T0313798

Photovoltaic (PV) System Delivery as Reliable Energy Infrastructure

Photovoltaic (PV) System Delivery as Reliable Energy Infrastructure

John R. Balfour and Russell W. Morris

This edition first published 2024

Registered Offices
John Wiley & Sons, Inc., 111 River Street, Hoboken, NJ 07030, USA
John Wiley & Sons Ltd, The Atrium, Southern Gate, Chichester, West Sussex, PO19 8SQ, UK

For details of our global editorial offices, customer services, and more information about Wiley products visit us at www.wiley.com.

Wiley also publishes its books in a variety of electronic formats and by print-on-demand. Some content that appears in standard print versions of this book may not be available in other formats.

Library of Congress Cataloging-in-Publication Data

Names: Balfour, John (John R.), author. | Morris, Russell W., author.
Title: Photovoltaic (PV) system delivery as reliable energy infrastructure / John R. Balfour, Russell W. Morris.
Description: Hoboken, NJ : Wiley, 2024. | Includes bibliographical references and index.
Identifiers: LCCN 2024000314 (print) | LCCN 2024000315 (ebook) | ISBN 9781119571193 (cloth) | ISBN 9781119571209 (adobe pdf) | ISBN 9781119571223 (epub)
Subjects: LCSH: Photovoltaic power systems. | Systems engineering.
Classification: LCC TK1087 .B353 2024 (print) | LCC TK1087 (ebook) | DDC 621.31/244–dc23/eng/20240206
LC record available at https://lccn.loc.gov/2024000314
LC ebook record available at https://lccn.loc.gov/2024000315

Cover design: Wiley
Cover image: © mphillips007/Getty Images

Set in 9.5/12.5pt STIXTwoText by Straive, Chennai, India

Printed and bound by CPI Group (UK) Ltd, Croydon, CR0 4YY

C9781119571193_180324

This book is dedicated to Karl, Jennifer, Vickie, James, Megan, Jonathan, and my six grandchildren. I have a special dedication for my wife, for whom Alzheimer's has taken her memories.

Russell W. Morris

I would like to begin with my father, Ronald B. Balfour PE, who acquainted me, early in grade school, with the scary and functional impact of engineering, science, and the laws of thermodynamics; my uncle Charlie, Charles Balfour (John Charles Balfour), who was my environmental science and energy mentor; and my brother Rick (Richard Anthony Balfour) and his wife Mary, who provided a variety of support on many occasions. In addition, I would like to acknowledge my life partner Patty (Patricia Ann Brown), who supported and assisted me during the editing phase of the manuscript.

John R. Balfour

Contents

Preface *xix*
Acknowledgments *xxi*
Abbreviations *xxiii*

1 Assessing PV Industry Challenges *1*
1.1 Introduction *1*
1.2 Terminology *2*
1.3 Preventive Analytic Maintenance *3*
1.4 Current State of the Industry *5*
1.5 Defining Failure and Success *6*
1.5.1 Failure *7*
1.5.2 Success *8*
1.5.3 Contracts *9*
1.5.4 Perspective *11*
1.5.4.1 The Catch-22 of Common Industry Assumptions and Subsequent Questions that Arise *13*
1.5.4.2 Stakeholders Are Often Held Back by the Simplest of “Assumptions” *15*
1.5.5 Loss of Corporate Memory *17*
1.6 Application of PAM *17*
1.7 Cost Control Considerations *18*
1.8 Project Versus System Delivery Process *21*
1.8.1 Project Delivery Process *21*
1.8.2 System Delivery Process *21*
1.9 PAM Concept *24*
1.10 Challenges Today with the Bidding Process *25*
1.10.1 Bidding Process *27*
1.10.2 Challenges Specific to the Project Delivery Model *28*
Bibliography *35*
Notes *36*

2 PV System Delivery Process *39*
2.1 Introduction *39*
2.2 PAM PV System Delivery Process *42*

2.2.1 Technology Fatigue *45*
2.2.2 Data Collection, Communication, Curation, and Their Effective Use *47*
2.2.3 Industry Language: Usage and Consistencies *51*
2.2.4 Safety *52*
2.2.5 Stakeholders *57*
2.2.5.1 Internal Stakeholders *57*
2.2.5.2 External Stakeholders *57*
2.2.6 Site Capability Survey *63*
2.3 PV Plant Commissioning *64*
2.3.1 Commissioning and Benchmarking *65*
2.3.1.1 Commissioning *65*
2.3.1.2 Benchmarking *67*
2.3.1.3 Electrical Testing *70*
2.3.1.4 Mechanical Testing *71*
2.3.1.5 Performance Evaluations *71*
2.3.1.6 Corrections Reporting *71*
2.4 Universal Real-Time Data (URTD) and Data Sharing *71*
2.4.1 Initializing the Site URTD Database *71*
2.4.2 Implementing the URTD *74*
2.5 PV Plant Lifecycle *76*
2.6 Standard Test Conditions *77*
2.7 Capacity and Capability *78*
2.7.1 Design Capacity *79*
2.7.2 Design Power Capability (DPC) *79*
2.7.3 Actual Plant Capability (APC) *80*
2.7.4 Raw Power Capability *80*
2.7.5 Reactive "Repowering" Threshold *81*
2.8 Addressing the Gaps *82*
2.8.1 Gap 1 – Design Power Capability to Actual Power Capability *83*
2.8.2 Gap 2 – Design Power Capability to Raw Power Capability *83*
2.8.3 Gap 3 – Losses Between Actual Plant Capability and Raw Power Capability *83*
2.9 Masking and Its Impact *85*
2.10 System Design Assumptions Drive Plant Fiscal Performance *87*
2.11 Conclusion *89*
Bibliography *90*
Notes *90*

3 Current PV Component Technologies *93*
Key Chapter Points *93*
Key Impacts *93*
3.1 Component Selection *93*
3.1.1 Lessons Learned *96*
3.1.2 Design Considerations *96*
3.2 Present State of Technology *97*
3.2.1 Technology Risk *97*
3.2.2 Cost of Technology *99*

3.3 Manufacturing Risk *100*
3.3.1 Variability of Quality among Manufacturers *101*
3.3.2 Decision Approach *103*
3.3.2.1 What Got Us Here? *105*
3.3.3 Integrators *111*
3.3.4 Primary Components *118*
3.4 Primary Technologies Discussion *121*
3.4.1 Cell Efficiencies *121*
3.4.2 Module Efficiency and Effectiveness *122*
3.4.3 Hail *126*
3.4.4 Module Degradation *128*
3.4.5 Module Defects *130*
3.4.6 Segregation of Safety and Performance Defects and Failures *131*
3.5 Inverters *134*
3.6 Equipment Removal, Disposal, and Recycling *142*
3.6.1 Equipment Removal *142*
3.6.2 Equipment Disposal *142*
3.6.3 PV Module Recycling *144*
3.6.3.1 What Is PV Module Recycling, and Why Is It Important? *144*
3.6.3.2 What Is the Present State and Need for PV Module Recycling? *145*
3.6.3.3 Who Ensures that Modules Are Recycled and/or Disposed of Appropriately? *145*
3.6.3.4 What Are the Costs Involved, and What Is Keeping Those Costs so High? *146*
3.6.3.5 What Are the Long-Term Impacts of Not Resolving Module Recycling Issues? *146*
Bibliography *147*
Notes *147*

4 SE/Repowering™ Planning Process *149*
Key Chapter Points *149*
Key Chapter Impacts *149*
4.1 Introduction *149*
4.1.1 Why **SE/R**epowering™? *150*
4.2 What Is the **SE/R**epowering™ Process? *152*
4.3 There Is a Continuous and Contentious Complaint about Lifecycle Performance *156*
4.3.1 Risk Reduction *160*
4.3.2 **SE/R**epowering™ *161*
4.4 Cannibalization *162*
4.4.1 Additional Requirements for Success for Commercial & Industrial (C&I) *162*
4.5 Impacts of **SE/R**epowering™ *163*
4.5.1 Improved and Higher Asset Valuation *163*
4.5.2 Long-Term Energy Production *163*
4.5.3 Plant Viability *163*
4.5.4 Revenue *164*
4.5.5 Operations Analysis and Decisions *164*

4.5.6 Commercial & Industrial (C&I) and Utility Economics *164*
4.5.7 Financial Viability *165*
4.5.8 Insurability *165*
4.6 Types of **SE/R**epowering™ *166*
4.6.1 **SE/R**epowering™ at Concept *167*
4.6.2 Existing Plant **SE/R**epowering™ *169*
4.6.2.1 New Ownership *172*
4.6.2.2 Major Financial and Legal Issues with Current Plant *172*
4.6.3 Distressed Plant **SE/R**epowering™ *172*
4.6.4 Relocation Plant **SE/R**epowering™ or More Likely Common Repowering *174*
4.7 Preemptive Analytical Maintenance **SE/R**epowering™ System Planning *175*
4.7.1 Planning Element Requirements *175*
4.8 RAMS for **SE/R**epowering™ *176*
4.9 **SE/R**epowering™ Considerations *184*
4.9.1 Plant Design at Concept *184*
4.9.2 Existing Operable Plant **SE/R**epowering™ *184*
4.9.3 Photovoltaic Module Recycling: A Survey of US Policies and Incentives Distressed Plant *185*
4.9.4 Plant Relocation *185*
4.9.5 Stakeholder Need – Requirements *185*
4.9.5.1 Substantively Better and More Accurate Asset Valuation *185*
4.9.5.2 Long-Term PV System Energy Production *185*
4.9.5.3 Plant Viability *186*
4.9.5.4 Revenue *186*
4.9.5.5 Ownership Operations Analysis and Decisions *186*
4.9.5.6 Bill of Materials (BOM) *186*
4.9.5.7 Spares *188*
4.9.5.8 Education and Training *188*
4.9.5.9 Codes and Standards *190*
4.9.5.10 What Does Conformance/Compliance Verification Entail? (Benchmarking, QA/QC) *196*
4.10 Technology Fatigue *196*
4.11 Data Collection *197*
4.11.1.1 How Are the Metrics Being Measured Today? (Data Collection & Analysis) *199*
4.11.1.2 "Who Should Gather the Data?" *200*
4.11.1.3 What Metrics Are Being Taken, and How Are They Being Measured Today? *201*
Bibliography *202*
Notes *202*

5 System Engineering *205*
5.1 Introduction *205*
5.2 Why Systems Engineering *206*
5.3 SE Process *210*
5.3.1 Success and Failure *212*

5.3.2 Stakeholder Requirements *213*
5.3.3 The Initial Concept Development and Feasibility *214*
5.3.4 Management of PV System Delivery *216*
5.4 Project Phases Overview *218*
5.4.1 Concept *218*
5.4.2 Detailed Design *218*
5.4.3 Manufacturing, Build, and Test *218*
5.4.4 Installation and Commissioning *219*
5.4.5 Operation, Upgrade, and **SE/R**epowering™ *219*
5.4.6 Site Restoration *219*
5.5 Systems Engineering Tools *220*
5.5.1 SIMILAR *220*
5.5.2 SMART *221*
5.5.3 Risk Management *222*
5.5.4 Project Management Tools *227*
5.6 System Versus Project Delivery Method *227*
5.6.1 Concept *228*
5.6.1.1 Determine Stakeholder Success and Failure Definitions *229*
5.6.1.2 Establish Concept of Operation, Usage, User, and Environmental Profiles *230*
5.6.1.3 Plant Stakeholder Requirements *234*
5.6.1.4 System Lifecycle Optimization *235*
5.6.1.5 Daily Variance in Power Generation and Load *238*
5.6.2 System Specification and Architecture *239*
5.6.2.1 Specification *239*
5.6.2.2 Maintenance Hours *239*
5.6.2.3 Variation in Solar Irradiance *240*
5.6.2.4 Acquisition Costs *241*
5.6.2.5 Energy Loss Budget *244*
5.6.2.6 Additional Environmental Effects for Concept and System Design *251*
5.6.3 Safety *253*
5.6.4 Common System Engineering Problems *254*
5.6.5 System Design *255*
5.6.6 Detailed Design *255*
5.6.7 Build and Test *255*
5.6.8 Installation and Commissioning *256*
5.6.9 Benchmarking and Certification *256*
5.6.10 Operation and Maintenance *257*
5.6.10.1 Maintenance *257*
5.6.10.2 Levelized Cost of Energy *258*
5.6.11 Upgrade and **SE/R**epowered™ *259*
5.6.11.1 Upgrade *260*
5.6.11.2 PAM **SE/R**epowering™ *261*
5.6.12 Site Restoration, Equipment Removal, Disposal, and Recycling *261*
5.6.12.1 Site Restoration *262*
5.6.12.2 PV Removal through Disposal *262*

5.6.12.3 Recycling *262*
5.6.12.4 Restriction on Hazardous Substances (RoHS) *263*
5.6.12.5 Manufacturing Compliance to Waste/Recycling *264*
5.7 Conclusion *264*
Bibliography *265*
Notes *267*

6 Reliability *271*
6.1 Introduction *271*
6.2 Why Reliability *272*
6.3 Success/Failure *274*
6.3.1 Stakeholder Metrics *275*
6.3.2 Lowest Level of Repair *278*
6.3.3 Failure-Free Operation *279*
6.4 Overview *280*
6.5 Reliability *282*
6.5.1 Synthetic Reliability Data – Bayes Theorem *282*
6.5.2 Interpolation *283*
6.5.3 Engineering Estimate *283*
6.6 Stakeholder Needs *284*
6.7 Reliability Predictions, Analysis, and Assessments *287*
6.7.1 Reliability Specifications *292*
6.7.2 Plant Reliability Drivers *292*
6.8 Reliability Program Plan *293*
6.9 Reliability Mathematics *295*
6.9.1 Weibull *295*
6.9.2 Bathtub Curve *296*
6.9.3 Normal Distribution *298*
6.9.4 Failure Rate *303*
6.9.4.1 Confidence Level/Confidence Interval *305*
6.9.4.2 *P* Values *307*
6.10 Reliability Block Diagrams (RBD) *307*
6.11 Fault Trees *311*
6.12 Failure Modes and Effects Analysis (FMEA) *311*
6.12.1 FMEA Procedure *315*
6.12.2 Risk Priority Numbers (RPN) *317*
6.12.3 Failure Modes and Mechanisms *319*
6.13 Failure Reporting and Corrective System (FRACAS) and the PV SCADA *324*
6.14 Root Cause Analysis *325*
6.15 Data Analysis *326*
6.15.1 Example Weibull++™ Analysis *326*
6.15.2 Praeto Analysis of Data *333*
6.16 Reliability Predictions *334*
6.17 Derating *337*

6.18 Reliability Testing *338*
6.18.1 Electronic Components *339*
6.18.2 Mechanical Components *340*
6.18.3 Test Time – MTTF *340*
6.19 Summary *341*
Bibliography *342*
Notes *345*

7 Maintainability *347*
7.1 Introduction *347*
7.2 Responsibility for Maintainability *350*
7.3 Types of Maintenance *350*
7.3.1 Unscheduled Maintenance *350*
7.3.2 Scheduled Maintenance *350*
7.3.3 Software (SW) Updates Maintenance *351*
7.3.4 Ancillary Maintenance (AM) *351*
7.4 Maintenance Cost *355*
7.4.1 Run to Failure *356*
7.5 Typical Maintenance Flow *357*
7.5.1 Fault Detection and Acknowledgment *357*
7.5.2 Work Authorization Delay *358*
7.5.3 Mean Time Till Onsite (MTTO) *358*
7.5.4 Equipment Delay Time (EDT) *358*
7.5.5 Fault Verification (FV) *359*
7.5.6 Fault Isolation (FI) *359*
7.5.7 Mean Logistic Delay Time *359*
7.5.8 Repair Time *359*
7.5.9 Repair Verification Time *360*
7.5.10 Overhead Time *360*
7.5.11 Minimum Maintenance Time *360*
7.5.12 Time to Repair *360*
7.5.13 Mean Corrective Maintenance Time *361*
7.6 Additional Maintenance Metrics *364*
7.7 Available Maintenance Time *364*
7.8 Maintenance-Driven Availability *365*
7.8.1 Crew Size *366*
7.8.2 Plant Maintenance Time *367*
7.8.3 Accessibility *367*
7.9 Preventive Maintenance (PM) *371*
7.10 Customer-Generated Maintenance *371*
7.11 Energy Storage *373*
7.12 Spares *374*
7.12.1 Spares Calculation *376*
7.12.2 Spares Storage and Availability *379*

7.13 Testability 379
7.13.1 Function of Testability 380
7.13.2 Design 381
7.13.3 Percent Coverage 381
7.13.4 Requirements 382
7.13.5 Special Test Equipment 383
7.14 Maintenance and Testability Specifications 384
7.14.1 Recommended Specification 384
7.14.2 Specification Notes 385
7.15 Conclusion 386
Bibliography 386
Notes 387

8 Availability 389
8.1 Introduction 389
8.2 Why Measure Component Availability 391
8.2.1 Capability and Capacity 393
8.2.2 Force Majeure 394
8.2.3 Annual Solar Irradiance 394
8.2.4 Customer Demand 395
8.3 Information Categories for Plant Availability (Unavailability) 395
8.4 Types of Availability 395
8.4.1 Energy 396
8.4.2 Power 397
8.4.3 Inherent 399
8.4.4 Operational 400
8.4.5 Achieved 401
8.4.6 Raw Availability 402
8.5 Confusion With Availability Metrics 403
8.6 Grid Availability 404
8.7 Specifications 404
8.7.1 Allocation 405
8.7.2 Requirements 405
8.7.2.1 Suppliers and Third-Tier Vendors 405
8.7.2.2 O&M 405
8.7.2.3 Owner 405
8.7.3 Standards 406
8.8 Conclusion 406
Bibliography 406
Notes 407

9 Energy Storage System (ESS) 409
Key Chapter Points 409
Key Chapter Impacts 409
9.1 Introduction Energy Storage Systems (ESSs) 410

9.2 Applications of Energy Storage *412*
9.3 Batteries *414*
9.3.1 Primary Versus Secondary *414*
9.3.2 Selection Criteria *414*
9.3.3 Types *416*
9.3.3.1 Battery Metrics *416*
9.3.4 Flow Batteries *421*
9.3.5 Battery Configuration *421*
9.4 Components of an Energy Storage System *423*
9.4.1 Battery Interface *423*
9.4.2 Storage Architecture *423*
9.4.3 DC- and AC-Coupled PV plus Storage Systems *424*
9.4.4 DC Coupling *424*
9.4.5 AC Coupled *425*
9.5 Battery Management System (BMS) *426*
9.6 Battery Thermal Management *427*
9.6.1 Housing Batteries *427*
9.6.2 Containment of Failures and Safety *428*
9.6.2.1 Fire Suppression *428*
9.6.2.2 Operating Ambient *429*
9.7 ESS Cost *429*
9.7.1 ESS Installed Cost *430*
9.7.2 O&M Costs *430*
9.8 Reliability *432*
9.9 ESS Maintenance and Operational Considerations *433*
9.9.1 Misconceptions, Myths, and Assumptions About ESS *433*
9.9.2 Plug and Play – PV+ESS *433*
9.9.3 ESS Dispatched Grid Service *434*
9.9.4 Stack Value – Value Stacking *435*
9.10 Considerations *437*
9.11 Electric Vehicles as Grid Storage *439*
9.12 Summary *441*
Bibliography *441*
Notes *442*

10 Data Collection *443*
Key Chapter Points *443*
Key Impacts *443*
10.1 Introduction *443*
10.2 Reducing Risk Begins with Data *446*
10.2.1 Component and Equipment RAMS Data Sharing *446*
10.2.2 Current Mandatory Reporting *447*
10.2.3 Proposed Data Format and Elements *448*
10.3 Shared RAMS Data *450*
10.4 Stakeholders *451*

10.5 Anonymized Plant Data *452*
10.6 Stakeholder Business Case for Sharing Reliability Data *452*
10.7 The Level Necessary to Control Costs and Improve PV Systems *455*
10.8 Monitoring for Better Data, Security, and Plant Cost Control *455*
10.9 Data Analysis *457*
10.9.1 Where Will the Responsibility for O&M and Data Decisions Reside? *459*
10.10 Data Presentation *459*
10.11 Process *461*
10.12 Implementation *463*
10.13 The Monitoring Plan *465*
10.13.1 Management Goals *467*
10.13.2 Establish Initial Data Requirements *467*
10.13.3 Define Assessment Strategy *468*
10.13.4 Identify Additional Data and Sensor Requirements *468*
10.14 Warranty Issues *469*
10.15 Synthetic Data *470*
10.16 Conclusion *471*
10.A Appendix *471*
10.A.1 Tropical *472*
10.A.2 Dry *472*
10.A.3 Temperate *472*
10.A.4 Continental *472*
10.A.5 Polar *472*
Bibliography *472*
Notes *473*

11 Operations and Maintenance (O&M) *475*
11.1 Introduction *475*
11.2 Safety *477*
11.2.1 Identification and Prioritization of Safety Issues *477*
11.2.2 Arc Flash Potential *478*
11.2.3 Job Hazard Analysis *479*
11.3 Reliability *480*
11.3.1 Reliability Data *481*
11.4 Availability *482*
11.4.1 Peer-Based Analysis *482*
11.4.2 Inverter Availability Guarantee *482*
11.5 Maintainability *483*
11.6 Testability *484*
11.6.1 Module Warranty *485*
11.6.2 Inverter Warranty *485*
11.7 Project Development *486*
11.8 O&M Plan *486*
11.8.1 O&M Philosophy and Strategy *486*
11.8.1.1 Self-Performing *487*

11.8.1.2 Third Party 488
11.8.1.3 An Alternative Commercial PV O&M Strategy: Local Electricians 488
11.8.2 Operations Scope 490
11.8.2.1 Construction Oversight 491
11.8.2.2 Substantial Completion Punch List 491
11.8.2.3 Capacity Testing 492
11.8.2.4 System Performance 492
11.8.3 Maintenance Scope 492
11.8.3.1 Preventive Maintenance (PM) 493
11.8.3.2 Scheduled Maintenance 493
11.8.4 Corrective Maintenance (CM) 493
11.8.4.1 Opportunities for Process Improvement 494
11.8.4.2 Preemptive Maintenance 494
11.8.4.3 Condition-Based Maintenance 494
11.8.5 Ancillary Maintenance (AM) 495
11.8.6 Pricing – Determining a Budget 496
11.8.6.1 O&M Price Distortions 496
11.8.6.2 Scope 496
11.8.6.3 Licensing 497
11.9 Conclusion 497
11.A Appendix A: Photovoltaic Fires Calculation Methodology 498
11.A.1 Rate of Deaths Per Fire 499
11.A.2 Total Cost Per Fire 500
11.B Appendix B: Operations Scope Example (Source: Courtesy of Higher Powered LLC) 500
11.B.1 The Operator's Role 500
11.B.1.1 Corrective Maintenance 501
11.B.2 Preventive Maintenance 501
11.B.3 Ancillary Maintenance 501
11.B.4 Tracking and Reporting of Technician Key Performance Indicators (KPIs) 501
11.B.4.1 Tracking and Reporting of Maintenance Provider KPIs 502
11.B.4.2 Tracking and Reporting of Equipment Reliability 502
11.B.4.3 Contract Management of the Third-Party Maintenance Providers 502
11.B.5 Curation of Service Maintenance Documentation 502
11.B.5.1 Record Keeping 502
11.B.5.2 Data Aggregation 502
11.B.5.3 Corrective Maintenance 502
11.B.5.4 Follow-up Issues 503
11.B.5.5 Data Sharing 503
11.C Appendix C: Maintenance Scope Example 503
11.C.1 Introduction 503
11.C.2 Maintenance 503
11.C.2.1 Safety 503
11.C.2.2 Site Access 504
11.C.2.3 Job Hazard Analysis 504

11.C.2.4 Training 504
11.C.2.5 Records 504
11.C.3 Preventive Maintenance (PM) 504
11.C.3.1 Administration 504
11.C.3.2 Measurement and Control Systems 504
11.C.3.3 Gen-Tie and Switchgear Maintenance (If Applicable) 505
11.C.3.4 MV Transformer (If Applicable) 505
11.C.3.5 AC Panel or Combiner Box (If Applicable) 506
11.C.3.6 Shelter/Skid (If Applicable) 506
11.C.3.7 Energy Storage Device(s) (If Applicable) 506
11.C.3.8 Inverter(s)/Converters 507
11.C.3.9 DC Combiner Box (If Applicable) 508
11.C.3.10 Trackers (If Applicable) 509
11.C.3.11 PV Modules and Other DC components 509
11.C.3.12 Grounds/Rooftop 510
11.C.3.13 PM Report 510
11.C.4 Corrective Maintenance (CM) Scope 511
11.C.4.1 Corrective Maintenance (CM) 511
11.C.4.2 Administration 511
11.C.4.3 CM Work Order 511
11.C.4.4 CM Field Service Scope 511
11.C.4.5 Key Deliverables 512
11.C.4.6 CM Service Report 512
11.C.5 Ancillary Maintenance (AM) 512
11.C.5.1 Administration 513
11.C.5.2 Types of Ancillary Services 513
11.C.5.3 AM Service Report 513
Bibliography 514
Notes 515

Glossary 517
Index 527

Preface

For the photovoltaic (PV)/energy storage system (ESS) industry to provide more (available) systems with lower energy costs, we must deliver PV systems as reliable energy infrastructure.

Our focus is on applying well-known systems engineering (SE) practices with integrated reliability, availability, maintainability (including testability), and safety (RAMS) engineering requirements. Doing so extends the useful system life and reduces operations and maintenance costs, which delivers more consistent energy production. This significantly improves the "actual" levelized cost of energy (LCOE); e.g., which determines the market value (i.e. USD $/kWh) that customers pay for the energy they use over time. For defined and traceable success, the initial investment costs must be balanced with the "all in" lifecycle costs and activities. These more robust, resilient plants deliver more energy for longer periods of time with greater consistency at a documentable lower cost of energy.

The critical elements in initial design and development are achieved when the project develops measurable success and failure criteria and applies them to the detailed specifications, requirements, design approaches, and processes. This includes RAMS attributes similar to that of current historical utility, aerospace, automotive, marine, medical, and other industries. For PV systems, these can be upgraded, maintained and can be functionally **SE/R**epowered™, thus extending the useful life of the plant while improving output and profitability.

By the beginning of 2024, PV systems and modules appeared to have reached their low point in first cost. At the same time, rising interest rates, wages, materials, increased demand, resource shortages, and government monetary policies led to increasing costs to design, manufacture, build, install, and operate systems or plants. Limited access to such resources as lithium and cobalt also affects the ESS sector.

Historically, most of the effort has focused on the PV system's initial investment (first) cost, which often treated RAMS as an afterthought and/or as attributes to be traded to minimize those costs. While many myths and assumptions are still rife in the industry, a growing sector understands that today's technology, applied effectively using **SE/R**epowering™ planning methods and processes, can produce more energy reliably and deliver that energy at a lower lifecycle cost.

An unfortunate fact: The existing business "least cost" model encumbers that process, as it drinks from a well of common myths and assumptions, i.e., PV+ESS is simple, cheap, and requires almost no maintenance. These, they are not! The sheer quantities of components alone result in potentially significant costs in maintenance and spares.

This book has been in development for over the last 10-plus years. It is based on our combined decades of experience with PV power system (PVPS) products, proposals, designs, reviews, installation, operations, maintenance, and activity in standards working groups and development.

Over the last 40–50 years, the industry skills, knowledge, and capability of PV professionals have not kept up with what the evolving technology demands. This fast pace of technology development creates a lag or delay in the training and educational curriculum to support the ever-widening variations in effective design and maintenance practices. This lag results in a lack of information or, worse, misinformation being used to both major and minor decisions on the design requirements of PV plants. This has resulted in inconsistencies in the ability to deliver reliable energy systems. The people involved include marketing professionals, manufacturers, engineering, procurement and construction companies (EPCs), hardware and software designers, developers, installers, maintainers, peripheral workers, and owners in the PV and energy storage fields.

Complicating the above is the growing impact of climate change on PV operation, maintenance, and profitability.

These consequences are not inevitable! They are simply the result of poorly supported decisions! As an additional consequence, many projects do not provide sufficient funding to fully address infrastructure robustness and longevity requirements, resulting in the current state of the industry.

Our goals and objectives are to offer a proven duplicatable approach with methodologies that are based on SE RAMS **R**epowering™ planning efforts. By defining a project using proven existing SE methods, it is possible to understand and focus on the cost benefits of this approach as they apply to the LCOE and the plant's entire lifecycle cost.

January 2024, Phoenix, Arizona — *John R. Balfour*
January 2024, Norfolk, Virginia — *Russell W. Morris*

Acknowledgments

Russell W. Morris, BSEE, MSSE, SM-IEEE, M-INCOSE: I would like to give special recognition to Dr. Ron Carson for his review and comments, Lawrence Shaw for his insight into the O&M operations, and Lawrence Montrose, a fellow Boeing Reliability engineer, as he provided a good counterpoint for ideas on software reliability. Lisa Owen gave me encouragement based on her experience as a writer. My three children, Jennifer, James, and Jonathan, have always supported my efforts, and I thank them deeply.

John R. Balfour BS MEP, PhD: My personal and professional experience has been shaped by my family and colleagues, each of which had an influence on my thinking, work, and focus. They all contributed to my awareness, understanding, and knowledge, which hopefully will make our lives and our planet better.

In addition there are those that would knowingly or unknowingly include my teachers, friends, and colleagues who shared their ideas, wisdom, insights, and perspectives that helped me better understand the many elements of the solar energy and technology – John I. Yellott, David Tate, Dr. Michale Boyle, Mike Brennan, Sunil Shaw, Jennifer Granata, Liang Ji, Govindasamy TamizhMani (Mani), Mark Wilhelm, Lawrence Shaw, Peter Mitchell, Stanley Mumma, Philip Allsopp, Roger Hill, Matt Marroon, Daisy Chung, Devarajan Srinivasan (Srini), Andy Walker, George Kelly, Sumanth Lokanath, Pramod Krishnani, Jon Previtali, Thomas Sauer, Patrick O'Grady, Sarah Kurtz, Craig Palmer, David King, Gerald Robinson, Ward Bower, Rue Phillips, Tyler Pearce, Natasa Vulic, Raginee Yadav, Jiawei Wu, Eric Chan, Yutao Zong, Steven Zylstra, Chris Henderson, Jaime Kern, Gord Petroski, David Prince, Jennifer Granata, Nelson Greene, Robin Gudgel, Bob Hammond, Tassos Golnas, Michael Borden, Kaushik Roy Choudhury, Steven Croxton, Eric Daniels, Thushara Gunda, Sam Vanderhoof, Joe Cunningham, Kent Whitfield, Erika Hanson, Thorsten Hoefer, Larry Freeman, K. Dixon Wright, Scott Sullivan, Abhishek Rao, Ankil Sanghvi, Aravind Venkatesan, Mudit Bareja, Nick Esch, Andrey Bednarzhevskiy, Chris Henderson, Jaime Surrette, Geoff Klise, Ryan Mayfield, Brad McKinley, Shane Messer, Craig Palmer, Ernie (Gilbert) Palomino, Nadav Enbar, Sam Walsh, Tom Broschinsky, Davide Grande, Liang Ji, David Devir, Gina Binnard, Mike Borden, Jeffrey Wehner, John Wohlgemuth, Leigh Zanone, Joe Brotherton, Andrew deRussy, Mahesh Morjaria, Stephen Coury, Scott Canada, and Cecilia Nedelko who provided great support during the disruptive Covid plague.

25 January 2024

Abbreviations

A	availability – generic term
A_A	availability, achieved
AC	alternating current
A_E	availability, energy
A_i	availability, inherent
ANSI	American National Standards Institute
A_o	availability, operational
A_{RAW}	availability, raw
As	arsenic/arsenide
ASNT	American Society of Non-destructive Testing
ASTM	American Society for Testing and Materials
BESS	battery energy storage system
BLAST	battery lifetime analysis and simulation tool
BOS	balance of system
C	centigrade
CAD	computer-aided design
Co	cobalt
CT	current transformer
DAS	data acquisition system
DC	direct current
DoD	depth of discharge
DOE	US Department of Energy
EAM	enterprise asset management
EPC	engineering, procurement, and construction
EPDM	ethylene propylene diene monomer
EPRI	electric power research institute
ERP	enterprise resource planning
ESS	energy storage system
EU	European Union
EV	electric vehicle
EVA	ethylene vinyl acetate
$F(t)$	failure as a function of time
FD/FI	fault detection/fault isolation

Fe	iron
FEMP	Federal Energy Management Program
FERC	Federal Energy Regulatory Commission
FR	failure rate
FRACAS	failure reporting and corrective action system
Ga	gallium
GFI	ground fault interruption
GHI	global horizontal insolation
h(t)	instantaneous hazard function
IBTS	Institute for Building Technology and Safety
IEC	International Electrotechnical Commission
IECRE	IEC Renewable Energy
IEEE	Institute of Electrical and Electronics Engineers
IGBT	Insulated-gate bipolar transistor
ILR	inverter loading ratio
IP	internet protocol
IRR	internal rate of return
IT	information technology
I–V	current voltage
KPI	key performance indicator
kVAR	kilo-volts-amps reactive
kWh	kilo-watt hours
LC	lithium carbon
LCC	life cycle cost
LCOE	levelized cost of energy
LFP	lithium iron phosphate
Li	lithium
LMO	lithium manganese oxide
LTO	lithium titanate
M_{ct}	mean corrective maintenance time
MDT	mean down time
MLDT	mean logistic delay time
MOA	Memorandum of Agreement
MPPT	maximum power-point tracking
MTBCF	mean time between critical failure
MTBF	mean time between failure
MTBM	mean time between maintenance
MTBSM	mean time between scheduled maintenance
MTBUSM	mean time between unscheduled maintenance
MTTF	mean time to failure
MTTR	mean time to repair
MWac	megawatts alternating current
MWdc	megawatts direct current
MWhr	mega-watt hours
NASA	National Aeronautics and Space Administration

NCA	nickel cobalt aluminum
NCU	network control unit
NERC	North American Electric Reliability Corporation
NFPA	National Fire Protection Agency
Ni	Nickel
NLE	normal loss expected
NMC	nickel-manganese-cobalt
NREL	National Renewable Energy Laboratory
O&M	operations & maintenance
OEM	original equipment manufacturer
OSHA	Occupational Safety and Health Administration
P	phosphorus
P(s)	probability of success
$P(t)$	probability as a function of time
PAM	preemptive analytics maintenance
Pb	lead
POA	plane of array
PPA	power purchase agreement
PPE	personal protective equipment
PR	performance ratio
PV	Photovoltaic
PVC	polyvinyl chloride
PVPS	photovoltaic power system
Q(t)	probability of failure
QA	quality assurance
QC	quality control
R&D	research and development
$R(t)$	reliability as a function of time
RAM	reliability availability maintainability
RAMS	reliability availability maintainability safety
RCRA	Resource Conservation and Recovery Act
RE	reliability engineering
ROI	return on investment
SAM	system advisor model
SAPC	solar access to public capital
SCADA	supervisory control and data acquisition
SE	systems engineering
SEIA	Solar Energy Industries Association
Sf or f^2	square foot
Si	silicon
Sm or m^2	square meter
SOC	state of charge
STC	standard test condition
TOD	time of day
TPO	thermoplastic polyolefin

UAV	unmanned aerial vehicle
UL	Underwriters Laboratories
UN	United Nations
UPS	uninterruptible power supply
UV	ultraviolet
V&V	verification and validation
Vdc	volts direct current
VOC	voltage open circuit
VRLA	valve-regulated lead acid
W/m*K	watts/meter-Kelvin
W/m^2	watts/meter-squared
Wac	watts alternating current
Wdc	watts direct current

1

Assessing PV Industry Challenges

In 2017, it was noted in a number of publications that there were ongoing and significant problems with photovoltaic power plants. There had been a number of plant failures, bankruptcies, defaults, and fire sales. Chapter 1 provides an overview of the industry as of early 2024 with some historical notes. The Preemptive Analytic Model (PAM) is introduced as well as the differences between what we call the project delivery model based primarily on cost and the system delivery model based on optimization of the cost and benefits for all stakeholders. Throughout the book, there is an emphasis on establishing success and failure definitions and criteria. This is in part due to the widely varying views held by the industry today.

1.1 Introduction

In this chapter and throughout this book, we use a Socratic method for discussion by pointing out many obvious and not so obvious industry challenges. Among them: are a lack of systems engineering (SE) and **R**epowering™ planning, product inconsistency, underperformance, insufficient/underfunded Operations and Maintenance (O&M), and a lack of long-term reliability and availability to mention a few. We address how many of these can be corrected. We ask some pointed and uncomfortably tough questions while critically looking for real solutions while stimulating discussion. Until the industry publicly discusses these positions critically, aggressively, and thoroughly, while proving those assertions correct or incorrect and providing/delivering better solutions, future industry-wide system performance, output, and fiscal results will remain underwhelming.

It is clear to us, the authors, that with over 60 years of PV development experience history, the long-term industry resolution of critical photovoltaic (PV) technical issues can be addressed in a more cost-effective manner. This requires adopting an SE approach while identifying risks to the stakeholders including plant design, component selection challenges, repair, and lost energy. This necessitates greater usage of operational information and failure data to identify, track, and analyze a plant's faults, failures, and service interruptions.

Our Premise:

The least-cost PV system, with or without energy storage that you can buy today, may in all probability, be the most expensive system you may ever build or own. Therefore,

Photovoltaic (PV) System Delivery as Reliable Energy Infrastructure, First Edition.
John R. Balfour and Russell W. Morris.

our focus is on the specification and delivery of reliable, available, maintainable, and safe (RAMS) profitable PV infrastructure not based primarily on the least initial cost.

Our Concern:

If we as an industry do not effectively address the real issues of today, we may find ourselves in a decade or so with a substantial number of worldwide PV plants that are underperforming and incapable of meeting the "real energy" requirements necessary, especially when called upon!

Our present PV industry business model and approach are inconsistent. Over the last 30-plus years, the industry witnessed a litany of project failures, bankruptcies, electrical fires, fire sales, and more. These flawed projects are often the result of poor management judgment and decisions using a least-cost process. The process comes without sufficient substantiation or rational data, other than assumptions or a sales pitch that promises to minimize the investment costs with a good return in the short term.

1.2 Terminology

As is discussed in Chapters 4–8, terminology and definitions drive understanding the capabilities of PV plants as well as their risks. At the technical level, there is a real/substantive lack of understanding of what reliability, maintainability, and availability terms mean. Of all the terms used in this book, and the industry for that matter, two tend to be consistently misunderstood and incorrectly applied. Capacity and capability have been abused as marketing terms and at the writing of this book, ill-defined and not applied accurately or effectively in the industry. Based on our research, we provide the following common definitions of capacity and capability.

Capability; noun: the ability to do something

Capacity; noun: the total amount that can be contained or produced

(https://dictionary.cambridge.org/us/dictionary/english/)

Using these we offer the following:

- Module capacity = The module nameplate power output as tested and certified at standard test conditions (STC) (UL1703/UL 61730 – PV Module Safety Standards Updates: Making the Transition).
- Plant capacity = The sum total of the module nameplate power.
- Plant capability = The plant health, condition, and status at any point in time to provide output power from the plant with the equipment that is available, while accounting for impact of all plant environments, equipment degradation, and failures.
- Performance capacity = The maximum power output of the plant for the equipment that is operational, available, and accounting for all degradation and failures at a specific site and environment (see IEC 61724-3).

Plant or STC capacity as a metric does not equate to performance or profitability. Capability in and of itself defines a state of the plant's ability to generate power but does not equate to energy production. In other words, the capability of the plant may be above or below what is demanded at any particular time. An example of this would be curtailment.

1.3 Preventive Analytic Maintenance

Preemptive Analytical Maintenance (PAM), Balfour and Morris (2018), is a philosophical and proven PV system engineering process and approach, which is practical in its implementation that includes embedded reliability, availability, maintainability including testability, and system operational safety. It focuses on a *system lifecycle planning approach*, which addresses plant **R**epowering™ planning as a "lifecycle" system delivery process. It includes a substantive set of engineering processes and procedures as compared to the presently discussed and practiced repowering approach, which is inadequate to extend system life for the delivery of cost-effective energy.

PAM is the application of a detailed lifecycle systems specification based on commonly accepted industry-derived nomenclature and metrics (see Orange Button, the IEC Reliability and Availability Standards IEC TS 63265:2022 and IEC TS 63019:2019, respectively, and Institute of Electrical and Electronics Engineers [IEEE] and Military standards) for new and existing plants based on all stakeholder's wants, needs, and desires. Identifying, defining, and specifying these early in the system delivery design phase, prior to Engineering Procurement and Construction (EPC) bidding, assure their adherence through the entire PV system delivery process. Doing so delivers plants that express improved performance, energy output, and asset value. This also improves O&M and cost stability over the life of the plant as these costs can be significantly reduced in the contract out years, while effectively optimizing a far more accurate levelized cost of energy (LCOE).

To achieve this in both existing and future PV power plants requires a consistent and continuous approach to performance and failure data collection, curation, analysis, and root cause analysis with clearly defined corrective actions to be taken.

PAM, illustrated in Figure 1.1, provides the framework and capability to deliver accurate operational risk and cost analyses. This approach, based on data acquisition, more than pays for improving the way PV plants are specified, designed, delivered, operated, maintained, and retired (Chapters 5–7). Using the acquired data to further understand the SE/RAMS (Systems Engineering INCOSE-TP-2003-002-04, 2015, IEC TS 63265:2022 and Reliability, Availability, Maintainability, and Safety) of operating systems provides the foundation for future improvement in lifecycle cost (LCC) (IEC 60300-3-3:2017).

There are precedents for this view as you read further.

"The great enemy of the truth is very often not the lie – deliberate, contrived and dishonest – but the myth – persistent, persuasive, and unrealistic." John F. Kennedy.

To quote Dr. Richard Feynman after the *Challenger* space shuttle disaster in 1983, "For a successful technology, reality must take precedence over public relations, for Nature cannot be fooled."

Table 1.1 is a summary of the various stakeholder's areas addressed in this book. Of special note is the broad range of users/stakeholders who can benefit from a well-defined set of practices that we are advocating.

Table 1.2 provides a brief summary of the elements of the book and the level of technology familiarity or difficulty of each chapter. It should be noted that regardless of the level of technical expertise, each chapter contains a broad range of material, which is understandable and beneficial to all readers. PAM as a model is foundational to all stakeholders in the PV industry. It is our goal to provide all readers with essential information and value to

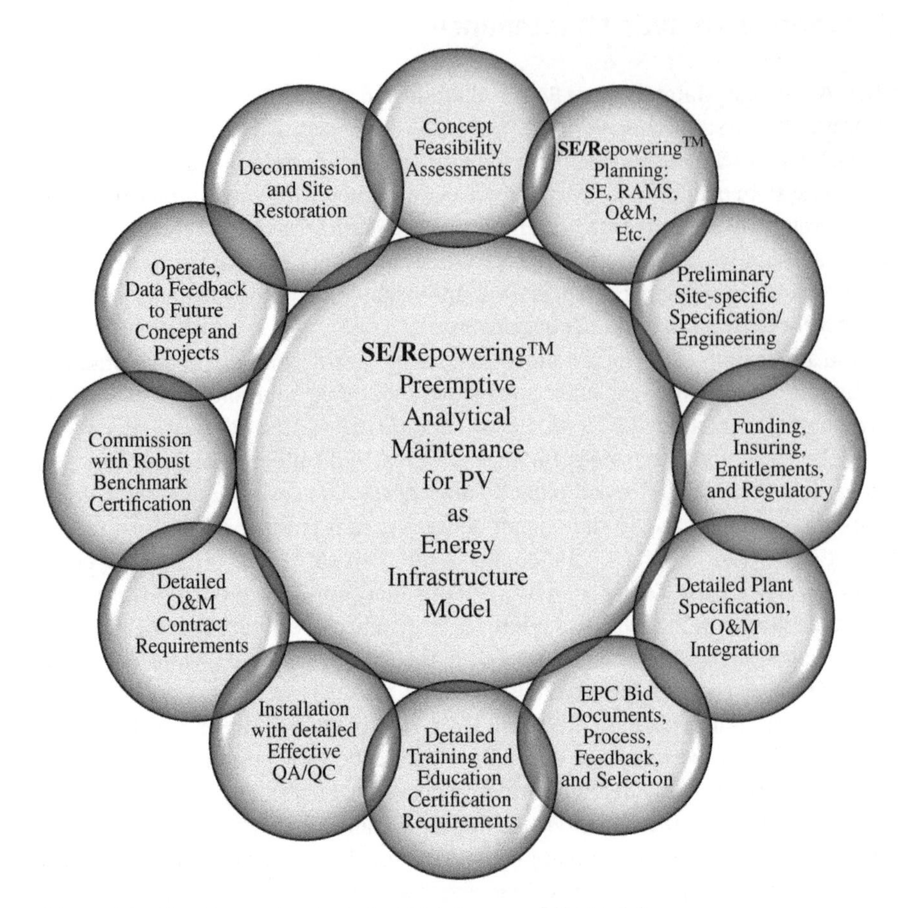

Figure 1.1 The Preemptive Analytical Maintenance, PAM Model.

Table 1.1 Summary of Stakeholder Vested/Critical Areas of Interest and Concern.

	Target	**Primary Areas of Interest and Concern**
Introduction	Student, new solar professional	Major elements of a PVPS (PV power system) and past and current problems
Major stakeholder	Vested financial, insurance, EPCs, utilities, owners, operators and maintainers	More detailed discussion of current situation and addressing the greater detail, cost, and planning needed for success
System engineering, design, and O&M	OEMs, Project SE, EPC, owner, operators and maintainers	Functional and operational discussions on needed improvements with SE, RAMS, and O&M; mathematical, statistics, physics, and chemistry details

Table 1.2 Breakdown of Information by Chapter.

Chapter	Title	Technical Difficulty	Target
	Acknowledgments		
	Preface		
	Introduction		
1	Assessing PV Industry Challenges	Low	All readers
2	Current Technologies	Low to moderate	All readers
3	Current PV Technologies	Low to moderate	All readers
4	**SE/R**epowering™	Low to moderate	Includes material for all readers
5	Systems Engineering	Moderately technical	Includes material for all readers
6	Reliability	Technical	Nontechnical foundation with supporting technical methodologies
7	Maintainability	Technical	Nontechnical foundation with supporting technical methodologies
8	Availability	Moderately technical	Nontechnical foundation with supporting technical methodologies
9	Energy Storage	Moderate	Includes material for all readers
10	Data Acquisition	Moderate	Includes material for all readers
11	Operations and Maintenance	Low to moderate	Includes material for all readers

participate in making better, more informed decisions regarding the delivery of PV plants. This may be accomplished incrementally or rapidly.

1.4 Current State of the Industry

Contrary to well-ingrained PV industry myths, assumptions, and common beliefs, the **delivery of PV systems (with or without storage) is *not* simple**.

The core of these PV myths lays within the industry itself. Since the first attempts to provide "Green Energy" (mid-1970s), there has been a practice to *not* publish, make public, or share detailed operational or critical failure data. This has been done, in part, by claiming that field operational and reliability data is intellectual property or proprietary information to reduce the likelihood of using this data in lawsuits for failure to perform to contract.

We believe that data, which is accurately collected, curated, and effectively anonymized, has little to no negative value on its producers when that data is not directly linked to a system, owner, or singled out to create potential embarrassment. The vacuum that has been created by this missing critical data has been replaced with myths and assumptions that tend to be based on marketing and greenwashing hype. The hype has led to an ignorance or a lack of understanding of PV plant physical attributes necessary for long-term profitability.

This has impacted how technologies are delivered to the industry that has been built into the present PV "project delivery" model.

Limited reliable and timely public information has created a set of assumptions that have led to a growing number of false and distracting conclusions and advertising by those inside and outside of the industry. There is an almost universal belief that photovoltaic power systems (PVPSs) are cheap, require little or no maintenance, and will last for at least 20–30 years as designed and built.

Stakeholders often lack sufficient information, consideration, and awareness of the PV metrics needed to understand the projects and, more critically, their risks. Failure to track detailed performance and accurate RAMS information and other critical data is generally not well documented nor well understood by the developers, EPCs, and the O&M team. This does not mean that industry participants are unaware of the impact of limited or incomplete data and its negative consequences over time.

Plant data are seldom appropriately tracked or understood by stakeholders either. Beginning at the concept phase, the initial stakeholders, such as the developers, owners, financiers, insurers, and SE teams among others, must establish their *failure and success* criteria. As new stakeholders come on board, without these definitions and without resolving the differences in interpretations for all stakeholders, the project is bound to fail, finding success elusive!

1.5 Defining Failure and Success

It is not uncommon for PV professionals to casually discuss the issues of failure and success among themselves and even occasionally with other stakeholders.

Yet, when discussing these terms with regard to a specific PV system being considered, it is essential to determine each stakeholder's definition(s) of success or failure and what significance it has for them. Elements of this are found rooted in a variety of well-developed project management skills and practices. However, there is much more to it. Because the definition and measurement of success should be clearly defined metrics in good project management practices, the same set of success guidelines developed should then be used as a risk management tool and further translated into meaningful monitoring metrics.

All stakeholders may wish to consider the following:

If you take the entire project team and ask them to define the elements of a project's success and failure, you will generally get different answers from each stakeholder. Those varied interpretations are signs of a lack of awareness of differences among your team and that often translate to a series of impacts on PV system delivery, costs, and results. Only by careful examination – and group consensus on what those answers translate to – can the project team begin to build a shared, defined set of goals and objectives. Doing so will define system and organizational value as a set of requirements to measure success and failure. Seldom is this examination process done thoroughly, however, most stakeholders have different views, requirements, interests, and perspectives regarding project success or failure.

Just "dictating" them as a company or organizational policy can generate unnecessary and complicated challenges. This lack of openness, awareness, clarity, consistency, and agreement can muddle everything! Addressing the team's questions and assumptions about success and failure can become key elements in organizational buy-in.

It is just as important to define failure and its levels of impact on the project. It is necessary to understand what the consequences might be when limiting factors[1,2] (project and environmental constraints) have not been properly addressed. When all the stakeholder's failure and success definitions are understood, the communication foundation will improve dramatically. The project characteristics, including the actual health, condition, performance, output, and revenues, will take on clearly defined meanings and measurable results. With discussion and agreement, the project team and all other stakeholders can work toward the same set of targets. The criticality is in addressing stakeholders by using common metrics and understandings that define how to effectively interpret perspectives and take actions on them. Chapter 5 goes into more detail on the process of defining success and failure, as well as the methods for performing the assessments.

1.5.1 Failure

The classical definition of failure is: "The event, or inoperable state, in which any item or part of an item does not, or would not, perform as previously specified."[3]

The emphasis here is on "specified" – to establish exactly what constitutes failure and the consequences to the stakeholders. Hence, it is **strongly recommended** that the developer, initial owner/s, and systems engineer complete the PV plant specification *prior to* delivering the request for proposal (RFP) or request for quote (RFQ) to the EPCs for bidding.

The application of the SE/RAMS processes (during the design phase of the failure modes and effects analysis [FMEA], reliability predictions, and fault tree analysis) supports the safety review and the development of specific requirements and metrics. It provides the basis for understanding potential flaws in the design and identifying critical corrective actions to be taken.

To define PV module, combiner, and inverter failure rates with availability requires the use of supplier and field data. However, analysts who do not clearly understand the supplier information or have misapplied it may grossly misclassify equipment failures. As an example, many can quote the basics of STC but may not grasp how actual performance varies under operating environments and, therefore, may warp its use and results. An example of this warping appears in the FMEA, where understanding the tool and its misapplication becomes critical. A PV module's capability and characteristics change between STC and actual environmental conditions that have a relational, voltage, current, and wattage shift that requires further analytical determination. In short, PV plants seldom operate at STC. However, plants are rated using only STC to define capacity, not performance or output (capability). While the stated STC nameplate "capacity" of the system may not change, the capability to produce energy and revenue over time is subject to derating, degradation, and the results of poor component selection, operations, and maintenance. This example highlights why we must focus on the health and condition of plants to control all present and future costs.

1.5.2 Success

Success is often defined as the opposite of failure; but this definition is not necessarily true for PV and especially between the various system stakeholders.

The PV plant's success lies in meeting the stakeholder's expectations of energy delivery when called upon by the utility, customer, or other stakeholders. It is critical that such expectation should be clearly defined in the specification, designed into the plant, and delivered to the customer/s. Any assumptions must be strictly defined and supported by technical data that is measurable and testable.

For example, a plant may be incapable of meeting the lifecycle nameplate power requirement with currently assumed degradation rates. This detail needs to be addressed early during specification prior to EPC bidding, not after installation, or during O&M.

Most would agree that if the requested energy demand is for 80% of the contractual energy capacity of the plant performance requirement (IEC 61724-3[4]) and the plant can generate 85% at that specific point in its service life, then the plant is probably successfully meeting the power demands and probably also with acceptable availability as defined in (IEC 63019[5]).

This capability to meet demand raises a set of important questions:

- What does this mean in the short, medium, and long term for stakeholders?
- What is the consequence if the plant is not meeting those contractual targets?
- What do we actually need to know to understand plant performance?
- And if the projected metrics for success are poorly calculated, documented, and/or underestimated, what could have been done to address the shortfall, i.e. failure?

This scenario is only considered an actual "success" if the stakeholder metrics have been defined, are real, measurable, and verifiable, and agreed upon previously, i.e. during system lifecycle planning and specification. During early project stages, success for the initial stakeholders does not necessarily adequately address the sustained success over time for the life of the plant and later stakeholders. Not fully addressing the criticality of "time" as a metric before and during project development – most specifically through the need for effective **Repowering**™ Process planning – sets up later stakeholders, investors, and owners for failure. As many stakeholders only consider their brief time with the project, their decisions may tend to be substantively flawed in the longer term. This attribute of time is a major concern for reliability (success over time) and maintainability (shortest repair time), and therefore affects the availability of the plant and consequent profitability. The determination for the owner's success is heavily based on contractual availability, which to date is primarily measured as the energy output of the plant at the point of interconnection to the grid. The measured availability is modified by what is demanded versus what the name plate (capacity) power should be as set by the contract.

Examples of success metrics for different stakeholders may include:

- **Owners:** Profitability and asset value over the planned period of ownership
- **Financiers:** Return on investment (ROI) within the specified terms of the loan/investment
- **Operators:** Required energy delivery as demanded within the capability of the plant

- **Operators and maintainers:** Are provided sufficient budget, staff, components, and tools to perform contract-based and specified metrics to address scheduled and unscheduled maintenance
- **Insurers/underwriters:** Require accurate clearly defined data and criteria to address claim events to avoid excessive payout for the insured portion of the plant
- **Safety (local):** Reduce or prevent any events that threaten, damage, or injure people and property
- **Logistics support:** Support for inventory and delivery availability of components, support equipment, and tools
- **Customers (utilities/end users):** Reliable delivery of contracted power when needed

1.5.3 Contracts

Common beliefs have compounded a broad range of cost challenges by the extensive use of restrictive, limited, and inconsistent O&M, Power Purchase Agreements (PPAs), and availability contracts that often contain extensive inclusion and exclusion language. Those contracts seldom provide the basis for an accurate picture of the necessary support costs required to properly and effectively maintain a plant, using defined O&M tasks (Chapter 11), sparing,[6] logistics, equipment, and manpower delay times (Chapter 7). Many organizations do not effectively collect this information, and thus tend not to have the essential information to make better decisions. This method creates a broad range of systemic issues that can be avoided by application of fundamental SE practices in a cost-effective manner. Nevertheless, all too often, it is the case where simple, generally inexpensive decisions that are made properly and early on, would save many projects thousands or possibly millions of dollars throughout the system lifecycle.

Most O&M and availability contracts reviewed by the authors and from industry-published reviews[7] have been written with very limited system maintainability requirements and specification in terms of actionable details, language consistency, identified tasks, and clarity. In addition, due to a substantial number of contractual *exclusions*, owners and customers (including utilities) end up paying more than expected or planned for O&M. Many of the corrective actions related to failure are primarily outside the scope of – and effectively excluded from – the signed contract. This is illustrative of, when not having sufficient detailed and documented field information, results in the LCOE becoming substantially higher than was ever projected in the project's concept, planning, finance, and operational phases.

With the introduction of newer PV technologies – such as module/panel level inverters, converters, bifacial modules, energy storage, integration of data acquisition systems (DAS), supervisory control and data acquisition (SCADA), and one- or two-axis tracking systems – PV power plants have become even more complex. With increased capabilities and complexity, stakeholders will see an enhancement and escalation of the type and number of risks for power production and commercial plant value (Chapter 5). The escalation results in substantially greater O&M costs as well, requiring increasingly higher maintenance costs, more specialized maintainers and support equipment, a greater type and number of spares, and all the attendant support costs, again increasing the cost of energy. This scenario is true whether discussing residential, commercial, industrial, and/or utility-grade systems.

While there has been continual technological improvement, it is not consistent and not being employed as well as it could. As a result, many opportunities to reduce operating costs and improve revenue are not yet accessible.

With or without energy storage, PV systems can be specified, designed, built, and operated for an optimized LCOE and LCC. It is important to do so by including initial or sunk investment along with predicted or projected O&M costs over the **entire** lifecycle of the plant. Accuracy will be defined by the identification of the "all in" plant costs from cradle to grave. Success requires an effective SE process and a program focused on the system's attributes of plant RAMS. That process must address both the technical and organizational issues, the crux of which has been holding back the PV industry for decades. In this book, we address the technical aspects with additional recommendations for the management of these systems.

The SE and RAMS approaches used within this book are neither new nor magic. They are based on over 100 years of industrial experience, with well-established physics, mathematics, and chemistry that have been implemented in a wide variety of industries. These industries applied systemic implementation of SE/RAMS to reduce silos that create internal and external communications barriers and inconsistencies. The results include reduced development time, initial and ongoing costs, and organizational "angst."

Organizational angst is often seen as too many internal or external stakeholders becoming needlessly involved, underinformed, disrupted, disruptive, stressed, given conflicting goals and or objectives, or overworked. It is the result of unresolved, simple, and complex system challenges that can be solved with good, thorough lifecycle planning. It is also the result of unplanned, unbudgeted, and poorly identified system physical or financial impacts on the system (project delivery model) due to a lack of comprehensive lifecycle planning, specification, and management.

The darker side of organizational angst can display itself as mental health issues and problems. Triggers range from excessive overtime work schedules, product, and supply chain issues and an all too common emphasis on reducing costs without clearly addressing the results of those reductions and their impacts. Many PV professionals are required to take steps that are contrary to how they have been trained, educated, or experienced with and that are counterproductive or even personally and ethically questionable. With often growing uncertainty and reduced consistency in many areas, anxiety can creep into a once confident team. Some of these may be serious, while others may be a continual erosion of good specification and design practices through O&M practices or other contradictory conditions at other levels of the project delivery process. This has resulted in substantive frustration for many PV professionals leaving them to deal with unresolved conflicting stakeholder requirements and internal personal and organizational conflict. For many exceptional and talented people, this has led to depression and/or leaving the industry. One author has personally known two skilled and talented PV engineers who committed suicide, in part due to the ongoing stress of following directives that were technically and/or ethically inappropriate. Though in extreme cases, uncontrolled organizational links can result in depression, behavioral incidents, and walking away from their jobs.

With the application of the PAM methods, once these issues are identified and resolved at the concept and SE phases for new or phased **R**epowering™ planning stages, less time and money are wasted at many levels within the project and the organization. This approach reduces unnecessary pressure to deliver and operate a weak system, while providing greater consistency, results, and profitability.

1.5.4 Perspective

Most of us would agree that everyone has a perspective based on personal experience, education, or available references, and in their (our) minds, their perspective is "**right**!" (i.e. the best or most accurate). These "assumptions" often drive the decisions of individuals. Our position is not to argue with that "fact"! However, we wish to point out that unless stakeholders have common agreed-upon definitions and perspectives, it is very difficult or impossible to achieve common goals and objectives. This is where the efforts to address all stakeholder needs reduces conflicts, misinformation, gaps in planning, and specification on through to project site restoration.

We are in a "post-fact-based era" or more succinctly, in the era when repeated business, political, or marketing statements (and myths) or actions are granted credibility without substantiation as facts. Even agreeing on basic scientific and technical concepts can become a challenge. Why waste time getting tied up in all that boring academic, detailed science and engineering stuff when the marketing information should tell us what we need?

...Right?

Yet, how can we achieve much of anything, if we cannot agree on a series of common facts, concepts, perspectives, and definitions? Basic notions – like agreeing on metrics, taxonomies,[8] ontologies,[9] and what are the project plans, goals, technical assumptions, and objectives – are essential. These agreements are used to make basic decisions to reach a common knowledge base, which is then used to make further decisions to improve the viability of a project. Something as simple as how we define project success or failure must be clearly identified and seen as an element of a common perspective. The understanding of the stakeholder perspective drives what is required, expected, and completed, and how it is accomplished. The fact is that unless a common set of perspectives is agreed upon, chaos becomes an underlying element of the project and results in unnecessary costs and risks.

When we go into a meeting, the focus is primarily to have our personal, professional, or corporate perspective drive whatever process or project we are working on, recognized, and addressed. Much of this drive is motivated by image, winning at all costs, or forming a certain reputation. It can be rather divisive. It seems to be a useless result to win the battle and develop and install a low-cost system – yet lose the war, with reduced long-term profitability. When a project team goes into the planning processes, if they focus on the variety of common perspectives in the gathering, they can begin to see and construct a clearer path to commonality and agreement.

One point of commonality among current projects is that the industry often wants to make the most money (profit), while making the lowest possible investment in time and capital in the process. That perspective would appear to be very simple and easy to share. Achieving it in reality is where the variation in perspectives begins to break down and creates challenges. Understanding the actual short- and long-term risks and addressing them should be of foremost importance to all stakeholders. As a result, addressing all the facts is the first step toward success.

Time plays a major role in all the perspective-related decisions and the eventual results! Consider the following example:

> If I am the financier, my goal will be to make the greatest ROI as quickly as possible at the least risk and to get out of the "risk pool"[10] before anything messy, bad (or substantially unprofitable) happens. Therefore, for a 20- or 25-year project, my efforts and perceived responsibilities are focused only on the first year or two, or in some instances the first five years of the investment or payback period. Essentially, my strategy is getting my payback and ROI while shifting/**shedding** the accruing risks to other downstream stakeholders by selling the plant off after a few years.
>
> Meanwhile, the risk-shedding process damages the system and results in much higher risks being assumed (whether known or not) by the next owners and operators.
>
> If you are the developer, you may play a number of roles, including those as a financier and/or EPC. Today, there are many combinations of these and other stakeholder roles. Or, as the EPC, you want to complete the terms of the contract, get paid, and wash your hands of any future responsibilities, challenges, and obligations.
>
> If you are an O&M organization, you certainly want to be profitable. In many instances, however, you are having to *bid for contracts below cost to win them* to stay in business, and you hope that opportunities to charge additional "time and materials" to the contract will bring it back to profitability over time. Your challenges are, in good part, the result of a ripple effect from poor decisions made far in advance, during the phases of concept, planning, or specification. The problem is that you are not alone.

What is being done at this point in the history of the industry to address the continuity and stability issues in the PV universe? In other words, if things are out of whack at specification and design, it flows through the whole project and its lifecycle. This conundrum is not a result of effective and healthy competition; it is the result of ineffective or incomplete process and management practices, the need for greater awareness for industry sustainability and growth, and in some cases counterproductive greed.

The above perspectives show how short-term thinking may result in the looming of a series of long-term financial disasters for the *Dumb Money*! This is especially risky for undereducated or underinformed stakeholders and future plant owners.

No one ever plans or wishes to be the Dumb Money!!!

Yet, as a result in many instances, they do not ask the right questions nor do they have adequate knowledge to fully understand what those right questions are to give them usable information. The dumb money tends to be those who focus on what *might be* the assumed financial outcome, return, and total value proposition. They tend not to see the necessity to carefully plan for and cost out all eventualities and have no objection to fly by the seat of their pants. As a result of accepting many of the common industry myths and assumptions, they make business decisions without looking at all the details or understanding the reality of how a plant functions *throughout its entire lifecycle.*

Effective examination of the risks includes understanding all the costs, or what is called the "all in" costs, and the entirety of all the potential project challenges. Risking to look dumb by asking an uncomfortable question is a lot more intelligent – and valuable – than looking smart, failing to ask critical relevant questions, with ugly fiscal results from a lack of details. Asking those dumb questions may get your organization critical information that will keep you out of the category of Dumb Money. In other words, asking and answering those critical questions may make you profitable and sustainable over time.

Unfortunately, they often identify these problems after the fact.

When we see companies that are doing unprofitable business, it impacts both the company and the whole industry. News about organizational bankruptcies and lawsuits is not a good thing for the industry and its stakeholders. With today's model focused on the *incomplete least-cost model*, stakeholders will never get a system that is "whole" in meeting most if not many stakeholder wants, needs, or expectations for the long term. The industry must take a closer look at what it takes to make the industry healthy and wholly sustainable. It must no longer support or tolerate those participants who have mastered shedding the risk to someone else. It also means all stakeholders must be educated and trained to consider a more global view beyond their primary – but often siloed – skills to understand the system, its dynamics, and the financial decisions that must be made to reduce the potential number of failures. This path will guide the industry toward maturity!

1.5.4.1 The Catch-22 of Common Industry Assumptions and Subsequent Questions that Arise

Risk shedding, or shifting, may allow some stakeholders to meet their primary goals and objectives in this calamity-laden PV project delivery development approach. This approach creates a series of catch-22s[11] that leaves the vast majority of project participants behind, or in some instances, damaged. Stakeholders must no longer be willing to meet the requirements placed on the organizations to deliver a project that works ineffectively during the system lifecycle. When individual stakeholders' goals and objectives are inconsistent or counter to the other project stakeholders, there is always a downside. The results tend to be unexpected systemic problems and losses resulting from the specification, design, construction, or overall delivery of a subprime PV system. Because their goals are not stakeholder group goals, we all too often see the vaguest minimums in quality, reliability, and functionality that can be negotiated contractually, to reduce short-term perceived risk. This perspective is often driven by the view that *it is not their job to be concerned about what happens after the fact*, i.e. within the warranty period or system life, even if their process creates or delivers unnecessary problems and unanticipated expenses. This situation creates a serious set of ambiguities where on the one hand, we need to make money at the least risk, while on the other, we would still like to be seen as having performed an outstanding job to ensure future work. With our PAM/**R**epowering™ planning process, the system stakeholders can achieve those goals.

We have many friends and colleagues who have tortured themselves over these issues. This approach, nevertheless, has led to the PV industry today where there is an ongoing conflict with a series of unexpected dilemmas or financial losses for the stakeholders, especially owners and end users. This scenario is true for many of the stakeholders in the process of developing or delivering projects: "get in and get out as quickly as possible." This time-constrained perspective creates a further series of internal challenges for the project.

There should be no expectation of perfection! Nor do we believe that the whole industry will immediately move toward this type of PV system delivery model and approach! It will take time and effort! However, as more of our colleagues move toward identifying these risks and challenges, it will improve the overall PV industry and power infrastructure. Our approach is not a simple solution – think of it as a roadmap. As industry participants begin to broaden their perspective of industry needs, the whole industry benefits.

If one is buying a project right after it has been constructed, the risk is generally, although not exclusively, perceived by the buyer to be minimal. The buyer generally believes:

(a) The design "as built" contracts account for all operational requirements and environmental concerns, conditions, contingencies, and issues.
(b) The components selected are reliable and of good quality with acceptably strong warranties from companies that will be around to service them.
(c) The plant is properly specified, designed, constructed, and installed by professionally skilled and well-trained personnel.
(d) Its operation will meet all the legal, regulatory, and other requirements whether built into the purchase contract or not.
(e) The plant delivered meets "all" applicable standards, and specifications including those listed in the RFP and others essential for project success.
(f) After the first year or two, the bugs will have been worked out of the system.
(g) The plant will generate power and revenue based on the carefully engineered, detailed, and accurate design. That design is based on accurate lifecycle costs (as per LCOE), as represented in the financial analysis whether before construction or after purchase.
(h) Our contracts protect us from a broad range of potential negative outcomes, thereby lowering our risk.

The above list forms the basis for questions that a buyer might consider asking prior to buying an operational plant. After experiencing over four decades in the industry, we believe everyone can do well, in fact much better, with an effective "systems process" with less waste in time, money, resources, and frustration.

Yet, in the present boom and bust mentality and industry cycles, there is still an incredible lack of consistency and adequacy at many levels during the project delivery process. Part of that contractual inadequacy is the assumption that *skilled experts* have designed and built this plant for an owner's long-term (20–30 years) use and profit. Therefore, as a long-term owner/player, "I have reduced my risk based on *their* due diligence and by hiring an independent engineer or series of experts, often our inhouse Pros!"

Nevertheless, additional questions arise over time:

- What if there was an insufficient specification budget for the independent engineer to be able to effectively complete the system evaluation?
- What happens if the data supporting that due diligence was incomplete or flawed?
- What if the essential information and details were never considered or addressed?
- What if the business model did not include all the attributes needed to meet all long-term expectations?
- Did a few days (or weeks) and a few thousand dollars of independent engineering effort provide sufficient information to actually make the most important fiscal decisions?
- What level of due diligence was required, and did it provide an effectively accurate evaluation of the plant, its ability to meet the lifecycle requirements, and a profit while doing so?

- What is the likelihood that the plant will meet its stated and projected (modeled) performance numbers throughout its lifecycle? (see P-Values in Chapters 6, also (https://en.wikipedia.org/wiki/P-value))
- To meet its performance numbers, have the health and condition of the plant including its RAMS metrics been effectively and accurately collected, curated, assessed, considered, and included?
- Were these RAMS metrics realistically developed, modeled, and evaluated to include relatively accurate long-term "revenue versus cost" financial projections?
- Is the evaluation process sufficiently robust to result in data that is informative enough to provide a high level of confidence that the system is working to specification?
- If it is not meeting expectations, how and when will we know?
- What was included in the original group of specifications and are they sufficient to meet lifecycle needs?
- Is there a clear understanding of the variability of the selected components RAMS attributes and supporting metrics? Note: Factors or conditions that stakeholders do not understand are often labeled as "academic." On a number of occasions, we have seen the client's eyes glaze over as they're thinking, "Why do I have to hear all of this academic garbage; it's not my concern?" Addressing those unknown or unvalued attributes and issues regarding RAMS will drive down costs by uncovering measurements required to make effective, cost-benefit tradeoffs and choices.
- What are the guarantees, and who has the responsibility to see they are protecting us?
- Are those guarantees and warranties actually sufficient, and who is managing them?
- What are the consequences if the information provided is not sufficiently accurate, reliable, or specific enough to meet my needs?
- What is the temporal (time-based) confidence level regarding the reliability of the plant and components as designed?
- Will it be delivered and performed per specification as contracted over the life of the plant?
- How will my organization and my investors be assured those good decisions have been made?
- How are success and failure defined and what additional steps are needed to ensure success?

1.5.4.2 Stakeholders Are Often Held Back by the Simplest of "Assumptions"

Stakeholders assume there is an additional cost, some even perceive it as excessive, when they delve into topical questions, resulting in the default assumption that all of this fact-finding is "too expensive"! However, stakeholders are actually creating problems for themselves when not closely examining these topics: the need for additional specification, due diligence, collaboration, and planning.

The use of the term "too expensive" is simply a throwaway line to effectively kill topical discussions, drown a series of critical questions or issues, or imply there is no interest or desire to gather the substantive information or data (Chapter 3). In today's market, it is still common to hear that the solution is to leave the important decisions up to the "experts" and/or other professionals to resolve them later. The next time the issue of "too expensive" is interjected into a conversation or process, the following questions need to be asked and answered:

(a) "If it is too expensive, what are those actual costs?"
(b) "Can you please provide those numbers with the detailed documentation so we can make more effective management and financial decisions with the facts, metrics, and supporting analyses?"

There are a growing number of organizations in the industry that have gained some of these capabilities, yet they are still very much a minority. If a business is buying and selling systems, their perspective will support their long-term goals, which may not and *probably will not* be yours.

Unfortunately, a conflict often appears between the desire and expectation of the stakeholders when meeting project goals and objectives. On one hand, some stakeholders lack the desire or need for clarity in communications, accurate information, and decision making. On the other hand, many stakeholders are beginning to question what they are seeing and are looking for improvement. Hence, with insufficient skills or knowledge, too many choices and decisions are left to *others*, often without even knowing who these decision makers are. The *others* may bear little or no responsibility for the consequences of their incomplete or faulty advice and decisions, and they may not have a clear understanding of what owner's want or need. This gap is the real reason why we focus on stakeholder perspectives in their definition of success and failure!

The greater challenge is the difference between expectations that are not defined and the realities of flawed business models that encompass a faulty, incomplete decision-making process.

The result of the above is that even at commissioning, the plant has not been specified or designed to meet the existing or future service life expectations or needs of owners and other stakeholders.

This scenario boils down to stakeholders believing in marketing claims and industry myths and assumptions, mistaking them as legitimate, while failing to perform sufficient due diligence for higher profit and reduced risk. This common approach has not carefully taken into consideration the differences and perspectives of the project participants. With failure to gather and use significant data and documentation in the contracts and specifications, the expectations of these stakeholders have an extremely low probability of being realized.

This problem is not new. It has been endemic in the industry for over the last three decades. Although for the most part, most stakeholders know about at least some of the problems. However, knowing about the problems or challenges and not acting on them is not the road to industry maturity or profitability. So, the question is:

What is more important? Buying the miscommunications, hype, and assumptions? Or guaranteeing that the plant being specified, designed, built, and can be operated to meet the long-term project goals in a meaningful manner that can be clearly defined and enunciated between all stakeholders?

We focus on the concept of building better plants that last longer and produce more energy over time at lower "all in costs," which results in far lower LCOEs. It is written for all the stakeholders and not just a portion of those interested groups.

With the incredible reduction in PV system delivery cost and a subsequent decrease in unexpected O&M cost, we suggest the industry consider taking a few percentage points of those reductions and reinvesting them into building better systems that will actually meet the needs of all stakeholders throughout the plant lifecycle.

1.5.5 Loss of Corporate Memory

The reality is that over time, all humans and industries tend to fall back on old bad habits. They often develop and gain great knowledge, wisdom, skill, and ability, then lose sight of its value, or get just plain sloppy. This loss of knowledge, interest, or corporate memory results in the failure to learn from past mistakes and/or the inevitability of making the same mistakes repeatedly. This slippage is why effective processes, procedures, and lessons learned must be ingrained through institutionalization and *continuously updated*. Accurate analysis and adoption of improvements must become a part of the organization's culture. At the same time, the organization must continue to validate and modify those processes. If management does not make this a priority, success is at risk.

Loss of corporate memory tends to exacerbate budgets that do not include sufficient training, internal mentoring, and mentor shadowing. This, along with the tendency to rely on younger staff while retiring more seasoned professionals, saves money but loses technical expertise.

An example of this is Boeing's attempt to reduce overhead by offering older experienced engineers retirement packages and relying on younger engineers to perform at the same level. The Federal Aviation Administration (FAA) has subsequently challenged the independency and capability of these engineers to certify system/aircraft safety.[12,13] This budgetary approach tends to raise medium and long-term costs while not passing on the benefit of previous lessons learned. Procedurally, this lack of institutional knowledge application weakens the whole team's ability to meet all stakeholders' needs and expectations.

The PAM **R**epowering™ set of processes, procedures, and approaches (as presented in Chapters 4–8) requires effective program/project process planning, requirements, specification, and risk analysis documentation and reduction – prior to EPC bidding. Preemptive planning addresses the key (and most of the minor) elements of delivering a "high-performance" PV system that meets expectations and specification. As such the PV system maintains its asset value longer, while providing greater predictable energy output over time. It requires shifting priorities by *focusing on value over time*, and the plant's lifecycle! This mentality is the true meaning and value of an *effective cost-benefit analysis*!

1.6 Application of PAM

The key to the PAM approach and process is to specify, design, build, and operate plants that include a series of plans and contingencies within a complete **R**epowering™ planning process to extend the PV plant life and its commercial value. It results in plants capable of lasting over 50–75-plus years, about the same historical time frame as other utility power generation infrastructure. These results are based on a focused and well-defined, upgradable/replacement plant **R**epowering™ process and design (Chapter 4). This design process includes a recognition that every component fails and should not be expected to operate across the entire lifespan of the plant.

PAM requires the EPC and its suppliers to follow contractual requirements and specifications for components that are more robust, interoperable, interchangeable, available, replaceable, and maintainable. Those components will have fewer replacement cycles because the plant has been specified, designed, and built to be more flexible for later modifications based on a well-developed and defined **R**epowering™ plan. It requires collecting and using better, more actionable data. This improves decision-making and results.

The reality is that the PV modules themselves are complex due to the chemistry and physics of PV cells and the sheer numbers of cells, devices, and interconnections. Inverters are becoming increasingly more complex, with evermore exotic technologies, complicated by the implementation of new power and computing technologies and software. All the intricacies and complexities are further complicated by how the system might be designed and installed in the environment where the components are placed. Even DAS and SCADA add to complexity when security, fault detection, fault isolation and detailed analysis, and cable or digital communications are included and used in the system. With the addition of energy storage, battery management systems (BMSs), and complex energy distribution systems, all with daily and independent variations, the myth that "PV systems are simple" should completely evaporate and disappear.

Yet, the Myth of PV Simplicity persists!

1.7 Cost Control Considerations

A persistent problem lies with the sourcing of components. In the PV component industry, the last 40–50 years have demonstrated an increasing acceleration of new and often improved technologies. The rollout of these new technologies has gone from approximately every two to five years to new technologies introduced about every 12–24 months. This shortened cycle creates problems with an ever-changing bill of materials (BOMs) for modules, inverters, and balance of system (BOS), especially after equipment or component certification (Chapter 3). This timeline has also accelerated component planned and unplanned obsolescence, causing energy and equipment unavailability. This significant demand places stress on finding spares in a timely manner to address voltage, current, and "form factor" changes (component physical geometry changes). It has often made fit and function replacements extremely difficult to find (Chapters 6–8). The addition of this "technology fatigue" is another critical factor in shattering PV myths while driving up unanticipated costs.

PV history is rife with plant challenges (Chapter 3) due to the assumption of its simplicity and failure to learn from past mistakes. One lesson is the belief that components and equipment will be delivered as "reliable products." When the product reliability is researched, oftentimes one is referred to marketing information rather than researched test and analysis results. In addition, lifecycle component failure data is seldom available or made available.

Many industries have fallen prey to the marketing term "reliable product," see Chapter 5. First, what is the definition of reliable? Second, what is the supporting data?

To procure a truly reliable item, one needs to identify its requirements and then specify the item, including its reliability with respect to site environments, users, usage attributes, and how they are tested and measured. The attributes of plant SE, reliability, and maintainability affect all aspects of construction, operation, production, availability, and finances (with positive and continual profit). By addressing fiscal and physical realities early, most of the plant concept and design decisions can be made to preempt previously under-identified and/or known but downplayed challenges. This approach can result in the avoidance or mitigation of substantive short- and long-term and consistent power generation challenges. Doing so is a major risk-reduction strategy.

By failing to address preemptible and controllable issues early, the industry has created a "high risk, speculative PV power generation, and energy market environment." This market has been demonstrated by a significant history of plant and corporate failures, closures, liquidated damages, unanticipated costs, and bankruptcies. This path drives up the cost of energy over time by paying too little attention to controllable costs or properly and accurately weighing all cost benefits with valid information. It can be an operational budget buster when components last a fraction of their predicted life. Low reliability and difficult maintenance procedures lead to higher maintenance rates, hours, manpower, and spare costs, which lead to significant O&M budgetary shortfalls (Chapter 11). As of 2022, efforts to provide more PV-based energy have risked billions of dollars of energy investment over the next decade or two in the United States alone. Further, the international market is exploding with a growing number of communities relying on PV technologies to effectively solve energy and climate change issues.

From our experience, to date, no one individual or organization in the PV industry has published or provided an accurate projection of a plant's LCOE.

This is based on decades of work in the industry, questioning our colleagues in all sectors of the industry and doing a more extensive series of internet searches of the available literature over the last two decades. We suggest that, as it is commonly accepted, the LCOE must be based on the "all in" cradle-to-grave costs and updated annually. "All in" must include costs that are generally ignored or overlooked today because the current project delivery models don't appear to include all cradle-to-grave costs essential for the generation of an accurate LCOE. Nor do they often accurately identify and include the cost of inflation tied to an appropriate Cost of Living Adjustment (COLA) over the working life of the system, effective labor and management training and education, the impact of shortages, and product availability.

Until this is done effectively with far greater accuracy, the least-cost project delivery model looks inexpensive because, with all the unidentified costs, it appears so. Yet, as a result of insufficiently addressing real and documentable lifecycle costs, the first cost/least-cost model will deliver the higher cost of energy as ignored costs overwhelm budgets early in the system lifecycle.

The perspective of an accurate "all in" cost provides the data and trend analysis to improve existing and future plants.

For accuracy, it must also include calculations for the Cost of Living Adjustments (COLA).

The effective service life of the PV plant's electrical energy generation and storage are primarily determined at the time of the SE concept and requirements development, and their specification prior to EPC bidding. SE-based decisions are a key part of the effective development process that defines the "all in" plant costs, energy output, and the performance of the plant over its defined lifecycle. It must also include inflation and real or manufactured price pressures. In our calculations, we have been using a 3% COLA. However, to address a worst-case scenario, using a range from 2% to 5% will provide possible insights to real costs throughout the system's life.

When tracked annually, these calculations are used to address the actual cost of the plant as constructed versus the modeled plant. Again, decades of internationally documented evidence support this position. It is borne out by excessive O&M costs, failed systems, imposed liquidated damages, and other unbudgeted or unaccounted items that directly increase the LCOE.

As an example: With few exceptions, we as an industry have not effectively addressed the costs of module and equipment disposal and site restoration. This topic is an emerging issue requiring new standards by standards organizations, city, county, state, provincial, regional, or national regulatory agencies. With some technologies requiring special processes and handling procedures or are labor-intensive at disposal, the consequent cost can be substantial, yet generally are not accurately accounted for either in budgets or LCOE calculation.

Through hundreds of conversations over the last decade and more, we know that *many experienced PV professionals want to see changes*. Yet, they have the sword of Damocles hanging over their heads. That professional angst says to do it properly, while the sword hanging over the industry mistakenly says, "It's too expensive!" The lack of published data indicates that few organizations have gathered sufficient detailed data or performed the analyses to make the case for the lifecycle perspective. To address this situation, the industry must adopt a more functional systems model that examines all issues and not just the convenient ones. However, breaking old habits is difficult, and yet critical to industry's success and viability. We have reached the junction where the whole stakeholder chain must act to shift priorities.

All too often new plant owners discover a raft of previously undisclosed and undiscovered issues following their due diligence. In essence, losses tend to increase and will continue to do so until there is an industry-wide system rating and certification standard to address a plant's asset value and its real health and condition. As of 2021, *International Electrotechnical Commission – Renewable Energy* (IECRE)[14] is in the process of developing such a standard.

But let's be clear! LCOE is and will continue to be a key financial and performance indicator. It was developed to define the actual and real cost of energy provided to stakeholders, especially retail end users. To fully understand the historical variability of LCOE, it requires a dedicated DAS or SCADA with measurements of the true plant health and condition down to the "repairable item level."[15] Currently, many plants have minimal, if any, significant health and condition monitoring and fault detection for maintenance below the inverter level. The systems often have no ability to provide fault isolation to the lowest level of repair unless it is built into the inverter. Providing a reasonable accurate LCOE at the concept and design phases is dependent on the depth, quality, accuracy, and detail of health and condition data from existing plants. This assessment can only take place when there is an industry-wide data sharing and analysis process (Chapter 10). That critical information must be included in the system specification. It is grossly affected by how faithfully the total system delivery process has been employed.

Thus, to track and project the system LCOE, the method requires an annual update of the initial model parameters from plant data to provide a more accurate basis for the operating cost, and ultimately for the plant revenue.

However, much of this is being addressed with new remote-sensing technologies. A number of emerging technologies from other industries can fill that gap! We address some of those noninvasive technologies in Chapter 10, Data Collection and Analysis.

1.8 Project Versus System Delivery Process

1.8.1 Project Delivery Process

The existing process is termed "PV project delivery." It begins at the project concept stage and usually ends at the commissioning and commercial operations date (COD) or the end of EPC warranties. It is based on the lowest or least-cost approach. It is damaging in four major ways:

1. A poorly specified, designed, and controlled bidding process;
2. The reliance on a project delivery model as opposed to a system delivery model;
3. The failure to provide an effective system delivery process;
4. The failure to collect detailed failure and operational data, to curate that data, and to distribute the analysis and pertinent information to stakeholders in understandable and actionable reports.

This information is required to build functionally better and more cost-effective plants.

We have seen some of the outcomes of PV-sourced fires in western states including a wildfire that resulted from improperly installed and unmaintained wiring. Numerous ignitions atop warehouses such as Walmart and Amazon highlight examples of systems not performing as their investors anticipated. The RAMS processes should have been followed to achieve and maintain reliable long-term energy production. Unfortunately, in many cases, in the rush to deregulate and restructure, **they were not!**

The PAM process focuses on "PV system delivery," which is a service lifecycle process that begins at the concept and ends with site restoration.

1.8.2 System Delivery Process

PAM differentiates today's PV delivery in two ways:

(1) PAM implements and uses SE plans and processes from concept to site restoration.
(2) It is firmly based on the acquisition, curation, analysis, and assessment of current and past plant data as a fundamental foundation for new plant modeling and predicting plant performance.

In the PAM approach model, we provide a PV systems delivery process that focuses on the full lifecycle of the plant, from concept through site restoration, with in-depth attention paid to the "all in" costs. It is based on a framework defined in the **R**epowering™ plan (Chapter 4) categories and subcategories that govern what type of planning must be completed to deliver a cohesive set of system and RAMS specification with associated requirements (Chapters 5–8). "All in" costs include *all* the lifecycle project costs and the losses from each project. PV system delivery presents a complete concept to site restoration approach that includes the processes and procedures to establish and meet successful short-, medium-, and long-term power generation and financial objectives.

Note: Today's system removal, abandonment, or deconstruction is not site restoration!

Site restoration requires that the site be returned to its original useful state or agreed-upon long-term usage after system life state.

The criticality in all of this is particularly related to the issue of planning, which addresses or underaddresses PV plant ownership. Most planning today does not effectively consider the "chain of ownership." It primarily focuses on the first owners. The first owner/s are generally an internal or external developer who may sell the entitlements or the completed plant. With a lack of foresight past the initial ownership, the needs of future owners and the customers are often ignored. This type of planning has created an overwhelming dilemma for the industry that is still not clearly understood or widely discussed today.

This oversight ignores the long-term interests, needs, and requirements of a functional PV plant throughout its lifecycle.

The PAM **SE/R**epowering™ systems delivery process approach has as its foundation, the design and development of historic utility electrical generation systems based on coal, oil, gas, nuclear, and hydroelectric energy sources. It is focused on "total project value" while delivering better performance and revenue at a more clearly defined, optimized (included) "all in" cost. This methodology has also been found in the industries of marine, rail, auto, medical, aerospace, and many more.

The minimal additional costs of construction and delivery that go into SE-RAMS and **R**epowering™ planning are financially justified (Chapter 5) by the added value of a PV system infrastructure that is designed and built to be updated. The actual LCOE is dramatically lower than if plants are designed and built for 20–30 years, as assumed in the PV industry today, yet not sufficiently reliable enough to operate effectively through most of that proposed life.

Due in part to deregulation, much of what was considered "organizational fat" at many utilities was reduced, including the historic focus on reliability, availability, maintainability, and safety. The restructured energy production and delivery resulted in disjointed systems and grid O&M practices to artificially lower energy production costs.

Nevertheless, the electric utility infrastructure model was abandoned around the 1990s during the utility industry restructuring and deregulation process. Deregulation was rolled out with the assumption that with less regulation it would cost less to build plants of the same quality and reliability to deliver lower-cost energy. We will point out that without adequately addressing fundamental long-term lifecycle needs, the long-term viability of the PV plants themselves falls apart. Though fiscally appealing, politically attractive, and rich in marketing promises about lower costs, the rollout has provided three decades of underplanned, unexpected fragility in the grid and energy production infrastructure, and in most instances has led to a game of catchup.

However, we are not alone!

Although aerospace and many other industries have used lifecycle cost in the past, the shift to a least-cost, fast-tracked development processes has led to a series of significant failures. A part of the LCC often overlooked is the cost of failure – one example of which is the Boeing 737 Max. As a result of the initial shift away from the fundamentals, in this case, Boeing has lost over $20 billion[16] over what may have been a change of $5–10 million with a requirement of an additional two to four months process for FAA flight certification for new safety of flight hardware.[17]

In the automotive industry, the use of unreliable airbags[18] resulted in tens of millions of vehicles recalled, billions of dollars in replacement airbags, more than 15 deaths[19] and associated lawsuits, loss of consumer trust, and ongoing legal cases of related safety issues. This shift in many industries has led to other examples of a collapse in goods and services, along with financial ruin for companies. For the PV industry, it has resulted in a number of solar power system fires with significant financial costs and safety risks for people. The result of this shift away from an LCC initial effort has delivered a negative impact on LCOE and on the stakeholder economics of finance, human health, and safety.

Other energy sources and technologies have stringent reliability and safety requirements developed over time. With utilities, their failure could take out electrical power, leaving millions of users in the heat or cold. Following hurricanes in Texas, in February 2021, and Ida in New Orleans in September 2021, there were major grid failures with hundreds of thousands of people without heat, power (conventional and Solar), and significant damage to the infrastructure. These failures created safety, financial, and property losses as well as personal hardships. These issues are happening at an accelerating pace as a result of least-costs programs and the growing impact of climate change. They will continue if not accounted for in their design!

Historically, electrical utilities had diligently, adroitly, and heavily focused on reliability and safety, accomplishing them with great pride. Yet, they are still subject to *force majeure* on top of today's growing challenges, lack of effective O&M, and resulting deregulation while reducing the emphasis on RAMS, SE, and far more intensive planning management. This has resulted in the utilities' depletion of their energy and O&M reserves – their rainy day funds.

Avoiding and reducing these challenges required their power plants and distribution complete specifications prior to EPC bidding, so that all bidders were bidding the same plant specifications and details. Substantial levels of specification and engineering were thus completed prior to the bidding process. Those requirements allowed owners to do a comparative assessment of each bid, resulting in a set of reliable and optimized costs for the delivered plant. In that process, the owners collected sufficiently valuable information with analysis to improve all future plants.

To properly address the issues above requires beginning at the concept and specification phase.

All stakeholders should contribute to and understand all their costs throughout the lifecycle of the system. This approach does not mean they need to know or will necessarily understand every detail. It is a path to start learning from history (as an industry) and begin to collect the appropriate information to avoid future problems – and having to respond to them when they could have been prevented. To make the needed improvements in plant efficacy, any plant issues arising between the initial concept and site restoration must be studied with the intent to update new and existing plant requirements, which are to be used for development, specification, and delivery for future projects. This assessment must include project, component, and field data metrics that are addressed and accurately factored into the plans of project specification, engineering, finance, regulatory, site, and other planning iterations.

1.9 PAM Concept

Our thesis is simple:

> The existing PV project delivery model and process are fundamentally flawed.
>
> The industry will only improve dramatically with a recognition by all stakeholders that the present "least-cost" approach is doomed to failure while costing the industry billions of dollars. The most effective redress is to adopt and accept a process of optimized **SE/R**epowering™ for plant development. It must be based on detailed system specifications and statements of work that define development, specification, operations, maintenance, disposal, and site restoration. This solution requires remolding the entire PV system plant delivery process and its practices.

Throughout the concept and planning phase, detailed specifications are developed to provide information to support the EPC's bidding process. A comprehensive specification contains the elements of the project concept, goals and objective, and all component specification within a set of appropriate operational and environmental site factors, which are addressed by an extensive set of "limiting factors" (see Appendix 11.A). These are a broad range of constraints that need to be addressed early.

Addressing these limiting factors will define the plant and its total lifecycle energy production, costs, revenues, and viability. The less detail provided in the specification, the greater the risk that substantial, unanticipated, and unbudgeted cost increases could occur in each phase between concept, repowering, and final site restoration.

Substantiation must be based on current, detailed plant performance, plus RAMS and O&M data, without which many projects are doomed to repeat history. Long-term success is, in good part, measured by adequately addressing all PV system limiting factors. Addressing these factors aids in defining plant life, its energy production, and its financial viability from concept until site restoration. Failure to address one of these factors, for example by underaddressing RAMS, accelerates negative growth over time in system viability, power generation, and energy availability. At the same time, there are still many people who say that we have too much data, that it is unnecessary or suspected in value.

Then, others see the unrealized potential. A colleague, Steve Croxton of Ameresco, noted a few years ago (December 2016): "We have all of this data we don't know what to do with. We know it must have some additional value, so how do we use it to make our plants work better, last longer, and produce more energy while lowering O&M cost?" This question and further discussions have resulted in his company beginning the process to take much better advantage of existing data while determining what additional data is required for long-term improvement.

1.10 Challenges Today with the Bidding Process

Initial design and installation costs are driven down at the bidding process, setting up a qualitative "race to the bottom." Bidding has often become an effort to win a poorly defined and constrained (least or lowest) cost competition. This dilemma is created by building an incomplete set of specifications on another poorly defined set of organizational/stakeholder and system goals and objectives. The process has been inappropriately termed "system optimization" when it is in reality a "least-cost plant capability minimization."

Alternatively, the term "lifecycle cost optimization" is the optimization of both the system cost in capital expenditure (CAPEX) and operational expenditure (OPEX) over the lifecycle of the plant. This method accomplishes a clearly stated and defined set of project and system goals and objectives. Details include all costs over the lifecycle, not just those leading up to the COD or the first 5 or 10 years. It is not focused on an optimal, lowest CAPEX cost as presently defined at commissioning or COD. Today's "system optimization" does not effectively address the total critical lifecycle and especially environmental costs and their impact.

Another abusive euphemism refers to today's system or cost optimization as "value engineering." The abuse of this engineering language results in a lack of attention to and consideration of essential information needed to make sound system lifecycle specification, procurement, and design decisions. Value engineering tends to result in the "gutting" of system RAMS characteristics. It becomes a race to minimize the CAPEX to achieve the lowest cost of project delivery at COD, which historically tends to dramatically complicate and raise OPEX over time, thus shortening plant life while reducing energy production and profitability.

Consider again the example of the Boeing 737 Max. It was designed and delivered based on lowering development cost, meeting a tight schedule, and with what ended up being incomplete levels of testing to verify and validate the hardware and software design changes in performance as documented in the US national and international media.[20,21,22,23] The perceived need to rush the product to market overruled and undermined the issues that define an effective RAMS process. In Boeing's case, it appears management, as with many other organizations and industries, had temporarily lost their way in meeting their historic focus on RAMS. Doing so cost Boeing a substantial amount of money (more than $25 billion as of 2021), reputation, and no doubt it will have a negative impact on the company and economy for many years. This outcome is primarily based on poor financial management decisions focused on short-term stock value and not long-term organizational health, survival, and growth. As a result, they have created grave concerns and doubts in the minds of their commercial customers, organizations, regulators, government, acquisition agencies, and ultimately their passengers.

This similar mentality of choosing the lowest sunk cost is presently rampant in many PV organizations today. It is driven by developers, financiers, owners, operators, and utilities pushing for the lowest initial delivered cost and LCOE as they attempt to force the construction loan size, financing, and projected O&M costs down. Their goal is to reap the greatest and earliest possible ROI and profits while ignoring the consequences of such limited methodologies. We do not question their right and potential value to the industry to be profitable. The questions we ask you to consider are: "What is the true value of the product being delivered?" and "How much profitability and revenue are left on the table with this model?"

These project delivery practices have thrived as a result of the broad length and depth of the industry buying into what we refer to throughout the book as well as ingrained PV myths and assumptions. This continues because there is a lack of critical data shared in the industry thereby allowing this business model to thrive and continue.

Developers and owners tend to reduce LCOE by slashing O&M scopes of work. O&M companies primarily make their best money when there is significant profit in providing the necessary maintenance support services based on cost-plus contracts. This formula falls apart if there is no budget or revenue to do the necessary work or buy the necessary spares before and after the system begins operations, thereby undermining system reliability and availability. This latter point is driven in part because no representative on the plant operations side, or elsewhere in the project, appears to have a responsibility for RAMS during system specification, commissioning, and/or before warranty periods end. Operators may have detailed responsibilities, yet in the overall plant operation, RAMS is not a key element of those efforts. This situation leads to putting off maintenance issues until they become a critical performance challenge that could result in liquidated damages (LDs), breach of contract, or bankruptcy. Under this scenario, a major challenge prevents the system from meeting its contractual performance, while there is insufficient or no budget for appropriate plant upgrades or repairs. The result is a substantial loss of energy output and revenue, leading to increases in the real cost of energy that define the "all in" costs of LCOE.

O&M teams are saddled with meeting the milestones, performance, and availability requirements for a plant without effective systems built in. They are now expected to resolve issues where there was never sufficient budget (and potentially never will be) to meet those essential requirements without a corresponding and substantial increase in LCOE.

Past and current practices have been to provide as much latitude in the specifications as possible for the EPCs to bid at an increasingly *lower Capital Expenditures (CAPEX)*. It results in the specification being very loose and open to a broad range of interpretations and changes. This latitude also results in insufficient checks and balances with little or no requirement to verify and validate the specifications, contract, installation, and operation at COD and over the lifecycle of the plant (Chapter 3). It exposes a serious lack of information on the components, plant health, and condition. These bidding process habits and practices drive up the *actual OPEX*, all too often substantially, thereby raising the total cost of energy produced and delivered, further deteriorating the realized LCOE. It is critically important to owners as these costs kill profit.

Little, if any, of the potential cost savings from plant or component cost reduction translate to improved specification or component selection, nor do they provide for effective reliability or lifecycle testing. While we see system costs dive, we do not see corresponding reinvestment in support of system RAMS or O&M. Commonly, the lifecycle mission is not adequately considered, defined, or enunciated.

As an example, RFPs for large utility-grade systems exacerbate this issue due to its sizable scale – there may be little to no accurate lifecycle consideration proposed for the life of the plant and the total operating costs. From the owner's perspective, this consideration can dramatically alter the proposed plant configuration and technology offers a broad range of choices, specifications details, and designs provided by the EPCs for evaluation. This factor in and of itself makes EPC selection incredibly important, difficult, and unnecessarily expensive. The EPCs are not provided with, nor are they providing modified or suggested project drawings and plan changes that can be easily compared against a common specification. This gap makes comparing the RAMS attributes, system availability, performance, plant functionality, energy production, revenues, and other lifecycle impacts of each proposal almost impossible to accurately assess. The selection is simplified short term with a least-cost approach, compromising decision making results. In essence, the project delivery model today is not only in trouble, it appears to be backward.

Yet, with this widely accepted industry approach, it is "assumed" (undefined, unspecified, and unsupported) that the plant being delivered is reliable and will somehow meet its proposed 20–30-year life and LCOE. This position is substantively and provably questionable!

1.10.1 Bidding Process

In today's common PVPS project delivery process, developers substantially reduce their system costs by pulling out what they perceive as nonvalue-added attributes from the project before it is constructed. This model allows the removal of value based on a marketed and assumed philosophy that "delivering PV projects is a high risk, speculative process and investment," where the developers are bearing all of, or most of, that risk. The fact is, as the existing process takes place, it is high risk, but not necessarily to initial investors. They are in and out of the risk pool quickly while marketing that their efforts have reduced risk and that they have diminished major project risks for future owners, operators, and other stakeholders.

As a result, the assumed risks upfront justify a substantial profit to the initial investors, who actually reduced their risk and system cost and quality at the expense of long-term profit, reliability, customer satisfaction, and stable energy generation for later stakeholders.

The PV project delivery model is the product of primarily competing for the lowest price at any cost, whether it is unidentified, undiscussed, and/or long term. It is wrapped or masked into the initial "assumed" LCOE. It brings us back to our question to the industry: "Has anyone ever delivered a plant with real or accurate LCOE project cost targets?"

Often, this insufficient, underlying process prevents us from addressing the real long-term systemic and LCOE issues under masked[24] conditions, issues hidden from simple visual observation. Most of these challenges are common and often right out in the open. Those challenges include a PV project approach that:

- Is founded on a process that makes decisions with incredibly limited product and reliability information;
- Makes least-cost choices a technology and investment priority;
- Predetermines the cost of goods and services that go into the project without clearly defining detailed specifications, life usage, and availability of those goods and services;
- Is slow to address existing industry problems, and often are easily swayed by a good marketing story;
- Allows a persistent lack of awareness on the complexity of issues that follow COD over time.

However, project stakeholders often do not perceive these broad range of issues as sufficiently real or important. The issues are either inadequately addressed or tend to be glossed over. Foundationally, many decision-making PV professionals still see the industry from a myths and assumptions perspective – which is often inaccurate – which focuses on the lowest price, bells, and whistles, the coolness and simplicity of the technologies, and so on. Finally, although many PV professionals do see the challenges in relationship to these issues, they do not wish to rock the boat. The situation makes it easy for the industry to keep doing the same thing for the same, uncomfortable results. We see these misconceptions in every sector of the process where the focus is on the first five, and up to seven years after commissioning. The plant life may often be discussed, but the focus is not on delivering RAMS during the complete lifecycle. After the initial period, one can expect the system's health, condition, and performance to begin showing signs of rapid decline, diminishing its operational and performance capability. Meanwhile, those initial project stakeholders have already moved on to other projects.

1.10.2 Challenges Specific to the Project Delivery Model

Let us examine the list of problems below presented by the bidding process within the project delivery model.

Problem 1.1 ***Size, Price Tag, Capacity, or Capability?*** Many who read PV literature in online magazines, other media, and advertising will notice the emphasis on a project being the biggest, least expensive, i.e. the cheapest per watt. It appears to us, however, that the comparison of these two attributes can be completely counterproductive. A lot of this is about organizational ego and marketing, which is normal; however, it tends to be short term in thinking, action, and lifecycle results. Reading those same sources, one would think that the most important thing is system capacity, or how many kilowatts or megawatts a system is rated for. Yet, capacity that does not work well results in a system capability that does not perform well, while producing less energy and revenue than is expected or required. A knowledgeable business approach mentality would not place the focus on *how big the project is* but *how much energy that plant will produce for the lowest cost over the longest time* (LCOE).

System capacity and *capability* are not the same! PV system capacity is based on the STC DC power rating of the plant (while production varies all day long as temperatures and other conditions shift and change hourly and seasonally). It is primarily an initial design metric for system sizing. *As a result, "All DC watts of rating vary outside of STC!"*

PV plant and component capability is the attribute of providing power when demanded. Capability includes the plant and its components meeting the defined mission of supplying energy and income. It addresses the issues of supplying (lifecycle) usable power over time. Capability is a system health, condition, and status set of metrics and outputs.

Problem 1.2 ***Warranties*** Often, simple issues – like squeezing more out of the warranties – are not adequately considered. This is because many, if not most, warranties can be negotiated *prior to purchase.* Details must be defined in those warranties initially negotiated with component suppliers and EPCs before the contract is signed. However, these details are often not effectively negotiated and applied to reduce long-term O&M costs.

The warranty period for some components is only a fraction of the plant life, typically two to five years, inverters may be 10 years and, in some cases extendable, while assuming a planned life of 20–30 years. Without sufficient OEM reliability data, a gap results when under-identified issues or changes begin to emerge after the termination of EPC's support and product warranties. These issues could have been more adequately identified but were not and may continue to plague the plant's remaining lifecycle. By negotiating stronger warranties for projects, substantial amounts of dollars will be kept out of the O&M budget, during and after the warranty period.

Previously, this negotiation flexibility required extremely large volume purchases. However, a rising buyer, developer, or EPC company showing growth potential or strong backing may be able to access better warranty conditions and pricing by addressing warranty issues with OEMs and suppliers. Sometimes it just requires having a good conversation with the manufacturer, which seldom takes place, to get a better deal. Often the focus is strictly on *negotiating lower costs while bypassing the warranty issue details*. This can be shortsighted, missing an important fiscal and operational savings.

Unless extended warranty issues are directly covered in the contractual product purchase agreements, the project may end up having to buy extra spares years ahead of need, in order to reduce operational risks for any planned lifecycle challenges. By this, we mean that by purchasing better products, they will require fewer spares especially if they have a stronger warranty with a more amenable warranty claim and return merchandise authorization (RMA) language in their contract. For success, it is essential to purchase from a strong OEM by looking beyond exiting short-term financial rating projections.

All PV stakeholders need to understand that, "Any detail which is not clearly defined in the system specifications and the contract (and its appendices), does not exist and cannot be expected to be delivered."

Problem 1.3 ***Risks Perception*** As investors perceive relatively limited risk in a PV system, partly due to tax credit and other financial incentives available to date, there is little incentive to build a better project. The initial owners assume that risks are reduced based on the perception that the project is less expensive, hence there is less exposure for default or financial failure. Future owners and downstream stakeholders must take into consideration the fact that their purchases will not be covered by those same incentives, and therefore, will have greater fiscal risks. They miss the point, which is directly related to asset value, by not requiring effective specifications in the initial deal, and go along with the assumption that the EPC experts are looking out for them throughout the lifecycle. However, they should realize their assumption is wrong! This is because it is *not* the EPC's job to look out for all owner's interests, as it is *not* in the contract! EPCs are referred to as "contractors" for that reason.

Another disconcerting assumption is that products specified for the project are adequate for the rigors of the local environment and will meet long-term performance and revenue requirements. This assumption, more often than not, is a strategic and tactical "error" for existing and future owner/stakeholder interests. It is a strategic error from the perspective that it is not validated from effective research, data collection, and analysis. It is a tactical error because it has set up the whole project and system for a series of expansive, unexpected, and unnecessary failures that could have been dramatically reduced by making better decisions initially (Chapter 3). As a result, this incomplete project delivery model continues its dive to the bottom.

From our perspective, the underlying myths and industry-wide assumptions are not being addressed through the present processes, which are actually creating, propagating, accelerating, and increasing project risk. There is an extremely high probability that if there were less real risk, such as due to greater project delivery safeguards, the actual finance and insurance costs would also go down. They may not be reduced significantly, but they would offset future risks, including project and O&M cost overruns.

Problem 1.4 ***Unsupported Operations and Maintenance*** Yet, in a more insidious manner, many O&M contract bidders today are lowballing their bids to win the contract. They run into a catch-22 by assuming that the low bid revenues can be made up by additional "time and materials" work to address the contractual exclusions or problems with the original contracts. As a result, they are well aware of the fact that those exclusions will need to be addressed. They count on the supposition that such work will be part of the additional and essential future services they provide.

This awareness includes many shortcomings of how the system was specified, designed, built, and operated. It also includes the fact that additional work will need to be added to meet the PPAs and other contractual agreements due to often incomplete scopes, specifications, or design.

Yet, the reality is that the tight budget and management have not accounted for the actual need or cost of such critical work; as a consequence, *much of that work is never funded*!

This systemic planning gap creates a series of further challenges, which would have been addressed if the O&M specifications were complete, detailed, effective, and based on industry-wide standards and contractual language. If those specifications were included during the O&M phase of planning, many of the mentioned problems would never exist in the first place.

At this time there are a number of new standards and certifications that are being written and many existing ones are continually being improved to address these types of issues.

Those standards, organizations, and language include building codes, *IEC*, *IECRE*, *American National Standards Institute*, IEEE, American Society for Testing and Materials, and a variety of other national and international codes, and standards that are in continual development and improvement.

It is not good enough just to have the standards. They need to be required, applied, and confirmed (verified) at the specification and on through the project during benchmarking and commissioning and when sold or reinsured.

Or as our colleague Roger Hill stated:

> **Standards are like icebergs. They are visible above the surface, but below, they came from extensive efforts in supporting research, technical development, guidance documentation, best practices, and testing.**

And yet today, it is not perceived as being in the best interest of many project delivery participants to have clearly defined standards and certifications. Standards and certifications provide a format so that the elements that they cover use consistent language, metrics, processes, and procedures to achieve certain ends. Doing so puts pressure on a multitude of marketing claims that are not fact-based while claiming to be the best or having the most experience or meeting all the codes and standards. While this has been a leverage issue for competition, it has become a major barrier to improving system delivery and addressing the broader range of RAMS attributes and stakeholder concerns. When competition pressure drives down the ability to deliver PV as energy infrastructure, it robs the entire industry and its stakeholders of value and eventually credibility.

Problem 1.5 ***Emphasis on Power Purchase Agreements*** Further exacerbating these challenges is the preference for utilities, owners, and other funding parties to concentrate on a PPA and its bidding processes. As mentioned, the present bidding process model is based on an aggressive, initial least-cost pricing (capacity and assumed cost per kilowatt hour). Developers encourage this approach as it is manifested as a *safe* least-cost PPA pricing but does not include all the real project costs as addressed in this book. Yet (what consistently amazes the authors over the last decade or so is that) the vast majority of PPA signers believe that they are at little to no risk. They often underestimate or fail to consider what happens if that contractual energy is not available, especially when they need it the most. They under-evaluate what happens to them if that PPA system owner cannot afford, or is unwilling to address, the unexpectedly rising costs.

Many buyers of PPA energy primarily focus on the lower initial LCOE without adequate understanding and evaluating overall built-in risks. In particular, limited information and awareness of the specification, design, and O&M challenges exist to help establish a plant's physical health, condition, status, and lifecycle known as plant or system availability (as defined in IEC 63019) or its capability to provide the expected energy on demand. A miscalculation in plant availability could result in energy under delivery, resulting in massive reliability and financial risks. These excessive risk elements have arisen primarily due to a lack of understanding. It is also partly due to the belief that PV is different enough from other energy generation technologies, thus "the same rules don't apply to PV." This thinking and its effects could not be further from reality. Without understanding or addressing the history, lessons learned, and lack of availability of operating and failure data, there is a tendency to repeat history, which has proven expensive. Over a few decades, following the introduction of PPAs, it is still not uncommon to hear participants on both sides of the PPA gleefully claim that "there is little to no risk inherent in the deal or process." The facts speak otherwise that includes remedies such as LDs and/or other legal challenges.

Problem 1.6 ***Project Value Determination*** The historic electrical utility infrastructure, with its previous 100-plus years of successfully safe and reliable energy generation and distribution was founded on and grown through the application of SE/RAMS. This illustrates that profits were being made by addressing and adopting SE/RAMS considerations early in the specification and design process. This methodology and approach allow for more consistent profit and reduced LCC and LCOE. The reduction in LCC and LCOE includes *all* the actual costs of ownership. Nevertheless, deregulation tended to reduce the consistency of those commitments resulting in shifting a substantive amount of risk to the private sector.

The PV industry was also grossly impacted by the deregulation of the energy production industry. Deregulation removed some of the emphasis on SE/RAMS and as a result the energy industry shifted to a lowest or least-cost model. This result continues to affect the quality of energy infrastructure and components. Since deregulation in the 1990s, reliability improvement in components and systems has slowed. This requires a return to SE/RAMS methodologies for system specifications and manufactured parts such as PV cells, modules, and some BOS lifecycles.

Typically referred to as the "price per kilowatt hour," the price for a plant to deliver energy is determined by market pressures and energy product availability. The price of energy is continually shifting through each sales tier beginning at production, distribution, and retail user energy sales. This is regardless of whether substantive project specification and planning have taken place or not. This point raises two very difficult questions:

- "What is the true value and price of electrical energy at any particular point in time in any market?" and
- "While looking at the 'all in costs' for delivered energy, what is the difference in cost and value with a well-planned SE/RAMS process, which includes an extended lifecycle **R**epowering™ program?"

These questions must be addressed to reduce risk!

This existing PV business model is fanciful, expensive, and dangerous! It is also more sensitive to global events such as climate change, COVID, and supply chain problems. There is insufficient planning to address the existing practices much less provide for additional contingencies, as the industry is already operating on tight margins.

The underlying assumptions with the industry tend to be that a PV plant will be or is "reliable" without defining what that means over time. Reliability is not and should not be used as a marketing term. Reliability is a measurable attribute of all the components of a PV system. LCOE is highly dependent upon having a demonstrated (measured) reliable system. We address the topics of reliability throughout the entire book from a series of industry perspectives. Chapter 6 covers the topic in detail.

Many companies (buyers and sellers) take great pride in building a plant considered to have the "lowest cost of energy" (a marketing claim), instead of addressing a more realistic "*levelized* cost of energy[25]" or LCOE. We counsel the industry to focus on a *real, documentable* LCOE with the lifecycle costs including contingencies specified upfront. It always comes back to the true asset value of the plant and the true "all in" cost per kilowatt hour produced throughout its life.

The lowest/least-cost approach, generally though not always, involves the cheapest components not necessarily built to appropriate standards and with minimum or no certifications. This business model often relies on bare bones due diligence especially when assessing variability and most likely no substantiating lifecycle or reliability testing. This irrational approach results in excessive costs due to required corrective actions. This leads to an increased LCOE, potential for LDs, plant derating or devaluation, and potential or real bankruptcy. A literature review indicates that the odds of this approach being successful are probably comparable to those of betting at a roulette table in Las Vegas: 35 to 1. Not everyone loses, and people keep spending more money with the same results. And, they keep coming back for more.

Together these problems above define the ceiling for the *assumed* project value, but they do not necessarily capture the capital infusions necessary to deliver the project throughout its *assumed* (unspecified) design life.

Please note that project available funding and the necessary lifecycle funding have a growing disparity that begins early in the plant's life and continues to increase. This difference is a product of early cost calculations that were done – prior to the actual bidding evaluations – without directly comparing the EPC and O&M bidding details to a full set of specifications. As a result, concept and SE decisions were and are based on disparate EPC bids. These decisions tend to deliver ineffective and incomplete results that later become unplanned and budgeted O&M costs throughout the plant lifecycle. The emphasis is primarily on the initial cash flow and the *assumed* long-term cash flow of the project. This process is further undermined by the fact that the initial owner will probably "flip" the system to an uninformed and unsuspecting buyer.

Problem 1.7 ***System Flipping and Overbuilding*** To "flip" a project or system, the existing practice is to develop a least-cost PV system/plant by generally following the assembly of entitlements, EPC bidding, construction, and/or commissioning. This plant may be sold at any phase but primarily within the initial five years of operations. The cost-determining process does not include a **R**epowering™ plan, the essential tasks necessary to deliver a far more reliable, robust, resilient, and long-term productive plant. The lack of considerations will become even more obvious and critical as the due diligence on plants (which is often minimal to start) is somehow increased to uncover numerous previously unidentified or under-identified issues (see Chapter 3). Astute buyers will want/need a complete data set to address performance, production, and persistent problems with the PVPS. Unfortunately, that depth of due diligence is often hampered by lack of incomplete data.

Due to the inconsistencies, exclusions, and potential omissions of data, future owner(s) will be unlikely to receive a suitably high commercial valuation on a plant. This is especially the case when due diligence is limited, incomplete, and inconsistent. As additional deficiencies become more apparent, the plant will be discounted and derated at all future sales. It is basically the building and dumping of a substandard project, usually to a relatively assumptive and unsuspecting buyer or investment team. As practiced today, the risks are shifted or shed onto the future owner(s) and away from the original developer, owner, financier, and EPC. If potential future owners and long-term investors understood these facts clearly, they would demand less risky, demonstrably more reliable, and profitable systems. As more systems are valued in the same fashion, this model has become increasingly problematic.

Our concerns about the manner of "flipping" are that people are planning and building substandard, subprime PV projects and continuing to reinforce a flawed and damaging business model.

They are able to sell it to the next owner based on existing performance metrics that look healthy because the system was overbuilt, masking many issues. This overbuild has two facets: the first is a sort of short-term performance insurance and the second is to produce a contractually compliant energy production curve. The overbuild, which is usually designed to broaden the daily energy production, masks a series of systemic health and condition issues. In time, it becomes more readily obvious. When energy storage is included, it becomes obvious far more rapidly indicating the presence of design and specification challenges and under-identified system problems including partial and total component and localized failures. Both of which have significant early and future consequences.[26] These problems are addressed in Chapters 3–7 and 9.

A more effective system delivery model provides greater capability, resulting in greater output delivery with additional energy and profitability for a longer period of time.

Most if not all experienced PV professionals know that systems being delivered today have the overbuild problems and performance challenges with an inability to determine the status and health of the system. This mindset has been accepted almost universally by the industry and is often referred to as a necessary element of *good* competition. The results are a growing series of financial and business decisions that can have calamitous results by ignoring the realities, as the willful ignorance has been accepted as *good* business practices. To reverse this continuing dilemma requires moving to a PV systems infrastructure reliability-based model.

As ignorance is not bliss! The reality facing the industry today is that "caveat emptor" (let the buyer beware), must be continually in play and considered throughout all PV system and financing efforts.

Problem 1.8 ***Lack of Specification, Including Requirements, Needs, and Expectations*** Many PV projects are delivered with a lack of essential planning, data, specification, understanding of and addressing new and changing technologies, and how rapidly those technology change. As a result, critical details are often glossed over.

For example, manufacturers tend not to adequately address the full range of worldwide climates and site environments the equipment is expected to operate in. A major consequence of this oversight is that many plants have components that are being stressed by operating outside of their designed environmental specifications. They may or may not also do limited reliability testing during design and build. In addition, they may only perform limited sample testing of the production lots. This limited level of assessment is enough to lead to earlier than expected plant failures. Even after a successful commissioning, there can be a significant loss of energy production and negative financial results, e.g. inverter performance decrease and/or excessive degradation of components resulting in LDs and excess unplanned O&M.

By sending RFPs or RFQs that are not based on a detailed "common system specification," every EPC is bidding on their unique system and may not deliver the plant envisioned or needed. Each EPC is proposing their own schemas based on loose specifications or their own interpretation of the specification. Based on individual system designs, attributes, and components, each schema has its own way of complying with the contractually least-cost requirement. The owners are left to try to select an EPC from disparate proposed concepts and designs without a common specification. This lack of an apples-to-apples bid comparison results in the inability to properly perform a comparative evaluation of competitive bids. The owners shift their focus to the least cost, without clearly understanding the consequences of accepting the bid that has been differentiated from the rest primarily by the first cost. These also include inaccurate LCOE calculations and poorly addressed and understood lifecycle performance.

The existing approach provides most of the upfront profits to the initial project financiers and developers. They carry the relatively least amount of actual risk and make a substantive profit, yet market to the industry about how risky their contribution is. As previously noted, the associated PV projects and systems could begin to experience serious failure issues after five to seven years, when those initial owners, financiers, and developers are often already out of the risk pool. This timing is a key element of that particular business strategy.

The disconnect in that business model and the challenges listed above are underpinned by the lack of detailed system specification and failure to include sufficient or appropriate SE and RAMS attributes in the specifications and plans. Without specifying the need for certain specific levels of reliability, for example, leads to an environment where nobody in the EPC process is incentivized to build more robust and resilient plants. If properly addressed, the reduction in kilowatt hour cost over time would be lower and more accurate. It would also result in lowering the LCOE to a more realistic and accurate value that is optimized for the fixed and variable cost elements. For the affected stakeholders, it provides for improved predictability for profit and ROI. The financial case would be more profitable with more consistent energy production and availability, while maintaining a substantively higher plant asset value.

Nevertheless, additional growth and understanding that leads to better cooperation is being forced upon the industry internally and externally.[27] The facts point to a growing number of experienced, trained, and educated PV professionals who understand the issues and solutions.

Until now, a problem/solution approach is often ignored or drowned out by a number of interests intent on repeating the mantra of the PV industry's least-costs approach and inconsistent application of SE practices. Therefore, we must focus on educating and improving the knowledge of best practices for PV stakeholders.

Therefore, we must focus on educating and improving the knowledge of best practices for PV stakeholders. These are some of the challenges we have identified and will address in Chapters 5-7. We see a bright future for the industry by addressing these issues. However, change is often resisted yet necessary!

The chapters in this book are arranged by titles of processes and methods that need to be used to provide for reliable, maintainable, available, and safe[28] PV power plants that produce energy at the best, i.e. lowest LCOE.

Bibliography

Balfour J., Morris R.W. (2018). Preemptive Analytical Maintenance (PAM), Introducing a More Functional and Reliable PV System Delivery Model – A Reliability Precursor Report, Academia.edu, March 2018, https://www.academia.edu/36904693/Preemptive_Analytical_Maintenance_PAM_Introducing_a_More_Functional_and_Reliable_PV_System_Delivery_Model_A_Reliability_Precursor_Report

IEC 62308:2006. Equipment reliability – Reliability assessment methods.

IEC TS 63265:2022. Photovoltaic power systems – Reliability practices for operation.

IEC TS 63019:2019. Photovoltaic power systems (PVPS) – Information model for availability.

INCOSE-TP-2003-002-04 (2015). *Systems Engineering Handbook*, 4th ed. Wiley July 7, 2015, ISBN-10: 1118999401, ISBN-13: 978-1118999400.

Elsayed, E.A. (1996). *Reliability Engineering*. Reading: Addison Wesley Longman, Inc.

SEBoK Editorial Board (2020). *The Guide to the Systems Engineering Body of Knowledge (SEBoK)*, v.2.2, v.2.2, R.J. Cloutier (editor in chief). Hoboken, NJ: The Trustees of the Stevens Institute of Technology.

IEC 60300-3-3:2017. Dependability management – Part 3-3: Application guide – Life cycle costing.

IEC TS 61836:2016. RLV Solar photovoltaic energy systems – Terms, definitions and symbols.

Notes

1 SEPA, Balfour, Resource Guide: Utility Solar Asset Management and operations and Maintenance, 2017.

2 Balfour, Morris, Preemptive Analytical Maintenance (PAM), Introducing a More Functional, and Reliable PV System Delivery Model – A Reliability Precursor Report, 2018.

3 MIL-HDBK 338B, *Military Handbook Electronic Reliability Design Handbook*, 5th ed.

4 Photovoltaic System Performance – Part 3: Energy Evaluation Method.

5 Photovoltaic power systems (PVPS) – Information model for availability.

6 Sparing is a process for the determination of the number of repairable or replaceable parts needed based on the reliability and maintainability attributes of the parts.

7 "A Best Practice for Developing Availability Guarantee Language in Photovoltaic (PV) O&M Agreements," SAND2015-10223, https://www.osti.gov/biblio/1227340-best-practice-developing-availability-guarantee-language-photovoltaic-pv-agreements.

8 **Taxonomy (general)** is the practice and science of classification of things or concepts, including the principles that underlie such classification. https://en.wikipedia.org/wiki/Taxonomy.

9 In computer science and information science, an **ontology** encompasses a representation, formal naming, and definition of categories, properties, and relations between concepts, data, and entities that substantiate one, many, or all domains of discourse. More simply, an ontology is a way of showing the properties of a subject area and how they are related, by defining a set of concepts and categories that represent the subject. https://en.wikipedia.org/wiki/Ontology_(information_science).

10 "Risk Pool" with regard to this work: While all stakeholders may be gathered into a risk pool in order to deliver a project, many stakeholders tailor their agreements to leave the risk pool as soon as possible. As many stakeholders (in some cases end users) are in for shorter time periods, and may not have much or any risk as they finish their contacted obligations, they leave the risk pool with little present or ongoing responsibility. Many downsteam stakeholders including energy users, later owners, O&M organizations, and others do not have long term protection over time. Part of the challenge is that those downstream stakeholders will be left with the consequences that develop later.

11 Catch 22: https://www.merriam-webster.com/dictionary/catch-22.

12 Seattle Times, 11 Nov 2021, FAA says Boeing Safety Engineers Lack Expertise To Certify Airplanes.

13 https://www.faa.gov/foia/electronic_reading_room/boeing_reading_room/media/737_RTS_Summary.pdf, p. 62, Recommendation 5.

14 IECRE – International Electrotechnical Commission Renewable Energy. This is a system for certification of standards relating to equipment use in renewable energy applications.

15 Repairable item level is the level at which maintenance can occur and repair the affected by repair or replacement to return the system to full specification.

16 CNBC, AEROSPACE & DEFENSE, Wall Street expects Boeing to take another big, ugly charge on 737 Max. B of A estimates total cost of crisis as high as $20 billion, PUBLISHED FRI, JAN 17 20207:00 AM EST.

17 Authors' assessment.

18 Department of Transportation, Takata Recall Spotlight.

19 https://www.consumerreports.org/car-recalls-defects/takata-airbag-recall-everything-you-need-to-know/

20 https://www.justice.gov/opa/pr/boeing-charged-737-max-fraud-conspiracy-and-agrees-pay-over-25-billion

21 Leggett, Theo (May 17, 2019). "What went wrong inside Boeing's cockpit?" BBC News Online.

22 "System Failure: The Boeing Crashes". Al Jazeera Media Network. October 16, 2019.

23 https://www.thetimes.co.uk/article/is-the-boeing-737-max-safe-ml59n3lhk

24 Balfour, JR, Walker, Robinson, Gunda, Masking of photovoltaic system performance problems by inverter clipping and other design and operational practices, 2021, Elsevier.

25 Levelized must indicate that the costs are spread out over the entire life span of the plant, aka system lifecycle.

26 "Masking of Photovoltaic System Performance Problems by Inverter Clipping and Other Design and Operational Practices," John R Balfour, Roger Hill, Andy Walker, Gerald Robinson, Ammar Qusaibaty, Thushara Gunda· Jal Desai.

27 New NERC reporting requirements and new IEC standards for wind and solar.

28 Safety is a topic in this book as some of the safety requirements drive reliability, maintainability, and consequently availability. Safety, however, is generally regulatory for the specific city, county, state, province, region, or country, referred to as the Authority Having Jurisdiction (AHJ). As such we do not go in depth to address the myriad of regulatory requirements for power generation that must be met and accounted for during permitting and periodic safety inspections.

2

PV System Delivery Process

As presented in Chapter 1, the "PV *Project* Delivery" process or model has and continues to inadequately provide for a successful PV-based industry as energy infrastructure. This chapter provides an overview of the element of the system delivery process covered in greater detail in Chapters 4 to 8.

2.1 Introduction

We propose to deliver PV as Energy Infrastructure. **What is meant** by **"PV as reliable infrastructure"?** Infrastructure is "... the basic systems and services, such as transportation and power supplies, that a country or organization uses in order to work effectively" (Cambridge online dictionary, https://dictionary.cambridge.org/us/dictionary/english/infrastructure). Our approach has four main elements:

1. Founded on systems engineering (SE) methods and practices using data acquired from past and current systems used to specify, design, manufacture, install, operate, and maintain a PV plant, thus extending its lifecycle to be equivalent to historical power generation.
2. Requires that project stakeholders; financial/insurance, owners, operators, maintainers, etc., are adequately educated/trained on relevant PV concepts, physics, and limitations.
3. Structured on clearly defined/agreed criteria of success and failure. This approach uses **SE/R**epowering™ planning processes and procedures, that are based on SE with reliability, availability, maintainability (includes testability) and safety (RAMS) requirements integrated into decision-making, component selections, and designs of the plant and its interfaces as a complete system.
4. Requires sufficient O&M budgets and reserves that are maintained to cover maintenance exigencies, spares, COLA, and force majeure for example, from cradle to grave.

Delivering reliable energy infrastructure assumes nothing about the technologies applied, plant size, essential lifecycle production capability, site environmental conditions, execution of an O&M plan, or the energy to be delivered to meet the customer needs. It is based on both current and future energy needs as supported by curated information (data) essential to meet stakeholder defined goals and needs.

Photovoltaic (PV) System Delivery as Reliable Energy Infrastructure, First Edition.
John R. Balfour and Russell W. Morris.

Results: An operationally robust/resilient plant that is defined, specified, designed, delivered, maintainable, and rebuildable to achieve more consistent energy production and delivery throughout its documented and sustainable lifecycle, with an accurate levelized cost of energy (LCOE).[1]

This "PV Systems Delivery" process addresses the major drawbacks of the existing practices. We address the who, what, when, where, why, and how to support the evolution of system delivery by addressing the old model issues. This is accomplished by shifting toward methods that are intended to address more accurate long-term lifecycle costs (LCC) and LCOE issues, not just the short-term financial success of the initial developers and owners. This preemptive analytical maintenance (PAM) process takes a wholistic view of the plant, including all stakeholder interests to provide for more reliable, available, maintainable, safe generation, storage, and use of PV energy.

The decisions made at concept and specification drive all of these choices. The results accrued from the decisions and actions made and taken at specification through commissioning, benchmarking (Kounev et al. 2020), system operations and maintenance (O&M), and site restoration grossly affect the initial and long-term costs of the system (Harper 2019).

Much of the initial work on a photovoltaic power system (PVPS) is to determine the power/energy (number of modules) requirements and physical size of the plant to meet the customer's needs. As part of the design, a large number of considerations and technical assumptions must be documented and agreed to by affected stakeholders. At issue is the difference between what was designed and assumed versus what the plant's capability really is at the time of commissioning and benchmarking. One result is that the plant's output seldom meets expectations, as the ambient environment is rarely at standard test conditions (STC) values.

That aside, by using a pyrheliometer and or pyranometer, one can calculate the expected plant energy output. A pyrheliometer is an instrument for the measurement of direct solar irradiance, also known as direct normal irradiance. Pyranometers are used to measure total hemispherical radiation – beam plus diffuse – planar with the modules (Chapter 3). More often than not, the output is less than calculated as a result of unsubstantiated technical assumptions, unaccounted for power losses in the system, overdesign (in percentage) with or without storage, and the degradation rates for the plant. Sources of unsubstantiated assumptions include:

- Modules, inverters, and other components will always be exposed to specified environments.
- Modules, inverters and other components are operating at specified efficiencies.
- Environmental measurements are taken from questionable locations.

Historically, over the last four decades, a key question asked and seldom fully addressed is: **"Where did the missing energy go?"**

Missing energy is the difference between what the plant is capable of and producing versus what is available and measured at the output!

Missing energy is the actual energy losses that may or may not be accounted for in systems planning and design assumptions or specifications. One of our key points is that some of this lost or energy (gaps) is recoverable. Accounting for all losses during concept and system design is difficult, and yet both financially valuable and instructive for all stakeholders throughout the plant lifecycle.

How much of that energy can be recovered consistently and reliably, and at what cost? This will depend on:

- The planning effectiveness begins with the stakeholder involvement process to establish wants and needs, the established, constructed, commissioned, and benchmarked requirements.
- Depth and completeness of system planning, specification, and design.
- The capabilities (knowledge and awareness) of the specification, design, and management team.
- A complete bidding process (RFP, RFQ) inclusive of additional feedback and recommendations from bidders.
- The Engineering, Procurement, and Construction (EPC) response to the system specification and design.
- The skills, knowledge, and awareness of the installation crews from top to bottom.
- The depth and completeness of the O&M team capabilities, their planning, support, and funding.

To answer these questions, all of the known and potential losses must be accounted for to determine plant health, condition, and status output over time (plant life cycle), not just at commissioning. This requires understanding the variability (extreme values, both high and low) of the loss elements. With the requisite skills, most of those losses can be identified, listed, and responded to.

Addressing the questions effectively allows stakeholders to determine what the actual measured fiscal and performance impacts to plant output will be as considered, designed, built, and operated through site restoration. By addressing these issues across the entire system delivery process, missing energy opportunities can be identified, then accounted for at each stage of the system delivery from concept through site restoration. In other words, many of these losses are avoidable through application of standard engineering practices, planning, oversight, training, and installation workmanship, then confirmed through rigorous commissioning and benchmarking. In essence, missing energy translates into missing revenue and higher O&M and LCOE costs.

The fiscal questions that are seldom addressed in detail begin with:

What does it cost to recover an additional 3%, 5%, 10%, or 15% of that energy over the initial 5, 10, 15, 20, 25 years or more if we include detailed and effective **SE/R**epowering™ planning?

At this time, different organizations and stakeholders across the industry have the greater or lesser capacity, capability, or awareness to address these issues. Those abilities are driven by organizational leadership generally based on faults and failures provided from existing plants and applied lessons learned. However, improvements and energy recovery will not effectively take place unless there is a management, plan, and budget commitment to do so.

2.2 PAM PV System Delivery Process

The PAM PV "system delivery process" model begins at concept and ends at site restoration! This directly supports the IECRE 04 3[2,3] approach for utility and commercial grade plants using and expanding on the elements of the system lifecycle.

A fact in the life of a PV plant is that those who learn from their mistakes (lessons learned), and make those changes and improvements needed, tend to have better results and greater levels of success with lower costs. This route requires that a number of key tasks are performed over the lifecycle but especially during the concept, specification, design, and development, *as core elements of the PAM process.*

PAM focuses heavily on the delivery of detailed specifications and planning upfront, to address the life of a plant with the associated lifecycle issues and costs. As a result, for example, many component decisions are made and resolved with little to no difference in the purchase price. Where the item costs more because of its reliability (Chapter 6) and system quality, it more than makes up those costs in planning, and components, which are usually offset within five years or less.

Quality in this context implies that the system is designed and built to the stakeholder wants, needs, requirements, and specifications of the SE RAMS effort. This also implies that the reliability of the plant relies on the manufacturing quality of the components, and how they are handled, installed, and operated through the lifecycle. Other design and construction attributes add minimal cost and are based on better planning, while offsetting wasted time, energy, expenses, and corporate angst due to missed opportunities. From experience and as documented by industry failures, it is in the 7th through 10th years when PV plants historically begin to display and experience serious, often unexpected issues. These include more unscheduled (fault and failure) maintenance (Chapter 7) than expected, the need for shorter intervals between preventive and ancillary maintenance, and greater difficulty finding form, fit, and function replacements for replaceable/repairable items. The PAM **SE/R**epowering™ goal is to preempt and *reduce* many of those issues and costs.

However, many in the industry do not fully understand that building better systems to stabilize Capital Expenditures and drive down Operating Expenditures results in higher availability of the plant, greater measurable energy output, and reduced variation in the power component degradation curves. Reducing these and other availability-disrupting events dramatically, improves integration with energy storage (Chapter 9).

In the PAM system delivery model, overall financial risk is better understood and managed. This is because many risk-laden elements are being carefully identified, addressed, and preempted, reduced, or eliminated. There are fewer surprises, much lower unexpected costs, and disruptions in plant performance, reliability, availability, energy production and revenue flow. This method changes the economics, in some instances, dramatically.

Best Practices For Project Lifecycle Cost, Risk Reduction, and Control:

- "If not clearly detailed in the system specification and contract *Prior to EPC Bidding and Design,* it will not be in the final product; therefore, it *Should Not Be Expected*!"
- All stakeholders must communicate effectively, continually, and accurately, based on consistent definitions and language to eliminate undocumented and undefined expectations."

In other words, if you are not concerned about, avoid, or are afraid to address additional risk by what you are doing today… "don't blame anyone else."

Do not expect other entities to protect your interests, profitability or provide system reliability, availability, maintainability, or safety if it is not clearly specified for all components/subsystems and addressed in the PV delivery contract specification. The assumption that the "experts" will take care of it for you is an industry issue where many participants appear to be hesitant or afraid to make the first serious steps toward resolving long-term industry issues. *In short, the best advocate for you (the stakeholder's) wants and needs is yourself*. Change can be slow, sometimes painful, and comes with an incremental cost. Change requires dramatically improved PV systems specification and design processes, education, training, stakeholder community involvement, and action! Success is far more likely when stakeholders understand the systemic, production, and consumer risks and consequences of their decisions and address them during the EPC bidding specification development.

One of the most effective modifications in the process is how we look at developing a more thorough specification upfront that includes substantive pre-engineering efforts.

Yes, there will always be modifications to cost, schedule, and, occasionally, decisions that require shifting and reallocating resources.

This can make the bidding process much more effective, far more accurate, less costly, and more reliable. Doing so accelerates the process of dramatically improved PV systems, with greater production and profitability, while saving time and reducing costly resources during operations. The process reallocates some costs and actions, resulting in greater project cohesiveness.

As a result, the EPC's will bid on the same specification thus providing additional value and information from their experience and skill sets. This results in improving plant design, construction, and PV system outcomes as infrastructure. Those changes that functionally improve the system will be primarily resolved in the initial contract, requiring fewer change orders, cost, time, and complexity. Nevertheless, this systems approach has seldom been effectively incorporated in the PV/energy storage industry.

Yes, as a whole, the industry is making progress through incremental steps, two steps forward and one step back. They are little baby steps because existing business models perpetuate a variety of PV myths and assumptions that continue to stymie progress. Those myths have been and can be easily dispelled with accurate and factual information. This progression requires the desire of wanting to improve and then taking the essential steps to gain that improvement.

The PAM approach directs the EPCs to be far more technically and specification oriented, and creative during the bidding process. EPCs must be rewarded for designing, developing, and installing more cost-effective and reliable systems. Care should also be exercised when bids appear outsized, larger, or smaller than expected, for the contract and specification supplied. The owner should have an estimated range of the bid ranges for the specification and the contract, which will be supplied to the bidders. The owners can expect a learning curve, but with the ability to compare proposals against a set of detailed specifications, there will be a better understanding of what the EPCs are actually going to build and where the costs are distributed.

For example, buying and installing more reliable parts or components may have as much as a 30% cost penalty for those components. Yet, that value may represent only about 5–7% of the actual cost of the end item (e.g. inverter). Should the supplier/EPC indicate that having reliable components is a 15–50% cost delta, the bid has to be seriously reviewed and questioned. If the extra 5–7% improves the item mean time between failure by 50%, that implies as much as a 30% cost savings each year in maintenance, spares, and fewer surprises in the 5–15 year range.

A fundamental issue that compounds today's results are the ongoing focus on initial performance requirements, the first five years or so (during initial finance), "not" long-term energy output goals and objectives of energy infrastructure. Too many assumptions are based on incomplete or insufficient data. Predictions established using available *marketing* literature typically do not adequately address the true attributes of the in-situ product, which affects the true operational capability. Or, if addressed, available information has often not been updated from prior versions or generations of equipment, making that data little more than a weak guess. Projection of costs should be based on current and relevant engineering testing and field data for proposed designs. Examples supporting data requirements include choices and decisions made about the rates and levels of deterioration, deration, degradation, anomalies, faults, failures, and defects.

These unchecked assumptions greatly impede lowering lifecycle project costs and risk.

"I had," he said, "come to an entirely erroneous conclusion, my dear Watson, how dangerous it always is to reason from insufficient data," Sherlock Holmes – *The Adventure of the Speckled, Band* (Sir Arthur Conan Doyle)

"Data! Data! Data!" he cried impatiently. "I can't make bricks without clay," Sherlock Holmes – *The Adventure of the Copper Beeches* (Sir Arthur Conan Doyle)

The miscalculations often begin with out-of-date default data used in modeling and assumptions related to the project. The term *garbage in, garbage out* (GIGO) comes to mind. This approach to modeling does not represent accurate results or expectations over time!

Understanding and clearly defining while controlling "all" technical assumptions and data input for more accurate systemic modeling is critical. Doing so requires having "control" of that data going in based on accurate well documented technical field data and manufacturer testing.

Data is too often cherry-picked, optimized, or in extreme cases, synthesized (fabricated) to make the results match expectations or requirements. Such data can also be incorrectly entered or misinterpreted, greatly affecting the model outputs used for project and system assessments.

We continually advise stakeholders to **Never** *lose control of the modeling data inputs and their output values*!

There are few, if any, true industry standard inputs. This complicates the comparison of bids from EPCs or suppliers! Therefore, there should be a clearly focused eye on the data being used, that it is current, relevant, plainly, and clearly listed, accurate, and then compared with the outputs. When someone refers to using "standard inputs," **beware**!

Always confirm the model inputs and outputs (Chapters 5–8 and 11) and that their iterations are accurate, supported by valid data, documented, meaningful, and applicable. By doing so, you will better understand the peculiarities of the model and the validity of the data going in, which are key to being able to produce valid data output to determine and improve future decisions. In addition, all models have strengths and weaknesses. Therefore, it is often best to compare two or more data models to better understand the range and differences in output.

This level of due diligence is initially taxing until it quickly becomes second nature as a more precise set of practices with improved results. It requires additional time, experience, knowledge, and attention to the ever-changing details of products. The quality, reliability, and capability of a product can vary from manufacturing batch to batch. Therefore, it is best *not to assume* product consistency without a robust manufacturing quality control (QC) system and supporting documented data. Technology fatigue can be as simple as using out-of-date component data or failure to recognize that technologies have expiration dates, although this issue is far more complex.

2.2.1 Technology Fatigue

Technology fatigue is the electronic industry's uncomfortable little secret and is seldom broadly discussed publicly. On average, new technology will be brought to market every 18–24 months. Current technology may only be supported for five years or less by the manufacturer. And, there should be no assumption that the new products are drop-in replacements for current products. This latter item is related to what is called form, fit, and function replacement, i.e. no other changes need to be made to the assembly to accommodate the replacement part.

How does one tell the actual difference between products and manufacturers when you may not have sufficient or only have questionable information?

Project schedule constraints may also play a major role in the impact of these product selection decisions. This dilemma is especially illustrative when product and manufacturer procurement decisions are turned over to the EPC with insufficient and poorly documented early specifications. Until technology fatigue is fully realized and addressed, it will continue to lead to uncontrolled system costs and project/organizational disruption. For many stakeholders, it is still a rather taboo, difficult subject if they have not used trained, skilled, knowledgeable, and experienced subject matter experts and staff. It is easier to leave the tough decisions for someone else to resolve later. As such, the industry generally does not make the tough lifecycle specification and design decisions necessary for long-term success. They all too often opt to trust without verification and to minimize the investment through perceived and assumed least cost!

This is reminiscent of the "Wizard of Oz" attempting to terrify Dorothy, the Tinman, Cowardly Lion, Scarecrow, and Toto. That fear of the unknown was exacerbated by the smoke and mirrors of what "might" be happening, when the truth was simply behind the curtain. Yet, it was the little dog who saw through it. Toto was not caught up in all the theatrics (marketing), extraneous noise, confusion, and misdirection. Toto being curious, just wanted to know what the facts were, making everything simpler and a lot less scary by simply pulling back the curtain.

The source of these "behind the curtain" issues is often the ongoing lack of sufficient, current, accurate, and effectively analyzed product operational failure information and data. Behind the "curtain" of proprietary information[4] (PI) or trade secret (intellectual property [IP]) is the actual measured and measurable rates and levels of deterioration, anomalies, errors, faults failures, defects, and in many cases, their causes. It is possible to acquire this information provided the owner will perform or budget for their project's reliability and maintainability of field data to be anonymized and made publicly available. This requires a standardized data acquisition format.

Progress today, with limited RAMS improvement, will take decades to come close to achieving product and system results as compared to incentivizing those efforts for more immediate implementation. We, the PV industry, cannot afford those slow decades-long improvements!

The critically missing field information is the actual reliability data on the components that underlie the system health, condition and status. This is further exacerbated if the product selected is not adequately designed for the environmental use cases where it is applied. Real but often outdated reliability field data can be found in published technical papers or other publicly available sources. Manufacturers have data that PV specifiers and owners often do not have, indicating the actual capability of their products. However, unless system performance is unacceptably low or problems manifest during warranty, there is little incentive to develop solutions to retrofit products in the field. This information is necessary for the project teams and owners to improve business and technical decision-making and positive results. For success, these issues must be discussed within the organization and other stakeholders, especially competitors who may be running into the same common problems. In addition, they need to demand the actual testing and field reliability data, including operating conditions and environments, from OEMs including their technical assumptions.

A result of not addressing data by the unprepared, the often simple solutions are elusive!

Not demystifying and understanding these key data elements that define plant health, condition and status, drives technical, performance, and revenue **expectations** askew. This situation is controllable by practicing basic, well-documented RAMS, SE processes, data acquisition, and sharing (Chapters 5–7 and 10).

The actual process is further complicated by a disturbing penchant to leave "risk shedding" unaddressed, which dumps inherent specification and design problems on others, specifically future stakeholders (Chapter 5). Risk shedding as a business priority undermines "risk reduction," usually by deprioritizing, understating, or ignoring the consequences. It is sometimes expressed as "... leave those details to the experts!" without specifying or funding the solutions for those details, or "leave the unresolved issues to the O&M guys, and it is their job to solve them!" without identifying those issues ahead of time. The resolution requires a cohesive project team with strong leadership, project management, communications, and effective processes to address both.

Present business models and practices all too often do not adequately address or consider the totality of "All" project stakeholder needs. As a least-cost project model, they result in masking, veiling, and or ignoring real yet often resolvable issues and solutions. When several issues are not adequately resolved, the reality is that critical project challenges drive up lifecycle costs. The common approach focuses on maximizing initial profits with little or no consideration for the consequences affecting the downstream project stakeholders and the industry. This old practice "sheds" known and unknown problems to the next stakeholder/s. It slows progress and provides little or negative value to industry maturity. It is a questionable unrealistic long-term competitive business strategy. In reality, it seems little more than churning people, money, and poor results, as disinformation, and may, in some instances, be illegal. It is sometimes sarcastically expressed in the statement, "It's nothing personal, its only business!" Yet, in reality, "**There is nothing more personal than business!**" This reality is especially true when the physical and mental health, safety, and investor/owner's or other stakeholders time and money are suspected of and/or knowingly being abused.

We have been told that these industry concerns are none of our business and, in some instances, have paid a price for it. The question that keeps coming back is: "Who's business and responsibility is it?" For long-term success, the real path to industry maturity requires addressing and mitigating risk as competitive goals and objectives, and not just dumping them on later stakeholders.

As we begin to deal more seriously with climate change, as a reality, the criticality of addressing stakeholders needs, wants, and expectations becomes even more important. Failure to do so eventually negatively impacts everyone, when resolving these common issues would benefit all stakeholders. This approach has become a harsh reality for all, not just businesses, and it impacts entire populations internationally.

2.2.2 Data Collection, Communication, Curation, and Their Effective Use

Compounding all the above is the concentration on initially stated performance requirement goals and objectives without performing sufficient data collection on actual plant health, condition and status measurements.

Unfortunately, data is primarily focused on how much energy was delivered to and measured at the POI (point of interconnection), the utility meter. The POI, because of its inaccuracies, may indicate that the system delivered more or less energy than actually consumed. The physical plant capability is often based on many unsubstantiated assumptions with insufficient measurable data. The result is inaccurate projections established through initial product sales literature and marketing hype intended to sell the product. This information is not the factual basis for a practical or profitable contractual buy. In addition, decisions are often made using "average values" without the necessary information or underlying statistics for associated levels of deterioration, degradation, anomalies, faults, failures, and defects and what they mean to all stakeholders. These practices lead to a significant reduction in reliability and increased O&M costs. Consequently, it reduces plant availability (see IEC 63109, PVPS–Information Model for Availability).

The exclusive use of contracted energy delivery as the primary required metric results in overlooking serious flaws within the system that are degrading its future capability for reliable energy production. The major concern for primarily using POI for the plant health, condition, and status is that it only measures the output power/energy and can have a measurement error of ±3% or more, for example. When seriously addressing these data issues, real temporal losses from various conditions tend to overestimate true system long-term performance and output.

As a result, it should be of no surprise that when systems are not producing the energy they are expected to deliver, the source of the problem cannot be immediately identified due to a lack of sufficient usable information to determine the current state of the physical plant. Much of the challenge is the industry's focus on PV system capacity (nameplate module DC watts), and weather-corrected performance, while ignoring the plant and its component "capability" to complete its lifecycle mission. It is further complicated by the complexities of calculating the impacts of climate on monthly energy performance. Making the monthly, quarterly, or annual system report numbers buries challenges for a short period of time. Note that the plant performance is not just energy delivered; it includes electrical frequency accuracy, the ability to absorb rapid changes in solar irradiance, and other under-addressed conditions or results!

This situation is analogous to owning a car and planning a trip. When considering travel, you may not be effectively focusing on the current health, condition, and status of the car, while expecting (assuming) it to achieve results similar to those published in the original marketing specification information, i.e. the EPA mileage numbers, rated performance, range, etc. Because there has not been a problem with vehicle performance before the trip, there is the expectation that there will not be a problem on the trip. The vehicle marketing literature indicates:

- **Capacity rating:** The vehicle is rated at approximately 190 HP (Horsepower) under factory test conditions.
- **Assumed energy resource input value:** The delivery is approximately 25 mi/gal or *X* km/l highway.
- **Maximum fuel load capacity:** The vehicle has a 16-gal (60-l) fuel tank.
- **Modeled mileage performance:** The new vehicle has an estimated EPA mileage range.

With these assumptions, you head off on your road trip. However, there are a few things you may not have considered that are a reality for any journey. They include considering the health, condition, and status attributes that define and affect the vehicles "capability":

- Low tire pressure and wear will affect mileage;
- Inadequate antifreeze mixture can raise engine temperatures impacting engine performance and mileage;

Rated vehicle HP capacity of the engine (Note while many people focus heavily on the engine horsepower as a capacity reading, just like with PV STC, during its operation, the car may never operate at that HP capacity. Automobile operational HP capacity is measured at a defined speed with a defined load and under certain other operating conditions. Similarly, PV modules/panels are sample measured at the manufacturer, under STC. Depending on location (site), the PV system may only pass through those STC capacity rating figures once or twice a day, and in many instances, not at all.)

Usage profile: The mileage and speed will be heavily impacted by traveling through cities, mountains, and valleys, which will also affect the performance and mileage.

Environmental profile: All of the environmental conditions that could impact the drive and the results including rain, hail, snow, temperature, humidity, altitude, etc.

User profile: What is the driving style? Does the driver use cruise control, exceed the posted speed limit, ride the brakes, drive very slowly, erratically, and so forth.

Load: Mileage reduction includes additional passengers and their luggage.

Unexpected/variable conditions: Don't forget the traffic and road conditions.

Safety risks: Unknown and uninspected brake wear, excessive metal particulate in the lubricating oil, or any one of a number of unrecognized and noninstrumented vehicle performance attributes that are degrading with the potential to cause significant damage or result in a safety event.

The trip costs twice as much and takes 20–40% more time due to needed repairs and additional fuel stops. These issues are also driven by how well the system, in this instance, an automobile, is maintained. Some cars are maintained very well, while others are not.

Part of the challenge is a lack of consistency in communications language, beginning with extremely basic notions of how the industry and organization define project success and failure(s), which are addressed in Chapters 4–6. Undefined, these two basic metrics impede the quantitative and qualitative growth in system RAMS. After all, the questions that should be continually asked are: "How do we know when we are being successful and to what degree?" "What information indicates failure?" and "How are we going to address it early?"

The real issue is not having too much data but using verified and validated data that is both accurate and timely! The solution is to combine sufficient, accurate, and quality data (Chapter 11) with the processes to properly analyze that data for dramatically improved results (Chapters 5 and 6).

Synthetic data is made up, copied, or fabricated to replace damaged or incomplete data.

Synthetic data (Chapter 10) is not a substitute for real and accurate data to fill in the gaps. It changes the analysis in a manner that is often poorly understood, if at all, or properly applied. Good analysis requires understanding variations of raw data and how it is interpreted. The PAM process uses the data's statistics to provide a broader understanding of what the data is telling you and where it is going, i.e. trending. Raw data that is incomplete and "filled in" using synthesis may tell two completely different stories about the health, condition and status of the system. That information should be noted and considered to see a broader and richer picture. It provides a more refined "capability" view.

Approved data planning and accurate data acquisition, QC, and analysis are critical steps and procedures. Their value is a cost control element of the overall LCC of the system.

One of the greatest challenges in the PV industry today is where incidents, defined as faults, failures, derating, or degradation, are not adequately documented and/or acted upon to improve the plant (existing or future plants) in a meaningful manner. The lack of an effective root cause analysis (RCA) negates the opportunity to find and build on many potential lessons learned. The information must be accurately documented, curated, and reconsidered throughout the system delivery process. Just indicating that something was replaced or was "broken" or when, is not sufficient. It does not allow the team to plan in a manner that would avoid similar situations, the cause, or provide the reliability information needed to make better choices in the future. Addressing the issues in a timely manner reduces soft and unbudgeted costs.

It is also foundational in the efforts of building PV and energy storage as infrastructure, or not!

These effects ripple through the whole project and/or organization. With a relatively small investment, there is an improvement in profitability by decreasing time and money spent on repetitive errors, failures, and their resulting challenges. Doing so delivers improved reliability, availability, output, and profitability, while reducing corporate angst. Yet, today in many organizations, there is insufficient attention paid by many stakeholders to the significance of those mistakes and failures that could generate valuable and constructive lessons learned.

Completely eliminating a culture that overlooks, ignores, and/or "covers up" (accidental and/or purposeful) design, manufacturing, and installation mistakes are some of the most valuable (fiscal) management strategies that can be incorporated. Without moving the culture away from the current practices, the industry will continue a cycle of incident repetition, which is unproductive, costly, and unnecessary. The project team must be rewarded for identifying any and all issues for whatever reason, then providing a solution as a group to improve the potential for success (i.e. lessons learned). Turning these issues into a blame game undermines the whole organization and results in personal, corporate, and project failure.

In looking at the readily available research, contemplating the results of thousands of technical and/organizational discussions, and understanding that every organization has made meaningful steps with improvement, is not sufficient. Only by communicating those issues and solutions on an industry-wide basis allows us to share the value of the improvements made to reduce our own systemic risks. *And Yes, we all have them!*

Having reread the notes from a few EPRI/Sandia industry meetings around 2013–2015, during two breakout sessions, many if not most of the issues we discuss here were identified, and defined. Now, about a decade later, we still wait for a resolution on many of these same issues.

Without sounding too anachronistic, earlier in the PV industry history and our experience, our competitors were often our good friends and colleagues, not just our competitors. In other words, we were all an "industry" of co-stakeholders.

We often formally or informally communicated about our problems, technical challenges, mistakes, and improvements. This had the effect of educating each other. As a result, much of what we learned we learned from our competitors. "Our" problems were "industry" problems, as were the solutions! Over time, this network evolved into a growing series of working and standards groups across the industry, creating new and more effective practices and standards. Those taking this tact grew our businesses by providing a smarter, better-educated group of industry professionals. We understood the clear difference between sharing industry information versus corporate business hype and marketing. To be clear, proprietary business information was kept proprietary, while common challenges were discussed and often resolved.

The results tended to address issues that impacted a broad range of users, raising all boats with the tide. Our relationships with manufacturers were also much closer, and many of those manufacturers listened and responded to those concerns. We did not have to be a huge organization to be taken seriously because we had real conversations with those organizations at all levels and their technical teams who designed and manufactured equipment.

Most of those industry characters who hoarded or ignored information, especially manufacturers, for the most part, all too often disappeared from the PV stage. Unfortunately, this often happened after their product had been delivered, found wanting, warranty claims increased, and the company reputation/s diminished.

This level of communication was disrupted by a shift toward more corporate-oriented business practices, holding that each company had the "magic" formula, and the bottom line was the number one priority, often without having a number two. Consequently, even within those companies arose segmented and siloed organizations whose sole focus is driven strictly by short-term bottom-line results. All too often, management becomes disconnected from the details that drive the business and success. This is why, with PAM, we focus on processes that address the disconnect between fact-based data and assumption-based management. This drives critical business decisions versus trying to mold organizations that have not addressed technology and/organizational inconsistencies effectively.

Much of the historic research has been codified through the ongoing and rapid development of new and existing international electrical codes, PV, energy storage, and other standards. These efforts move us toward better, more reliable, and profitable plants. They include many new or modified IEC, IECRE, IEEE,[5] ASTM,[6] ANSI[7] standards, and the evolving NFPA[8]/NEC[9] codes as well as NERC[10] and FERC[11] reporting requirements and others. Several of the most important newcomers are in the form of further development of the evolving IECRE "PV Plant Rating and Certification Standard" (in process); IEC TS 63265:2022 Reliability Practices for the Operation of PVPS; and the IEC TS 63019:2019, Availability Information Model Standard.

These standards allow for a more effective and comparative set of metrics and recommendations to document and support the assignment of a true commercial asset value for a plant throughout its lifecycle. Yet, these are but a few critical standards completed, while many are still in process, and others are being updated, based on the efforts and volunteerism of many in the industry. One example is the IECRE PV Plant Rating and Certification Standard, which, when completed, will be a tool similar to the Kelly Bluebook for autos that standardizes asset value assessment. Having that kind of information can stabilize an industry. It does so while providing a standardized data set to improve bankability (the ability to finance and insure) and by providing data to differentiate a good product from one that is not as good or is without sufficient value. This results in real meaningful fiscal risk reduction and a reduction of the cost of energy.

Nevertheless, it is important to understand that electrical codes and a broad range of standards are in the process of development and continual improvement. It is especially true as technology changes. A standard gains effectiveness through its evolution and iterative improvements as the industry amasses new experience, knowledge, and wisdom. It is also important to note that these codes and standards are generally based on technical consensus among a variety of PV industry professionals. All of this collaboration does not just create physical value; it improves systemic, organizational, and project value, especially at resale, refinancing, and reinsuring a plant.

2.2.3 Industry Language: Usage and Consistencies

Addressing these issues and many others can usher in a vastly improved PV industry as people adopt the same taxonomies and ontologies. Taxonomy is "the practice and science of classification of things or concepts, including the principles that underlie such classification."[12] It defines the usage of consistent industry language. It includes the ontologies:

> In computer science and information science, an ontology encompasses a representation, formal naming and definition of the categories, properties and relations between the concepts, data and entities that substantiate one, many or all domains of discourse.
> Every academic discipline or field creates industry-specific ontologies to limit complexity and/organize information into data and knowledge https://en.wikipedia.org/wiki/Knowledge. As new ontologies are made, their use hopefully improves problem-solving within that domain.[13]

The use of consistent language and classification – in combination with common and agreed-upon definitions, system metrics, and standardized calculation methodologies expresses common system realities and provides greater communications accuracy. For PV, this work began to be codified in the Orange Button project funded by the US Department of Energy (DOE). Orange Button defined more than 4500 solar industry terms and more than 15,000 US Generally Approved Accounting Practices terms. The established terms are readily available to the industry[14] and continue to be updated.

The final piece of the communication challenge of the PV Industry today is that failures are not adequately assessed, analyzed, documented, and/or inconsistently addressed for feedback. The lack of a RCA to provide consistent feedback fails to improve the plant and future plans in a meaningful manner. These inconsistencies could include technical, procedural, organizational, and financial elements. To make effective use of failure event data requires the performing, documenting, and curating of the RCA.

During discussions with other colleagues, we raised the importance of completing an RCA. Of interest is that several of our colleagues have said, "Root cause analysis is a legal term we don't wish to address." Although surprising, it is fascinating that they reject a well-established and accepted critical engineering reliability technical assessment tool and process: "… and the benefits of an RCA by using this as a throw away excuse." It indicates how estranged many facts in the industry to improve systems are lost, misunderstood, or abused.

From our perspective, RCA is demoted to a legal term when there is a deliberate or under-addressed failure to address this risk. In the long-term, it becomes a legal issue at the heart of lawsuits. Lawsuits, liquidated damages, and penalties become the result of a lack of planning, action, and leadership. From a reliability perspective, RCAs preempt legal issues. An RCA is a major cost containment tool. However, it requires time, training, and a dedicated budget!

2.2.4 Safety

Today's standards and codes are not sufficient for a robust solar and energy storage industry! Like the ever-changing technologies, standards, and electrical codes are continually in the process of development and improvement as they adapt to changes in technology and awareness. A safety standard becomes increasingly effective through this iterative process as the industry evolves and matures with additional data, analysis, knowledge, wisdom, and experience.

For example, the NEC evolved from the NFPA, which can be traced back to the World's Fair in Chicago in 1893 and is updated every three years. Unfortunately, it is often inappropriately applied as a PV design tool, as is clearly stated in the NEC, **ARTICLE 90, Introduction**.

"ARTICLE 90, Introduction, 90.1 Purpose, clearly states that code addresses:

(A) **Practical safeguarding:** The purpose of this code is the practical safeguarding of persons and property from hazards arising from the use of electricity.
(B) **Adequacy:** This code contains provisions that are considered necessary for safety. Compliance in addition to that and proper maintenance results in an installation that is essentially free from hazards but not necessarily efficient, convenient, or adequate for good service or future expansion of electrical use.
(C) **Intention:** This code is not intended as a design specification or an instruction manual for untrained persons."

The NEC code was and is strictly for electricity usage safety!

John Wiles, a great industry friend and supporter from the Southwest Technology Development Institute at New Mexico State University, has been an effective teacher, friend, and trainer of code to our industry. John indicated in several of his trainings, "Meeting the NEC does not mean that the system actually has to work, it means that the system is safe for and during operation!" That comment about the code not requiring the system actually to work always stuck with many of us as important. It also indicated there was a divide between the requirements in the NEC, industry awareness of the actual role of safety, and the knowledge that PV professionals must exceed the code for the systems to actually work appropriately.

Many PV design engineers today take great pride in designing code, yet are ill-prepared to address system reliability, availability, maintainability, or long-term viability and output. Addressing system lifecycle requirements is often not effectively a key element in the PV plant delivery process. This results from stakeholder assumptions and a lack of industry awareness that time and environment are seldom adequately addressed.

This reminder speaks volumes as to design based on NEC code. We always told our clients, designers, and electricians that if they specify and design properly, a system should already meet or exceed code. This ideology was often hard for our new engineers to grasp; ... many never did!

If the specification and design teams' efforts are strictly to meet the local, outdated, or the latest NEC code, they will not have delivered a low-risk system. The PAM process requires that PV owners, specifiers, and design professionals continually strive toward improving systems by designing them beyond the adherence to existing code for the long-term system and project success! Our philosophy included two major positive impacts on our business. Our systems were first safer, with fewer operational and maintenance challenges. The second was doing so allowed us to build higher performing systems delivering more energy per nameplate watt (measured under STC conditions).

Therefore, the stakeholders must focus on delivering a system using an effective PV System Delivery Process that goes beyond those codes and standards. It is our opinion that many companies and design teams claim to incorporate existing codes and standards, yet may have not correctly applied them or fully understand what they indicate and how to effectively apply them. Worse, they may be operating off of an out-of-date or superseded standard or code, resulting potentially in a considerable loss of time and money.

Many Authorities Having Jurisdiction (AHJs) policymakers and elected officials, often lag in adopting or updating new codes. It is always best to work with the latest published code versions, then discuss your reasoning for designing beyond the current code during a SE plan review to reduce confusion. When exceeding the code, provide the reliability information that supports your additional efforts. This communication can be incredibly important during any PV System Delivery and **SE/R**epowering™ process. In some instances, avoiding, under, or reinterpreting codes and standards becomes a management strategy. It is generally a poor business decision. Like everything else, it is imperative to monitor the codes for changes that must be incorporated into a design, while focusing on continual lifecycle improvement.

Our experiences led to the discovery that many PV organizations do not even have a copy of the latest national or local codes or standards available. Yet, many sign project contracts indicating that they are following those now contractual/legal requirements. These practices create a series of interesting challenges explaining how contractors deliver potentially deficient systems in safety, reliability, and performance, which in many instances may still pass inspections.

An example of code value was following changes to the 2008 NEC code. There were modifications that made for some important changes affecting the design and installation of PV systems. One would have resulted in a $50,000 reduction in cost for a PerfectPower™ client. One of us, Dr. Balfour, President of PerfectPower™, had asked Bryan Gernet of Arizona Public Service (APS), with whom he talked about the city of Phoenix and whether he thought it was a good idea for the city to accept the new code rapidly. He had met Bryan in 1994 while getting the APS Environmental Showcase Home PV system operational. He also worked with him until Mr. Gernet retired. Bryan reviewed PV drawings for a proposed PV and storage system in the APS service territory.

He wrote Lanny McMahill, Electrical Inspections Supervisor for the city of Phoenix. He was informed of a change to the NEC code and asked how difficult it would be to have the city of Phoenix adopt the new code. Lanny suggested that a letter of support from one of our utilities would be helpful if they were comfortable accepting the change. Bryan obliged.

Many AHJs do not adopt codes quickly, sometimes for years, following their releases. In most instances, as with the city of Phoenix, for example, the city council would have to vote to have that take place up to 10 years after the code had been published. Because of the cooperative nature of our efforts, the city fast-tracked adoption, which was a positive step for PV in our community and much of the state. Yet, there was another point of view from local competitors. They were very unhappy with the city's acceptance of the change of code, even though it would save them and their clients' money. This was because they now had to learn and implement the new code requirements.

It was not until he researched the story for the book that he realized Lanny had been part of the team that had updated the 2008 NEC code, representing the International Association of Electrical Inspectors. They both hoped for and or required a request from the outside to offer the city council an opportunity to review and accept the new code.

While many of our colleagues can often be reticent about and sometimes downright hostile toward the NEC, they may not understand how much effort goes into getting a national agreement on making PV plants safer. Part of it, to be sure, is because it is just one more thing they had to do and learn, while many owners and managers just see code changes as an unnecessary expense. Nevertheless, a missing element in many organizations is that when they do follow code, instead of the team being cognizant, it is often only one or two people in the organization who understand its application. It is not unusual, for the installation team to not fully understand the application of the code on a daily basis or what it means to address these safety standards properly in the field.

Dr. Balfour's occupational experience as a PV EPC provider for over three decades developed a sincere concern about safety for people (his teams and his clients) and property and for the people responsible for ensuring those rules and requirements are met. He also learned substantially more about the code from Mr. Gernet. Mr. McMahill also taught him much more about the need for critically serious inspectors and what went into their responsibilities, interests, and concerns.

Those lessons and relationships prepared us for the next steps needed to address the issues of safety better. They were the foundation for looking beyond just meeting the basic minimal requirements to seeing what can be done if you go above and beyond in thinking and actions.

Many of our colleagues will tell you that they already do everything necessary for success and cannot understand or explain the reason for the many demonstrated industry project failures.

There is frequently a lack of in-depth, consistency, and continuity of education and training, not only within organizations but across the whole industry. Yet, they freely provide language about their compliance with codes and contractual standards in their marketing statements.

In some instances, when asking marketing about how they or their companies define safety and success, the response will be little more than a 1000-yard (meters) stare, as if they don't quite grasp the importance of the question or what that might have to do with success. Much of this is delicately wrapped in the common excuse for lack of detail, "this data is intellectual property" (IP). Yet, one might think or conclude that not applying lessons learned or acting on our mistakes could be just the opposite of IP.

There has been so little discussion of what makes a project work properly to achieve success or failure that it sometimes becomes difficult to determine an appropriate project and design philosophy. As we have shown, the concept of a turnkey system is, at best, a weak approach when not addressing RAMS, and at worst, it may be a future source of unacceptable losses.

When first contemplating this book over a decade ago, most of the industry would not openly, or even under some NDA conditions, discuss these issues or what the project stakeholder's criteria for success or failure were. Attending the 2012 Solar Praxis "Inverter Conference" in Phoenix, Arizona, there were panelists that clearly stated that "if their management were aware of them discussing these issues, especially on a panel, it could result in reprimands and possible dismissal." Most, if not all, of those panelists are now leaders in the industry who are beginning to address many of the issues; however, their management, in many cases, is still not organizationally sufficiently addressing the PV plant as a system.

To address the lack of meaningful industry communications about these sensitive issues, some of our best information was gathered using the "two beer rule." This is often accomplished with honest, brutal, and straightforward conversation over a beverage and/or a good meal. These truly meaningful and confidential conversations require a level of trust, with honest discussion outside of the workplace or organization. However, that trust must be established, honored, and then maintained personally and professionally.

About 10 years ago, the first response from our publisher, John Wiley & Sons, and half of the industry expert reviewers evaluating these approaches, included comments such as "all of these problems and challenges raised in this book proposal have been resolved years ago." Yet, many, if not most, of the problems and challenges, for which solutions were found years ago, have remained.

This was due to a lack of actually applying lessons learned as they were not well documented and were seldom included in any of the then-current training. In fact, many lessons learned, as a consequence, were never passed down within an organization, let alone the industry. The same problems exist today. One proposal reviewer's response was that nobody had any real problems indicating that "everything is great, the industry is growing, and companies are making money."

The published and unpublished (internal documents) literature on the large numbers of closed, rebuilt often referred to as repowered or bankrupt plants, however, continues to contradict these rosy assessments. Today, many of our colleagues are still struggling to have frank and open conversations within their own organizations, let alone the industry, in order to supply the information necessary to management for resolution. Many talented people have moved on to other jobs outside of the PV industry due to growing frustration with the lack of progress.

We are continually told that the discussion regarding the analyses of the source of the plant failures is academic or theoretical and that one should not be caught up in "the paralysis of analysis." Yet, to avoid "paralysis of analysis," it is necessary to actually perform analyses of vetted data, while avoiding making misleading assumptions! This process of avoidance is often a clear indication that there is little cross-organizational awareness of the real technical issues, a siloing challenge that is an ongoing source of many systemic deficiencies. This internal risk of not shedding the throwaway attitude "it's not my problem or responsibility" is expensive.

"Siloing results in limiting key communications, accuracy, and collaboration through segmented contractual and organizational activities. Information handed off from one silo to the next is often assumed to be accurate and complete. Valuable information may be available at the next cubicle, yet not part of the process."[15]

Several of our references noted in this book are 30 or more years old. Many of these referenced issues have been creating turmoil in the industry for decades, and many of the published solutions are not new. Turmoil in part because they are directed at a lack of data, documentation, and analysis that helped define the functional solutions for the automotive, marine, medical, and aerospace industries. It is amazing how much of it has been published in papers over the last 30 years, yet remains fairly well ignored. Many research papers, reporting topics, and problems in which solutions were found are accessible on a broad range of public and private institutional websites at no charge. There are also numerous international standards and documents available on the web behind paywalls.

Nevertheless, as we attempt to discuss the problems and solutions with clients and others, there is a substantial segment that still holds these issues to be "academic"! That characterization seems to fly in the face of the well-documented fact that many companies have lost billions of dollars with numerous losses in the last 20 years by not paying attention to these basic issues. Nevertheless, based on the marketing hype that PV is simple, green, requires little or no maintenance, and has a solid earnings history, investors continue to invest in component manufacturers and projects that have not addressed all of these "academic concerns." It begs the question, "How is this physical and fiscal rupture simply academic?"

Many of the current and previously accepted myths, assumptions, perceptions, and perspectives for today's PV reality are broadly seen as inevitable "facts" of an academic nature. This makes it difficult for them to be effectively challenged and replaced. For many, it's hard to believe there are any substantive problems in the present approaches because the industry is successful, growing so rapidly, and obviously, "if things were really in such poor shape, that wouldn't be happening." As a result of the deep-seated practice of believing the present industry model is functional, it is difficult to explain to someone how the choices they made for their plant will result in underproduction, failure(s), and/or that many of these current issues had been recognized, and resolutions developed decades ago. "After all, everybody does it this way, right?"

It is in addressing the uncomfortable conditions and situations that improvement begins to take place. This is where the greater value in technology, stakeholders, their time, and their money are actualized in a more meaningful manner. Doing so builds a stronger industry with fewer problems and far better results, especially when things evolve and change so rapidly. If the real goal is a profitable business for the long term and the reduction of organizational/corporate angst, PAM and **SE/R**epowering™ provides a more functional path.

2.2.5 Stakeholders

Stakeholders are the individuals and organizations affected by, and that have an impact on the PV projects, systems, and industry. Many tend to go unnoticed either by benign neglect or overtly ignored. Although they have a role to play in the PV plant's lifecycle, they often work separately in a siloed manner, where one participant may not even be aware of the involvement or needs of the others. This compartmentalization often leads to expensive inconsistencies. There are generally two types of stakeholders, internal and external.

2.2.5.1 Internal Stakeholders

Internal stakeholders are primarily organizations and individuals involved in the specification, engineering, procurement, construction, operations and maintenance, and general delivery of the system on through to site restoration, as seen in the list below. They tend to have a direct financial and/or legal stake in the project and tend to carry much of the risk.

For PVPSs, they include the stakeholders identified in Table 2.1.[16]

2.2.5.2 External Stakeholders

External stakeholders tend to be regulatory or informational types of agencies that may participate in or oversee codes, laws, reporting requirements, grid owners or managers, or

Table 2.1 Table of typical internal stakeholders.

Developers	Financiers
Insurers	Utilities
Owners	System and Design Requirements Specifiers
Investor/Shareholders (Publicly or privately owned, traded, or PPA)	Customers (Usually buyers of energy, energy services, and/or end users)
Design or Engineering, Procurement, and Construction (EPCs)	Construction Labor/Work Force
PV System Asset Managers	Operations and Maintenance Providers
Project Managers	Legal Personnel/Firms
Manufacturers	Suppliers
Logistics/Shipping Providers	Certification Professionals

potential individuals or groups that may be impacted by the project including neighbors and even competitors. The list below references only US stakeholders, while there are many similar organizations internationally.

1. Government and nongovernmental large-scale PV or power purchase agreement (PPA) bidding agencies (In the US can include local, state, and national governments)
2. **Grid operators:** CAISO, PJM, MISO, NYISO, ISO-NE, and ERCOT[17]
3. **National regulators:** NERC, FERC, FEMP,[18] and others
4. Additional agencies can vary from site to site and country to country
5. Neighbors, including communities, property owners, and environmental groups
6. Regulatory organizations, such as Public Utility Commissions (PUCs)
7. AHJs
8. Competitors
9. Residential, commercial, or industrial buyers and users (of the energy produced)
10. Educators, undergraduate, postgraduate, OEM, and other commercial

An example of fully addressing external stakeholder's concerns is the following based on Dr. Balfour's professional experience:

> At HighPerformance PV (HPPV), Dr. Balfour had a client who brought his team a project that was in the post entitlement development phase. His team was hired to address issues that would lower the cost of operations and maintenance, resulting in LCOE.
>
> The HPPV team pointed out that if they were not following the project site County code and specific contractual entitlement requirements regarding their permit, they could have serious negative time and financial impacts. The county requirement for the property included details about the site to "be returned to its original agricultural state" upon site restoration. If this issue and others related to their permit arose, there could be mandatory shutdown, serious costs, schedule consequences, legal fees, and penalties until the deficiency was corrected.
>
> We were asked, "What does that have to do with our contract on O&M?"

We indicated that their O&M planning must ensure they had provided adequate checks and balances to avoid the "out-of-county requirement" situation or issues and not drive up project costs and energy production losses. This was foundational to proper planning for our client to reduce O&M costs, especially as it addressed environmental issues and conditions. They needed to address all requirements that came before and after the system has been commissioned and during operation.

Based on their project permit, the county could come in at any time to inspect, review, and determine if they were in or out of compliance. By doing so, they could determine if the permittee had created issues that needed to be resolved before they could continue to produce energy. If they were out of compliance or created a nonpermitted set of conditions, those issues would have to be addressed and reinspected prior to having any potential shutdown lifted.

Our process was to protect them from a potential loss of generation, legal issues, a hit on any insurance that might cover such an issue, and the fact that all of those costs would go into the O&M budget. We indicated that if there was an incident *they had planned for*, it would be much easier to negotiate with the county and avoid any contracted service (production) interruption.

In essence, a stakeholder, the county as per the permit agency, required the permittee to address any issues that impact restoration. To do so, it should be addressed in their budget and planning. If they preempted all those potential issues in their planning, there would be one defined cost, if not, any corrections and the loss of energy production would come out of the operations and maintenance budget. Our position was that it was less expensive to plan and avoid the problem; however, neither of those potential outcomes had even been considered. Mind you, we never even got to the part about financial penalties that would be added including shutdowns, costs for actions, inspections, and legal costs.

As for the cost of doing this, we indicated how this level of planning could be used in their future plans of this type, which would be minimized in the cost per future plants.

Failing to address this during O&M planning, these issues can throw a monkey wrench into the project plans, finances, and revenues if they did not do as they had contractually and according to regulatory requirements, agreed to. Our position was that addressing these details could avoid driving up the costs. It would also avoid bringing a variety of other unidentified or defined stakeholders into the fray, which would also cost them additional time and money. After all, when a PV plant is shut down for a week, a month, or a year, it has devastating impacts on the financials, including the LCOE, and PPAs, which were not being considered.

The only thing that could have been worse and more expensive was if their team did nothing.

The result was that the HPPV contract was ended early, although they were paid the full contract amount. What we had included in our contract specification was not what they actually wanted, though this is what they asked for. They wanted a sign-off indicating their cost would be about half of the real costs, thereby shedding risk to future stakeholders.

Obviously, or maybe not so obviously, our client's lawyers read and accepted the county requirements when our client bought the project as reasonable pro forma requirements. It was clear in the total contract, and permit, that the site requires rigorous site restoration.

However, was it clear to our client's development team? They now controlled the project, while the internal or external O&M providers had not addressed all of those requirements or costs that they were contractually legally committed to provide. This is a common organizational shortcoming in most siloed organizations. It may be the result of one silo understanding the requirements, while another assumes that all contracts, as per entitlements, allow the same responses or actions.

We discovered the team we worked with, *who had not been involved in the early development and entitlement phase*,[19] had left those details to the EPC, internal experts and others. The EPCs may not have read the county permit requirements after the purchase, possibly just noting them on their project checklist. They may not have understood the later consequences. It may not have been in their contract, assuming that it is the projects lawyer's or O&M's responsibility. They may have failed to fully fathom what the project permit requirements covered and or the fiscal consequences. It appeared that all of these present stakeholders assumed that someone else was responsible and supposed that it had been taken care of! Yet, it was not!

It is important to remember that the EPCs will only do what their contract specifically requires, and if these issues are not addressed in the specification, with appropriate tasking in the contract clauses, they cannot be held responsible. Our client was the owner/developer, and they did not initially want to hear about these kinds of issues and may or may not have addressed them.

Based on available information, we presumed "someone" on the new ownership team had read and understood what all this meant and had determined the legal and fiscal implications, yet had never passed it on as requirements, in context, to the rest of the team. The owner's contact was stunned and amazed at the potential risks! Having identified the risks, we presumed that they would have to go back through the process of analysis for the plant, as planned. Failure to do this may cause the county to act punitively, to address the issues. However, we do not know if the reviews actually took place, as the contact quit within a month, and no further communication was received. Consequences of not addressing these questions included shutting down the plant, which could drag on for days to months, resulting in legal and other costs that were never included in the initial project costs, O&M budget, or LCOE! This outcome would be the same whether it was labeled as a design issue or an O&M issue.

HighPerformance PV simply flagged a critical O&M requirement that could result in hundreds of thousands to millions of dollars in legal fees, additional changes, penalties, new oversight by the AHJ, and a new permit and inspection process that was never considered in the original budget or LCOE. We learned that our project contact moved on to another organization shortly after we were paid.

This example addresses the importance of defining and practicing detailed identification of stakeholders, their wants, needs, and requirements, documenting this in the specification, effective processes, and procedures for the SE and RAMS processes.

Another example includes PerfectPower Inc., a company Dr. Balfour founded in 1998 and left in 2010.

During its profitable years and as a result of his company instituting five-year warranties with annual inspections on all systems and other processes and procedures, a service tech could review and check for minor and major problems, which helped preempt unscheduled outages. For small systems, a technician could address up to five or seven residential/small commercial systems in one day. When Dr. Balfour left his company, the first thing the company did was to eliminate most of the carefully established processes and procedures that had been developed over more than a decade. These were eliminated on new and existing systems as they were considered too expensive in their internal, yet incomplete, cost–benefit analysis. They were, however, already locked into future cost and profit through existing contracts.

Most of those processes and procedures were considered by PerfectPower management to be inconsistent with the rest of the industry approaches and, thus, "an unnecessary cost burden" to the projects they now supported. The intention was to save time and money without addressing the implications of omitting effective and proven specifications, processes, procedures, and methods. With these eliminated, installation labor and O&M costs went up dramatically. They also quit evaluating the existing monitored system performance resulting in the loss of requisite data to assess plant health, condition and status, an additional budget hit.

One of the reasons for the choice to automatically provide 5-year warranties and annual cleaning was to have control of the projects and their information, in order to address any installation or equipment issues early. It was a functional lessons learned process that also had the side benefit of identifying problems created at installation, thereby keeping all segments of the company in line with meeting a very high level of quality and customer service. It also ensured that installers did not cut corners and followed well-established and documented processes and procedures.

In a short time, the number of unexpected project warranties, truck rolls, and associated costs increased. Within a year or so, the cost of commissioning increased dramatically because of installation schedule delays, cost of construction errors, and unplanned service calls for new systems, and excessive truck rolls for existing systems.

This influx of problems for PerfectPower created confusion and wiped out the profit for many projects. This also generated unexpected and avoidable costs for owners, owners, and operators. These customer impacts of reverting back to common industry practices and other poorly informed choices and changes killed the company in a relatively short period of time. Yes, the first few months looked good, *on paper*, and showed fewer expenditures before the company's savings and profits were eviscerated. These results occurred after the company had been consistently profitable for the prior 6 years.

PerfectPower shut down a year or so later after he left, when project management considered the fundamentals of PAM policies and procedures provided more costly over value to the company. Nevertheless, prior to shutting down, the CEO indicated to him that they attempted to reinstitute those changes. It was too late! The other outcome was a rapid loss of goodwill from customers and referrals. The company has previously made a substantial number, if not most, of its sales through referral and word of mouth. That loss of credibility and of income was never considered and quickly dried up.

The greatest lesson learned from this company's failure was that organizations working under a misguided rush to cut costs for improving near-term profits can and generally will backfire. When a cost–benefit analysis was completed, many of the important factors that went into the impact statements for specific decisions were not even considered. This is fairly common throughout many industries, not just PV. The assumption and associated action are often that being profitable and increasing profits are simple, "Just get rid of or cut the fat or waste."

But what is the definition of fat? How is waste defined and measured? This is one major problem that many organizations have: defining waste before performing cost–benefit analyses.

Too many assumptions go into processes and designs based on little or no information, especially since many organizations actually had the data and information upfront in the project but failed to use it to make better decisions. What appears to be common in the PV industry is that upper management is often so divorced from the realities of the technologies and practices in their organizations that they make ineffective, damaging, and costly decisions. Unfortunately, all too often, those poorly supported decisions are self-inflicted.

Pressure was built by looking at or setting a goal without taking the time to understand what would be needed to achieve it. This is especially true of organization teams or individuals who consider themselves to be "experts." They have taken industry gossip, myth, and assumption as gospel instead of addressing these issues by researching the facts and adjusting their business or cost models instead of considering the consequences. However, often their attitude and approach only results in hiring cracker barrels or hearsay experts. Their expert status relied on common "standard industry myth, assumptions and practices" that had already proven to be poor at best.

With all the marketing hype and excitement about the advances in the PV industry, it is not unusual for organizations to devolve like this. Part of this is because of the intense marketing and the massive amounts of useless and distorted information that are continually being pumped out, much of which is advertising "flash," not science or engineering.

Over the past decade, a growing number of PV professionals have become clearly aware of and making progress on eliminating issues by addressing some previous myths and assumptions. However, overcoming bad habits in an industry's management and execution can take decades to complete.

The flow of information, documents, data, and processes requires consistency and accuracy; it requires constant communication between stakeholders where information is carefully selected, curated, and made available to the appropriate parties. Access to information must be carefully managed and distributed by a specific individual or group.

PV is not the only industry to have these challenges. In fact, it is ubiquitous among all forms of organizations. It often occurs within an organization's work centers, such as sales/marketing, design, QA/QC, manufacturing, and other departments. It increases in complexity where they tend to use their own tools and or language for specific communications and data sharing among themselves while distributing only a selective set of data/information. This phenomenon is not a conspiracy but a lack of effective and practiced systemized management and communications.

Siloing creates serious communication challenges and has internally disruptive impacts, as one hand does not know what the other hand is doing. It slows the improvement process down while veiling and exacerbating underlying issues. This situation cannot improve when siloed stakeholder individuals and or organizations working on the same project fail to communicate effectively with each other. As long as owners continue to treat projects as "one-off," e.g. not keeping consistent stakeholders and individuals with well-developed PV skills, while not sharing essential project information, they create problems that are unnecessary. The result will be the loss of critical project management skills that go into addressing the lack of information which can and does quickly complicate things. There is seldom a data traffic cop to direct the flow of information, much less continually confirm and secure its truth and accuracy. Therefore, siloing is often matched with the attitude that "No one is to blame!"[20] This sentiment is used by many individuals, organizations, and industries to avoid responsibility.

2.2.6 Site Capability Survey

In many initial site surveys, there is a primary focus on the site with little emphasis on the areas adjacent to the site. This can result in unaddressed issues, from shading to environmental damage.

> For example:
> In one project, a major portion of the north segment of one array was 10–12 ft lower than the southern array. The primary solution was that fill would be brought in to dramatically reduce shading.
>
> The assumption was that additional runoff would flow to the east to drain the area. The initial site survey overlooked the fact that both sides of the freeway to the north also drained into where the fill was to be placed. The existing drainage looked substantial; however, it was insufficient.
>
> The expressed concern, which was poorly documented, was that after a substantial snow event, followed by warmer weather and rain, could flood the entire area. The initial response was, "That's why we are using the fill and raising the array!" This resulted in the opinion that the hazard had been addressed and due diligence on the site was complete.
>
> It was pointed out that the array would probably not be damaged; however, because there was limited drainage, the flood water could back up to flood the substation. Although the issue would hopefully be discovered by the AHJ and or the utility, it was argued that the fill would resolve the array issue, and the scenario was unlikely. The result of this was that the problem was minimalized (trivialized) and thus ignored at that time.

This was a case for the need for understanding the potential variation of environmental conditions or a combination of conditions that, although rare, are potentially damaging environmental events that would cause major damage to the plant. This applies to green field, brown field, or exiting plant sites.

The **Site Capability Survey** (SCS) process includes gathering and documenting all current and relevant site conditions:

1. Elements, all local potential site human and environmental elements/conditions.
2. Details, documentation for discussion including plan development or adjustment.

3. Actions/activities, defined steps to be implemented to identify, and address impediments.
4. Punch list before and after, not initiating Commercial Operation Date (COD) until the punch list is completed. (Many punch lists are not completed resulting in construction costs being moved to the O&M budget.)
5. Results as reported for this particular moment or segment of time and including:
   (a) A clear listing of all intended applied processes, practices, procedures, goals, and objectives. (Why?) If the actions to be taken are not clearly identified, procedural actions can be overlooked, unbudgeted, or ignored.
   (b) The specific actions and activities to be taken, identified, and documented. (What?)
   (c) The metrics which will be used to determine plant and component condition details. (Where?)
   (d) It includes "new" sites for potential plants before and "after metrics" following a touch-up repowering. (Note: **R**epowering™ planning delivered systems include the collection of specific physical and operational data collection changes over time.") (When?)
   (e) The Team members review all PV plant conditions and capabilities as they exist at that point in time prior physical operations including what was specified, nominally to changes. (Who?)

2.3 PV Plant Commissioning

One of the weakest links in the present PV project delivery model approach and a major contributor to PV plant health and condition masking is a lack of **SE/RAMS/R**epowering™ and its requisite tools. This uncontrolled and under-controlled void conceals energy, time, and fiscal loss with the inability to adequately quantify the health, condition, and status of a plant beyond the COD. It is akin to launching a highly technology-driven sophisticated ship without accurately determining its initial condition, tracking system, or O&M challenges.

No matter what universe of choices is made prior to launch, these must be clearly identified (measured), addressed (corrected), and documented (benchmarked) before the plant is accepted. It is where the concepts, practices, health, condition, status and documentation of the plant are confirmed, for operational fiscal reasons and needs.

Our position:

> All PV plants, with or without energy storage, require a thorough and effective benchmarking and commissioning process. Our research and experience as an integrator and EPC confirm that this practice helps control system delivery, and O&M costs when begun at specification. This is true for on or off grid residential, commercial, industrial, and utility grade systems. Different types and grades of systems require diverse levels of documentation, data capture, curation, and evaluation. This gives stakeholders critical information to maximize system life, cost control and energy production, while providing specific and accurate information regarding what was designed, built, and operated. We continue to recommend sharing that reliability data to aid the industry and all stakeholders.

This practice provides a specified (at the initial contract) set of requirements and documentation. It must be available to the plant owners, for sales contracts, permitting, construction, insurance, O&M, and eventual sale for evaluation and continuing energy production. It allows for maximizing plant lifecycle operational, energy production needs, and lifecycle cost control value.

2.3.1 Commissioning and Benchmarking

2.3.1.1 Commissioning

For commissioning, as of the date of publication, the most recent source of practices, metrics, processes, and procedures can be found in IEC 62446-1[21] (2016). Unfortunately, commissioning is often whatever the system contract specifically calls for, whether it be IEC 62446-1 for commissioning or part thereof. Without detailed follow-up documentation, the level and quality of that commissioning may never be complete or effective. While this IEC standard is the primary standard to refer to and use at the time of our publication, in a world of rapid technology evolution, it is dated and not always complete for long-term service life.

Commissioning and benchmarking are used to provide a documented handoff from specification/design and installed by the EPC team to the owners and O&M team. The handoff is not successful unless all performance evaluations and acceptance tests are satisfied by the project's completion. The same is true for grid-connected solar projects that must meet any number of local or utility regulations.

There's another benefit of PV commissioning that varies with the level and quality of commissioning and benchmarking. It indicates to what extent and depth the systems have been optimized to meet the delivery of documented expectations at the time of COD. It is designed to confirm the engineering, physical, and financial condition of the plant to provide a snapshot, not a lifecycle picture of the plant. It is designed to meet the initial expectations for potential profitability, capability to meet payback projections, PPA requirements, and give direction and comfort to investors regarding the return on investment (ROI).

The greater depth, quality, and thoroughness in its evaluation and benchmarking, the greater the level of confidence participant stakeholders may have with this baseline.

Nevertheless, the value of the commissioning and benchmarking is commensurate with the quality and thoroughness of the specification, design, and operational (O&M) aspects of the planned, designed, and built plant. The variability of its effectiveness reflects how long initial fiscal stakeholders have a financial interest in the project. This can be or is, in many instances, only measured to the length of time for the initial bank or PPA funding and contracts generally two to seven years.

Yet, as a baseline, it must also be updated throughout the system life. It needs to include the original and annual commissioning, benchmarking, O&M information, and reporting to stakeholders regarding the plant's ability to produce energy currently with a prognosis of future production. This process defines the real "All IN" costs for LCOE, which should also be updated annually for operators/maintainers/owners/utility – financials and basis for financials. Doing so creates a template for future decisions, specifications, and designs.

What are the specific requirements for completion of the contract? Those requirements may be minimal or provided in great detail and may include:

1. What detailed level of commissioning and benchmarking data is required for lifecycle improvement evaluation?
2. What are the processes, procedures, metrics, and documentation requirements?
3. How will they be measured, and against what?
4. How much of the plant will be measured and documented in detail? (As compared to a higher level statistical confidence evaluation.)
5. How will this data be curated, made available, and to whom?
6. How will it be shared with stakeholders within and without the system delivery process?

Today's sample commissioning and benchmarking, though less expensive than Monolithic or Sectional sampling (see below), is an incomplete process that is assumptive of the initial COD health, condition and status of the plant. It would be similar to practice with modules and inverters: Where only a limited portion of PV system modules are flash tested, then assuming that the rest of them are statistically equal to the average of those tested. Or, that just because some of the tested manufactured inverters met specifications for operation, the rest will also as be acceptable as long as the green light turns on, then giving them a pass also.

Sampling Types during Commissioning:

(a) **Monolithic:** The entire plant is commissioned prior to start up COD including the area physical and environmental conditions surrounding the site. This is where the entirety of the plant is documented in great detail and a detailed "benchmark" is established for future comparison.
(b) **Fractional:** A specific percentage/fraction of the plant and its components are documented. Fractional Commissioning may include a specified percentage of the modules, strings, and combiners by number of plant segment to reflect a representative sample of the entire plant.
(c) **Sectional:** A utility-sized plant may be commissioned and or benchmarked, in large multi-megawatt sections, such as 20–100 MW or larger blocks as they are built and completed over time and brought on line as sub COD segments.

To improve the commissioning and benchmarking process, additional new technologies and methodologies (processes) will be required, which generally adds a small amount to the investment and or operating costs. With the application of **SE/R**epowering™ from the beginning, these added costs are quickly recovered.

When commissioning/benchmarking is combined with independent third party assessment and inspections of manufacturer's facilities, products and performance of installation testing/evaluation, a number of benefits accrue to the stakeholders:

- Owners have greater success receiving products that meet or exceed specification and reduce the defects, damage, faults, and failures from manufacturing and shipping.

- Many independent evaluation lab services include the contractual ability to facilitate manufacturer replacement prior to COD. This avoids a variety of costs for time, replacement, shipping, and labor from being applied to product warranty.
- Reducing these problems prior to COD, pays for itself throughout the product lifecycle by reducing the impact to O&M budgets and simultaneously improving energy production while reducing these issues and costs as O&M budget items.

Repowering™ and repowering? Repowering™ is a detailed PV lifecycle planning "system delivery process" designed to effectively extend plant life from two to four times greater than a functional lifetime today.

The common term repowering is often akin to or an emergency set of O&M actions or activities to bring the plant back into contract availability compliance. It is a plant event and condition triage, not a system specifying planning and lifecycle operation process beginning at concept. While it may require an evaluation and specific component or software replacement, it is not designed to keep a plant at a high level of energy production. As a result, it is not capable of discovering or addressing all of the hidden challenges in the system.

To control costs and output for PV/Energy Storage (Ramasamy et al. 2021) as infrastructure, O&M contracts must include an annual benchmarking update. This is especially critical for performance, revenue, and warranty purposes. It gives owner/operators critical tools to identify performance deficiencies that are driven by defects, degradation, faults, failures and damage, and document them for corrective action. It acts as a set of managed practices to keep energy production up and unwanted costs down. It is becoming a more cost-effective set of practices as additional automation technology lowers costs and improves accuracy.

Unfortunately, commissioning is often whatever the system contract calls for whether it be IEC 62446-1 for commissioning or part thereof or some variation of the same. With the unfounded concern about initial cost, commissioning, and benchmarking tend to shortchange the later stakeholders over the remaining life of the system.

2.3.1.2 Benchmarking

Benchmarking, Kounev et al. (2020), Ramasamy et al. (2021), is based at a specific point in time data, collection, and documentation of the capability of the system where details are measured, confirmed, and curated. Measurements include plant initial condition, health, status, and documentation of any upgrades, total and annual degradation, any known defects, force majeure and or incidental/induced damage. Based on benchmarking (data points), commissioning should establish the accuracy of any anomalies that must be addressed prior to actual COD.

In the quotation below, four different forms of benchmarking are identified, presented, and defined: Performance, Practice, Internal, and External benchmarking. You will notice that we include the organization to deliver critical information about who does what and how. It can identify gaps that exist, which may not be obvious, though in need of evaluation and modification to enhance system and organizational effectiveness. In mature industries, benchmarking is the mapping of the system processes, hardware, operations, and results that deliver continual improvement especially when clearly documented and fed back to the organizations involved.

As noted in "The APQC Blog," there are four basic types of benchmarking that can and should be applied to new and **R**epowered™ systems and organizations. These are:

1. "**Performance benchmarking**[22] involves gathering and comparing quantitative data (i.e. measures or key performance indicators). Performance benchmarking is usually the first step organizations take to identify performance gaps.
   What you need: Standard measures and/or KPIs and a means of extracting, collecting, and analyzing that data.
   What you get: Data that informs decision making. This form of benchmarking is usually the first step organizations take to identify performance gaps.
2. **Practice benchmarking** involves gathering and comparing qualitative information about how an activity is conducted through people, processes, and technology.
   What you need: A standard approach to gather and compare qualitative information such as process mapping.
   What you get: Insight into where and how performance gaps occur and best practices that the organization can apply to other areas.
3. **Internal benchmarking** compares metrics (performance benchmarking) and/or practices (practice benchmarking) from different units, product lines, departments, programs, geographies, etc., within the organization.
   What you need: At least two areas within the organization that have shared metrics and/or practices.
   What you get: Internal benchmarking is a good starting point to understand the current standard of business performance. Sustained internal benchmarking applies mainly to large organizations where certain areas of the business are more efficient than others.
4. **External benchmarking** compares metrics and/or practices of one organization to one or many others.
   What is needed: For custom benchmarking, you need one or more organizations to agree to participate. You may also need a third party to facilitate data collection. This approach can be highly valuable but often requires significant time and effort."

For greatest value and system lifecycle success, all elements of the benchmarking should be completed, accepted, and curated following the completion of EPC punch list issues prior to benchmarking documentation acceptance and COD.

A more effective and measurable methodology would begin on a foundation requiring the introduction of the **R**epowering™ cradle to grave planning through site restoration processes and procedures. This approach begins with exploratory or preliminary benchmarking based on a detailed plant corrections process and a preferable new annual operational benchmarking and commissioning. (Note: Once the initial **SE/R**epowering™ commissioning and benchmarking are complete, the annual benchmarking becomes a much less time and labor-intensive process.)

With this approach, PV systems with or without storage are required to be commissioned as you would a high-tech marine or aerospace vessel or vehicle. This is generally a detailed set of clearly defined metrics or conditions and practices made a key element of the contract to build and operate the plant. When effectively administered, commissioning, and benchmarking, the plan clearly identifies technical and nontechnical issues and assures a map indicating if and how they have been corrected.

A benchmark is crucial to future owners and stakeholders. It is critical to the plant value assessment and understanding the plant is condition at handoff. An effective benchmark process includes:

(a) An organizational set of processes and procedures to make corrections to keep production and revenue on course.
(b) A qualitative and quantitative set of clearly defined and documented set of testing and evaluation criteria must be complete and consistent with contractual requirements. This includes:
    (i) Mechanical Testing and Documentation
    (ii) Electrical Testing and Documentation
    (iii) Assessment of plant solar conversion efficiency and how it measured
(c) Testing must indicate which system, component, or subsystem is being evaluated. Does it meet the manufacturer's and plant's specifications and requirements, and is it operating within the specified and designed system, voltage, amperage, and environmental temperature range?
(d) Clarity for how any corrective actions will be carried out.
(e) It must include the construction ground work, conduit, wire, and cabling, system racking and tracking, safety, and accessibility.
(f) The benchmark report or summary:
    (i) Performance Testing, Evaluation, and Documentation
    (ii) Correction and confirmation of items listed Reporting, Completion, and Documentation

Many plants operate without those corrections, pushing those items and their costs into the O&M budget. The reality is that many incomplete plants are placed in operations prior to completion.

Benchmarking is based on the details as confirmed with a complete set of measurements. It begins with the initial plant metrics, data collection, and analysis prior to COD. A regular or later benchmarking event is applied at a period or schedule to collect that data and compare later changes throughout the plant. It includes plant condition, health, status, upgrades, degradation, defects, site initial and operational environmental (storm, climate change, and flooding), or induced damage (casual, purposeful, and accidental). Insurance claims are a good example. Based on benchmarking (data points), commissioning should identify and confirm any anomalies that must be addressed and retested:

1. This is the plant condition and what we have at a specific point in time!
2. These are the actions and activities we plan to take to gather, analyze, and curate that data.
3. These are the metrics for what we will find or found to determine if we meet our intended goals and objectives.

There must be a complete and detailed Site Survey before and after completion metrics, especially when initiating or following an emergency touch-up similar to a common repowering.

The primary function of commissioning is to guarantee that the facility meets the operational requirements of the investors, whereas others are listed below. More specifically, commissioning allows one to:

1. Evaluate, measure, document, and verify that the installed components are complete and installed per manufacturer and contractual requirements.
2. Confirm/certify that the operational PV system has been installed as specified and designed per specification including and meeting all engineering, operational, AHJ, and requirements.
3. Identify, classify, document for correction of any potential defects, energy losses, or performance issues within the solar plant.
4. Make a prima facia or enhanced case for the financial security and ROI predictability of the entire clean energy investment.

The initial finance package stakeholders will focus on the minimum cost and schedule for commissioning and benchmarking. Their project responsibility ends when that period and time, usually five to seven years, is up, and they have recovered all of their investment, profits, tax incentives, and other project-level incentives.

Review and confirmation of plant specification, design, engineering, and completed construction for operations begin with the contract documentation. This is to confirm that the plant is specified, designed, and completed as per system contract requirements. It must also include a confirmation of any additional system elements or other energy production elements, whether the system is grid-connected, a microgrid, or a standalone non-grid-connected system. It must answer a series of questions, including:

1. Does the plant represent all of the requirements as defined in the contracted planning, specification, and design elements as completed to initiate completion for grid operation, i.e. COD?
2. If there are unresolved issues, what are they, and how must they be addressed and or corrected?
   (a) Some variations may be unacceptable,
   (b) Some may be renegotiated based on shifting requirements or product availability. (Although, all of those changes must be documented and signed off, preferably prior to commissioning and benchmarking. This documentation should result in agreement on what the acceptable modifications are and lay a foundation to address the unresolved issues.)
3. What steps or actions need to be completed to move ahead for commercial operations?
   (a) Our position is that the final payments are only made once all of the commissioning corrections or modifications are confirmed and benchmarked.

2.3.1.3 Electrical Testing

This involves inspecting all the electrical components in the PV system, including the wires, panels, inverters, and even substations. The primary goal of electrical testing is to verify that every component has the correct current, voltage, and amperage – both individually and across the entire facility.

2.3.1.4 Mechanical Testing

Most PV facilities have relatively few moving parts. But mechanical inspections of all racking, mounting, and other structural components of the system are essential. This is especially true for solar projects installed in areas with temperature extremes, high winds, or seismic activity.

2.3.1.5 Performance Evaluations

With performance evaluations, you measure the expected (or modeled) output of the facility to its actual output (after adjusting for things like weather). This performance ratio (or "PR") helps to establish a baseline when forecasting both the long-term energy output and financial returns of the solar project. It is also used to meet a set of contractual requirements at operational inception.

2.3.1.6 Corrections Reporting

The final step involves providing the facility manager or system owner with a report detailing recommended corrective actions for any issues discovered during the previous testing and evaluation stages.

2.4 Universal Real-Time Data (URTD) and Data Sharing

In addressing data siloing, documentation, and communication issues, we have spoken internationally over the last few years regarding the concept of a URTD system (Chapter 10). The URTD data for an organization of a site is initialized using historical environmental information. It should be noted that the environment is inclusive of the site-specific climate conditions and the impact that terrain has on modifying those conditions. Things such as shading from trees, hills, and buildings, and the elevation change over the site with soil conditions that may erode under significant rainfall or snow melt.

This system requires the collection, securing, cleaning, anonymizing, and curation of data while sharing the appropriate information to project stakeholders across the project and or the industry. URTD and its real-time capability include "as built" impacts and all lifecycle plant data (see Chapters 5–9). It is based on using a detailed URTD with a skilled curator. The curator is much more than an information traffic cop, as they oversee the collection, creation, analysis, quality, and disposition of data to the right stakeholder's attributes for the other stakeholders. They also ensure that the data flow is bi-directional between the sources and the repository.

2.4.1 Initializing the Site URTD Database

One of the most aggravating, difficult to acquire, and critical data issues is the understanding, modeling, specifying, and designing for appropriate site environmental conditions (Chapters 5–9). Long before the documented reality of climate change,[23,24] PV professionals often failed to account for site historic and the potential for extreme climate conditions. It is generally a result of a lack of understanding that environments can have variations and result in significant overstress of components resulting in early failures. The site's natural environmental variability over long periods of time (20–50 years) should be assessed for maxima and minima, and the statistics applied to handle potential once in 100, 500, or 1000 year or more events.

However, with the reality of climate change providing additional extremes, it is in the best interests of all of the stakeholders to enhance or expand the systems environment extreme robustness.

These assessments should include temperature, humidity (duration), sand and dust, and precipitation. For the longer-term events such as once in 100 years, the effects on the equipment can be catastrophic. All the design elements need to account for the extremes. In 2020, in Texas, they had a major winter event that took out the power grid. More to the point, there was significant damage to the regional wind and PV plants as their design criteria did not include these types of events occurring.

Many projects appear to rely on the use of the average temperature (or statistical mean if used), with a 60% confidence level of the site environments. Some do specify a higher or lower temperature, but generally, this is based on a relatively short time span, say 5 years or less.

An interesting example of proceeding with a traditional view of limited environmental site research includes the placement of a few utility-grade systems, which were partially located on a dry lake bed. The project delivery/design team found the local southwestern desert weather data, which may or may not have been sufficient. It may have been sufficient at that time for finance energy calculations or possibly for too short of a full environmental picture over a period of time.

> … or they may have missed the indication that when it rains, the Dry Lake Bed became a wet lake of various depths depending on rain water inflows.

Having had an interest in climatology, both authors, with or without sufficient data, would have found sufficient indication of occasional flooding by the simple title including "DRY LAKE BED." We do not understand why the deeper parts of the lake were not avoided, or why the mounting system was installed so that in the event of a rare potentially damaging storm, flooding was factored in. At this time, due to a growing increase in extreme weather rainfall events, raising any and all electric components might have been a more reasonable approach.

We suggest that climate change impacts on-site requirements research must become more effective and detailed in how climate risk is addressed. For system reliability and grid capability, these and other climate events may have serious negative impacts on large segments of PV as energy infrastructure for grid stability.

Greater concern and attention regarding time and confidence levels for infrastructure must look at a longer temporal window with a higher level of confidence to improve lifecycle results. At this time, there is still a lack of research and interest in the PV community to look much closer at changes/variations in climate models, which drive potential risk and temporal vulnerabilities.

As has been demonstrated over the last decade, we are seeing a substantial increase in climate variability in the number and types of *extreme* weather events. This requires a mental, philosophical, economic/fiscal, and engineering shift from designing plants as they have been historical. This delivers a more functional approach to meet requirements during the finance period and beyond. By using historical weather data 20–25 years and longer, the industry will be more effective in addressing the costs associated with the lack of climate change specifications. That shift will not protect against all climate change consequences. It will provide a more solid foundation for the dramatic reduction of damage and its causes.

As an example, some U.S. locations have had four 1000-year events – referring to the severity – in a decade. In 2016 …"the United States logged no fewer than four deadly 1000-year floods in states as widespread as Texas, West Virginia, Maryland, and Louisiana." South Carolina suffered a 1000-year flood in October 2019. The same area experienced no less than five once in a 1000-year event.[25] In the Southwest US, in 2018, Phoenix had a once in a 1000 year flooding event, and in 2017, Texas was hit by Hurricane Harvey, which left devastating flooding over a large portion of the state. Also, in 2017, PV systems in Puerto Rico, the US Virgin Islands, and Barbuda suffered major damage or failed completely due to solar modules being ripped from the supporting structure, balance of plant (BOP) equipment damaged by flying debris, and twisted, bent, and broken structures. North Carolina suffered from Hurricane Florence in 2018, and a hail storm in Texas damaged over $1 billion of PV modules. Historically Arizona had severe weather events from a super storm that was documented for Maricopa County. "Maricopa County's largest flood occurred from February 19 to 26, 1891, when the Salt River swelled to 18 ft deep and 3 mi wide following days of rain, *The Republic* reported. A railroad bridge over the Salt River in Tempe collapsed, destroying homes along the river bank."[26] This last event was recorded in Maricopa County, but was felt all over the Southwest. "The Southern California region received 33″ of rain was reported in Descanso in a 60-hour period. 2.56″ in San Diego. From 2.16 to 2.25, a total of 4.69″ fell in San Diego."[27] These would also be construed to be once in 1000-year events.

Bright spots in the weather damage story have occurred. During Hurricane Harvey (2017), many communities in Florida that were dependent on solar did not lose power after the storm. This can be attributed to system/plant designs that accounted for potential weather extremes.

As we go to press (2023), in the USA, we have seen excessive California Pineapple Express rainstorms, growing numbers of tornados across the mid-west United States and records for extreme weather continually broken. The same anomalous patterns are taking place worldwide and are occurring more often and with greater extreme results.

As shown above, failing to take into account the actual history of a site or region can potentially result in significant losses. Today, insurance companies are either charging more or not covering these types of events due to the cost of repair.

Consideration should also be given to microclimates on or near the site, as these can make the extremes worse. "A microclimate is the distinctive climate of a small-scale area, such as a garden, park, valley, part of a city or subregion. The weather variables in a microclimate, such as temperature, rainfall, wind or humidity, may be subtly to extremely different from the conditions prevailing over the area as a whole and from those that might be reasonably expected under certain types of pressure or cloud cover. Indeed, it is the amalgam of many, slightly different local microclimates that actually makes up the microclimate for a town, city or wood."[28]

Microclimates in the same general geographic area can provide startling different numbers. Selecting the wrong location to acquire a site's environmental data may result in potentially significant errors. These include the microclimate conditions of prevailing winds, temperature, rain, snow, hail, precipitation, altitude, geography, soil type or chemistry, and moisture conditions.

Given today's technology, there is little excuse for not obtaining more realistic and accurate site data. It is not unusual for environmental site datasets and information to be "cherry picked." Some developers and/or others apply climate data to improve performance numbers in their modeling that provide outputs that are better than the actual site may produce. For example, a plant operating in the open desert or a semi-tropical clearing can vary significantly from a plant within a city's boundaries even though they may be only 5–10 or less miles apart. Concrete or pavement absorption, retention of heat or cold, and altered air flow patterns can make a city site considerably hotter or colder than an open-air site that is located on rangeland. A plant in the mountains can have a dramatic difference within 1–3 mi based on microclimate or elevation.

Although this may not be as common as it was previously, there are problems with using average high and low data measurements over only five years or so for design specification. Using these average measurements to design the plant will result in marginal operation during critical energy demand (maximum output) periods, even in nominal conditions, or before considering extreme climate change conditions. There is little documentation to indicate how many times this type of data has been used to get financing for projects that would not have been penciled out with proper data. However, the results are that many times, performance based on average (predicted) climate data do not match up with the reality of performance-measured site data.

Some projects make interpolations of weak data sets and yet treat them as if they are complete and historically accurate. When providing the first PerfectPower specification for a General Service Administration project, one competitive bidder indicated that our numbers could not possibly be met. However, their bid modified the specification by using site weather data from a location about 94 mi away, with a different climate type and 1500 ft lower in altitude. It also appeared that when they ran their modeling, some of their other performance assumptions were inaccurate.

The competitor's claim was that our analysis could not be correct. Our system climate performance data indicated higher energy output was based on weather data within a few miles of the site, and the modules that were selected were more sensitive to that environment. This was unfortunate primarily because that competitor at that time did not fully grasp the differences between site data from two separate locations that looked similar yet were far from it environmentally.

As an industry, we have yet to prepare for and address these issues adequately. We have often been ill-prepared for yesterday's climates and plant design, much less addressing those for the under-addressed tumultuous future climate and site changes.[29]

2.4.2 Implementing the URTD

By consideration of all the stakeholders, their challenges, and the conditions they create, we can begin to substantially improve, while adequately delivering information support in

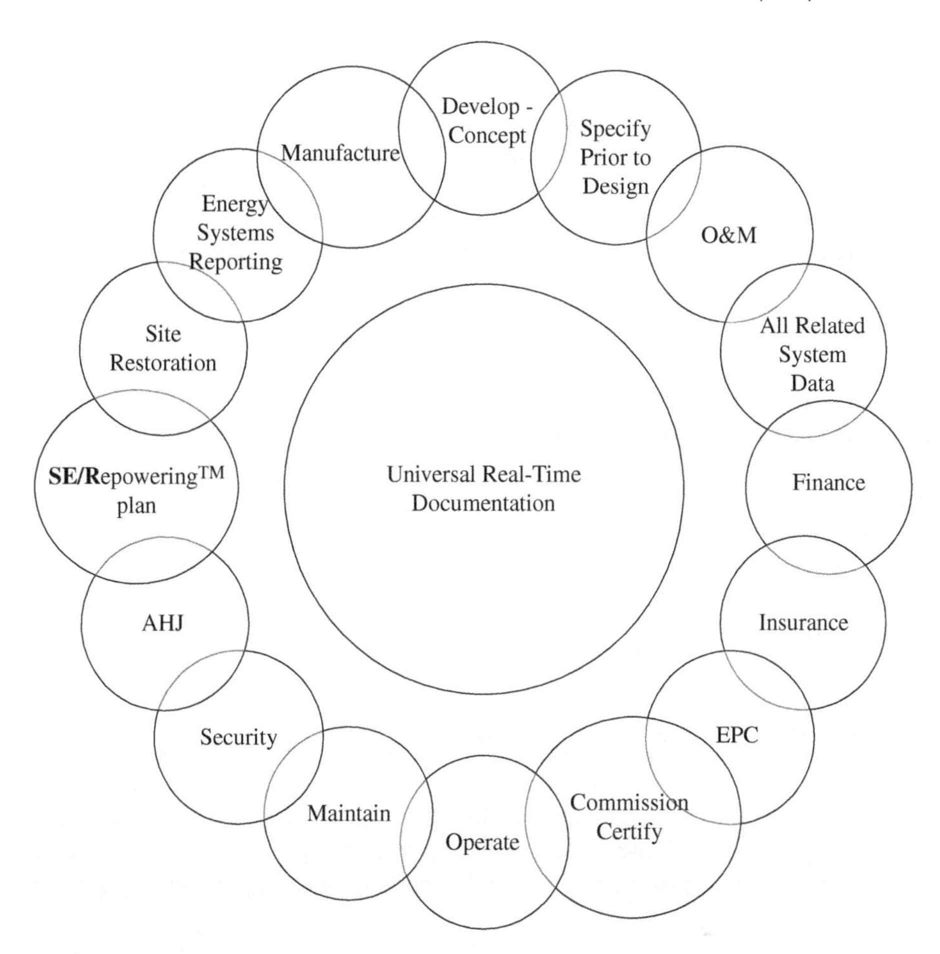

Figure 2.1 (URTD) Data Sources and Users.

a consistent manner for the reliability of the grid and all stakeholders. A well-designed and implemented URTD, Figure 2.1, can be, for example, a portal or gateway to providing all the FRACAS (Failure Reporting, Analysis, and Corrective Action System)[30] and other reliability data input (see Chapters 5–7, and 11). This approach can also provide an almost immediate link to and from a remote handheld device in the field. Its integration begins at the concept and continues beyond the life of the plant, the fleet, and the organizations involved. It is the "ipso facto" repository for system, fleet, organizational, and other pertinent details that tend to be collected by several stakeholders but often not shared, even among those who need the information to successfully support their project responsibilities. The URTD also focuses on updating drawings and other visual and multi-electromagnetic frequency data to illuminate the common challenges with "as built" drawings.

"**Computerized Maintenance Management System (CMMS)**, also known as **Computerized Maintenance Management Information System (CMMIS)**, is a software package that maintains a computer database of information about an organization's maintenance operations (Cato and Mobley 2002)." Chapter 11 discusses and recommends best O&M practices for utilizing a data management tool. A CMMIS provides the basis for aiding technicians to perform their tasks better while providing management with the ability to track all aspects of maintenance.

A CMMS can offer multiple core maintenance functionalities. It is not limited to manufacturing operations but expands to facilities, utilities, fleets, hospitals, sports arenas, and other industries sites with equipment/assets subjected to inspection, repair, and maintenance. With continually improved technology and increasing competition, more and more companies are switching to CMMS versus using manual methods to track and/organize information. The different components of a CMMS include, but are not limited to:

- Equipment Data Management
- Preventive Maintenance
- Predictive or Scheduled Maintenance
- Labor
- Work Order System
- Scheduling/Planning
- Vendor Management
- Inventory Control
- Purchasing
- Budgeting
- Asset Tracking
- Wireless access for rapid reporting and updating

A URTD system includes a CMMS[31] or CMMIS, which is the foundation of much of the ongoing system data collection that supports the whole process. It includes project, system, and all other project-related data, most of which is digitally updated by a remote system technology component with the ability to provide reporting to each appropriate stakeholder.

2.5 PV Plant Lifecycle

The PAM **SE/R**epowering™ view of a PVPS starts from concept to retirement, disposal and includes site restoration, often to the original condition or use, to be complete. The tracking and defining of all required activities and costs sum up the lifecycle or “all-in costs.” Leaving anything out will result in an inaccurate, low, and misleading LCOE. It will also hide important challenges that will need to be dealt with earlier in the plant operation than planned, many of which could have been avoided in the first place.

Wherever there is an inconsistency or a break in the lifecycle chain, the results are, almost without exception, additional substantial unbudgeted costs. These unforeseen expenditures result in reducing plant health, condition and status along with lowering performance eventually reducing revenues. Strategically, it is less destructive, disruptive, and expensive to preempt these cost issues.

For the most part, PV professionals tend to focus on their specific portion and responsibilities of the project/process without taking a more complete (systems/SE) view of the entire evolving lifecycle. As a result, there is seldom continuity throughout the project, especially when including external stakeholders such as suppliers, contractors, and others. These gaps create substantial inconsistency and/or confusion, which may lead to turf battles. The struggle stems from the incomplete and/or unavailable details of essential information and the "assumption" that the information provided is complete, accurate, and consistent with defined project goals and objectives. At times, even a subtle error, misspecification, or misunderstanding of the impact, may have substantial time and cost impacts.

The entire concept of a lifecycle is an organic "cradle to grave" view of the PV plant life. As with humans, health issues, defects, or diseases ripple through the body throughout their entire life, while some have a greater impact than others.

When evaluating the whole lifecycle and how any single player will impact the others, a series of questions emerge. Those questions make it easier to begin to deal with the irregularities that can harm or disrupt the project performance and revenue. *We are presuming – not assuming – all stakeholders are looking for a successful project outcome.* A critical goal (and result) is where everyone does a great job, they are proud of, and are paid financially and emotionally for their efforts. We are also presuming that the vast number of negative issues that exist in plants today do not or should not benefit anyone! What we seek is to have the entire project team able to clearly see the big picture and the linkages between project members and segments. They must understand their roles in the process, have the information and tools to resolve issues early, and confirm that all of the required steps have been taken as a team. Under those conditions, the results will be well within their expectations or beyond.

The PV system delivery "Limiting Factors" address all the critical factors including their metrics and clarify issues that are easily overlooked or ignored. We believe the next stage of project development is improved by "continual improvement" of systems, components, subsystems, O&M. Addressing these issues effectively enhances system viability and profitability. The more complete and the earlier these processes are applied, the better and more valuable the long-term results. Issues may be weighted differently while addressing the details, which are a matter of applying a coherent preemptive process.

Lessons learned, for example, can occur at multiple phases, such as during team assembly, site selection, civil and system engineering, construction, commissioning, benchmarking, O&M, and even when upgrades or effective **SE/R**epowering™ are considered or performed. These lessons can drive the concept phase for the next plant. Data developed and shared in lessons learned from other projects and plants can help influence the current and future plants by resolving identified faults and failures and implementing the learned lessons.

2.6 Standard Test Conditions

PVPS plant design consists of hundreds of specifications and or requirements. These include requirements of power generated, which is based on PV module certification testing at STC.[32]

It is important to acknowledge the fact that although modules are certified at STC, STC is a single point on an *IV* operational curve. (For cell or module (*I*) current (*V*) voltage curve.) The module operational *IV* curve varies at each temperature and irradiance, altitude, air mass, and non-STC condition.[33]

Standard Test Conditions (STC):
Irradiance: 1000 W/m^2
T: 25 C Cell Temperature
Reference Solar Spectral Irradiance,
Air Mass 1.5 (AM1.5)

Figure 2.2 Standard Test Conditions (STC).

Module design and performance data are based on OEM suppliers performing qualification testing at STC.[34] It is important to acknowledge the fact that although modules are certified at STC, it is only one point on an *IV* curve, and it is not necessarily the one needed to assess the site reliability and power generation capability of the module and, as a result, the plant. Module *IV* curves continually vary at each different operating condition impacting irradiance rated and measured at 1000/m^2, cell temperature at 25 °C and with an air mass of 1.5, Figure 2.2, or other conditional variation.

It is critical to be aware of and understand that each cell/module design and chemistry is different. Those differences result in a broad variation in energy delivery under the myriad of difference environmental conditions.

Misinterpreting something as basic as STC can result in surprisingly small, medium, or large failures in output and revenues. Although this supposedly will be automatically included in the modeling for the system, if it is not correct, everything ripples downhill from there.

Extreme operating and storage conditions, for example, accelerate the degradation process, which stakeholders often misunderstand. Additional consideration and review harken back to the importance of understanding the *actual* site conditions and its environmental history.

2.7 Capacity and Capability

Capacity is what the power/energy rating of the plant was designed to provide under STC conditions. Capability is what the plant can actually produce based on the daily irradiance and health, condition and status of the plant through its lifecycle.

It is critical to assess how well we understand whether the module is operating normally (within design parameters), at what level of degradation (age or other related), or if it is marginally operating or if the module has failed. These conditions continuously change throughout the life of the equipment in the plant and are often not considered to be so variable. This results in over estimating plant capability, although the plant capacity numerically will always stay the same.

An example of current practices is shown in Figure 2.3. What is planned, specified, designed, installed, and operated is often very different than many stakeholder expectations. Although throughout this book, we point out that expectations are not specifications, and wants and needs are two different things, which in this instance, expresses serious underperformance during the first 10-plus years (Figure 2.3). Although the plant was designed and constructed for an assumed 20–25 year life, it was not specified for the rigors of the site, environment, or the components essential to meet that target.

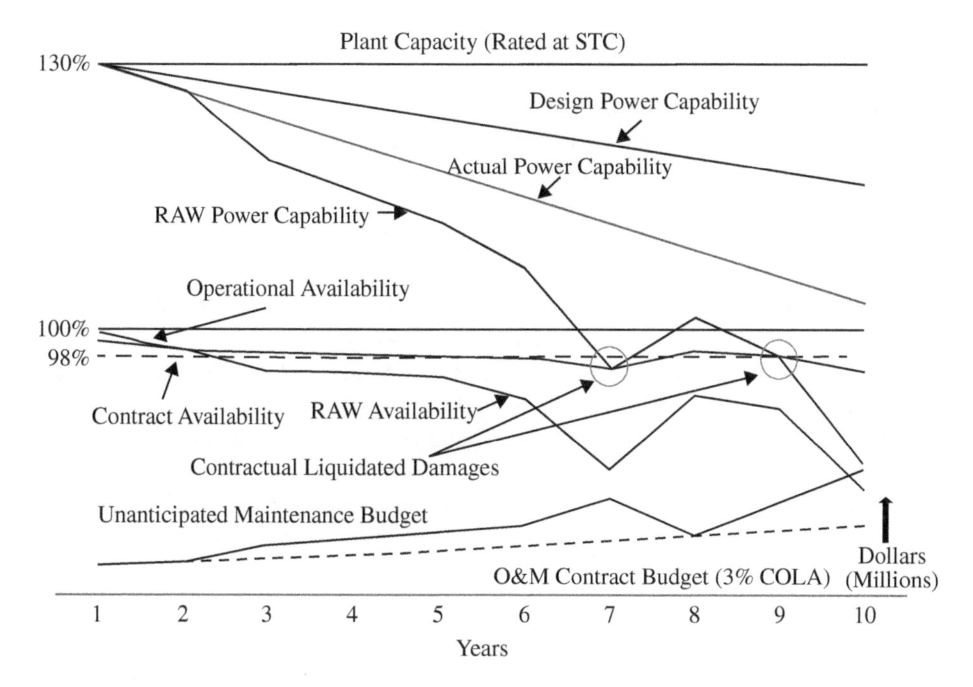

Figure 2.3 A Simple Model of Current Industry Practice Results. (Colour rendition of this image is available on the book's landing page on wiley.com).

2.7.1 Design Capacity

This is the certified module DC STC nameplate power plant capacity in DC watts as designed.

(a) The design capacity is strictly the module DC rating times the number of system modules for a specific system, design as rated under Standard Test Conditions.
(b) By itself, this metric provides an energy rating for DC output of the system modules at COD.
(c) The capacity of the plant is fixed at design/installation and does not vary over the life of the plant *unless* the plant is derated or rerated for fiscal, or for example, different modules have been introduced or installed as replacements at **SE/R**epowering™ or legal reasons.
(d) For this graphic example, the plant design is based on a 130% overbuild with inverter clipping. This overbuild is in part, to offset module degradation, defects faults, and failures and or broaden the performance curve shoulders.

2.7.2 Design Power Capability (DPC)

This is the plant's AC output over its lifecycle for modeling purposes as assumed. It is based on the original design assumptions of degradation, defects and installation quality, regardless of the source of that degradation. Current practices tend to "assume" that the modules are the only, largest, and prime source of degradation.

(a) DPC is the calculated "assumed,"[35] designed, and expected output of the plant over time consistent with its performance and output. It includes the project's technical and nontechnical assumptions that go into plant design considerations and modeling using any number of modeling tools.[36]

(b) Average component or system values or averages with selected variance metrics, are the design engineering assigned values that *may or may not reflect* a specific range of site and operational parameters accurately. Among the sources of variances are:
  (i) Limited site and operational consideration of modules and/or inverters only;
  (ii) Failing to account for plant wiring, connectors, other component manufacturing quality, shipping, installation quality, and monitoring accuracy;
  (iii) Initial installed and evolving faults, failures, defects, under addressed O&M, etc.

(c) The DPC is only as accurate as the sum total number of applied or considered lifecycle defects, degradation rates, mismatch, packaging, handling, shipping, and installation and maintenance assumptions, included in the plant modeling. These are often insufficiently understood, poorly defined, and sometimes not or under-identified, considered, and addressed. One of the weakest links is a lack of **SE/RAMS/R**epowering™ planning and its requisite tools. These void concerns and conceal energy, time, and fiscal loss with the inability to adequately quantify the health, condition, and status of a plant beyond the COD.

2.7.3 Actual Plant Capability (APC)

(a) Expected as initially designed and built. Often assumed to be at the published designed degradation rate once measurement starts with the POI. What is often found is energy delivered that is lower than expected much earlier than expected.

(b) Often measured at the POI meter, the Inverter/POI meter may have a ±2% variability from what is actually produced. Our position is that a ±2% or 4% potential variability is unacceptable and skews the APC and actual raw power capability of the plant as a critical and comparative value. This can result from meter quality or lack of calibration.

(c) Includes accounting for all sources of demonstrated and measured losses/degradation of the plant that includes all the sources of degradation, defects, soiling, etc., and quickly repairable failures that can take portions of the plant down, e.g. fuses that take out a string, an underperforming module that drags down string performance, circuit breakers or fuses that take out multiple strings, partial loss of a combiner, or an inverter which is under performing, etc.

(d) This is the energy the plant is capable of delivering to the customer at the output of the plant as installed. However, it does not necessarily fully or accurately address the status and or capability of the plant. Fully measuring and understanding the health, status and the condition of the plant is a key element of maximizing energy output to the grid and revenue.

2.7.4 Raw Power Capability

The internal measured and *potential* deliverable power/energy of the plant as in operation. It includes all sources of defects, degradation, inverter, balance of plant (BOP) failures, design and component selection errors. It also includes module and other component mismatches, all effects of site environments, soiling and manufacturing variance (lot to lot/batch to batch differences, for example), and a run to failure versus infrastructure approach to O&M.

The differences between APC and raw power capability are primarily the result of a lack of underfunded or ineffective O&M practices which grow over time. Those losses are in energy production and revenue lost.

This is what is produced moment to moment at the component level and throughout the lifecycle, measured as a detailed assessment of the plant's internal health, condition, and status, not necessarily what goes onto the grid. The use of the POI meter will generally not reflect the true energy output of the plant due to meter measurement errors. Nor will it indicate the temporal capability of the plant. The output may be minimally or well above or below the measured value. **Therefore, producer/consumer risks can be significant. If the meter is reading high, the consumer is paying for energy not delivered. If it is low, the producer is being paid less than the energy provided.** This lack of accuracy also masks plant lifecycle performance and output "capability" issues. Identifying and addressing these issues can be aided by applying IEC TS 63265, the Reliability Technical Standard to identify the reliability attributes of the plant and IEC TS 63019, the availability standard to better identify and understand the availability of those component sub segments.

Figure 2.3 provides a generalized look at a system over a limited timeframe, which will vary depending on the depth and quality of the initial plant specifications and construction. It illustrates the difference between the conditions the plant was designed for, and what actually happens over the first decade or so of general operation of a plant.[37] This difference in viewing plant health, condition and status has been clearly identified in the IEC TS 63019:2019 Availability standard and has been addressed in the Reliability Technical Standard (IEC TS 63265:2022). These standards allow PV professionals to clearly establish a set of guidelines, and metrics for addressing the health, condition and status issues, and the resultant Availability related to different levels of module, installation, plant degradation, site environmental factors, and other failures.

When including energy storage (Chapter 9), "masking" issues become evident quickly if not addressed during specification and design. Energy storage relies on the available excess energy generated in the plant to provide for charging battery cells or other storage media. Depending upon the configuration of the plant coupling, DC to DC to AC, DC to AC to DC, etc., with well-designed energy storage systems (ESS), more energy may be lost to the initial overbuild. In addition, system defects and irregularities can be discovered and addressed earlier following COD, which may impact warranty replacement.

2.7.5 Reactive "Repowering" Threshold

An unplanned set of activities triggered by poor production performance to replace some or all of the modules, inverters, cabling, connectors, etc., also known as repowering to get a plant back on track to an improved power generation capability. This is often triggered by not meeting O&M contact requirements, thereby resulting in liquidated damages.

This is a major reactive O&M activity that addresses specific plant output and conditions challenges. However, unless it is well planned (beginning at specification), monitored, and judiciously carried out, it often tends to be little more than a major yet temporary patch.

It is an O&M, management, and fiscal response to a lack of in-depth and unsuccessful initial energy planning, specification and ongoing system health, condition and status, data collection and activity for/on the system. This is seldom, if ever, included in the LCOE, i.e.

as the failure to include all of the all-in costs, to determine the real cost of energy produced over time. When we look at this approach, we see a "least-cost project delivery" model clashing with well-planned "system delivery model" business practices. It is like buying an expensive high-performance vehicle and limiting the essential and required maintenance, while still expecting it to meet expectations during its operational life.

As a result, due to incomplete specification and planning, the actual cost to produce energy over time ("the" or "a" defined lifecycle) is always more costly than initially calculated and planned. This results in a waste of time, money, equipment, labor, profitability, and may result in serious loss of income and essential energy availability when called upon.

2.8 Addressing the Gaps

What metrics need to address and quantify those changes, and what do they mean to long-term PV plants with or without energy production over time?

Figure 2.4 includes and highlights the areas where early and complete efforts to address the system engineering requirements can significantly improve performance and output.

The indicated Gaps (1, 2, and 3) are the metrics that define differences, which are built into a PV system that have not been adequately addressed at plant lifecycle specification. It indicates what is missing after the fact.

Our approach is to focus on the decisions and conditions that create these gaps, determine what they cost, and preempt them.

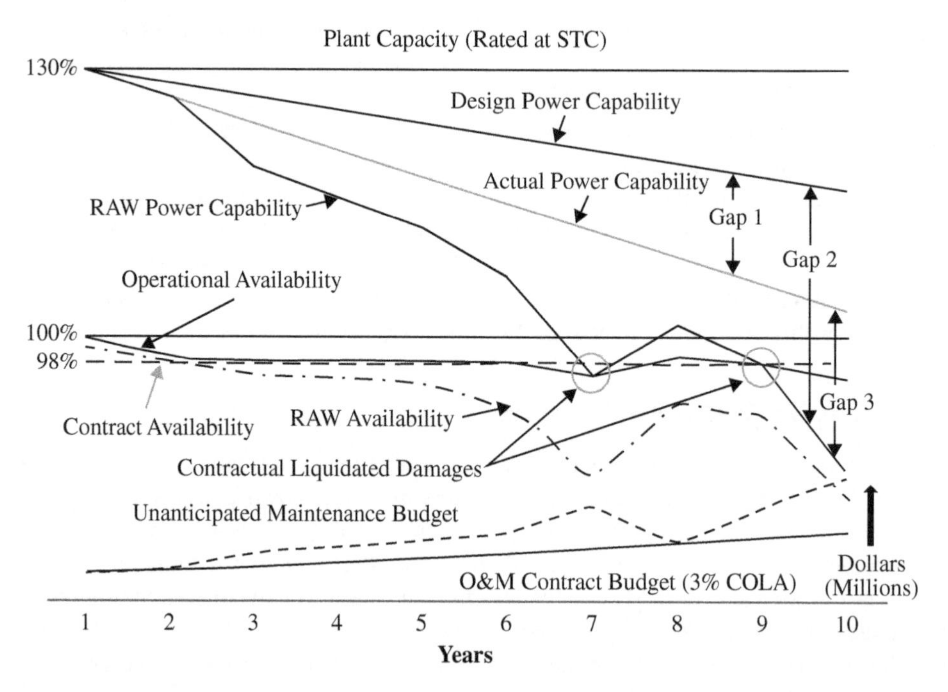

Figure 2.4 Identification of Performance Gaps. (Colour rendition of this image is available on the book's landing page on wiley.com).

2.8.1 Gap 1 – Design Power Capability to Actual Power Capability

This gap is the result of not accurately identifying and specifying all of the PV plant details such as attributes, product variability, and limitations. It includes "all" stakeholder wants needs, and requirements primarily through an abbreviated least-cost "project" delivery model. This tends to be driven by a lack of or incomplete understanding of the PV component's specification, manufacturing variability, and irregularity as delivered operational products.

To reduce the difference between design and actual power capability, Gap 1, we recommend the PV System Delivery model. The Project Delivery Model has failed to "invest" sufficient time and attention in detailing the general assumptions that are or may result in unsupported, inaccurate, and untraceable technical assumptions.

2.8.2 Gap 2 – Design Power Capability to Raw Power Capability

Gap 2 is the difference between what was assumed to be designed and what is actually built and operated, accounting for the system degradation and failures that result in reduced energy production.

It accounts for much of the "missing energy" expected over time. It indicates, especially when compared to the unanticipated maintenance budget, the total amount of energy lost and the additional O&M costs, which were not adequately budgeted and will not be recovered.

Not all of the missing energy and budget can be recovered in a cost-effective manner before or after the fact, i.e. following COD. However, when applying an effective SE/RAMS model at specification as we propose, a substantive/measurable amount of lost energy and lower O&M costs can pay for recovery when included in the financial model with an adequate O&M budget. Addressing these at specification provide a much more accurate LCOE.

Gap 2 *incorporates* the difference between the *plant* as a designed and installed system and the effects that account for *and include* component reliability, O&M maintenance philosophy, practices, and site environment variance over time.

2.8.3 Gap 3 – Losses Between Actual Plant Capability and Raw Power Capability

Most of the missing energy between the APC and the raw power capability can be identified and corrected when effectively addressed. The unrecoverable missing energy is primarily due to normal component inefficiency, aging and entropy.[38]

Gap 3 losses are often the result of poor specification, lack of training regarding component handling, and a lack of adequate O&M budget, service, and monitoring. (Example: On a 10-megawatt plant, nine inverters are operating nominally while one is operating at 72%. This plant delivering 97.2% with a single POI meter with an error bar of ±2% may well appear to be meeting contractual requirements.)

Control of the missing (masked) energy losses is established at specification, design, construction, and fully funded O&M approach. When effectively defined during the system specification process, potential sources of those energy and revenue losses, and the time

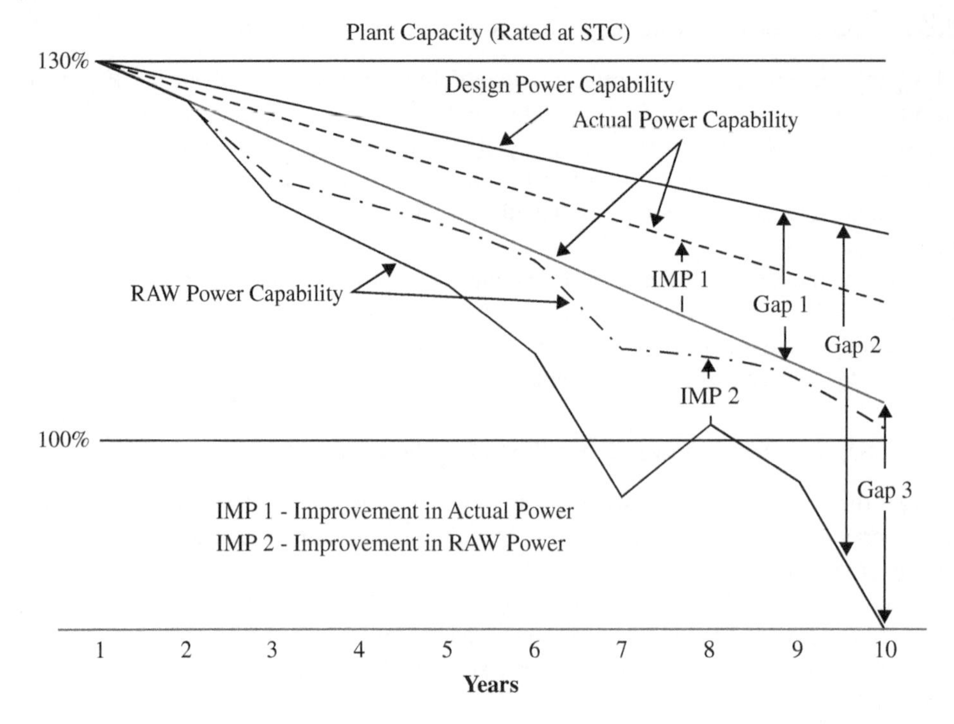

Figure 2.5 Example of Areas of Improved Performance. (Colour rendition of this image is available on the book's landing page on wiley.com).

and costs of recovering them become more controllable, while providing the threshold of and for planned active intervention. This requires not only monitoring improvements. It includes an improved, simplified awareness of basic physics through data collection, usage, and sharing. It provides the details of existing challenges, what to do, when, and the consequences of inaction on cost and revenue.

Most of these masking issues are or can be accounted for through an effectively specified, designed, constructed, operated, and maintained through an **SE/R**epowering™ Planning and Deliver process.

Gap 3 losses are like having a team of equally sized and powered horses to pull a load; however, one or more may be sick, injured, or dead.

Figure 2.5 illustrates the effect of improved specifications, inclusion of true component statistical variability, and more detailed, and fewer exclusion, O&M contract. Improvement 1, improved actual power, is predicated on overall better quality components, better specifications with actual site environmental profiles, tighter installation and handling, and a total site benchmark. This last item is needed to understand the COD's true plant capacity.

Improvement 2, improved raw power capability, is based on the improved reliability and quality of installation, high-quality O&M, detailed data collection with trend analysis, and funding and facilities for spares and repairs.

2.9 Masking and Its Impact

PV systems planning, specification, and design provide an interesting set of problems and challenges. PV plants today are generally overbuilt by 10–50% or more. We refer to this "overbuild" as creating a negative impact on assessing the true health, condition, and status of the plant by creating a "masking effect" (Balfour et al. 2021) or simply "masking." **While overbuild is not necessarily bad or negative, it can be critical in offsetting and hiding normal module and other component degradation, faults, failures, and efficiency issues.** If not properly monitored (measured), it hides the reality of what the plant is capable of delivering. We call that "masking" because it can hide the actual state of plant health, condition and status. Our masking effect paper provides several methods for calculating some of the impacts on power production that are often unrecognized or misunderstood.

One aspect of masking is that it is initially the difference between the DC power output capability of the plant, in excess of that needed to meet the AC power contract until it is no longer available for conversion. Masking includes all of the elements that hide, disrupt, or interfere with providing the contracted power that is not measured, monitored, documented, and reported. Additional sources of masking occur at the convergence of PV system business, political, and engineering assumptions, thinking, planning, decisions, and actions. If all of these factors are not addressed, they contribute to significant issues when the plant cannot meet demand years earlier than planned or expected. In reality, masking is seldom adequately considered as there is a lack of awareness of the topic and its lifecycle impacts.

For example, the plant DC may be potentially capable of delivering 122 MWdc to the inverters (accounting for degradation, failures, faults, reduced solar resource, etc.). However, the plant may only deliver ~100 MWac to the POI. In addition, masking occurs on the AC side at transformers, breakers, fuses, torqued connections, etc., related to component health, condition, and status. This difference is generally not measured and/or reported. The result is that there is 22 MW (minus the power loss due to inverter efficiencies) of power that is hidden or masked including from other causes. The SE/RAMS approach in this book notes that in order to understand how much solar resource is wasted, one must measure the power being generated at the source (strings) and at various points up to and beyond the input of the inverters to the POI.

What is the potential value of masked power?

Masked power, if and when recovered, is the power that could be available to supply to the customer if demanded and if the plant is capable of generating that power. A prerequisite for utilizing recoverable masked power is that the masking sources must be identified, measured, monitored, tracked, evaluated, and acted upon. This establishes how much-masked power may be recoverable. Today, in most, if not all, instances, plants delivered through the PV project delivery model tend not to consider nor do they provide the essential tools to identify masking elements, conditions, and potentially recoverable energy. This is a critical part of our focus on a **SE/R**epowering™ system delivery model, which requires that these masking challenges are addressed at specification and benchmarked by all stakeholders.

As energy infrastructure, we know that historic electrical energy infrastructure addressed many, if not most, of their masking issues early in the specification development to meet resilience, RAMS, fiscal requirements, and contracts to deliver energy. Historically, stakeholders did not let a coal, gas, nuclear, or hydro plant run to failure, hence the need for and value of redundancy!

In those technologies, additional power is supplied, when the plant has included redundant generating units (e.g. PV plants would include strings, combiners, inverters, and Balance of Plant (BOP) components) that can be brought on line as needed. In addition, with this criticality, recovered masked power can be stored in an ESS, and then used for example, to compensate for PV resource (irradiance) power transients at sunrise and sunset or provide additional power during nighttime. Both of these can be financially practical and beneficial; however, the latter application is the more functional adaptive alternative.

Plant overbuild based on initial assumptions, specification, and design to generally compensate for some levels of degradation, faults, and failures. Yet, these often do not adequately cover all potential sources and variations in condition and performance metrics. The effect of this is that as these conditions exceed the design assumptions, it often results in lower power generation while reducing the raw availability performance (measured at the inverter input and the POI meter). These tend to grossly affect energy availability much earlier in the plant lifecycle than planned.

"How critical is consistency of energy availability?" In answering this question, RAMS specified plants register as being of critical personal, community, financial, and organizational value and necessity as addressed in IEC 63019!

As shown in Figure 2.3, when the plant's capability drops below that needed to meet contractual requirements, the risk of liquidated damages and/or breach of contract can come into play much earlier than expected. This risk is indicated by missing performance targets triggering liquidated damages starting as early as year 7 and again at year 9. In this example, liquidated damages can also continue to occur after year 9, applying traditional repowering practices.

However, one of the more extreme long term concerns is that en mass, this loss of assumed energy delivery may have a negative impact on overall stakeholder energy availability, when needed, and falls short of meeting Climate Change wants, needs and necessities.

In some cases, these issues can trigger liquidated damages as early as year two or three!

Unbudgeted O&M costs rise unexpectedly as performance drops below contractual expectations and requirements. Though avoidable, if not specified and designed, these project delivery performance shortfalls contribute to unbudgeted or under addressed maintenance costs and fiscal complications.

It is of interest to know that once many of the issues of masking are better understood and measured, it becomes far easier to identify and address many of these challenges *prior to COD*.

Therefore, we focus on health, condition, status and capability during specification and plant operations. Masking allows multiple system failures to exist undetected or misclassified until a tipping point is reached. This is when the masked issues can force the performance level (based on the defined metrics in IEC 61724-3[39]) and its availability

(IEC 60109), below contractual requirements. At that point, while addressing the now visible effects of masking, stakeholders discover a substantial amount of the plant's underperformance, requiring additional large, unexpected, and unbudgeted maintenance. Avoidance can deliver a negative fiscal impact, which may result in O&M cost increases with a one-two punch through potential liquidated damages, potential loss of O&M or PPA contract, and/or bankruptcy.

2.10 System Design Assumptions Drive Plant Fiscal Performance

Over the past three decades, it has been unsettling to see PV with or without energy storage designers miss a 0.5% loss here, a 0.25% there, or a 0.33% to reduce the component cost of the system. This is all too often practiced without any financial cost–benefit or other analysis of how it ripples through the system throughout its life. If we look at these systems as being PV infrastructure, it would seem that the lack of accounting becomes both expensive and wasteful while being fiscally counterproductive.

Even a concrete bridge over a ravine requires maintenance to ensure reaching planned life and to keep it in safe useful service as infrastructure.

To address this unwieldy trend, greater thought and consideration become a hallmark of cost-effective and reliable PV infrastructure delivery. This could include:

1. Energy Planning and Accounting
   (a) To enhance system/component output, longevity, and maintainability, it is essential that the stakeholders:
      (i) Define and document the system stakeholder expectations for operationality as energy infrastructure including all time and cost elements.
      (ii) Define what the acceptable energy losses are from cradle to grave.
      (iii) Budget for these losses as energy infrastructure.
      (iv) Complete a fiscal analysis to quantify the tradeoffs between what are acceptable and unacceptable losses of energy production and revenue over time.
   (b) Address the lack of industry-wide data sharing that quantifies these elements.
   (c) Develop the international standards to provide the language, guidelines, quantifiable metrics, and practices necessary and require certification of those details and metrics.
2. Establishing Realistic Degradation and Defects Values:
   (a) Assumption that degradation is a module-only or primary problem for plant performance over time:
      (i) Assuming an average module degradation does not accurately address the implication of higher and much higher degradation rates for up to half the modules
      (ii) If the manufacturer states that the average degradation rate is only a 0.5% average, this only covers potentially one part of the installed modules. Variations in lots and or batches may result in much higher rates due to materials variability alone.

   (iii) Even using a 0.8% average has the same types of problems, yet still does not address an accounting of what levels of degradation are being passively accepted or even specified and designed into the plant.

(b) The gap between design and actual is the actual power capability accounting of the total degradation rate of each generation unit (modules through to inverters).
   (i) Degradation is also present in the DC (and AC) BOP and the inverters.
   (ii) Connectors have increasing contact resistance resulting in IR loss for each connection.
   (iii) Fuses fail at higher rates as time is accumulated (removes partial or complete strings that may not be known).
   (iv) Circuit breakers contact resistance increases.
   (v) Inverter capability decreases with time due to component aging.
   (vi) Mismatch (multiple manufacturers, use of different production lots, binning errors, etc.).
   (vii) Failures (Installation, DOA, induced during maintenance, weak components [early failures], etc.).
   (viii) Damage from packaging, shipping, handling, and installation.

(c) Finding the gap for degradation and defects in existing plants requires DC side monitoring to at least the string level.

(d) Reducing the gaps for future plants requires understanding what the degradation and defect rates are in the field, allowing for higher confidence in accounting for them and ensuring that sufficient margin exists to account for underestimated degradation and failures.

(e) It requires demanding greater levels of confidence in component RAMS from OEMs.

(f) Average degradation rates for plants have been found to be up to 1.4% or greater by NREL. The questions need to include: How do we determine at concept how much loss in degradation, defects, masking, and weak specification/design is acceptable? Addressing this requires a better understanding of the physics of those components in real environments, their RAMs, what is delivered, and how it is installed at the site.

3. Establishing realistic financial value to discover and correct (recovered capability) determine cost-effective elements of infrastructure:
   (a) Greatest value for current and future stakeholders (this addresses asset value)
   (b) Consistence (energy) and availability (when you want and need it)
   (c) Reduce wasted/lost income and profit – while minimizing and controlling cumulative losses
   (d) Addressing all stakeholder needs over time
   (e) Political – Address PV investment including waste – misapplication of funding, the misuse of funding, overcharging for product, use of wrong item
   (f) Paradigm shift of energy production at best effective cost versus the lowest cost
      (i) Stabilizes cost of electricity to stakeholders
      (ii) Dramatically reduces and or eliminates brownouts and blackouts

As an alternative to the throw-away and unsupported comment, "It's too expensive" we should come to terms with the fact that when Dr. Balfour first took the professional path toward the PV universe, modules cost about $24 per watt. Please save us from the histrionics myth and excuse of "It's too expensive!"

A critical and standard infrastructure practice should ask the questions:

- What is wasted (lost) income – how much money from waste is acceptable to leave on the table?
- How much energy does that lost income represent?
- As we move toward additional electrification of vehicles and replacement of other fossil fuels, can we afford as a planet to not reliably meet those rapidly growing electrical generation needs?

2.11 Conclusion

Our position is that sample commissioning and benchmarking, though less expensive than monolithic or sectional, is often an incomplete process that assumes that a sampling of the component performance represents to the plant as a whole. Production batches of PV modules that are partially flash tested, then assumed that the rest of the batch is equal to the average or level of those tested, creates unnecessary risk and cost.

For existing plants, utilizing a **SE/R**epowering™ program can be instituted during its future operational life and provides the basis for greater power generation, and more energy delivered at lower O&M costs with extended lifecycles.

A more effective and measurable methodology would require SE and introducing the **R**epowering™ cradle to grave planning through site restoration processes and procedures. For existing plants, this approach begins with exploratory environmental benchmarking, a Site Capability Survey, and for new plants, a thorough operational benchmarking and commissioning protocol.

The foundation is with an intensive SE RAMS, benchmarking, and commissioning process that requires and delivers greater specification and design attention, decision-making, and activities prior to EPC bidding. It may be completed as a single event or accomplished in segments, thereby bringing the plant to a far greater state of energy production, reliability, measurability, production, and cost control and profitability over time.

As illustrated above, the term *lifecycle* is the complete life of the plant from the cradle to the grave or reuse or recycling, and not, as currently practiced, the period covering the first owners. As this chapter ends with the topic of the lifecycle "failure versus success," the future of the PV industry is at the crossroads where you, the stakeholders, must begin to understand and implement the PAM PV System Delivery and System Engineering **R**epowering™ processes for substantive improvement. This is why we began the book Introduction by asking the questions: **"Where are we now, where do we want to get to?"** and **"Where do we go from here?"**

Failing to take a more systematic and structured approach to define, specify, design, and install an entire project will result in failing to meet contractual requirements. This can lead to liquidated damages, unplanned shutdowns, negative profit, and, in the worst case, bankruptcy. Those results are clearly business practices that are not productive or profitable.

We added some history to this text because many in the industry who are newcomers, i.e. less than 5 or 10 years of PV experience, do not have a framework as to where we came from, how things exist today, and what needs to take place for improvement for tomorrow. Our goal is to help you and your organizations deliver projects with RAMS processes that provide plants that are functional, cost-effective, and profitable throughout their entire lifecycles.

Bibliography

Balfour J.R., Hill R.R., Walker, A, Robinson, G, Gunda, T, Desai, J., (2021), Masking of photovoltaic system performance problems by inverter clipping and other design and operational practices. United States: N. p. Web. doi:10.1016/j.rser.2021.111067.

Cato, W. and Mobley, K. (2002). *Computer-managed Maintenance Systems: A Step-by-step Guide to Effective Management of Maintenance, Labor, and Inventory*, 33. Butterworth-Heinemann ISBN: 0-7506-7473-3.

Wang, D. (2017). Benchmarking the performance of solar installers and rooftop photovoltaic installations in California. *Sustainability* 9 (8): 1403.

Kounev, S., Lange, K.-D., and von Kistowski, J. (2020). *Systems Benchmarking, For Scientists and Engineers*. Springer Nature Switzerland AG https://doi.org/10.1007/978-3-030-41705-5.

Ramasamy, V., Feldman, D., Desai, J., and Margolis, R. (2021). U.S. solar photovoltaic system and energy storage cost benchmarks: Q1 2021. *Tech. Rep. NREL/TP-7A40-80694*.

Harper M, 2019, *What are the Four Types of Benchmarking?* Nov 13, https://www.apqc.org/blog/what-are-four-types-benchmarking

IEC TS 63265:2022. Photovoltaic power systems – Reliability practices for operation, 2022.

IEC TS 63019:2019. Photovoltaic power systems (PVPS) – Information model for availability, 2019.

Notes

1. The levelized cost of energy (LCOE) The LCOE of an energy-generating asset can be thought of as the average total cost of building and operating the asset per unit of total electricity generated over an assumed lifetime. https://corporatefinanceinstitute.com/resources/valuation/levelized-cost-of-energy-lcoe/
2. International Electrotechnical Commission – Renewable Energy.
3. Rules of Procedure for the Certification of Photovoltaic Systems according to the IECRE-PV Schemes.
4. https://www.inc.com/encyclopedia/proprietary-information.html
5. International Electronic and Electrical Engineers – Society.
6. American Standard Testing Methods.
7. American National Standards Institute.
8. National Fire Protection Association.
9. National Electrical Code.
10. North American Electric Reliability Association.
11. Federal Energy Regulatory Commission.
12. https://en.wikipedia.org/wiki/Taxonomy
13. https://en.wikipedia.org/wiki/Ontology_(information_science)
14. https://orangebutton.io
15. "Resource Guide Utility Solar Asset Management and Operations and Maintenance" a SEPA/HPPV publication, 2016 http://solarelectricpower.bmetrack.com/c/l?u=5ECD14D&e=8E338F&c=71B0E&t=0&l=23B5622D&email=%2FtPzXG1FWzWHOFjOzfsnyo3STdsmV9HICJ%2FwYbRnGP0%3D&seq=1

16 This includes variety of organizations that participate in benchmarking, commissioning, inspectors, testing, data collection etc. They may be short-time or long-time internal players in the process and may include Code, Standards, testing, and certification organizations.

17 California Independent System Operator (CAISO), Pennsylvania, Jersey, Maryland Power Pool (PJM), Midcontinent Independent System Operator (MISO), New York Independent System Operator (NYISO), ISO New England (ISO-NE), Electric Reliability Council of Texas (ERCO).

18 The North American Electric Reliability Corporation (NERC), *Federal Energy Regulatory Commission* (FERC), Federal Energy Management Program (FEMP).

19 The entitlement (permitting) phase is the foundation for a PV plant development project. Working with environmental, local, regional, state or government zoning, and other AHJ regulations, this phase determines if the site can be used and the appropriate documentation has been completed. Entitlements must be adhered to move forward on construction and followed throughout the entire operations and site restoration.

20 "No one is to blame!", *Little Dorrit*, Charles Dickens, 1855.

21 IEC 62446-1 Ed. 1.0 b:2016 – Photovoltaic (PV) Systems – Requirements for Testing, Documentation and Maintenance – Part 1: Grid Connected Systems – Documentation, Commissioning Tests and Inspection IEC 62446-1:2016 defines the information and documentation required to be handed over to a customer following the installation of a grid connected PV system. It also describes the commissioning tests, inspection criteria, and documentation expected to verify the safe installation and correct operation of the system. It is for use by system designers and installers of grid connected solar PV systems as a template to provide effective documentation to a customer. This new edition cancels and replaces IEC 62446 published in 2009 and includes the following significant technical change with respect to IEC 62446:2009: expansion of the scope to include a wider range of system test and inspection regimes to encompass larger and more complex PV systems.

22 https://www.apqc.org/blog/what-are-four-types-benchmarking

23 https://climate.nasa.gov/news/2951/climate-change-could-trigger-more-landslides-in-high-mountain-asia

24 https://climate.nasa.gov/news/2981/nasa-space-laser-missions-map-16-years-of-ice-sheet-loss

25 https://www.edf.org/blog/2016/09/01/we-just-had-five-1000-year-floods-less-year-whatsgoing

26 https://www.azcentral.com/story/news/local/arizona-weather/2018/09/30/history-arizonas-damaging-floods-hurricane-rosa-tropical-storm/1480276002

27 A History of Significant Weather Events in Southern California, Organized by Weather Type, PDF.

28 https://www.metlink.org/fieldwork-resource/microclimates

29 There are a number of different levels and forms of data to be collected beyond that being practiced today (see Chapter 10).

30 https://en.wikipedia.org/wiki/Failure_reporting,_analysis,_and_corrective_action_system

31 https://en.wikipedia.org/wiki/Computerized_maintenance_management_system

32 IEC 60904-3, EC 61215, IEC 61646 and UL 1703.

33 Module testing for variations at other environmental values are covered in IEC 60904-3, IEC 61215, IEC 61646 and UL 1703.

34 IEC 60904-3, EC 61215, IEC 61646 and UL 1703.

35 Bollinger, M., Will Gorman, W., Dev Millstein, D., and Dirk Jordan, D. (2020). System-level performance and degradation of 21 GWDC of utility-scale PV plants in the United States. *J. Renewable Sustainable Energy* 12: 043501. https://doi.org/10.1063/5.0004710.

36 Note: Based on and how the inputs are defined and applied, different modeling tools result in sometimes considerably different outputs. Accurately determining a reasonable output may require using and carefully evaluating multiple models/tools to determine which of the outcomes are more likely to be accurate or valid for the specific system and its site.

37 Preemptive Analytical Maintenance (PAM), Introducing a More Functional and Reliable PV System Delivery Model – A Reliability Precursor Report, 2018.

38 Entropy (https://en.wikipedia.org/wiki/Entropy) is a scientific concept, as well as a measurable physical property, that is most commonly associated with a state of disorder, randomness, or uncertainty. The term and the concept are used in diverse fields, from classical thermodynamics, where it was first recognized, to the microscopic description of nature in statistical physics, and to the principles of information theory. It has found far-ranging applications in chemistry and physics, in biological systems and their relation to life, in cosmology, economics, sociology, weather science, climate change, and information systems including the transmission of information in telecommunication.

39 IEC-61724-3, Photovoltaic system performance – Part 3: Energy evaluation method.

3

Current PV Component Technologies

This chapter addresses PV components and provides an overview of the current state of the technology. PV power system (PVPS) plant problems with early component failures due to poor selection criteria and the lack of specification of critical reliability and maintainability attributes are the main drivers. Implementing a system engineering approach that addresses stakeholder needs and resolves differences between them leads to a more cost-effective power plant.

Key Chapter Points

- Risks – Strengths and weaknesses of components and subsystems (particularly PV modules and inverters, and for energy storage [Chapter 9])
- Technology fatigue, what it is, and what its causes are
- Efficiency and testing
- What drives component selection? What should be done, and how can it be managed?

Key Impacts

- Better information results in better product selections, lower costs, and higher revenues.
- Technology fatigue, the impact of continual change on the industry in cost and reliability.
- Bankability, the missing fiscal responsibility, and the importance of testing and certification.
- How the decisions made today impact projects by system availability for tomorrow.

3.1 Component Selection

The component selection process affects the cost of a PV power system (PVPS) from manufacturing to end-of-life energy production and site restoration. The component in this context may be a resistor, capacitor, module, inverter, or other. Selection requirements include addressing service life, reliability, ease of maintenance, scheduled maintenance requirements, performance, and availability over the life of the system.

Some of us have or had the responsibility to make the technical/finance decisions in system specification, design, and construction, who prefer to look to the long term for PVPS's ability to produce energy reliably at an appropriate cost. In essence, what is the

Photovoltaic (PV) System Delivery as Reliable Energy Infrastructure, First Edition.
John R. Balfour and Russell W. Morris.

best component for the system where it is located and how it will be operated? In tandem, the issue of Rapid Technology Change Fatigue has been a growing issue for approximately two decades or so.

The starting discussion revolves around: "What is the best module, inverter, other component or which are the best, most reliable suppliers for those parts?"

The introduction of COVID into the mix created a broad range of additional internal and external stress issues on technology delivery and for actors within the industry. This placed additional physical and mental stress on the people both in the industry and outside. The impacts will probably not be fully known for a decade or more. In addition, the industry and the rest of the world are dealing with issues related to growing infrastructure supply chain challenges that further complicate things.

Today, it is more difficult to choose the right component and company than it was 10–30 years ago. Personally, and professionally, we have struggled with these questions since PV modules sold for US$24 per DC watt. Before 2010, as an industry, we were not fully aware of these issues because there was little or no detailed field data available. Today, many plants, O&Ms, and suppliers collect data, but most of it is publicly unavailable. It is treated as proprietary information (PI), or intellectual property (IP) and not released because it is believed to be potentially detrimental to a company's image, reputation, and/or success. This is the case even when we speak about anonymized data. As a result, improvements can be slow while buyers maybe purchasing products of unknown quality or capability. So, how do we address the growing complexity of not knowing basic and essential component information?

Our approach is to begin by asking some simple initial questions, including:

- Will this technology operate as advertised?
- How well will it operate on this site under the operating ambient and induced environmental conditions?
- Is there sufficient test and or field information to confirm its capabilities in the field?
- Is it designed or appropriate for the environment in which it will be installed and operated?
- If it can't meet the requirements for the planned usage environment, can the system be effectively adapted to accommodate it?
- Does the manufacturer have a good business and technical history?
- Will the manufacturer be around to support the warranty or warranties as well as available to support spares or repairs?
- Will the component be easily installed and maintainable throughout its life?
- Will the owner or O&M be able to make the required changes to accommodate form, fit and function replacement products in the future?

These initial questions may seem pretty simple. Yet, underlying fundamental requirements must be met, validated, and verified as needed to assure a successful product and PVPS operation. Many problems arise partly because there are far too many myths and assumptions (Chapters 1 and 2) driving the PV least cost juggernaut. As a result, it is insufficient to just ask around for advice on which components are the most appropriate for your project.

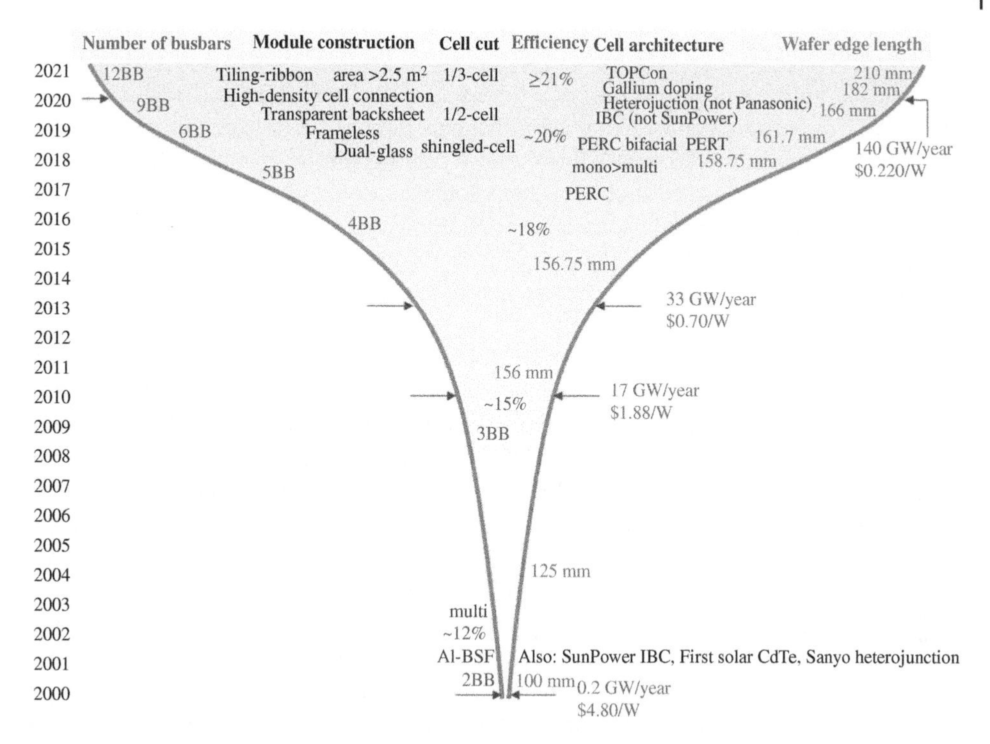

Figure 3.1 Growth of Solar Cell Technology 2000–2021. Source: [2]Hieslmair Henry/https://www.youtube.com/watch?v=M7BHcxxugwY.

A common saying in Latin, is "***scientia potentia est***," which translates to "Knowledge Is Power," or "***ipsa scientia potestas est***," "**Knowledge Itself Is Power**." To be successful, it is crucial that the stakeholders or teams make decisions based on sufficient data and knowledge. *Especially in reference to the PV industry, we would argue that neither of those classic statements are true in and of themselves!* If they were, then the PV industry would, today, be far more sustainably, robust, reliable, and profitable. It is like assuming that because you have a dictionary, you know and understand all the words and definitions it holds just because you own one.

Where is the power and wisdom in that?

Never the less, when it comes to knowledge, it does little good to know something yet do nothing about it! If you do not have the background or experience or have made the raft of mistakes we and our colleagues have made over the last few decades, you may be doomed to make those same mistakes. That seems rather wasteful! Avoiding mistakes requires change! One of the reasons we go into many details in this book while providing you with history and some perspective is to give you information that ties together a series of reference points.

Instead, consider this, ***Acting*** **on accurate information (Data) and Knowledge Is Power.**

There is an Astounding pace of growth and innovation (Figure 3.1), yet: "Field Data That Reveals How PV Modules Will Perform Over 25+ Years Does Not Exist!"[1]

3.1.1 Lessons Learned

Without learning and documenting our previous mistakes (lessons learned) and addressing them, we lose any of the effective "power" that experiential "knowledge" provides us. We suspect the same is true for you. We believe we can pretty much guarantee that without that depth of lessons learned, your PV systems and plants, with or without storage, will also lose energy production and the associated profitability that they could have provided. The decisions you and your colleagues make, based on the information you gather and use, will be the difference between being knowledgeable and ignorant, or between being potentially successful or failing.

When you take the time to choose the correct products for the site environment, with the simple presumption that you want to deliver a system that will function properly (effectively) with as little additional operations and maintenance cost, that is a great start. To do so, we must understand the basics of SE and RAMS (Chapters 5–8). Doing so requires specialty expertise to address the design and implementation questions. It requires technical knowledge, facts, good usable and accurate data, appropriate technical processes and procedures, with a developed ability to practice them and the ability to act on them to reach the stakeholders clearly defined goals, objectives, and requirements.

The use of the POI meter for the plant, for example, does not address plant health and condition, only energy delivered and generally with an error range of $\pm0.3\%$ to $\pm3\%$ or so. As a result, most commissioning may not clearly or accurately indicate the plant's actual health, condition and status before the owner accepts it. This tends to create an issue as the owners believe that the commissioning is viewed as an accurate picture of plant health and condition now. The actual effect would be determined by the number of modules on the string, their condition, and the gain or loss of energy results. This ripple effect and other similar issues are often under-addressed or estimated when looking at long-term plant operations, maintenance, energy production, and profitability.

The issues of failure must consider whether it is a full or partial failure and its impact. Therefore, the entirety of the failure causality and result must be considered. In the PV industry, a prevailing perspective results in the belief that PV failure is either 0 or 100%, which is misleading and contrary to the realities of component or subsystem failure in the PV universe.

3.1.2 Design Considerations

For a specific site, it requires understanding how the actual site environments impact the components over time or lifecycle. Different components have different lifecycles or service lives, and it is critical to understand those lifecycles to determine the appropriate components and how to get the most out of them. It is crucial that the life of the system be determined first. This is true whether applying Preemptive Analytic Maintenance (PAM), Balfour and Morris (2018), to a new system expected to last 50–100 years or more, or by applying **SE/R**epowering™ planning, processes, and procedures (Chapter 4) to an existing business model and system to extend its life another 30 or more years. This is where understanding the reliability of components gives you a realistic view of "how long." This is where applying well-founded science, engineering, and finance processes can produce the best results.

Once you determine the expected life of the system, then allocate those requirements to the component/s or subsystems that will meet those system requirements. For example, if an inverter must be serviceable for 20 years, the major assemblies in the inverter must have mean time between failures (MTBFs) in excess of "*n*" times the MTBF of the inverter to meet the above life need. Or, the component must be replaceable (like a filter) or rebuildable. As will be shown in Chapter 6, the statistical average implies only about a 60% confidence in reaching this value, thus the need for the maintenance attributes or ease of repair or replacement and the ability to fault detect this component. This generally requires paying a bit more for some components. Components that are selected for a project based on first cost/least cost are often the wrong components. Again, it may seem simple, yet it certainly is not. The sad part is that the better component is not always the more expensive one/s. You just have to understand what you are buying and the metrics that tell you what you are getting, if they are available. The question to ask is; What data is available and what isn't? And if that data is not, why not, and what can you do about it?

You may ask: How do I make better selections, and what can be done to reduce the pain of making poor ones?

For PVPS, the components often number in the thousands and more. So, what considerations must be made at the specification and not after discovering that a component you selected may be wrong?

In this chapter, we primarily address PV modules/panels and inverters. So where do we begin?

3.2 Present State of Technology

3.2.1 Technology Risk

PV technology has been changing rapidly over the last 20 years and accelerating. The result of this is that there have been significant improvements in component capability, such as 300 V IGBTs now operating at 1000 V. These new capabilities appear about every 18 months to 2 years. As good as this is for the industry, it has also led to technology fatigue, where companies discontinue product production in as little as every three to five years or less. So, no sooner are these components installed than they start becoming more difficult to find for repair and replacement.

As the PV market continues its rapid expansion, investors, developers, and owners are regularly challenged by manufacturing changes and supply chain constraints. Assuring them that reliability and long-term performance will remain intact is not solved by warranties and standards alone. The only proven method for protecting component quality is to qualify, verify, and validate the new components through performance qualification stress testing and reliability testing over the range of environments and operational stresses the product may experience. An odd accompaniment to this is that bigger is not always better.[3]

The present state of PV component technologies is one of the continual, minimal, or all too often extreme changes from product revision to product revision. These may be difficult to track since many may not have a change in part number, may only contain a revision number, or potentially have a completely new part number.

Whether we look at modules, inverters, racking/tracking, or other system component technologies, the rate of evolution is increasing. Modules, for example, have some of the most difficult changes to grasp and attempt to keep up with. Manufacturers continue to market new cells and modules with different form factors (shapes and sizes), variations of product performance, and in some cases different bills of materials (BOMs) for the module itself.

These continual changes challenge the industry to make effective long-term choices. Modules, racking, and tracking have the same challenges because they are continually being changed, often with improvements, but not always. This greatly impacts the reliability and availability of new high-quality products entering the market.

> When we use the term *quality* in this book, we are referring to clearly defined improvements to the measurable attributes of the system through the application of PAM. Yet, more importantly, it ripples through the whole system of components, their specification, selection, installation, and maintainability through its lifecycle.

As we proceed, we address PAM as a quality process and methodology for assuring and/or constantly improving system RAMS attributes.

Seldom have we heard anyone speak in detail about system lifecycle quality and reliability publicly on this aside from in a conference environment among colleagues and friends. Those of us who have been in the industry for any length of time are sometimes disparaged because we "talk negatively" about industry performance and reliability issues and realities. However, those of us who have been there, made the mistakes, and do not want to see others do the same, see it differently.

It is like knowing that the bridge over a great ravine or chasm is out on the highway up ahead. As we attempt to warn people that there is a problem, we are being told, "Quit being so negative." That does not change the fact that the bridge is out or that there are negative consequences of not addressing the rapidly changing reality!

Yet, the facts are very clear, even for those who do not want to talk about it publicly. To build a better industry, we must address our problems and issues, which are all resolvable!

This concern was clearly stated by Hongbin Fang – Longi's director of product marketing – when he announced, "Our development logic is to look at every aspect of PV project deployment with a holistic view and find a best overall solution: not only ingot, wafer, cell, and module manufacturing processes but also module deployment processes, such as transportation, installation, and system integrations."[4] This came with the announcement that Longi was not going to expand their production until they addressed what they referred to as "holistic issues." If all of those considerations have not been adequately addressed in the system specification prior to Engineering, Purchasing, or Procurement and Construction (EPC) bidding, the overlooked or ignored interactions tend to result in a variety of different kinds of defects, faults, and failures. These create financial consequences, which are usually damaging, expensive, and avoidable.

Though he was speaking about PV modules, the same is true for every component in the system. They are connected, interrelated, and rely on each other, and if the right combinations are selected, things work exceptionally well. Yet, when they are not, serious problems arise and multiply.

It is not good enough just to choose a product. One must completely and clearly understand the impact of that choice beyond the initial financing and construction of the PV system. They must understand the impacts through the system lifecycle, while sometimes committing to something that appears to be simple on the surface, yet creates problems that continue for decades.

It is further complicated by the fact that all of these changes require some level of specialization when addressing system specification, which is often either missing or insufficient to make substantive qualitative choices. This complicates the marketing versus factual information challenge. Yet, it impacts the ability to understand how these products will operate beyond the first 7–10 years and beyond (Chapter 4 **SE/R**epowering™).

3.2.2 Cost of Technology

While the drive for reducing component costs at any price still, there is a growing concern about issues that impact reliability, robustness, quality, and product consistency. The concerns for RAMS are recurring themes in broader industry discussions while driving niche conversations about specific problems, all too often – after the fact. However, the industry has been slow in taking action!

The new semiconductor technologies tend to reduce the die geometries for processors and other specialty chips. Power semiconductors (IGBTs[5]) improvements tend to be more along the line of improvements in dopants, better geometries for voltage and current stresses, and improved (lower) thermal resistance, reducing the temperature stress. Surface mount components have had improved strength to handle differential thermal stresses and improved performance of resistors and capacitors. PV cells have developed better substrates, larger cells, and improved interconnect (solder and welding). Efficiencies have improved at a relatively steady pace for the last 20 years.

All of these changes generally drive the cost of components up, while the large quantities used on PVPS allow for quantity discounts. PV modules, for example, have gone from US$24.00 a watt to less than US$2.00 to US$3.00 a watt. Some modules can be as low as US$0.30 a watt, but caution is advised for these low prices as we have pointed out that low cost may imply low quality.

However, actual system costs per DC watt are far more sensitive to buying power and actual use. Utility-size systems can, for example, get significant discounts for very large quantity buys, while commercial, residential, and off grids systems cost substantially more. This is where a challenge arises where people talk about the PV costs per watt, as compared to the actual costs that are sensitive to system type, size, and a broad range of other factors.

This appears to indicate that it is possible we are beginning to see that over the next few years and decades, there will be a much better balance between pricing and the ability/capability of the components to actually achieve what the manufacturers are claiming. Yet, more importantly, do so by delivering effective products that do what the stakeholders want, require, and need.

The relatively new standards for PV System Availability (IEC 63019), Reliability Practices for the Operation of a PVPS (IEC 63265), among others, will assist in component improvements in manufacturing requirements and results. They begin the process of raising the reliability and asset level bar and then meeting or exceeding that bar. These, along with an IECRE PV System Rating and Certification standard, will be a functional reality in the next few years. The reason for this is that those emerging standards provide awareness and guidance on how to see and measure the existing challenges, often hidden in plain sight. Though encouraging, they will not be perfect. However, they will provide substantially better direction and insight for the industry, providing a clearer set of requirements for those stakeholders who actively participate in the delivery of PV systems.

This chapter focuses on the component-related issues that are critical to consider when reviewing and selecting specific types of PV products, components, and warranties. The emphasis is based on understanding what is actually required for each component's mission, while meeting the systemic needs versus what may simply be "cheap cool technology."

It clearly brings home the message that the critical requirements go beyond the first cost of that component or subsystem, especially if you wish to control O&M, system ownership and management costs. It shifts the focus and adjusts to include the global issues of how that component will operate throughout its life as part of a system.

3.3 Manufacturing Risk

After four decades in the PV industry, when we go to a conference (pre and post-COVID), it is not unusual for someone to ask how excited we are about all the new technology/s and products. After all, if you look at all the manufacturing *marketing hype*, there are new "*miracle*" products that do amazing things and cost very little to do so. Or, as we have so often seen, the "*miracle*" is actually illusion, simply smoke and mirrors. Unfortunately, much of the marketing information we see today is long on promise but short on detail and, in reality, performance!

Although we focus on reading about the new products becoming available, our experience and the industry as a whole lead us to be far more wary. Our response to our colleagues and friends who ask about our lack of excitement over new products is generally going to hear, "I'm not all that interested in the newest or the latest until I've seen the reliability test data and conditions and it has been delivered and operating in the field with detailed and accurate data to support marketing claims. Look around you; many of the new products that you see here today will be gone in a year or two. If you don't factor that into your decision making, you will make many of the same mistakes we did!" All that being said, we still have to be aware of all of the new products and gimmicks coming into the industry and carefully sort out what is real and what is not real.

In short, I want to make sure that I am selecting the appropriate product to do the job and reliable enough to perform its mission throughout its life. And I want to know how long its life will be based on conditions in the local environment it will be placed.

A great place to begin with RAMS is with specification, design, manufacturing, and production:

> The quality of PV modules manufactured for a specific project is determined during the manufacturing process, not afterward. To assess and assure quality, direct oversight of all critical manufacturing processes is required including major steps such as cell stringing and lamination.[6]

Nevertheless, even before one evaluates the individual product at the manufacturing level, many can be eliminated by addressing basic quality, reliability, and ease of maintainability.

At a PV conference in Phoenix, AZ, around 2006 or 2007, Dr. Balfour had five of his employees' reviewing products. They rushed up to him as a group and said, "You've got to see this really cool product that we need!" He thought to himself that he had seen them totally swayed by remarkably effective marketing. As they all returned to the booth, the smiling salesman was excited to see that they had brought "the boss."

Dr. Balfour asked the salesman if he would mind if he spent time talking to his people privately. The salesman smiled, feeling as if he had already made a sale.

Dr. Balfour turned to his team and said, "OK, you know we have a process to evaluate new products. Let's go through the first five initial questions required to determine how viable a product will be for our needs and those of our customers." The discussion went on for about 30 minutes as they went through the following questions which included:

1. Does the product provide anything new that will improve its, and our mission? (i.e. delivery of a better, more reliable, easier to install, maintain, and productive system)
2. What is its manufacturer or product delivery, warranty, support, response time, and track record? (i.e. product quality, responsiveness, customer service, and transparency)
3. Is it easy to install, service, and replace over its service lifecycle of many seasons, years, and decades? (i.e. component quality, ease of installation, long-term maintainability, reliability, and warrantability)
4. Can it be easily or readily maintained using preventive maintenance (including cleaning, e.g., or subcomponent replacement at predetermined intervals (i.e. scheduled maintenance))?
5. If the design changes the form factor dramatically, is there a form, fit, and function replacement?

Still, the unanswerable questions are all too often about having useful detailed, and cumulative data to make better decisions. We know it is generally unavailable from manufacturers or the rest of the industry. Since the late 1990s, we were just beginning to see colleague concerns and new technical reports from the National Labs that raised serious questions. Those questions and issues continue today.

3.3.1 Variability of Quality among Manufacturers

> "Variability of Quality Among Manufacturers
>
> After more than 20 years of industrial-scale PV module manufacturing, the solar industry still shows significant variability in quality from module manufacturer to manufacturer – cost and volume often still takes precedent over establishing and maintaining high-quality management systems.
>
> PI Berlin has completed a significant volume of quality assurance on modules from 22 major PV module manufacturers – mostly Tier 1 and Tier 2 manufacturers. Figure 8 shows differences in defect rates between these manufacturers – ranging from under 1% for the best performers to over 5% for the worst. With an average defect

rate of 2%, around 60% of manufacturers demonstrate above average quality control in manufacturing. Only 5 manufacturers, or just under 25% of the total, have consistently achieved defect rates under 1%. These manufacturers would be regarded as having achieved excellent, consistent quality."[7]

"Tier 1 manufacturers are mixed throughout these results, thereby supporting the ongoing hypothesis that buying from a Tier 1 manufacturer does not guarantee Tier 1 quality - especially without proactive steps taken to manage quality."[8]

Figure 3.2 illustrates the variance in defect rates for Tier 1 and Tier 2 manufacturers.

As with most new entrants to the PV industry, the component marketing hype was exceptionally good and obviously effective. He knew his employees did not care much about his opinion, as they all considered themselves experts in the industry (with a few months to a few years' experience) and were swept up in all the excitement. They also had difficulty comprehending that such a new shiny product might have some drawbacks. For example, as with all new products, not all of the design limitations may have manifested themselves. Fortunately, by giving them the guidance to initially evaluate the product, they had the opportunity to balance fact from myth/fiction!

His approach was to simply have them develop their own opinion based on asking tough, detailed engineering questions about the attributes, how they were determined, and how they were verified and validated. At the end of their discussion, he asked each of them individually what they thought and what they would decide if they had to make the decision. One by one, they all agreed that this product did not meet our tough

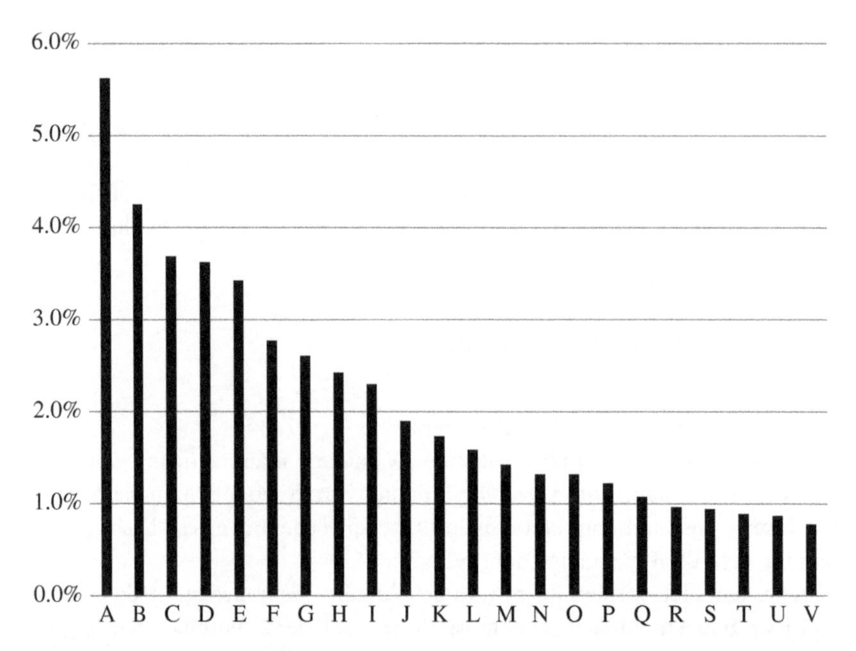

Figure 3.2 PV Module Defect Rates for Tier 1 and 2 Manufacturers (from 2016 to 2020) (Courtesy of PI Berlin from their Figure 8).

requirements and customer needs. Without exception, they all indicated that this product was not a good enough choice for our company or customers. They realized that there were suppliers that could not provide any reliability information or detail other than what the salesman indicated/claimed that it was "the most reliable," which meant nothing. No, the company did not have a supplier track record as it was a relatively new manufacturer, and although it had a very distinctive "look," it really had nothing to offer over existing products.

A few years later, after Solyndra went bankrupt, members of his team still referred back to that conference and the fact that "they decided," not the boss, that this was not a good product for the company or industry. Their choice was based on determining if there was enough information to make a realistic decision based on facts. The bottom line was there were not enough facts that took it out of the realm of simply getting excited about a "Shiny Object," a cool new product.

3.3.2 Decision Approach

Having observed massive change in PV technology arenas, where nothing stands still, we have serious concerns about how decisions are made regarding the following that may or may not be adequately addressed:

- Who makes the system specification choices, and when and how do they do it?
- What are their criteria, metrics, processes, procedures, data, and analysis?
- How is lifecycle reliability for components and systems seen, understood, and considered?
- What does "cost reduction" mean? In other words, *what are those real versus assumed cost benefits*?
- Who will be stuck with the wrong choices?
- How important are efficiency improvements in modules and inverters, resulting in greater energy density?
- Have all the limitations and weaknesses been identified and corrective actions performed?
- What are the actual improvements in performance, and will they be experienced in the field?
- As the manufacturer changes the BOM, how will the replacement parts operate, and can they be replaced without modification?
- How will we know what BOM version we are working with?
- How do we determine what the applicable stakeholder specifications are that should be included, and does the product meet those requirements?
- What is the near and long-term outlook for the availability of the product?
- What is the backup plan if any of the products are no longer available to replace failed units?
- Will the product original equipment manufacturers (OEMs) still be around to service their warranties or provide replacement parts?
- When will the market change where future system owners start asking for or demanding reliable systems that meet the project requirements over the whole specified lifecycle?

- Will this product be maintainable throughout its entire lifecycle?
- How do we make sure that our organization, including management, understands that all of these issues have tradeoffs in time and money as they impact the balance sheet beyond a quarterly or annual assessment?

The following are *areas* where *improvement* has lagged or fallen behind:

- In many instances, poor choices result in a loss of overall component, system reliability, and output or profitability consistency.
- Too many products have no publicly available manufacturer and competitor performance and reliability data over a broader range of environmental applications.
- The manufacturer or supplier has provided no track record information for comparison or technical evaluation. (This is complicated by the fact that most PV systems do not have an adequate O&M program to find product issues early, keeping EPCs and owners in the dark until there are, in many instances, major problems.)
- Many organizations in the project delivery chain do not collect and analyze sufficient data to make the decisions they need for long-term production and profitability. (This is the barrier we see that is created or self-inflicted, keeping the industry from moving to a system engineering delivery process with dramatically improved results.)
- An increasing loss of manufacturing organizations. (This is not necessarily because they had poor or bad products. It is often because somebody had a cheaper, all too often lower-quality, product. In many instances, this cheaper product provider was hyping their marketing, indicating they were providing an improved or superior product. This is frustrating, especially in those instances where manufacturers who produce a better product lose market share to inferior products and are gone from the market.)
- Customer service appears to be shrinking across all industries unless you are a very large customer.
- Product availability and stability in the market where by the time you have built your plant, that product may no longer be available for purchase in the future.
- Clear documentation of performance, availability, and reliability information, especially related to reliability data, indicates how a product will function at that site over time.
- The complete lack of documentation of lessons learned or awareness of what happens when choices are not effective or productive. In other words, who in the organization is actually and/or sufficiently tracking or mapping the results of these good or bad decisions?
- Is that information available to management to make better decisions in the future? (In many organizations, the answer is no, aside from limited documentation!)
- The real concern is where there is too much variation to select from, distracting from the actual ability to analyze the product/s, reinforcing the selection of generally the lowest cost product based on limited information. As a result, in a rush to get the project done, the cheaper product wins out almost every time.

- And finally, is the organization making decisions based strictly on customer or supplier relationships, present or past, that have absolutely nothing to do with an understanding of the quality and capability of the products being selected?

> However, module buyers, project owners, and investors need to remain mindful of the risks to quality, which growth creates. Poor quality can ultimately lead to solar power not meeting its economic and environmental sustainability promises, which in turn could negatively impact the long-term growth of the industry. It's therefore important we pay as much attention to quality as we do to technology and other commercial considerations.
> The good news is that risks to quality can be mitigated with relatively little effort and expense to the buyer, via contractual requirements and independent quality assurance at the factory.[9]

3.3.2.1 What Got Us Here?

The overwhelming theme behind products and selection is not: "What is the best choice to do the job?" It is all too often, "What is the cheapest product to meet the budget?"

With terrestrial PV technology being available for the past 50–60 years, why has PV-related RAMS not improved consistently and/or substantively and, in some instances, been reduced?

The issues of consistency are like waves washing onto a beach where a surfer wishes to get the best ride. If he or she picks the right wave, that could result in an incredibly exciting, uplifting, and rewarding ride and experience. If the surfer picks the wrong wave, it can result in a washout or a wipeout. If we are accurate in this assessment, what are the causes and solutions?

In the nonterrestrial (extraterrestrial) PV market, great strides have been and continue to be made. These have been far less sensitive to development costs and the cost of building the resulting products. RAMS were identified as critical, and as a result, products were very carefully designed and specified. However, many of the lessons learned by the aerospace market have not made their way to the terrestrial residential, commercial, and utility markets.

From a historical perspective, many of us who were in the industry during that time often saw this as a barrier to PV quality, reliability, resilience, and durability. They were not a foundational requirement as they were in their own technologies, which provided lopsided development of products while under-addressing RAMS. It allowed the utility industry to discount the technology based on a perceived lack of PV technology maturity.

Internationally many other organizations also focused on the same issues and provided a variety of different levels of funding.

While the market grew, the US Department of Energy (DOE), with some influence from the electrical utility industry, set an aggressive low-cost bar for PV electrical energy production. This sped-up PV product/component development focused on lowering costs yet ineffectively omitted the need/s for substantive improvement in component reliability and longevity.

As a proposal reviewer for a number of DOE funding applications under a series of funding programs and titles, an issue continually popped up in Dr. Balfour's mind. During some of our proposal evaluation efforts with DOE and National labs, we often raised a red flag asking why reliability was not a key required element philosophically, operationally, or technology delivery wise. We made a more critical element of DOE FOAs (Funding Opportunity Announcement). "Why was there seldom any mention of reliability in the FOAs and almost never in the applications themselves!" It was, at times, a frustrating and painful experience! This lack of attention to reliability (RAMS) of components in the PV industry has created a resilience gap that still has not been adequately addressed for systems and component funding announcements. Nor has it been adequately integrated into PV projects and systems that reap substantial tax benefits in the use of federal and state funds. We believe that by including RAMS attributes and resilience requirements in all governmental and other commercial funding, most of the problems we see in PV system components and systems would be dramatically reduced within 3 to 5 years. This would result in transformational progress for the entire industry.

Many, if not most, US and other utility companies pushed for the deregulation surge in the 1990s. The results of the deregulation process did not turn out exactly as planned. Many of the important lessons learned over the previous century or so were lost or neglected.

The process of introducing PV technologies into grid applications was taking place in parallel with the deregulation of the electric utility industry. The goal of deregulation was to reduce costs, but one of the effects was that the utility industry became segmented, whereas previously, it had been rather monolithic. In other words, the general process previously was relatively seamless, as the regulated utility industry focused on RAMS and took great pride in those technical and lifecycle attributes.

However, segmentation resulted in numerous inconsistencies that led to the utility infrastructure not being effectively or adequately managed, maintained, or repaired over the last few decades. Part of the infrastructure funding in 2021 was to address the degradation, of the utility grid. Under the old system, one primary organization (the utility company) was in many instances responsible for the whole energy system delivery process providing RAMS-based energy to the grid (Kim J. et al. (2021)). Planning, siting product through system specification and design, oversight of the energy generation technology, the infrastructure, and integration necessary to deliver power, operations, and maintenance were almost seamless. Maintenance was just as critical as good specification as both were directly linked to long-term success. As a result, these outcomes were sources of utility industry technological pride. Plants were designed as infrastructure to be maintained and rebuilt (**R**epowered™) over many decades.

That approach no longer exists! ... and unfortunately, to a great extent, the PV industry has followed that utility approach as illustrated by the failure of the Texas power system (2021 – grid, wind, and PV) during a heavy snow/ice storm. These are not isolated instances or issues, as this lack of energy/grid resiliency is found in many locations outside the United States. This is a great example where marketing of additional utility values and cutting muscle out of systems operations and maintenance continues to undermine all energy production and its delivery.

Within the PV industry, systems engineering thoroughness never began, so we have a gap that has yet to be effectively addressed. Until it is addressed, we will continue to see many of the current problems continue into the future. As a result of this understanding, we have used lessons learned from the older utility process that was very systems engineering and RAMS oriented.

As a result of poorly developed deregulation as practiced, after having over 100 years of incredibly successful technology development, priority in the new model was to cut costs and do so dramatically. To achieve this, budgets for operations and maintenance were thinned, often slashed to a point where RAMS was and still is jeopardized. The mantra became an approach to cut out the "fat" in budgets without adequately differentiating fat from muscle. This freezing of the cost bar and the requirements on PV equipment forced substantial improvements in certain PV technology areas, which were not expected. In the stampede to market, reliability and durability were not effectively part of these substantial and essential improvements.

In hindsight, the cost bar should have been paired or balanced with stakeholder system RAMS requirements, which would have delivered far better and more sustainable lifecycle-based products and systems.

During a 2012 conversation at a PV conference in San Francisco, one of the authors sat at lunch with a small group of regulators and utility executives. Part of the conversation was about their perception of the fact that the continual cost-cutting trend could not continue forever and that, at some point, the US government was going to have to step in and bail out the utility industry. "After all, it did work for the banks so why not utilities?" they agreed! The opinion of those executives at that lunch was consistent in supporting that position. The bailout discussed would be up to the US$2 trillion range to stabilize and update the US electrical utility industry to bring it back to a sustainable level. As a result of this strategy, the US utility industry has created some issues for themselves and their customers that are still unaddressed.

These include:

- A growing number of urban and rural fires.[10]
- The resulting need to shut down segments of the utility distribution system, thereby reducing or eliminating power to hundreds of thousands to millions of ratepayers during marginal or extreme weather conditions.
- Substantial growing losses and damage to public and private property.
- Growing numbers of personal injuries and deaths.
- A growing number of lawsuits.
- Bankruptcy for one of the largest utility companies in the USA, PG&E.
- A growing loss of confidence in the utility industry and its ability to deliver reliable, available, maintainable, and safe energy.

The conventional utility industry created its own nightmare scenarios, stimulating PV technology growth and competition while creating internal issues for both. Unfortunately for everyone, these risk-ladened practices also made their way into the PV industry.

This is where the focus continued on the first cost and away from RAMS. All of this has been exacerbated by a number of internet and data-related security issues, which will take time to resolve.

When we put together the pieces, we have a resulting "brutal or brutalizing cost model." It thrives on an overall reduction in the appreciation for the cost and safety benefits of RAMS. It is combined with the rapid growth in the PV market nationally and internationally, along with other challenges, that continue to strain the PV industry. Today, it favors developers who are very profitable while continuing to produce substandard and poor-performing plants internationally.

Much of the improvement prior to that was moderate and incremental. Funding all scales of PV projects for new technology products in the industry was difficult, and the market had not yet developed to support rapid growth. As a result, products evolved more slowly while there was a greater deal of individual and industry technical discussion and interplay going on between manufacturers and users. That process allowed for improvements in the reliability of products, which positively impacted availability and maintainability. In other words, if something did not work, it was not unusual to be able to talk directly to the people who designed it and those who oversaw the product development. This often resulted in effective feedback finding its way into the products.

From personal experience in that prior era, one could talk to the designers of a product and, if there were issues, determine a workaround prior to purchasing. That was because many of us understood that no product was perfect, and therefore, a cooperative informational approach was very important and made available. In instances where we could not have that relationship, we developed our own internal networking to address products, issues, and solutions. As growing masses of investment with little to no knowledge, understanding, or awareness of the technology and its evolution exploded, that cooperative environment dramatically reduced, although there are inklings of a resurrection.

Throughout this era, component and system research done in private, national, and international labs, and universities began to document these issues with limited resources. Although there was substantial input and funding from the private sector and governments internationally, the market tended to look at these studies and results as being strictly "academic." Because they did not fit comfortably into the existing business model, they were often discounted, ignored, or minimalized. The time for addressing this disconnect is well at hand! This is where science and marketing conflicted!

When we combine this with the pressing shift to an aggressive first cost only business model, PV industry players tend not to have the budgeted capacity to address nonfirst cost issues as thoroughly as they did in the past. We knew we were continually on a learning curve, today the criticality is that it is often ignored! The advent of substantial siloing has created more barriers than productivity. This trend has still not abated. As a result, in many instances and many organizations, one hand, does not know what the other hand is doing. That is, until there's a problem!

The fact is, today, if you are not a big buyer, OEMs seldom want to talk about these issues unless you are buying megawatts. Even then, it may be a hollow discussion. Yes, they may poll buyers for marketing purposes; however, there is a great need to provide shared data and address the void between products and reasonable long-term needs!

The exception to that is that bigger buyers have greater access to better product batches with better quality and reliability. Those bigger buyers can hire consulting professionals to perform inspections and evaluate the products being manufactured prior to purchase and during product production to receive a slightly to moderately better batch of products.

For any given batch of products, there is a quality and performance variation range that results through testing, grading, binning, and product rejection. Grading groups of module products into a series of output batches, rated in capacity DC watts per module. This is why we see products from a manufacturer with the same physical form being provided with different Standard Test Conditions (STC) ratings.

The better-graded products tend to go to the bigger customers and buyers especially those who are well educated as to technology development and its capabilities. Nevertheless, this has not raised the bar for sustained product quality improvement for the PV industry!

Yet, those inspection services are all too often ignored. When buying millions to tens of millions of dollars' worth of product, the cost of paying for those types of services are minimal compared to the results one receives. The issue of finding better-quality products requires understanding what is being produced nationally, internationally, and between all manufacturer's own plants. That is why having someone visit the production plant to identify and address any problems, especially in changes in the BOM, is critical to making the best decisions and getting the best results (see Chapter 6).

> It should be noted that the overall difference in quality by country is much less significant than manufacturer to manufacturer or between factories within a single manufacturer.[11]

> Among these countries, factories based in India have exhibited the lowest defect rates, although the relative volume produced in India is still small compared to other Asian countries. India does however have a long history of producing PV modules with the likes of Tata BP Solar being an earlier producer. Cambodia is a relative newcomer in PV module manufacturing, but thus far OEM factories based in Cambodia are performing reasonably well compared to others in southeast Asia. Factories in Thailand, on the other hand, have exhibited the highest defect rates at over 2.5% on average.[12]

Although many in the industry may deny the concerns we have seen grow and continue today, the results we see in industry bankruptcies, systemic challenges, and financial losses tend to support our position and assertions. For the most part, those issues are technically controllable.

During the late 1980s and 1990s, as growth in the industry accelerated, we began to see a series of consolidations, a harder drive to reduce/minimize costs, and the beginning of continual change from manufacturers offering new products at an ever increasing pace.

The industry focus had begun to shift toward a greater and greater need to compete with traditional electrical generation. Traditional generation put substantial pressure publicly on the PV industry in order to slow its growth and emphasized cost repeatedly. At the same time, the utility industry targeted a specific cost per kilowatt hour, and robbed their operations and maintenance budgets to meet those numbers. This left many utilities with dramatically fewer reserves and resources not just for emergencies, it also created future unfunded emergencies.

During this era as many utility companies were forced into accepting PV, the resistance was in providing marketing delay pronouncements. For example: "We are very enthusiastic and supportive of PV! We embrace PV as a new and exciting power technology. However, at this time, PV technology is still too expensive, risky, not well developed and is still very immature."

It may be of interest to note that they drank their own Kool-Aid, which resulted in swallowing the "PV is simple" approach. They also focused heavily on power purchase agreements (PPAs) believing that doing so reduced their risk which has in many instances come back to haunt many utility and nonutility projects. We raise this issue because the PV industry is in desperate need of utility experiential insights to make major project and component selection improvements. However, today those communications are extremely limited. That chasm resulting in that lack of effective communications must be bridged to improve the nexus between both industries and their infrastructure futures.

This was the beginning of an explosive growth in utilities signing PPAs with the mistaken myths and assumptions we have addressed throughout this book. That allowed their vendors to treat new PV plants the same as their existing energy deregulated technologies and strategies.

Today, while utilities are major players in the PV industry as far as system purchases and PPA's, we seldom see them in any meaningful numbers at the table when it comes time to improve and create new standards. As experienced PV professionals with the knowledge that many of our associates agree with this, we as an industry need them at the table and would welcome them.

In hindsight, every utility roadblock and impediment that was put in place for the PV industry, was generally overcome while products improved and as costs dropped. Again, in hindsight, we should be thankful for the pressure that was placed on the PV industry that resulted in better products. However, the sad part is how the opposite took place within the utility industry technologies that exist today and their treatment of PV.

Inverters became more efficient and effective as there was a critical development in Maximum PowerPoint Tracking (MPPT). In 1998, Trace Inverters were touting their MPPT technologies at a solar conference in Washington, DC. At that conference, their team was bemoaning the inconsistencies of their products and made it clear to many of us that there were still substantial issues. During that era, frank discussions allowed many of us to work around the shortcomings of the technologies as we provided that feedback based on what we all learned about the existing challenges. Those issues were soon resolved in their and other manufacturers products.

Yet today, although this is a critical component in most inverters, many PV professionals still do not have a completely clear grasp as to how effective those varied MPPT technologies operate under a variety of environmental usage circumstances. As we often rely too much on packaged energy modeling software, much of the industry has not learned how to maximize energy production in their plants, while continuing to leave additional energy production on the table. This is one of the shortcomings in our industry focusing on least cost projects and not effective system specification and development by applying practical systems engineering methods.

MPPT was and is still a substantive move forward toward better and greater energy collection while allowing systems to deliver more energy. It has still not been maximized, which can improve energy production, especially when energy storage is a key element of a system design.

As we have written numerous times over the last few decades, many designers and system buyers do not understand how to get the most out of their MPPT. In many instances, energy is wasted as a result of clipping (Balfour et al. 2021) while marginal energy production outside of the MPPT window can also be lost. With the advent of more energy storage, some of those losses can be recovered.

On the module side, a number of issues like cell browning were addressed and for the most part resolved. New cabling and connectors allowed the industry to get away from running conduit from module to module, which was both a cost, time saver and safety bonus. Mounting hardware improved and became more flexible, usable, and replaceable. And there is an incredible amount of variety of product in all of those areas, many of which are in continual levels of improvement. However, it is important to always keep in mind that these improvements have not made new or existing components or products "bulletproof" in spite of what the marketing may indicate. We continue to see many products that, when cheapened, create new unexpected challenges.

3.3.3 Integrators

During that era from the 1980s through the early 2000s, full-service PV companies were often referred to as "Integrators." They integrated a diverse variety of products and technology, then consolidated a project with EPC. Our evaluation of the PAM integrator approach as compared to the existing EPC process, relied heavily on exceptionally high-quality project communications, project-wide definitions, especially of what success and failure were and a minimal amount of system and project siloing. The process was integrated and not segregated.

Primary historic EPC tasks and responsibilities are included as compared to our PAM approach, yet are not limited to the following as shown in Table 3.1.

To deliver effective PV systems historically requires a broad range of knowledge and experience to make systems work effectively. This took place as continuing integrator education and implementing improvements to make systems work well, grew. It was as if everybody that was successful in the industry was experiencing the kind of education you get when you go into a rigorous graduate educational studies program or a professional journeyman training program.

A good example of the parallel between graduate studies and high-quality journeyman training was experienced with a utility grade system that was built in Eldorado Valley, Boulder City, NV.

During a review of plans, the master electricians for the project indicated that the utility feeding transformers were of the wrong type and would not survive a year. The response from the project engineers was that they were PV experts, professionally licensed engineers, and that the electricians should mind their own business because *they were only the labor*.

As predicted, the majority of those transformers, a very expensive system component, failed in the summer in less than a year!

Table 3.1 Comparison of Current Approach and PAM SE Approach.

EPC Engineering Functions: (per Wikipedia[13])	PAM PV (EPC) System Integration Engineering [14]
Basic engineering	Total project management
Detailed engineering	System and project planning
Planning	Basic engineering (usually for bidding purposes)
Construction engineering	Detailed engineering
	Component performance and RAMS data review
	Civil engineering
	Construction engineering
	Product quality improvement team w/wo O&M
The procurement functions	
Logistics & transport	ID most appropriate component procurement
Receiving	Purchasing
Procurement	Logistics & transport
Invoicing	Receiving and storage
Purchasing	Invoicing
	Subcontractor monitor and control
	Manufacturing and on-site product inspection
	Warranty responsibility and service
The construction functions	
Electrical installation	Project management
Mechanical erection	Civil site work
Civil engineering	Electrical installation
	Mechanical erection
	Installation inspection and Prepunch list inspection
The commissioning functions	
	Benchmarking
After-sales-service	After-sales-service
Testing & commissioning	Testing & punch list completion, benchmarking, Testing & commissioning (The last two are cooperative.)
Contractual warranty requirements	Contractual warranty requirements
The integrated operation and maintenance function	
Operate and maintain system	Data Collection, curation, root cause analysis
	Operate and maintain system

As qualified as the engineers were, they did not understand the use of those inverters with the transformers that they selected, especially in that environment, and how there were compatibility issues based on inverter electronics. Over the past few decades, we have continued to see this kind of professional and semiprofessional arrogance throughout our industry.

While the master electricians did not have advanced technical degrees, they had real-life experience and union training in both the classroom and certifications and had a far broader awareness of the realities of equipment operation in the field. This included a substantial depth of products in a specific design and environmental context including its lifecycle. They saw and experienced far beyond just the product specification.

We learned decades ago that when there was some inconsistency or disagreement, it was best to talk it out because both sides often had different parts of the puzzle that needed to be solved. This is how we learned that when going into a situation, it is best to have the mindset that in spite of our experience, "*we could be wrong*." By not listening, discussing, and working toward a common perspective, we often lose out on the benefits these learning opportunities present.

There are a few lessons learned from other projects that, if applied, could have avoided this and many other types of situations that are similar. They include:

1. **Data**: In effective PAM project development, all the information available must be carefully considered, evaluated, and confirmed or discounted.
2. **Siloing**: Siloing both internally and externally creates avoidable situations, problems, and costs that never make it into the LCOE.
3. **Unprofessional conduct**: Professional arrogance and discrimination are cost expansion issues that serve no one, while undermining the project and the entire team.
4. **Perspective**: Applying the integrator mindset, all parties have value even when they are wrong.
5. **Lack of general knowledge**: It is still common in the industry, that many project teams, do not take the time to understand all of the equipment they are using (specifying), nor do they fully understand the environments that they are designing and building systems into.
6. **Failure to document and apply lessons learned**: When teams do not go into each project questioning everything, they will continue to make poor decisions that may not become an obvious critical failure until after later owners or operators are stuck with often expensive problems with no budget available to solve them.

The integrator era was intense, ever-changing, and communications needed to be comprehensive. The reality is that most of those necessary elements have not changed, while many of the industry myths and assumptions still play a major role in decision-making. In other words, issues like the nonsense about the simplicity of plug-and-play since the 1990s, although discussed, required PV professionals to have a detailed grasp of the technologies, their application, and frailties, and how they actually worked throughout their entire organizations. We had to actually understand what would work and what would not. And, we made many mistakes, which resulted in many companies going out of business. Yet we, as an industry, tended to share and keep most of that knowledge in the industry until the industry exploded in new growth.

The specification was treated much differently. It included a balance between the most appropriate product and the most cost-effective product over time. The development of the system design and scope focused on addressing the balance between cost and long-term reliability. Integrators/EPCs had better access and experience with existing equipment and product information based on substantively closer relationships with the OEMs. This was even the case with large organizations that mandated silence. Problems were still discussed beyond the office walls. We learned, in many instances, what would work and what would not.

In the past, those skills generally were provided by a small, far more cohesive team than is often the case today. With the growth of organizations and projects, much of this work has become separated and siloed (Chung and Balfour 2016), which reduced PVPS delivery cohesiveness. Specialization and diverse interests among different siloed functions based on individuals often coming from different industries, created internal barriers to greater levels of success. These issues must be addressed if the goal is overall project success throughout its lifecycle and not just the first few years.

The National Electric Code (NEC), for example, was far from adequate and has continued to evolve. Much of that evolution process, however, has continued to be resisted by many in the PV industry as being a cost issue. Fortunately, most of us in the industry look at safety as a cost-reduction issue, which is a different mindset. Nevertheless, we are still fighting battles about safety.

The industry was such that we did not have many answers. For example, our first documentation for modules and inverters was on mimeographed sheets. Xerox copies later replaced them before the industry got more sophisticated. As a result, our best source for additional information was often our competitors. Most PV professionals do not grasp the notion that the competitor is still a stakeholder. Then, our competitors communicated on how to resolve issues. Even in the passion of competition, we more often than not worked together, a practice which must be rekindled in the industry. That is how we grew an industry!

It appeared very clear to those who were observing that the issues and the terminology for reliability and PV, generally outside of the extraterrestrial PV market, considered reliability to be little more than a "marketing" term! As of the writing of this book, that really has not changed. The usage of this reliability term continues, yet it is seldom based on sufficient metrics or comparative analysis that measure the component reliability itself, almost without exception. And unfortunately, all too often, when there is information available, it still tends to be ignored when it should be applied specifically to the project and the site conditions.

Since the 1990s, the rate of product change has grown and continues to accelerate. New product introduction cycles have become shorter and shorter without sufficient amounts of reliability or health and condition feedback as to how well the products work in the field. Some companies, of course, focused on reliability in part for image, pride, and professionalism.

Because there are assumptions that a longer warranty will protect people, economic interests, and projects, the final frontier in PV maturity appears to be that of actually addressing the reliability of those products. That warranty "fiscal security" notion has been undermined by the growing number of OEM's that have gone bankrupt, were bought out, abandoned the market, and or have not been sufficiently supported by large companies or simply dwindled away. And, of course, there are the gotcha warranties, too, that look good but really are rather empty.

All of this was further complicated by the fact that once a product has been certified, there is no requirement that that product will be fabricated with the same BOM. This is a major quality and reliability issue. As a result, batch consistency has continued to be predictably unreliable, especially as most purchasers do not pay sufficient attention to the potential differences in production of different batches until there are moderate to major failures. This continues while many lesser product problems are not usually even detected or tracked properly. This common practice has given the industry many marginal and minor failures, which have a major impact on product life and energy production.

> Based on the extensive independent oversight of production which PI Berlin has conducted, the manufacturing processes representing the top five sources of defects for modules are as follows:
>
> - Cell soldering
> - Material lay-up
> - Material control
> - Rework
> - Junction box application[15]
>
> Another interesting observation is that frame damage ranks as the second leading source of module defects. While a damaged frame itself may not present an obvious reliability or safety concern to the module, damage to anodization layers on the frame can lead to corrosion in the field. Frame damage rarely occurs during module manufacturing itself. Instead, frames are often damaged upon receipt by the module manufacturer and not well screened prior to use in production. In fact, component material defects represent over 60% of all module defects, including defects with:
>
> - Cells,
> - Frames,
> - Glass,
> - Back sheets,
> - Junction boxes.[16]

In essence, PV professionals who tend to buy products often have little to no clear or broad view available of exactly what they are buying. They know that prices have been going down, which on the surface, seems like a good thing. However, they don't know or have the documentation to know how well that product will hold up and perform during its normal in situ life cycle conditions. And besides, most people have no clear data regarding the real lifecycle for any particular set of climatological site conditions over time. All of this has been seriously exacerbated by the fact that as climate change becomes a major PV challenge, the industry has not effectively or fully addressed the issues existing prior to that awareness.

The major challenge is while the cost of components and systems has dropped dramatically, the reliability, availability, maintainability, and safety of many components and systems has not improved in direct or proportional relationship to price reductions. At the same time, very little of that drop in system cost has been turned into improvement in product and system RAMS. Most of the improvement has been based on resolving specific situational issues instead of looking at the process holistically.

While this seems like a massive conspiracy, the reality is that the existing model is cheap and not about critical details or RAMS. As it stands today, PV professional engineers and/or specifiers can specify a good product based on good information or experience, only to be cut out of the budget, or in some instances, replaced by a product sold by a friend or relative. This appears to continue because the whole PV project delivery chain is inconsistent with their knowledge levels, and, in many instances, choices are completely separate from project environmental realities. That is not to say that all companies are doing the same thing. There is also a continuum of good to medium results, which is all about industry progress. The problem is, if you are the ultimate owner, how will you ever know what you actually bought?

Products and manufacturers are changing so rapidly that the risk of buying basic components has continued to rise. This trend is the opposite of technology improvement and industry maturity. The issue of technology fatigue is complicated by the fact that this price squeeze and rush to the bottom, does not allow for the appropriate amount of research and analysis, while digging into the realities of these new products. Therefore, stakeholders will not see dramatic or even marginally improved system lifecycle reliability unless they take the time and money to invest in determining exactly what they are buying and whether it is the right or appropriate product. This must be kept in mind because there is seldom, if ever, a perfect product (Chapter 4).

Previously, this was a conversation that took place on a one-on-one basis and seldom ever made it to the open forum of conferences and technical working groups. That is rapidly changing!

Unfortunately, the mentality of this least-cost system/component business model has continued without providing the essentially effective support structures and specifications prior to EPC design or bidding, while stressing the industry and its members in a number of ways.

These include a broad range of products and components that:

- Do not sufficiently meet their basic warranty periods.
- Were not designed for and or do not meet the environmental requirements of a specific site. This is in good part because all modules are tested under the same defined conditions. Yet, different module technologies, BOMs and components, have different thermal and spectral sensitivity, form factors, outputs, and other details. Thus, they deliver different levels of energy collection and degradation rates in the field under real day-to-day, hour-to-hour conditions. Not every manufacturer provides a broader range of environmental testing beyond those requirements for STC certification. This additional testing, which would be necessary, requires additional financial resources; however, it is probably less costly than today's practices. A common perspective and position is that it is easier to take the hit on the warranty and industry improvement than invest in a

substantively better product. Inverter testing also has similar challenges, although those test requirements are improving.

- To complicate matters, operations, and maintenance practices today seldom have the equipment, the time, or the budget to discover, document, or address warranty issues early. This allows many warranty issues to never be addressed until it is too late.
- Result in excessive poorly and or undefined and unbudgeted operations and maintenance costs. While specifiers, EPC, and many engineers understand the underlying conditions and circumstances, their initial focus is on getting the contract and then building the system per contract. The long-term owners have a much different set of concerns.
- Those are of a lifecycle nature, which is generally under-addressed. The fascinating part of this process is that EPCs often have sufficient experience with components to understand there are better product choices. Nevertheless, the existing model makes that a change order and not a requirement of the contract, thereby seldom making it into the initial system specification.
- Are not effectively or easily maintainable. For modules, there are no spare parts and modifications, and generally, repairs cannot be made. Under warranty, you are replacing the whole module "if" you can accurately identify the level of fault, failure, deterioration, or form/type of failure (Chapter 10). With inverters, the challenges tend to be in the areas of effective testing equipment and procedures, along with the need for specialized technician services. Some manufacturers only allow factory-certified and/or trained technicians to work on their products, which positively improves O&M while in many instances delaying the repair. Warrantees often require that the owners do the basic required maintenance, which is essential, yet often delayed, under-addressed, or ignored. Due to the myth of PV systems being simple and not needing much maintenance, this tends to create additional issues. All too often, the labor to do so is not funded. Hopefully, there may or may not be spare parts that can be used to repair them.
- That is not available within a relatively short time period after they were introduced. (This is particularly of concern with subcomponents of systems and once any spares have been used up, it may be almost impossible to get replacements. A project that is completed a year, 2 years, or more out may have few options.)
- They have wide variability in their capabilities to operate within the same model number under different batches and sometimes under the same batch numbers.
- Where no replacement parts are available other than cannibalization, the cost of labor can become prohibitive and eventually force owners to replace massive sections of a system. Doing so is often referred to as repowering. However, it is more the result of following a poor business model and defective/short-sighted management practices. (See **SE/R**epowering Chapter 4).
- That isn't easy to or completely untestable. Often there are insufficient sensor and data points to determine if a product in the field is underperforming or failing unless it is observed directly or the failure is substantive.
- It should have never been specified in that product or system in the first place. Historically, a good example of this is plastic zip ties. They cost about a penny or less, then fail after a few years with a cost of over a dollar each to replace.
- Continue to exacerbate organizational angst because of increased frustration with the existing model and its results. If organizations took the time to understand the labor costs

and mistakes made due to organizational angst, they would begin to fine-tune their internal systems and processes and how they do things. They would treat systems and labor much differently in order to get a better ROI (return on investment).

3.3.4 Primary Components

This chapter is not to reestablish the science and engineering of these components but to address the issues of what should be considered and why, when specifying systems and components. The level of component design change has tended to create a market where one product may seem to be incredibly effective, while a similar product with minor changes may result in unexpected, unnecessary, and unwanted problems. Hopefully, within the next few years, the level of change and the repetitive parade of releasing new products without a track record will slow. At that point, stakeholders will benefit from choosing better products or components with fewer failures and lower failure rates.

The most basic considerations are: Which components are most cost-effective and capable of performing over the lifecycle of the system being designed and built for a specific location? Please note that we include the terms "cost effective and lifecycle" and not just a simple throw away version of an incomplete "cost effective" analysis. Operational conditions have a major impact on product performance and lifecycle. This is often overlooked.

The vast majority of the money invested in components will be for modules, inverters, and racking or tracking equipment. Yet, there are other products that are also evolving and rapidly changing. They include Met stations, SCADA and DAS systems, energy storage systems, cabling and cable connections, remote sensing devices, and technology, to mention a few.

When stakeholders focus primarily on least costs, it saves minimal time upfront, dramatically increasing system costs, labor, schedule, and risk over time. This tends to be true for all technologies where better decisions and selections result in better product serviceability and lower lifecycle costs.

This variability of industry-wide situations would become far more visible if a broad range of PV reliability testing and field anonymized data were shared. Doing so would address actual O&M lifecycle costs while also factoring in the millions to billions of dollars in underperforming assets. These include associated electrical infrastructure, plants, and systems that have been derated, rebuilt, or torn down prematurely where there is no publicly available data to actually compare. Because there is little to no industry data to be reviewed across the industry, many still live in a fool's paradise without understanding the real costs over time. This is where we see a real challenge with the LCOE numbers provided today. Those projections do not include the realities of large medium and small project failures that owners are stuck with after the fact and have to write off behind closed doors. Therefore, it is important that the project team understand the site environment based on the most accurate information they can find. This refers to all service life elements: including the physical geomorphology of the surrounding area, location, actual site climate zone, and varying conditions year over year. They must address all environmental conditions by taking a seriously critical look at the average climate and the extremes. This is becoming even more critical with climate change and the fact that trends from the past have a lower probability of being the same in the future.

This means that the process we propose with an integrated project team is continually factoring in higher levels of plant reliability and the ability to address changing conditions. All of this will pay off through the life cycle both qualitatively and fiscally.

Please keep this question in mind: "Will those system components work properly and as required in their chosen environment over the life of the system?"

For example, when selecting inverters; are their internal component attributes harmonized with the site environmental conditions of temperature, humidity, altitude, etc., and the induced environments due to equipment manufacturing, transportation, installation, and operation? In addition, it is critical to understand not only how often equipment, e.g. inverter, may exceed its design capability (and be physically derated), but how often it is overstressed close to those limits. This is often under-considered or addressed.

This means that during specification, it is important to understand what components went into that inverter, module, poorly worded or other components or subsystems. This must beg the question, "How much has the BOM changed since initial design component testing and certification?"

It is essential for the specification team to fully understand the component well beyond just reviewing a product cut sheet (marketing information). There must be a component review to see if there is reliability data available for that component or set of components. There are a growing number of competent organizations that have the skill, knowledge, and capability to do this in both the lab or in the manufacturing facility itself. This data must be compared to the planned operating environment and understand how those numbers were established and if they are valid for the specific component. If reliability values/numbers were questioned by users on a regular basis, this will hold manufacturers and component providers' feet to the flames to provide better and more accurate information.

A good example of that situation occurred during a working group for one of the IEC inverter testing standards committees. We looked at a condition where there was a capacitor "flash over." The capacitor directly shorted out to the cabinet through the air, destroying the capacitor and melting a hole in the cabinet. This type of condition is dangerous, unacceptable, and avoidable!

Over three meetings, as this issue was discussed, we requested and received the spec sheet for that capacitor. It was not until we reviewed it in the third pass that we realized the temperature stated for the capacitor was ***not the operational temperature, it was the storage temperature***. The realization was simply that that capacitor should never have been specified for that inverter application under the temperature conditions stated.

To make matters worse, once we dug into the standard that it was tested under, we discovered that the testing was also inadequate for the usage. Just having the marketing information was not and is not sufficient. Reviewing the information carefully and considering all the details would have meant that the OEM either would have never used that component or that the system purchaser would have, if the information were available, had access to all the documentation for the inverter BOM and bought a different product. However, the inverter product selected will plague that entire plant if it is not corrected after the fact by the OEM, or until a sufficient number of units have failed to where the owner replaces those products, or the plant is bankrupted or scrapped.

In this instance, there was little clear information about the operational temperature range for that capacitor. The fact that it took us a number of reviews to see this detail in the small print further addresses the issue of detailed examination and research of component documentation. This indicated that similar inverters in similar situations would have the same types of problems over time and that this problem was designed for the product. We did not assume that this decision was made consciously in order to save money; however, it did cross our minds. Nevertheless, the question that we could never answer was, "Why was that product cut sheet only indicating a storage temperature?" This raised a number of questions:

1. What level of reliability testing was done on that product with that BOM?
2. What were the technical design assumptions for that product and the inverter?
3. What other selections were made for this and other products without providing or having sufficient information and review, to make the kinds of technically critical choices to result in an effective and reliable inverter?
4. What level of design derating was applied, if any?
5. Where was the requirement/specification breakdown that led to this issue?

When reviewing situations like this, the question always comes to mind, "Was this a conscious choice or was it just an oversight?" The author's perspectives here are that neither was acceptable. Nevertheless, because this was probably not clearly addressed in the OEMs specification processes and procedures, that alone increased the probability of the error and the unexpected additional expense to the owner of replacing those parts.

As the working group agreed, *that capacitor should have never been used in that inverter even in milder environmental conditions*. Of course, all of this could have been preempted by actually looking at the details of the component being considered, asking the right questions, getting the properly supported data and analyses, and making a better design decision.

A further root cause analysis (RCA) assessment of the condition review and the effects of failure indicated that there were several additional factors that also contributed to the arc over. Due to its close proximity to the cabinet wall, a voltage potential exists between the capacitor and the cabinet wall. To have a flash over, the air characteristics, temperature, humidity, and altitude, and the separation between the cap and the wall contribute to exceeding the breakdown potential. This leads to a further consideration of moving the capacitor further away from the wall or installing a piece of dielectric material (Teflon, ABS plastic, etc.) on the cabinet wall to increase the voltage breakdown potential.

This is not an unusual problem. It is far more common than industry professionals are usually aware of or would want to admit. We will presume (not assume) that the manufacturer did not have a clearly defined set of design and test processes, procedures, and documentation to avoid a situation like this and that the price was right in the initial review of the specifications and that was good enough. They may have truly believed that they and their product were in good shape.

These challenges are common throughout the industry. This is often because the focus always seems to come back to driving down the price with insufficient or minimal consideration for reliability. The plant is expected to perform as required without addressing sufficient environmental testing or the potential cost of warranty failures. However, another issue to consider here is the cost to the inverter manufacturer to replace under warranty even a few percent of the number of inverters sold. In addition, there is the other consideration that issues like this one discovered by the industry tend to taint the reputation of the manufacturer. Nevertheless, there are few broadly required practiced existing checks and balances to address these types of situations industrywide.

In today's helter-skelter market, manufactures get away with this for a while. However, as the industry evolves, this type of black mark on their capabilities and reputation will impact their future sales. The authors have worked with teams throughout their careers that have a bad habit of using the terminology that "we don't need reliability – our engineers design reliable products" or worse, "we assumed" that.

> One of the themes of this book are to clearly say to the reader, **"NEVER ASSUME ANYTHING", GET THE FACTS!**

When engineering systems and determining specifications, it is not professional or acceptable to "assume" anything. That does not mean that management may not force you to, as we have seen all too often. We see that happen consistently throughout the many industries. The worst part of it is that management has not taken the time to educate themselves and, as a result, is unwilling to fund the design and test effort, they have bought into the myths and assumptions, or they just do not care. The focus of the work for industry maturity must be on documenting, as best as possible, what was found and why the decisions were made. As a result, in some of the work related to the IEC Availability Standard and the IEC Reliability Standard, a number of us developed a system mapping tool to define the issues and allow professionals to track the issues back to where a decision was made. This has a positive impact in being able to avoid the same types of problems in the future as it documents lessons learned.

3.4 Primary Technologies Discussion

Basic technologies for PVPS include cells, modules, combiners, inverters, and power components (transformers, contactors, and capacitors), which are considered primary components. Other components are wiring, connectors, nonstructural conduits, nuts, bolts, etc. Here we discuss cells, modules, and inverters.

3.4.1 Cell Efficiencies

The PV plant is based on the use of photoelectric cells, which seems obvious; however, it is also one link in a chain of cells that easily exceeds 1,000,000 links. The actual solar conversion efficiency is physically based on the chemistry of the cells as well as the environmental impact of temperature, irradiance, albedo, altitude, and airmass. Under STC conditions, Figure 3.3, the cell efficiencies are measured by the output under those specific conditions to determine capacity. The cell's actual power output at a specific site and at any particular point in time is determined by the cell size, chemistry, configuration, and irradiance when measured under site operating conditions.

> Standard Test Conditions (STC):
> Irradiance: 1000 W/m^2,
> Cell temperature, T: 25 °C
> Air Mass 1.5 (AM1.5)

Figure 3.3 STC.

The NREL graphic on the next page (Figure 3.4) tracks the Best Research – Cell Efficiency over the past four decades. It includes Multijunction (2-Terminal, monolithic), Single Junction GaAs, Crystalline SI, Thin Film, and emerging PV technologies. Note that these are laboratory trends; however, that is not what often reaches the field for energy production over time.

It is important to understand that cell and module efficiencies are often considered constant, while in the field, they are always shifting due to aging and environmental conditions. As a result, they seldom operate at STC.

PV plants are rated by total module name plate capacity. For example, a plant with 10,000 modules rated at 300 W each results in a nameplate *capacity* of 3,000,000 W (3 MW). However, as discussed throughout this book, the actual plant output is seldom operating at STC. For example, modules degrade from the nameplate power beginning at production through installation to the end of module life as a result of time, defects, degradation, and failures. Much of this is also driven by environmental exposure, handling, and maintenance. Note: It is not uncommon to not have the actual PV cell history from its manufacture until it is installed. The name plate number can be used for initial plant sizing and discussion, but it does not indicate the plant's capability to produce power over time. In essence, when comparing two or three PV plants of the same nameplate capacity and location while using different module technologies, it will result in different performance, outputs, energy production, and profitability while exhibiting different levels of technology-related degradation.

This should be very carefully considered when shopping for least cost modules products because the first cost will not address the ability/capability of those products to operate as expected in the field throughout their lifecycle or use cycle.

These realities regarding PV technology are often misunderstood or misrepresented to or by many stakeholders. This affects the day-to-day power generation *capability while addressing module effectiveness* and over time the asset value of the system.

Cell power decreases with temperature driven by the voltage, even though the short circuit current (I_{SC}) actually increases slightly due to thermal excitation. The reduction in the open circuit voltage of silicon photocells (V_{OC}), for example, is about 2.2 mV/°C.[17] This adds up to a fairly large change when all of the temperature variables are added.

3.4.2 Module Efficiency and Effectiveness

As to the importance of understanding module ratings versus installed performance, we begin by addressing efficiency and effectiveness. It would be a mistake to assume that cell and module efficiencies are the same.

As discussed previously, there is a difference between efficiency and effectiveness. We begin by looking at a tested module or cell efficiency, which is often idealized using testing at STC. What is targeted in the lab may not end up in the field. When assessing system capabilities, it is critical to address what takes place to that efficiency after we have included the defects, faults, and failures driven by environmental conditions.

From the discussion of temperature effect on the efficiency of solar cells, a 400 W module (96 cells) that is certified at STC may only produce approximately 367 W at 40 °C (104 °F) and even less at 50 °C (122 °F). This is where we separate to show the difference between efficiency and effectiveness. The fact is that efficiency is a standard measure of the

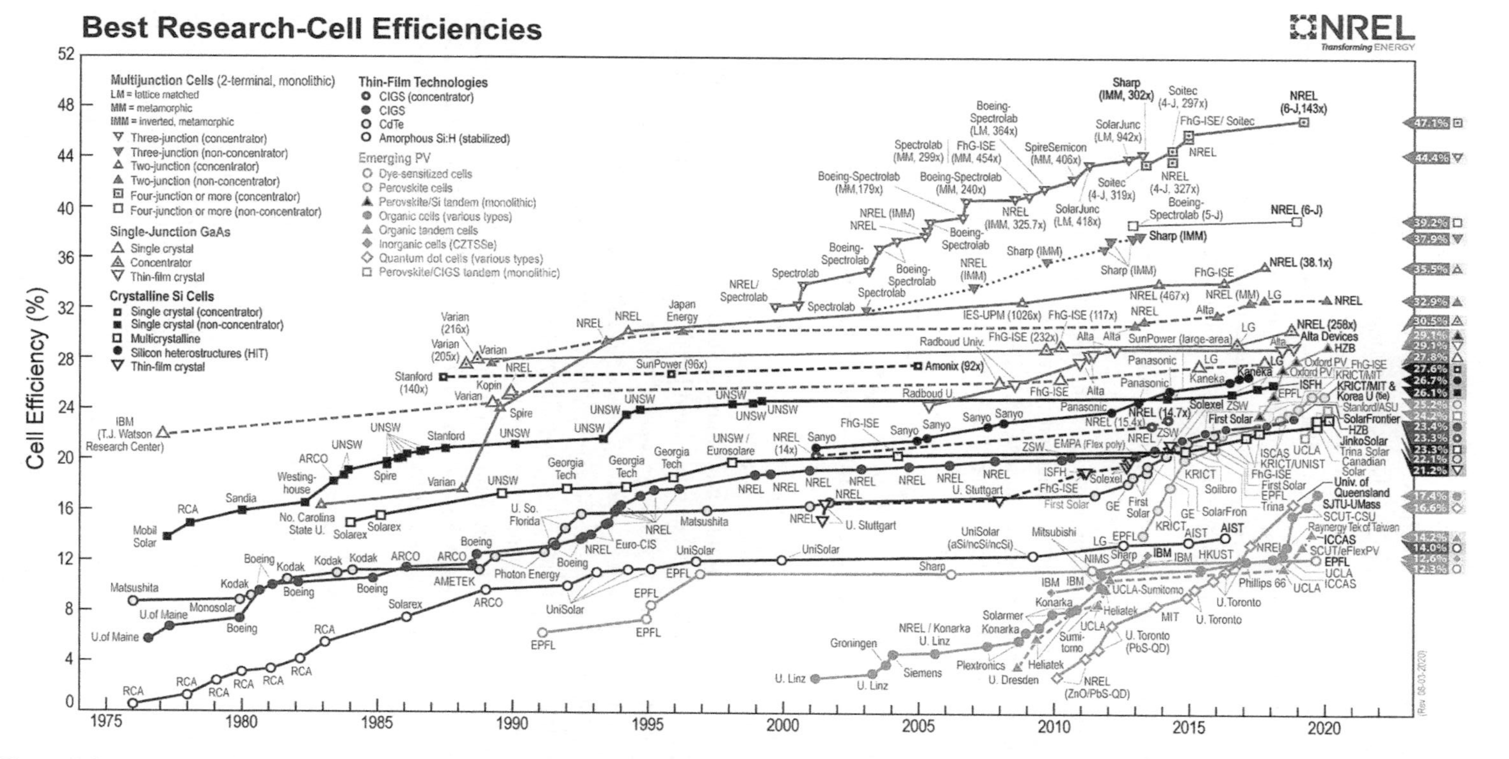

Figure 3.4 NREL Cell Efficiencies 1976–2021. Source: United States Department of Energy/https://www.nrel.gov/pv/cell-efficiency.html/Public Domain. (Colour rendition of this image is available on the book's landing page on wiley.com).

cell/module attribute under ***one specific set of conditions***. At the same time, effectiveness is the attribute of the module power output ***under all of its physical, temporal, and environmental operating conditions***.

Doing so requires that the product is properly packaged, shipped, delivered, installed, and commissioned in a manner that does not cause damage to it. This is the basis for the PAM **R**epowering approach, addressing module and plant effectiveness over time, as opposed to using initial component efficiencies as the basis for plant specification, design, and purchase. This lifecycle approach addresses most, if not all, of the questions and data necessary to dramatically improve system performance, energy production, and profitability over time.

From Figure 3.4, you will note there is a substantial difference in efficiencies over four decades.

When the cells are integrated into modules, the module efficiency will vary based on the cell and its type, chemistry, spectral sensitivity, number of cells, size of cells, and spacing. It includes the external size of the module itself, framing, type of EVA backsheet, glass used if the glass is being used, the composition of the front and back, BOM, quality of construction and assembly, and other subtle issues.

Looking at module efficiencies, Figure 3.5 tells a different story. It includes not only the cell size; it addresses nonenergy collection cell spacing and the frame on the face of the module. Installed in a plant, there is a further effect due to soiling that reduces the module efficiency.

Components and subsystems must be capable of operating in everyday real-world environments and during a variety of expected and unexpected extreme conditions! To achieve project/system operational and financial success, we must address the dearth of sufficient operable data! This is because tested lab data and the results found in the field are not the same, nor do they take into consideration a number of factors, as mentioned, from packaging to installation and testing if there is follow-up testing. While the trend is toward improving tested product efficiencies, effectiveness is what they will actually do in the field over the entire module and or plant life cycle.

This is seldom completely grasped by much of the industry that is not involved specifically in product testing compared to the field results. The results discovered are the divergence between what is promised in marketing and what is seen in the field is, product reliability.

Various technological advances introduce new risks. A comparison of recent developments is shown in Figure 3.6.

> Bifacial modules exhibit a clearly higher rate of defects compared to mono-facial modules. In PI Berlin's experience, the challenge with bifacial modules is a result of both the new cell design and the new module construction. Soldering the rear-side of bifacial cells is different to the soldering of the rear-side of conventional mono-facial cells and this can present the manufacturer with challenges in adapting the cell soldering processes. In addition, bifacial modules are typically constructed with two sheets of glass rather than one sheet of glass on the front and a polymeric back sheet on the rear. Additional inspections are needed to assure the quality of the rear-side of a bifacial module, and these are not always well conducted, or performed often enough. Rear-side power measurement of bifacial modules is infrequently performed at the factory.[19]

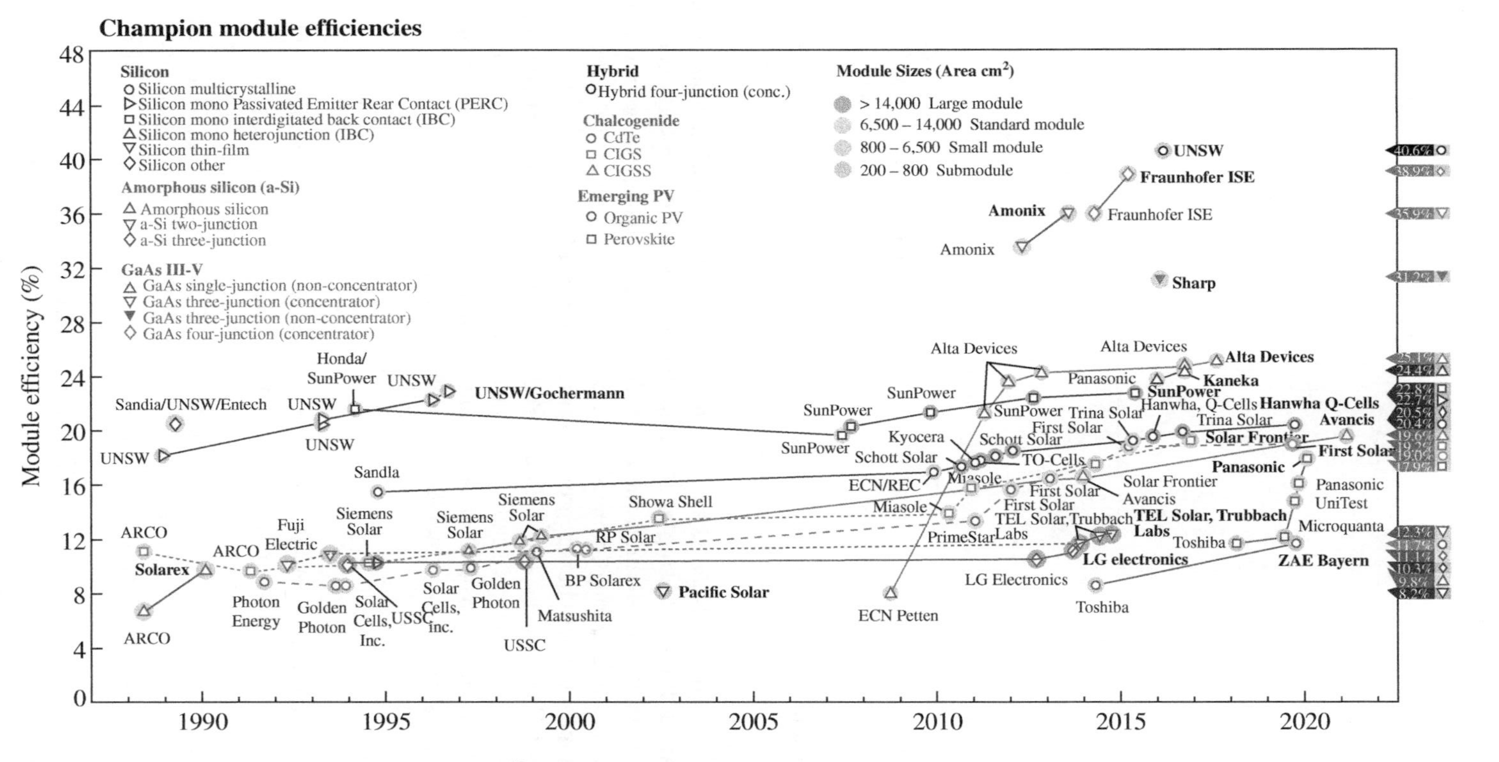

Figure 3.5 Module Efficiencies 1988–2021. Source: [18]/U.S. Department of Energy / Public Domain CC BY 4.0. (Colour rendition of this image is available on the book's landing page on wiley.com).

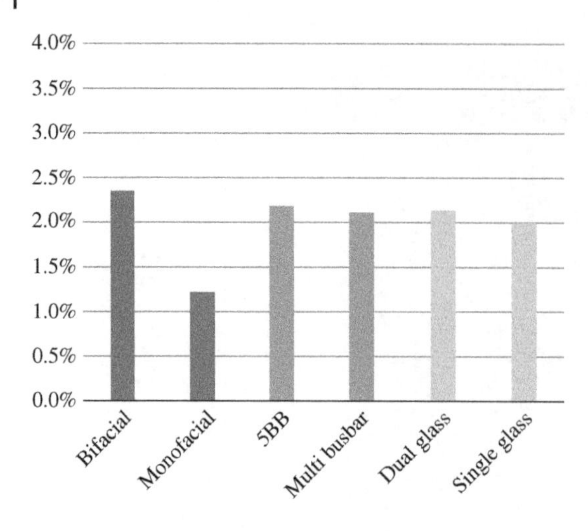

Figure 3.6 PV module defect rates related to technology changes (from 2016 to 2020). Source: Courtesy of PI Berlin.

Figure 3.6 illustrates the defect rates for various technologies in the construction of modules. Testing requirements are completed under the following IEC standards. Please note the different editions and years to indicate the latest updated standards. For effectiveness, good standards *evolve over time by reflecting the changes in technology and their design.*

Those standards are based on a consensus process, which means that a substantial number of professionals in the industry must agree on what the minimum requirements are as they are updated.

IEC 61853-3, **Photovoltaic (PV) Module Performance Testing And Energy Rating – Part 3: Energy Rating of PV Modules**. The purpose of this document is to define a methodology to determine the PV module energy output (watt-hours), and the climatic-specific energy rating (dimensionless) for a complete year at maximum power operation for the reference climatic profile(s) given in IEC 61853-4.

IEC 61853-4, Photovoltaic (PV) Module Performance Testing And Energy Rating, describes the standard reference climatic profiles used for calculating energy ratings. Part 4: Standard Reference Climatic Profiles contains an attachment in the form of zip files (climatic data sets), which are intended to be used as a complement.

3.4.3 Hail

Hail is one of the most damaging weather extremes a PV system can encounter. Past testing for hail focused on 2.54 cm (1 in.) hail at a "terminal" velocity of 106 km/hour (60 mph). Reality is that hail is now commonly reaching 5 cm (2 in.) in diameter, and if it is in a thunderstorm's down draft, the wind velocity can exceed 150 km/hour (90 mph). In addition, most of the testing was for single impacts, while in hail storms, there can be 10 or more impacts per square meter (approx. 10 ft^2) in five minutes. Chapter 6 addresses this failure mode in more detail.

IEC TS 63397, **Guidelines for Qualifying PV Modules for Increased Hail Resistance**, is a good example of a critical standard that is now (2022) in process and will probably not be completed until after the publication of this book.

Modules have to be certified to the above performance standards; however, there is a loophole. What many PV professionals do not clearly understand is that once that module group has been certified, there is no requirement to be recertified if the BOM is changed unless there is a major material change in the module. Different material is defined as "material that differs in its chemical composition, type designation, or specification from the material it replaces (including, e.g. electrical, optical, mechanical properties; the nominal values including tolerances shall be considered" (IEC TS 62915:2023 Photovoltaic (PV) modules – Type approval, design and safety qualification – Retesting)[20].

For hail resistance, we have always had storms that produced hail, sometimes large hail, but in the last 20 years things have gotten increasingly worse. Due to climate change variations in local, regional, and national weather patterns, the extremes of storms that produce larger hail occur more often and do not fit the historical patterns as they did previously. Understanding the effect of larger, higher velocity hail and the cost of exposure of modules that are not designed or built to this environmental hazard will require more robust designs to improve performance and reduce force majeure costs associated with these events. The size of hail stones also is not routinely measured and reported, compounding the problem.

The challenge for exposure to hail is to develop a set of test procedures that address different sizes and velocities of hail over a wide range of different module products and configuration designs made of different materials (Table 3.2). In addition, the capability of a module to survive hail impact changes as it ages. While hail testing focuses primarily on a single or a few module strikes, *to date the authors have never seen a hailstorm or the module that is only hit once or twice during a single storm.* Generally, they are pummeled repeatedly.

This evolving standard will have to go through a number of phases to provide all stakeholders with sufficient information to make informed decisions. A series of challenges are created by how those modules are shipped, installed, and maintained. From firsthand experience, if during installation, module mounting hardware is not properly torqued, damage may cluster around those improperly mounted modules. This is another realm of testing that will also have to be addressed, i.e. the effects of variations in installation mechanics. Following a moderate to massive hailstorm, for example, there are modules that will not show defects or failures for weeks, months, and possibly years. These can be compounded overtime due to wind-induced flexure or even ground movement or vibration. It reinforces

Table 3.2 Summary of Hail Damage and Testing.

Hail Issues	IEC TS 63397 Hail Testing Issues
Impact on insurability/finance	IEC standard requirements versus hail rating
Additional budget necessary	Industry acceptance and application of the standard
Impacts on PPA requirements	Hail ball size and density (confidence level)
Module replacement	Evaluation of single versus multi strikes testing requirements
Tracking stowage	Test for consistency
Modules/availably	Testing for hidden or micro-damage
Energy production flow	How will product results be presented?

a greater need for installers and all other PV professionals to be better trained and educated. The reason for this is to overcome the fact that just because a stakeholder in a PV project has not seen or heard of an issue does not mean it does not exist.

3.4.4 Module Degradation

Different module product technologies have different sensitivities to irradiance and temperature while degrading at different rates depending on the module, its BOM, actual site temperature, and all local conditions. As a result, module output is impacted by the basic design, materials, and components selected (specified), then all of the conditions found in its design application, which influence module health, condition, and performance. These impact actual output and revenue over time. Figure 3.7 illustrates a variety of degradation sources in PV modules.

Although plant power module degradation rates (Figure 3.8) are generally quoted at 0.5%/year, over time, the reality ranges between 0.5% and 2%. Another example of degradation impact and variation can be found in the NREL paper, *Photovoltaic Degradation Rates – An Analytical Review* (Jordan and Kurtz 2016). "This article reviews degradation rates of terrestrial PV modules and systems reported in published literature from field testing throughout the last 40 years. ...The distribution has a mean of 0.8%/year and a median of 0.5%/year." Using the mean or median will result in a substantial difference in expected module performance, output, and revenue throughout the plant's lifetime.

Later analysis by Kim et al. 2021 (Figure 3.8) noted that degradation rates run −0.8% to −2.0%/year. The principal sources of degradation for silicon cells are shown in Figure 3.9. There is a wide variation in degradation rates as reported in various countries (−0.8% to −4.0%).

There are a variety of fiscal impacts due to degradation including the ease or difficulty in identifying the causes and resultant effects over time. In most cases, the sources of degradation are not initially visible and barely measurable. As maintainers inspect modules and strings, additional tools are essential to clearly identify the cause and determine the effect as it continues. Many of the tools for evaluation and their discussion can be found in Chapter 10.

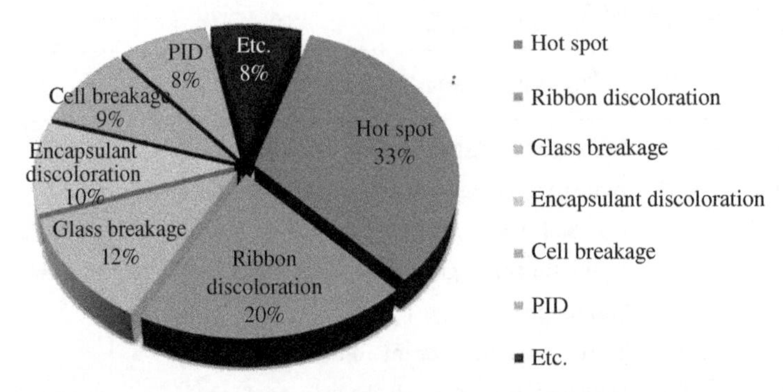

Figure 3.7 Sources of Degradation of Silicon Solar Cells Over Time. Source: Adapted from TamizhMani 2016).

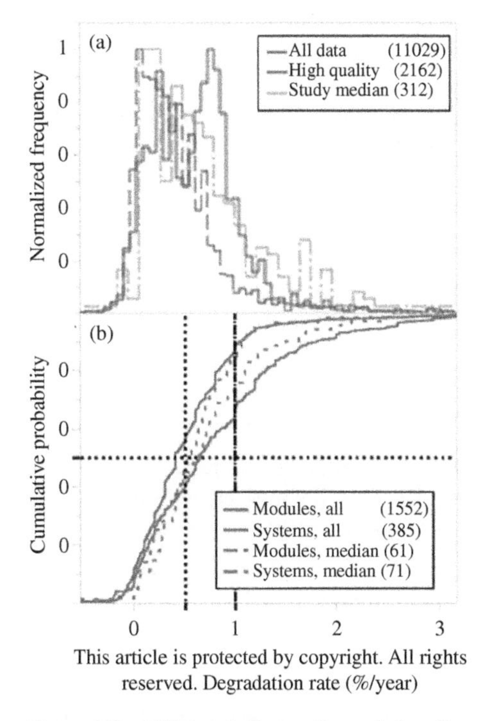

Figure 3.8 PV Module Power Degradation Rates Observed in the Field Range from 0.2 to 2%/year (2016).[21,22]

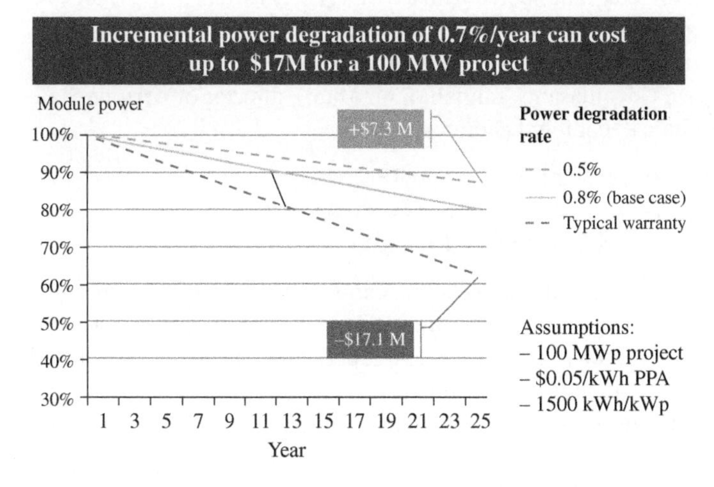

Figure 3.9 The Cost of Underestimating PV Module Power Degradation (Solar Buyer 2021).

Because there is not sufficient data – the information is all over the map and requires substantially greater levels of data sharing and diligence to ensure a far clearer view of product degradation, its rates, and its impact on system energy and performance.

Figure 3.10 resulted from a statistical analysis of "11000 degradation rates in almost 200 studies from 40 different countries." The light gray/blue distribution is from the PV

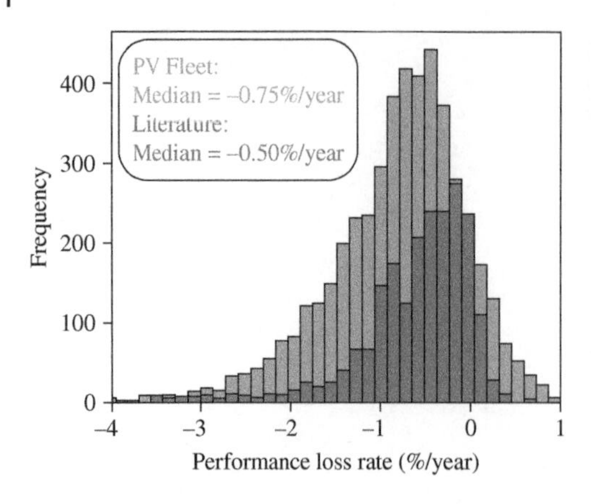

Figure 3.10 Statistical Analysis of PV Degradation Rates. Source: Jordan et al. 2022/John Wiley & Sons/CC BY 4.0. (Colour rendition of this image is available on the book's landing page on wiley.com).

Fleet analysis and is all system's degradation rate or performance loss rate. The literature distribution is the "high-quality data" from the compendium paper. "High-quality data," meaning they used at least two measurements to determine the performance loss slope. The literature distribution is much lower because 80% of the data points were modules, and only 20% were systems data, in contrast, all of the light gray/blue distributions are systems data.

3.4.5 Module Defects

As of this publication, there are 86 different identified defects that can be found in fielded PV modules as identified as of today and scored with a Risk Priority Number (RPN) process (TamizhMani 2016). In time, additional defects may be identified. These defects have been identified visually and with the appropriate equipment and addressed internationally by our friend and colleague Dr. Govindasamy TamizhMani (Mani), director of Arizona State University (ASU) Photovoltaic Reliability Laboratory.

- Some defects are purely cosmetic, with low or zero RPN scores;
- Some defects pose safety risks or performance risks with high RPN scores;
- Some defects pose degradation risks with high RPN scores.

This does not address all of the possible defects that can take place from the beginning of manufacturing until final recycling. However, these defects can be identified and documented with the appropriate field tools, knowledge, and experience. Others can be detected with different forms of attached and remote sensing.

Testing for the effects that cause problems with solar cells/modules includes yet is not limited to the following:

a. LID, light-induced degradation, is an effect found in boron-doped silicon solar cells. The exact source of this failure mode is still unknown. Some research by University of Manchester, UK, and University of New South Wales, Australia, appears to indicate that free oxygen atoms bond with boron and cause the effect when first exposed to light. This effect may be as much as −2% in addition to the efficiency.
b. LeTID, light, and elevated temperature induced degradation is a long-term effect that adds to the inherent degradation of 0.5–1.0%/year over the life of the module,

c. PID, potential-induced degradation factors include voltage, heat, and humidity. This is not a major effect but is dependent on the manufacturing process and quality. PID, if it occurs, can impact the energy output of a module by up to 30%.

3.4.6 Segregation of Safety and Performance Defects and Failures

Doctor Mani's research and team have identified 25 safety and 61 performance defects (TamizhMani 2016) that include visual detection in a standard set of O&M practices (Table 3.3).

Modules with safety defects should be replaced immediately upon inspection and are generally covered under warranty.

Performance defects impact performance and energy production, which impacts the bottom line. Identifying and addressing these issues impact all stakeholders throughout the system lifecycle.

Bifacial PV adds an additional level of potential defects, safety, and performance issues. Because of their design to collect energy from both sides of the module, they have an additional layer of glass, and cell connections and include added weight to the mounting system. That size and weight addition requires greater handling skill and capability from manufacturing, through operations and maintenance, and all the potential opportunities for damages in between.

Large-scale deployment of bifacial photovoltaics will take time to improve their reliability to today's level. R. Kopecek et al. note, "Low photovoltaic module costs imply that increasing the energy yield per module area is now a priority. We argue that modules harvesting sunlight from both sides will strongly penetrate the market but that more field data, better simulation tools and international measurement standards are needed to overcome perceived investment risks."[23]

The importance of determining the root cause of the defects is related to the system CAPEX and OPEX sensitivity which includes:

- If there is a warranty safety issue O&M should automatically make the replacement.
- Depending upon the warranty, performance issues may or may not qualify until they reach a certain level of degradation or performance deration.
- It is critical to understand the OEM requirements regarding warranty claims and the information required to successfully secure whatever remedies are in the warranty.
- It is important that the project team understands that warrantees can be renegotiated with manufacturers but only prior to the purchase.
- Often, the larger the purchase, the more responsive the OEM may be.
- Is there still a warranty or a company that can warrant that product and defect?
- If there is no warrantee, the product replacement costs for resolving the issue become an unbudgeted O&M cost unless it was addressed in project planning and O&M budgeting.
- If there is a warrantee, the costs for resolving the issue may become an operation and maintenance labor cost that was not budgeted unless it is in the initial project planning.

Addressing PV System Delivery fiscal risk with regards to warranties, project stakeholders should include a series of scenarios in their financial modeling and thinking to address "no warranty" and early failure of components. Doing so raises awareness as to potential

Table 3.3 Module Defects That Can Be Visually and Sensor Detected.

Total Number of Safety Defects: 25			
Glass	**Frame**	**Bypass Diode**	**Junction Box**
Front glass crack	Frame grounding severe corrosion	Bypass diode open circuit	Junction box crack
Front glass shattered	Frame grounding minor corrosion		Junction box burn
Rear glass crack	Frame major corrosion		Junction box lose
Rear glass shattered	Frame joint separation		Junction box lid fell off
	Frame cracking		Junction box lid crack
Wires	**Backsheet**	**Cell**	
Wires insulation cracked slash disintegrated	Backsheet peeling	String interconnect arc cracks	
Wires burned	Backsheet delamination	Hotspot over 20 degrees C	
Wires animal bites slash marks	Backsheet burn marks		
	Backsheet crack/cut undersell		
	Backsheet crack/cut between cells		
Total Number of Performance Defects: 61			
Glass	**Frame**	**Junction Box**	**Cell**
Front glass highly soiled	Frame bent	Junction box lid lose	Cell discoloration
Front glass heavily soiled	Frame discoloration	Junction box warped	Cell burn mark
Front glass crazing	Frame adhesive degraded	Junction box weathered	Cell crack
Front glass chip	Frame adhesive oozed out	Junction box adhesive loose	Cell moisture penetration
Front glass milky discoloration	Frame adhesive missing in areas	Junction box adhesive fell off	Cell worm mark
Rear glass crazing		Junction box wire attachments lose	Cell foreign particles embedded
Rear glass chips		Junction box wire attachments fell off	Cell interconnection discoloration
		Junction box wire attachments arced	Grid line discoloration

Edge Seal	**Encapsulant**	**Back Sheet**	Grid line blossoming
			Busbar discoloration
Edge seal delamination	Encapsulant delamination over the cell	Back sheet wavy	Busbar corrosion
Edge seal moisture penetration	Encapsulant delamination under the cell	Back sheet discoloration	Busbar burn marks
Edge seal discoloration	Encapsulant delamination over the junction box	Back sheet bubble	Busbar misalignment
Edge seal squeeze/pinch out	Encapsulant delamination near interconnector or fingers		Cell interconnect ribbon discoloration
	Encapsulant discoloration (yellowing/browning)		Cell interconnect ribbon corrosion
			Cell interconnected ribbon burn mark
			Cell interconnected ribbon break
			String interconnect discoloration
			String interconnect burn mark
			String interconnect corrosion
			String interconnect break
			Hot spot less than 20 degrees C
Wires	**Thin Film**	**Bypass Diode**	**Others**
Wires corroded	Thin film module discoloration	Bypass diode open circuit	Module mismatch
	Thin film module delamination – absorber coating		Solder bond fatigue / failure
	Thin film module – AR coating		

real-world conditions and can stimulate a far more accurate and selective view of what is or might be considered for purchase. This exercise will focus more on the components selected, their medium and long-term availability, and common future, seldom or unbudgeted costs contingencies that may be missing in LCOE.

The challenge for downstream stakeholders is that component and system degradation, including defect rates, are established as metrics prior to design and PPA contracts, then included in LCOE projections. They are often assumed based on minimalized OEM technical document sheets while not reflecting the experience of the industry over the last few decades.

While continuing a review of the historical foundation for module and systems defects, there is still a wealth of low-quality, unpublished, or unavailable information on thousands of PV plants that, if available, could help refine the evaluation accuracy. Additional data would positively impact system delivery efforts by clarifying the realities of plant health, and condition, and status plant health and condition realities as they relate to output and revenue. Nevertheless, the historical trends indicate that system lifecycle defects and degradation used in LCOE calculations for financing continue to overestimate the system's capability to produce energy throughout its life.

Traditional commissioning and benchmarking do not produce a set of measurements taken over time. It does not address the details of the entire plant. It only measures specific details and generally a small percentage of the plant. The details of component and latent defects and degradation tend to be under addressed for the products selected and particular site environment.

The results are simply that many plants, as presently designed and constructed, may never meet PPA contractual or other future stakeholder requirements, wants, needs, or asset values.

Therefore, to define the failure of a PVPS, an additional requirement is needed to establish fault, failure, and degradation for reliability and availability metrics. The initial requirement is to define the level to which the term *failure* applies. A solar component failure can potentially cause the loss of power from a module. If there are 10,000 modules, the loss may appear to represent 0.001% of the total power output of the system. However, if that module failure ripples through a whole string, the loss may grow to 0.05% or more. As installed, there may be as much as 0.5% or more of the modules failing or underperforming. This is typically undetectable unless every string is measured.

3.5 Inverters

There are many inverter types with a large number of components to consider when selecting inverters. Although it is convenient to leave review and selection to the "experts," it takes much more to understand the potential impact of inverter failure or under performance associated with the design. In short, writing a well-tailored and defined specification is necessary to ensure documentation of need/requirement, but just buying to a specification is not the best way to complete the purchase.

Inverters are complex systems with components ranging from control cards, MPPT, power control drivers, power cards (switching), high-voltage filter capacitors, high voltage transformers, cooling equipment, sensors, and safety circuits, among others. As an example, under ideal conditions, a 1 MW inverter may generate as much as 10 KW or more of heat due to switching inefficiency in the output driver and requires cooling. There are also inherent safety hazards for maintainers (high-temperature components and shock hazards). When operating outside of the main design point, many companies provide equipment that has built in MPPT circuits that will improve the overall efficiency from the irradiance on solar modules to the AC output.

There are a number of functions necessary for the inverter to reliably perform over its entire life. MPPT[24] is one such function as it may help reduce stress in the inverter and optimizes the operation to best use the solar energy input from the array. MPPT impacts energy collection and output including the effects of physical and environmental impacts on performance. Nevertheless, when designing carefully, additional clipping can have a negative impact on inverter life and its energy production capabilities. As DC energy enters the inverter, not all of it leaves as AC energy (Balfour et al. 2021). The majority of the energy becomes usable AC power, and some are consumed as conversion efficiency loss $(1 - \varepsilon)$ of energy due to the various resistive losses in the system. These latter losses are not necessarily controllable. Energy loss due to clipping, also referred to as shoulder, where some of the energy is lost because the total power available from the array is greater than the inverter can use to convert it into salable AC energy.

The failure of a component in an inverter resulting in the inverter shutting down can represent a loss of between 1% and 5% or more of the total plant power output. The actual impact on the plant power production is tied to the size of the plant and the number of inverters that are a fraction of the plant output. This may or may not create a specific problem for the plant on any given day, depending on the actual demand. Regardless, this should be defined as a critical failure as it may cause the plant to fall below the demand power requirement. This leads to the need to define success (IEC 63019).

When oversizing the array to account for the design margin[25] and variability of solar irradiance, clipping practices create additional heat within the inverter when too much energy is being provided by the array. If there is excessive clipping (DC/AC ratios greater than about 1.3:1), it will impact performance over time, resulting in additional degradation with reduced output and lifecycle from the inverter.

Figures 3.11 and 3.12 helped us understand better what was going on inside the inverter, allowing us to make the most out of the MPPT window. We are often trained to view module IV curves as a two-dimensional behavior of voltage and current and as a simple/clean representation of module output. However, it is critical to understand that energy-tracking is not simply a matter of overlaid IV curves similar to STC. All too often, solar professionals don't see the impact of MPPT as being a shifting multiple-dimensional set of elements and conditions. For each individual inverter (model, age, and unit), MPPT is ever-shifting. In the field, MPPT is continuously adjusted by time, and many previously discussed variables are listed below.

It should be noted that there is a need to understand the impact on MPPT of all the variables including voltage, current, degradation rates, failed modules, wind (cooling and heating), solar insolation, sun angle striking the module surface, temperature, health and condition, and other issues.

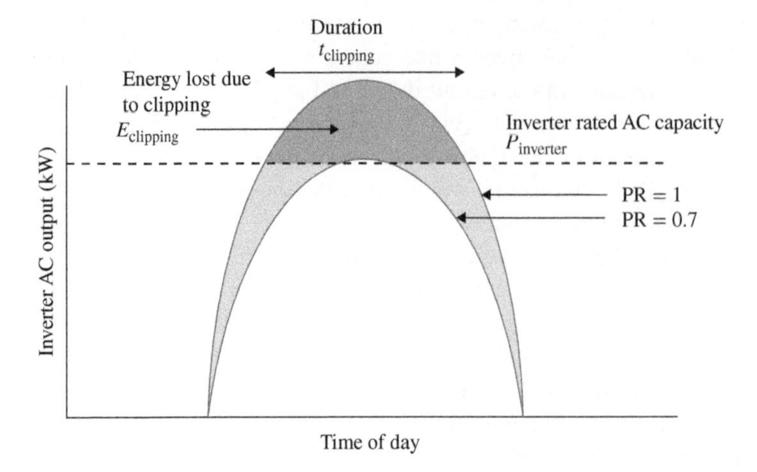

Figure 3.11 Conceptual Diagram of Daily Power Profile. Source: [26]/with permission of Elsevier.

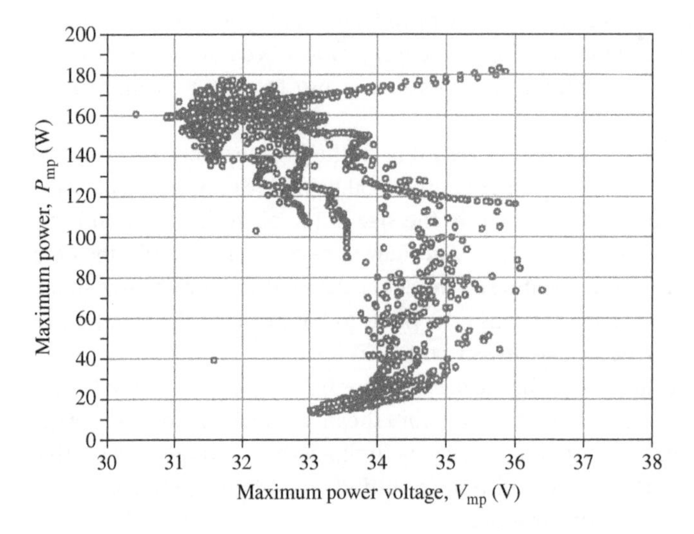

Figure 3.12 Plot of 3300 MPPT Power versus Voltage.

The challenge for specifiers, designers, and many other PV professionals is that until they see the impact of these multidimensional elements as they create voltage and current gyrations, they may have a slightly skewed or incomplete view of how each individual inverter will act as a part of the system as a whole. This is where tweaking the system between specification and design can assist stakeholders in seeing better output and performance throughout the inverter's life. It changes the perspective on how to get the most out of the plant. It also lays a better foundation for product selection and lifecycle results.

The work accomplished by D. L. King, Boyson, and Kratochvil, of Sandia National Laboratories in 2004, based on both the overlay of energy tracked within the MPPT window and the Wasp curve, was a breakthrough for many of us. It allowed us to graphically visualize and determine how to target the MPPT window to squeeze the most out of a PV inverter and system. To improve our design, performance, and output, we targeted system design with the understanding that module and inverter output was going to degrade over time. This allowed us to design such a module output that would pass through the MPTT window without leaving too much energy outside of the inverter window (dashed box), as seen in Figure 3.12. The "Wasp-shaped" scatter plot Figure 3.11 helped us understand how dynamic the energy collection is within that MPPT window. Please keep in mind that there are a number of different MPPT formats based on varying algorithms where the MPPT provides variations to what we see in Figures 3.11 and 3.12. As a result, this set of curves will vary for each different inverter type and environment. For specifiers and designers, the variation in modeling does not necessarily give us a very clear visualization and/or full understanding of what occurs daily or annually, even when all the data has been properly entered into that model. In its early work, the project team must look carefully to fully grasp these issues as they change over time and with multiple levels of system degradation.

For many of us, this was the first look and set of clues dating back to about 2006, into what was happening while addressing the industry-wide question of: **"Where does all the missing energy go?"**

We believe that solar professionals should very carefully begin to learn about these details to squeeze more energy out of a particular specification, design, and installed system. Even squeezing 3–5% more in energy lifecycle production over time is seldom carefully analyzed or understood, thereby leaving/wasting substantial amounts of cashflow and profitability on the table.

Clipping is only part of the mystery. In order to have more energy available in the morning and afternoon hours, the PV strings may have an excess design margin. However, once the modules reach max capability, there is excess energy that the plant is not effectively delivering.

What varies by inverter product and type is the input threshold for voltage and current at low irradiance levels that do not "Turn On" the inverter AC production side of the inverter. This may seem a small portion of "potential energy production"; however, it can be considered.

We also see much of that "lost energy" as being potentially recoverable if it can be drawn off for energy storage in Energy Storage System (ESS) or Battery Storage System (BSS) or possibly through the products in the case of hydrogen for use later. Doing so may avoid converter clipping almost completely with the benefits of longer component life and greater energy production capability throughout the life of the system.

MPPT is an attempt to address that problem and reduce excess loss. Still, MPPTs are not perfect, so there is a small energy loss for a variety of conditions which can add up over time. It is an essential part of addressing some of the energy losses based on product, system design, depending on which MPPT algorithm is specified and implemented, varying environmental conditions, and the defects, and degradation, including a number of other losses. By better understanding all of the losses throughout the lifecycle, decisions can be made to extend product and system life while generating far more energy.

"Figure 3.11 is a scatter plot of over 3300 performance measurements recorded on five different days in January in Albuquerque with both clear sky and cloudy/overcast operating conditions for a 165-W_{mc}-Si module" (King et al. 2004).

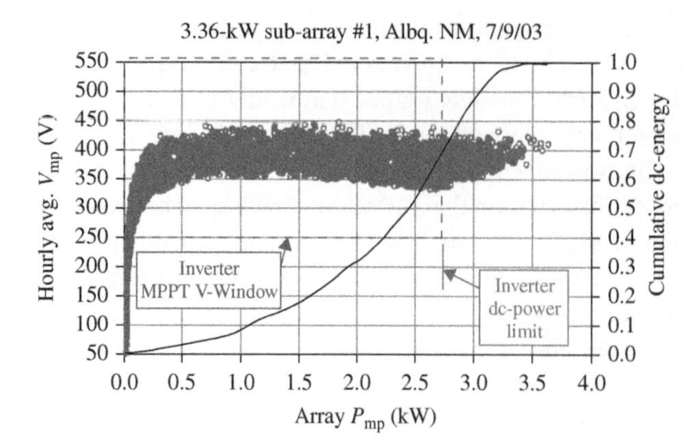

Figure 3.13 Hourly Subarray Volts and Cumulative Annual DC Energy versus Array Power. Source: [27]/with permission of Elsevier.

As you can see above, Maximum Power and Maximum Power voltage are all over the map. These normal system DC power and voltage fluctuations meet NEC requirements in the system tested and they meet the manufacturer's specs. They are based on a system design for Albuquerque New Mexico that show how performance changes, in this instance, in the winter. It is not static and is almost never operating at STC. In designing and building PV systems as energy infrastructure, understanding more about what is taking place within inverters aids in improving design and reducing O&M. While many may look at this as an academic exercise, the reality is that it reveals the mystery of what is taking place within the inverter, what is coming out while allowing specifiers and designers to squeeze more energy, i.e. improve efficiency, out of a system and lowering lifecycle costs.

Yes, the system may require additional inverters, however the cost benefit favors output and profit! It also reduces losses during an inverter total or partial failure.

Now let's plug power into the system and see what happens. The results will be presented as the input as laid over the MPPT window in an inverter (Figure 3.13).

Figure 3.12 is a "Scatter chart of calculated hourly-average performance values for a 3.36-kW array in Albuquerque, NM, over a one-year period. The 'window' superimposed on the plot shows the voltage and power constraints for the inverter used with the system. The fraction of the cumulative annual energy available from the array is also shown as a function of the array power level."

The paper's author states, "The vertical spread in the P_{mp} values is primarily caused by changes in the solar irradiance level, with lesser influences from solar spectrum, module temperature, and solar cell electrical properties. The horizontal spread in the associated V_{mp} values is primarily caused by module temperature, with lesser influences from solar irradiance and solar cell electrical properties."

What we see above is:

- Hourly Voltage at maximum power (X-axis) over array maximum power (Y-axis).
- Any DC power produced above the limit set by the manufacturers to physically or electrically limit to what the input maximum of the inverter is, will be wasted and or underutilized. This protects the inverter but reduces system performance. It may not negatively

impact the inverter in most of these conditions; however, there are inverters that turn some of that excess into heat. Heat and wasted energy equate to poor performance.

- A dashed line indicates the inverter MPPT window or MPPT box for the limits of the inverter converting DC to AC. The data points indicated by hourly readings inside the box are DC power that will be converted to AC energy by the inverter. The points outside of the dashed rectangular box are DC energy that is wasted and or underutilized either because it does not meet the parameters that the MPPT requires or, as in this case, the array is oversized for what the inverter can accept. Please notice that there is a waste on both sides of the MPPT window, the lower left and the far right.

By quantifying excess energy, one strategy for the system design is taking the excess energy off and storing it for later use. Storage can be accomplished through battery storage and/or converting the excess into hydrogen for later use in a fuel cell. This gives the waste additional value by providing higher revenues when sold to the customer at a higher rate.

Note the line based on the Cumulative DC Energy of the system. It indicates that out of 100% of the possible DC input, up to about 70% of the energy makes it out of the inverter to the other side as AC power. Yet, there is a 30% loss of energy going through the inverter. (Depending on component and system design, that 30% may be higher or lower.)

A note on clipping that you may wish to consider:

The idea of losing a little bit of energy for a small part of the day or year seems insignificant and both time and money-consuming. It isn't, it is what people think they are paying for with a professional PV EPC.

It is easy to clip a little, but how much is a little? Many of the designers who have actually grasped these simple concepts made an about-face in PV system design.

It is easy to overlook the details if you convince yourself it is insignificant. Please do not design or install it that way. It will catch up to you and or your customer.

Through our work with the American Hydrogen Association (AHA) in the late 1980s and early 1990s, considerations and lessons regarding hydrogen production have stayed with us over the years. Part of that lesson was that depending on how the electrolysis process was designed, hydrogen can be self-compressing at lower voltages and current. Applying that knowledge could skim off the wasted DC input when voltage and current were outside the MPPT window, as seen in the lower left side of Figure 3.12 above.

Combining that knowledge with what we see on the right side of the graphic, should the additional energy be diverted prior to reaching the inverter, additional energy could be used to electrolyze and store hydrogen for later use. If effectively done, a proper cost analysis may indicate that the benefits come in multiple areas of extending inverter life, maximizing the use of MPPT, and setting aside hydrogen for energy production at a later, more advantageous and profitable state.

While this is not the primary focus of our book, it may be worth the industry's interest to delve into these issues at a much deeper level than has been applied to date.

- This also tells us that even if the inverter operated at 100% efficiency (which it can't), it could not convert the DC energy due to a system with too large of an array or too small of an inverter. The system efficiency is lower as compared to what may be expected.

- Finally, while an inverter may be rated with a 95–98% efficiency as tested in the lab, it is not necessarily reflected in the field. That waste plus output that we see above is further impacted by the specification, design, and environmental considerations of that particular plant.

Today, as designers use software to make performance decisions, they seldom have the opportunity to consider this form of MPPT window graphic and scatter plot as it would appear for their specific site on an annual basis. Nor do they tend to have the opportunity to understand exactly what the boundaries are of the MPPT window for a particular DC/AC configuration to consider options over the system lifecycle, especially as system components degrade. In our previous work, when we could determine what these boundaries were, it allowed us to design more effectively to squeeze every AC watt out of the DC side of the system that the design could handle.

For those inverters that we used, where the manufacturer provided the information, we could make decisions that are now automatically made with software. Because, in many instances, this is just a matter of data input, there is a common assumption that software will allow designers *to get the most out of a system design*. However, without the additional considerations for parameters not contained within the software models, that design may not optimize energy production over time. If designers look at the system based on general assumptions about IV curves with maximum and minimum voltage and current, they may be missing some opportunities to squeeze 5 or 10% more energy out of a system array. It is through taking actions like this that we consider the results as true PV system infrastructure optimization.

Where this becomes incredibly important is when energy storage is applied, and excess DC power may be drawn off prior to reaching the inverter that previously was wasted energy. This means that one can design a system to extract the absolute most energy while matching under loading, not overloading the inverter to force a clipping situation. Today, inverter DC input may be at 120 to 150% (DC/AC ratio of 1.2:1 to 1.5:1) or more, whereas in the past, we underloaded inverters at 80–85% to allow the inverter to run more effectively, under less stress, more efficiently thereby extending its life when properly maintained. This approach developed more energy for use especially in extreme hot and cold conditions. However, as we learn more about the module and other systemic defects, degradation and the need for better and more detailed specifications, there needs to be a better balance between the issues of inverter performance and the need to provide handling of the multidimensional variables and component selection.

Below are some of the considerations that designers and specifiers of tomorrow may wish to consider very carefully as they choose components and ask questions of the manufacturers.

1. The system meets code and manufacture requirements, but it is not performing well, or as well as it could be. Why?
2. The array was oversized for the inverter. Or, as we discussed, the inverter was undersized for the system.
3. The software for sizing strings allowed this configuration to take place. As a result, the design meets the NEC Code. While system performance sizing software has become very sophisticated, it may not effectively address the effect of temperature extremes or

variability on sizing in addition to actual derating to meet performance expectations. Specifiers and designers must account for the actual site's extreme temperatures, whereas the Code is designed for safety, not performance. When designing systems properly, you will exceed the code and maximize performance. Note: Please read the introduction to the NEC Code. (Chapter 2) if you have any questions about that point.

4. In time the PV system modeling will indicate the waste condition, but at this time, most models do not consider all the details of the MPPT window. This is possibly because each inverter has a separate set of windows, or as we graphically introduced, the above Wasp curve (Figure 3.12) plot and MPPT window. However, there is still the issue of how those models hold up as different segments of the system degrade. These are questions that can be partially answered through a SE and RAMS process. In time, there is a high probability that the software that is being developed today for sizing systems will be able to include a more accurate view of actual performance over time once sufficient data has been made available to the industry.
5. The designer and the plant delivery chain:
   - If inexperienced or insufficiently experienced, the designer may simply go through the motion of string sizing and design. Often, we find that the inputs for the modeling software are standardized, which creates an individual site and system problem. Data are inaccurately applied and/or there is insufficient data to make an accurate selection.
   - A more experienced designer may think, "Yes we will be clipping a little energy but it is not that much so let's not worry about it." However, when you look at the long-term operation of the inverter and the total amount of energy that could have been produced, their existing process may not allow for addressing what are often under-evaluated inputs and selections of inputs. We believe an effective cost analysis, which uses more than one set of software formats for comparison and more clearly understood inputs with a variety of different choices, may be made in the specific design. This also requires that the actual site environmental conditions including the extremes are adequately addressed in the analysis.
   - The integrator owner may have thought, "We have to lower the cost of the system. It will cost less for an undersized inverter with the same string lengths. The Code and the manufacturer let me do it, so let's get the job done and get a check cashed before payroll comes due."
   - The inspector is responsible for safety and primarily focuses on the established and accepted AHJ-approved code version. As much as the inspector may care about wanting more productive plants, the focus is system safety and meeting or exceeding code. They generally have no jurisdiction over performance. At the same time, experienced inspectors could be consulted to reduce what they see as challenges. In an era when climate and PV system output results have a critical relationship to the future of the AHJ and the entire community, it would be worth considering looking at setting AHJ requirements to meet RAMS attributes by having energy available when needed.
   - The system owner thinks they have a great least-cost deal, until they realize how the system underperforms in the out years and will stay that way; therefore, they are likely to be unprofitable and unsatisfied.

- The system owner has been told that PV is a commodity, and this has translated into the false assumption that equipment, design, and labor are all the same, "So I will just buy the cheapest system to get the best deal." If you believe that before reading this, now you know better and have no real excuse for designing low-performance subprime PV systems. It is not that difficult to do much better as a designer or an integrator.

There is no effective lifecycle design or install standards in place today, so the natural default is to Code that is a safety standard or below.

3.6 Equipment Removal, Disposal, and Recycling

For decades there has been a myth that when the PV plant comes to its end of life, the cost of equipment removal and site restoration will be covered by the recycle value of the equipment being removed. That myth has resulted in zeroing out the cost of equipment removal, disposal, or recycling for the purposes of LCOE. It also assumes an unfounded value for the equipment to be removed. If one just considers for a moment the amount of equipment and labor that it takes to remove a substantial-sized system, the fallacy would be apparent. However, today we still have an industry that believes in this myth, which is one of the reasons we stay with our position that *no one has ever accurately identified or calculated the true LCOE for a project*.

Consider for a moment, if you will, that in the removal of equipment, just the equipment requires tearing out of modules and racking, digging up buried conduit and cable, breaking up concrete slabs, and removing concrete for footers if they are used or pads for inverters, removing all of the extraneous equipment and hauling it away. There is a substantial amount of labor involved and, of course, the use of heavy equipment.

Then, you have to haul it away and pay to have it landfilled after taking the bits and pieces that can be recycled. Also, consider that some of the materials will be treated as toxic or hazardous waste, which adds to the cost. That is today!

So how do we handle the equipment removal and disposal. Surely, most of it in time will be recyclable, some will be reusable, and some will be landfilled.

3.6.1 Equipment Removal

If we speak of removing modules and inverters, they will need to be disconnected, taken down, properly handled, evaluated, batched, sorted, packaged, and shipped for what can be used or repaired for spares and recycled. This will require the development and application of clearly defined standards, processes, procedures, and practices for the dismantling, sorting, inspecting and/or testing, packaging where necessary, and appropriate handling till any recycled or reusable product is reinstalled.

3.6.2 Equipment Disposal

The issues of equipment disposal are very complicated on several fronts.

PV components are made from valuable materials, some of which are rare and expensive. They cover the gamut from modules, inverters, combiners, racking, etc., and construction materials. Today, for the most part, they end up in landfills whether they are toxic, presently unusable, or recyclable.

While the arguments for disposal always focus on short-term fiscal discussions, there is far more to it. The results and real global cost for energy, the delivered LCOE, and the actual energy available will be determined prior to the end of this decade. Various options for businesses are to do full recycling, recycle certain materials (metal frames, copper wiring, etc.), or just take everything to the dump. How policy and business interests plan and act in the next few years can potentially deliver very positive or negative scenario results (profit and environmental). Those actions, or lack thereof, will also be seen in global environment results and in the real cost and availability of energy including who will have access to that energy.

All of this creates a dilemma:

Figure 3.14 is a projection of the growth of PV from 2000 to 2050, which may be understated based on more serious responses to climate change and the growth in EV technology. This growth is driven by "green energy" demands but has the side effect of creating significant waste in landfills. This shift toward dramatically reducing fossil fuels, e.g. replacing it with PV power generation, creates a bit of an energy and environmental conundrum. We want dramatic environmental improvement; however, if it is done improperly, it creates additional problems and expenses beyond what exists today. As previously noted, several dopants are used in producing solar cells, and some are harmful to the environment.

Yet, while much of the discussion talks about capacity growth, much of the industry is assuming 20- to 25-year lifecycles for these systems and products. This is especially true for PV modules. Yet, the data has been telling us for over a decade the reality that useful product life and acceptable performance can vary dramatically from expectations. Unfortunately, the global results have been a negative variance in a relatively short period of time, while the trend continues the same.

As a result, the graphic above and most projections that follow do not indicate what will actually be operable in, let's say, 2025. We discuss this in greater detail throughout this book.

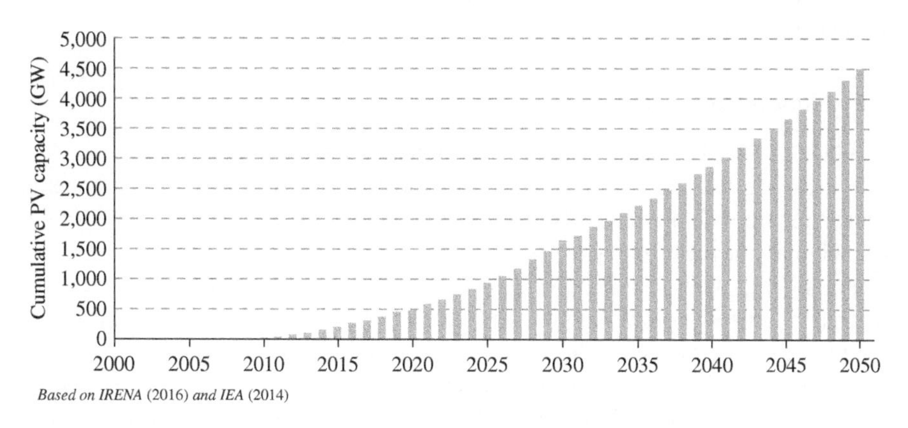

Figure 3.14 Expected Cumulative PV Capacity 2000–2050 (IRENA 2016).

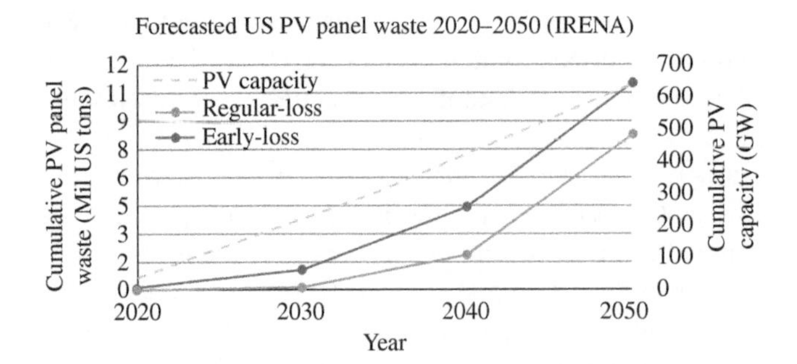

Figure 3.15 Forecasted International PV Panel Waste 2020–2050.

Hopefully, this will stimulate vigorous industry conversation, addressing what is presently being practiced as todays PV project delivery is ephemeral. Between issues of design, specification, product quality, and handling, we combined with a lack of shared data and effective operations and maintenance of their own story. The amount of energy projected and the reality of what will be available is substantively different. In short, what we think we will have in 2050 in the industry, as practiced, *could be dramatically less* than what is projected to operate and be available, providing undefined shortfalls.

Please consider this carefully when reviewing, Figure 3.15 "Forecasted International PV Panel Waste 2020–2050."

If under-addressed, this set of conditions, the early failure, repair, and disposal of a number of types of PV plant components, will have an undesired and unanticipated impact. The results could include:

- A premature global loss of trillions of dollars of products;
- Loss and replacement of components from early defects, faults, failure, and underperformance;
- An energy shortfall of massive essential electrical generation potentially in the thousands of gigawatts.

PV recycling will be critical in meeting future energy needs. Yet, it is only one crucial element. The decisions to be made or not made soon will define how we live, the quality of that life, and our physical and fiscal living conditions.

3.6.3 PV Module Recycling

3.6.3.1 What Is PV Module Recycling, and Why Is It Important?

PV recycling is the gathering and processing of PV modules into their discrete components to be reused to manufacture modules and other products. In other words, the recycling process is a remanufacturing process. The USA has more than 70 gigawatts installed today, and the number is growing. More than 140,000 PV modules are installed in the USA every day, and that volume will grow. All products have a useful life, and at some point, those products will experience an end-of-life event, and solar is no different. Solar energy is sold for its positive economics and its perceived greenness and sustainability. As a responsible industry, if we choose to merely bury our solar waste in our nation's landfills, then we lose in two ways.

1. By burying our solar waste, that practice is clearly not sustainable. It is also an expensive process that does not fully take advantage of those system elements that can be recovered, remanufactured, and used again for future energy infrastructure.
2. The materials used in manufacturing PV modules (like silver, copper, and silicon) have a value and can be mined directly from the decommissioned PV modules. Today, we are not quite sure what the value is, and that is one of the unknowns that really slows the process of determining what a true LCOE is or could be.

3.6.3.2 What Is the Present State and Need for PV Module Recycling?

In Europe, more than 90% of the PV modules that are decommissioned today are recycled. Because of the Waste from Electrical and Electronic Equipment (WEEE) Directive in Europe, PV modules are required to be properly handled at the end of their useful life. WEEE legislative requirements have created a condition for PV module manufactures and stakeholders to come together and cooperate to achieve common goals of collection and processing responsibly at the lowest possible costs. PV Cycle, a nonprofit, has been working since 2008 to organize the European solar industry, working on policies, regulations, coordinating, collection, and working with recycling industry partners such as Veolia to responsibly and economically process solar waste throughout Europe.

In contrast to Europe, in the USA, less than 10% of PV modules are currently being recycled. Because there are no national standards requiring solar panel recycling, nor a coordinated effort to solve the growing waste stream, various states and local jurisdictions are just now beginning to address the problem of discarded solar modules. The US EPA (Environmental Protection Agency) https://www.epa.gov/hw/universal-waste, along with the DTSC (Department of Toxic Substances California) https://www.environmentallawandpolicy.com/2020/10/california-classifies-solar-panels-as-universal-waste/, have issued guidelines and regulations calling out for the proper handling and disposal of solar modules (https://dtsc.ca.gov/solar-panel-faqs/).

Although the PV cells make up the majority of the parts in a module, the hazardous elements are primarily solders (Tin-Lead Solder if used), some plastics, and the dopants (mainly arsenic, indium, gallium, boron, and phosphorus) used to build the cells. The cell dopants and other materials actually make up much less than 5% of the cell, and the other 95% is silicon (glass). The cells themselves will generally have copper, silver, or aluminum metals, and aluminum is used primarily for the module frame.

Many states are in the process of drafting new legislation, recognizing the concern and responsible actions necessary from PV system owners, EPC contractors, and others. For example, North Carolina no longer allows PV modules to be dumped in state landfills; however, now, those decommissioned solar modules from North Carolina are being transported to other states that still allow landfilling of solar modules. Washington State has passed regulations that require any entity that imports solar modules into that state, must provide a stewardship plan to properly handle those imported modules at the end of their useful life, at no cost to the end user.

3.6.3.3 Who Ensures that Modules Are Recycled and/or Disposed of Appropriately?

In Europe, PV module disposal is regulated by Waste from Electrical and Electronic Equipment (WEEE) legislation and has buy-in from PV module manufacturers. It is also supported throughout the entire value chain, from installation to end users. The USA and much of the rest of the world have very few requirements for end-of-life processing. It is noted that when there are few regulations and the responsibility is totally voluntary, most end-of-life solar is either dumped into landfills (legally or illegally) or shipped to developing countries that have little to no environmental protection laws.

What does the future PV module and recycling need to look like over the next two or three decades?

3.6.3.4 What Are the Costs Involved, and What Is Keeping Those Costs so High?

The IRENA report (Figure 3.14) shows that in the International, in 2020, there was 1 GW of PV modules being scheduled for decommissioning. Because there is no national requirement to sustainably collect and process PV modules, most solar modules end up in landfills across the nation at an average cost of less than US$1 for each solar module. Because of the low volume of solar being properly collected and recycled, the current cost is between US$15 and US$25 to process each PV module. While the price is high and the lack of regulations keeps the price high, and the lack of subsidies to kick-start a national solar PV recycling program are nonexistent, the price continues to stay high unless the bulk of decommissioned PV modules is sustainably processed. With the greater volume to recycle PV modules, along with the development of innovative technologies and processes, the costs of recycling would then drop to be in line with what has already been proven in Europe. By following Europe's example, (less than US$0.70 per module) it would be less costly to recycle than to landfill (US$1 per PV module) PV modules.

Worldwide, e-waste disposal is a major problem. Many consumer electronics/products today have a recycling fee charged to end users at the time of purchase. Example: car tires US$2, flat screen TV US$11, car batteries US$15, bottles/cans US$0.05, and mattresses US$11. Many of these programs have successfully supported the collection, recycling, reuse, and "**circular economy**" concept. Today, 98% of car batteries are recycled and turned into new batteries, keeping millions of tons of lead out of landfills and groundwater. If solar modules were included in a subsidy program (US$0.013/W) like many other consumer products, this would easily support national collection and processing, and the subsidy could be lowered over time.

In addition, as new international standards become universally required and applied, the growth in recycling and reuse of those products will address the issues at clearly defined costs. The authors believe that a variety of standards must be universal requirements for applying and receiving tax incentives or other benefits (deductions, credits, or Feed In Tariffs [FITs]), permitting, and inspections to ensure those standards are being met throughout the entire PV system lifecycle. Beyond the appropriate environmental thing to do, it is good business and, in time, will be very profitable.

3.6.3.5 What Are the Long-Term Impacts of Not Resolving Module Recycling Issues?

If we do not have better ways and programs to recycle PV modules, then the prices will continue to stay high for recycling. Larger PV projects are beginning to face the rising costs of bonding and insurance over 20–30 years. Bonding to ensure solar plants are properly decommissioned can be severely affected by the high recycling costs. Insurance costs have risen substantially and dramatically as weather-related incidents have increased, and the high costs to dispose of modules is on the rise.

Modules contain both valuable and toxic materials. By landfilling solar, we lose both the good components such as silver, copper, aluminum, and silicon, but also we risk damaging our fragile water tables with lead, cadmium, and other heavy metals.

Bibliography

Balfour, J.R. and Morris, R.W. (2018). *Pre-emptive analytical maintenance (PAM), introducing a more functional and reliable PV system delivery model – a reliability precursor report.*

Chung D, Balfour J.R., 2016, *SEPA Handbook Resource Guide Utility Solar Asset Management and Operations and Maintenance*, SEPA 1220 19th Street, NW, Suite 800, Washington DC, 20036.

Jordan, D.C. and Kurtz, S.R. (2016). *Photovoltaic degradation rates — an analytical review.* https://www.nrel.gov/docs/fy12osti/51664.pdf (March 2023).

Jordan, D.C., Anderson, K., Perry, K., et al., 2022. *Photovoltaic fleet degradation insights*, First published: 24 April 2022. https://doi.org/10.1002/pip.3566.

Kim, J, Rabelo, M, Padi, S.P., et al., (2021). *A review of the degradation of photovoltaic modules for life expectancy, Energies*, 14, 4278. https://doi.org/10.3390/en14144278; https://www.mdpi.com/journal/energies.

King, D.L., Boyson, W.E., and Kratochvil, J.A. (2004). *Photovoltaic Array Performance Model.* Sandia National Laboratories.

TamizhMani, G., (2016). Risk priority number (RPN) for warranty claims, quantifying risk for every defect. *PV Module Reliability Workshop–Lakewood*, CO-22 February 2016.

Notes

1. Henry Hieslmair, *NREL Virtual 2021 PV Reliability Workshop*, https://www.youtube.com/watch?v=M7BHcxxugwY, DNV.
2. Ibid.
3. "2021 PV Module Quality Report," PI Berlin: Trusted Solar Advisors.
4. "Bigger is not always better," PV-Magazine, AUGUST 28, 2020, EMILIANO BELLINI.
5. Insulated-Gate Bipolar Transistor (IGBT) is a three-terminal power semiconductor device. Current devices have Very High operating voltages (>1.2 KV) and High current capacity (>800 A), which makes them ideal for inverter applications. https://www.mitsubishielectric.com/semiconductors/php/oSearch.php?FOLDER=/product/powermodule/igbt.
6. 2021 PV Module Quality Report, PI Berlin: Trusted Solar Advisors.
7. Ibid.
8. Ibid.
9. Ibid.
10. https://www.cnbc.com/2019/08/20/walmart-sues-tesla-over-solar-panel-fires-at-seven-stores.html
11. Ibid.
12. Ibid.
13. https://en.wikipedia.org/wiki/Engineering_procurement_and_construction

14 Functions include: (as per PAM specification prior to EPC bidding). Note that this presumes a prebid system specification.

15 Ibid.

16 Ibid.

17 Ponce-Alcántara, S., Connolly, J.P., Sánchez, G., Míguez, J.M., Hoffmann, V. (2014). Ramón Ordás Energy Procedia 55: 578–588, *4th International Conference on Silicon Photovoltaics*, Silicon PV.

18 https://www.nrel.gov/pv/module-efficiency.html

19 2021 PV Module Quality Report, PI Berlin: Trusted Solar Advisors

20 However, the same material provided by a different supplier or manufacturing location is not necessarily a different material as long as it can be clearly demonstrated.

21 Jordan et al. 2016, *Compendium of Photovoltaic Degradation Rates*, Progress in PV 2016.

22 Confirm Solar Buyer of NREL (Why does it say NREL?)

23 Kopecek, R. and Libal, J. (2018). Towards large-scale deployment of bifacial photovoltaics. *Nature Energy* 3 (6): 443–446.

24 The MPPT circuit forces the solar inverter input to work at a specified voltage by varying the resistance of the inverter input using power electronics. The higher the resistance, the higher the voltage across the solar panel. This also results in heat/energy loss and thus affects the overall efficiency of inverter.

25 Design Margin - the design headroom (difference between component rating and applied stress) or safety factor on a specific design characteristic or design requirement (design input). Design margin reduces stress on the components thereby increasing the life and reliability of the item.

26 Balfour, J.R., Hill, R., Walker, A., et al., 2021. *Masking of photovoltaic system performance problems by inverter clipping and other design and operational practices*. United States: N. p. https://doi.org/10.1016/j.rser.2021.111067.

27 Ibid.

4

SE/Repowering™ Planning Process

Key Chapter Points

- What is the **SE/R**epowering™ planning model and why is it critical to delivering PV plants as energy infrastructure?
- Why and how is **SE/R**epowering™ of Critical PV system delivery a lifecycle cost control and risk reduction tool?
- How do we address the many existing PV industry repowering myths and assumptions?
- What **SE/R**epowering™ process steps are essential to industry stability and what is its value?
- What does the process **SE/R**epowering™ as a system deliver model, its planning, and delivery consist of?

Key Chapter Impacts

- Understanding the four types of PV Plant **SE/R**epowering™
- Defining the critical planning elements for PV **SE/R**epowering™ planning and system delivery
- **SE/R**epowering™ economics impacts on Commercial & Industrial (C&I) and Utility PV
- Understanding and resolving environmental and process "limiting factors" to build and operate systems more effectively while delivering a lower lifecycle cost of energy (LCOE)

4.1 Introduction

One of the hallmarks of any industry's maturity is when it has mastered and demonstrated the ability to provide effective lifecycle planning, specification, product selection, and system delivery. To achieve this hallmark requires an industry to address all levels of specification, engineering, design, procurement, construction, benchmarking, commissioning, O&M (Chapter 11), site restoration, and the points in time and resources needed for **SE/R**epowering™. Lifecycle planning and specification determine the Capital Expenditures (CAPEX) – the cost to deliver the plant; and Operating Expenses (OPEX) – the cost to operate, maintain, and complete site restoration. OPEX includes lifecycle plant costs for

Photovoltaic (PV) System Delivery as Reliable Energy Infrastructure, First Edition.
John R. Balfour and Russell W. Morris.

operation and maintenance, from commercial operation date (COD) up to five decades or more of operation, and site restoration.

4.1.1 Why SE/Repowering™?

The short answer is about lifecycle energy production gained or lost, O&M, and initial system versus project delivery. There currently exist many plants in various stages of aging, derating, cumulative failure effects, and the cumulative effects of system degradation that will require a plant to review their plans and data for upgrades, repairs, and/or today's version of repowering. For the lowest LCOE, all of these require that a substantial effort be put into applying the principles of SE found in Chapter 5 to identify the plant's health, condition, and status, as you develop the **SE/R**epowering™ plan and its implementation. For **SE/R**epowering™, plans and processes include the actions taken to improve plant energy production (better quality and/or increased power) over the life of the plant. Today's processes do not collect sufficient reliability (TamizhMani, 2014), O&M, and maintainability data or integrate it to make effective decisions. Making any changes to the plan must be based on documented, verified, and validated information that is acquired over time. If there is sufficient detailed data collection and analysis, it will drive improved cost control throughout the plant lifecycle. This gives owners and operators the flexibility to determine when hooks and triggers are identified to address cost-effective plant **SE/R**epowering™ actions based on plant health, condition, status, and customer demands. It addresses often invisible component and technical challenges that exist.

This **SE/R**epowering™ process chapter's focus is on identifying and planning for issues regarding **SE/R**epowering™ a PV system and the considerations that must be addressed to achieve 25–100 years of successful energy delivery results. To date, we see the existing PV industry paradigm as incomplete and inconsistent with the long-term goals and objectives for a mature and profitable power generation industry, comparable to hydro and nuclear power plants. As we have demonstrated in Chapter 3, we see the PV industry is beginning the process of shifting from **PV as a Speculative – High-Risk Industry**, toward becoming an **Integrated PV System as Energy Infrastructure with Storage Industry**.

Changing the paradigm from "speculative PV" practices to one built as "energy infrastructure" through the development of an effective **SE/R**epowering™ plan extends the useful life to 50 years or beyond, delivering more reliable and profitable systems. Thus, it effectively reduces lifecycle costs and risks, while focusing on making 'all' project stakeholders financially whole.

Presuming that one has the appropriate entitlements, which may vary from project to project, and funding, etc., **SE/R**epowering™ provides the ability to upgrade the plant without a significant unplanned reinvestment in the infrastructure.

This book is focused on addressing uncomfortable, complex discussion topics of the attributes and realities of today's repowering. Until those complexities have been adequately identified and addressed with effective **SE/R**epowering™ processes, and structured lifecycle planning, our industry will continue to get what we have today. That is PV projects with substantively disruptive and underperforming financial consequences.

As a result of our efforts, we have redefined today's usage of "repowering" toward a more functional lifecycle set of **SE/R**epowering™ plans, processes, requirements, practices, and implementation.

What is "**SE/R**epowering™" as compared to today's version of "repowering" activities?

SE/Repowering™ is an energy infrastructure (with or without storage) is a lifecycle planning, implementation, management, delivery, and site restoration process. It relies on systems engineering, reliability, availability, maintainability (including testability), and safety (RAMS) practices and attributes. The process is designed to address as many issues as possible early on at the concept stage and resolve them to control all costs for both CAPEX and OPEX. This also means delivering a more accurate LCOE. An effective **SE/R**epowering™ approach extends the useful life from 25 to 100 years or beyond, making systems more rebuildable, reliable, and profitable. Having the capability to rebuild over the systems lifecycle is a complete paradigm shift from historic practices.

Current PV projects are often marketed as having a lifecycle of 25–30 years with appropriate derating (Chapter 6). Yet, with present components available, marketing and business practices, project delivery is directed toward the initial investment first cost/least cost while sacrificing RAMS. These practices continue to cast a negative impact on CAPEX and OPEX, which directly undermine Life Cycle Cost (LCC) and LCOE. LCOE goes up or down based on real costs, i.e. true LCC, not projected numbers based on significant and often unsupported assumptions and models without verification or validation. This continues to be problematic and expensive. As a result, throughout this publication, we have been presenting an improved model for delivering highly functional and cost-effective PV as infrastructure systems with or without storage.

Mature industries incorporate business critical links which include lifecycle processes put in place and carefully managed from project inception. The most critical of these processes is the "lifecycle plan." The plan should contain specific and realizable actions that are structured, yet flexible and responsive with triggers to changes that can occur over the course of the plant's life.

Current PV plants are designed and built, often based on marketing "myths" and "assumptions" (Chapter 1), of long-term costs or component capability. These decisions continue to be made with insufficient data (Hill et al. 2015), while primarily focusing on minimizing the cost to design, build, and operate the plant as it is the primary level to illustrate a lower LCOE. By using this "least cost" process with limited O&M planning, which includes many contractual exclusions, the projects result in plants with little or no adaptability and reduced long-term value. This predicament is true whether O&M is practiced internally by the owners or subbed out on a contractual basis.

PV as an industry must and can and will become a mature industry. The industry began in the 1960s and now has 60 years of growth. Unfortunately, while PV technology has been improving, many lessons learned for PV plants seem not to have been effectively learned, analyzed, documented, or shared. It is essential that this cycle is changed for industrywide economic stability. For the PV industry to mature, it must address substantively improved systems engineering and O&M specifications, costs, data, and modifications. Substantive progress takes place when planned at concept, implemented, and carried through to project completion (benchmarking and commissioning) on to the end of plant life, and site restoration details (Chapter 3). This process results from developing a complete **SE/R**epowering™ plan to extend system life, energy production, and asset value.

With the PAM model approach to PV, we refer to it as a "**SE/R**epowering™ planning, implementation, management, and delivery process." It includes broad considerations of "all" stakeholders, structured elements, and comprehensive planning that the owners or owner-operator tracks over the potential 50-plus years of life of a PV plant infrastructure.

Its value is derived from the flexibility to undertake systematic updates and modest-to-moderate course corrections over the entire lifecycle of the system. Success is derived from preparation and carefully timed actions with managed changes, as compared to the present PV project approach of persistently responding to major (and minor) events as they occur. The current PV project delivery approach has been demonstrated to be reactive and not effectively preemptive. Reactive is generally meant to be that the existing approaches result in the unplanned and misapplication of resources as a response by management and/or engineering to a problem in the plant. Had management applied well-known RAMS/SE methods and processes, a series of problems would have been preempted, eliminated, or ameliorated at the most cost-effective point in the lifecycle. The reactive approach costs unplanned funding of underaddressed plant and engineering management, development of corrective actions, testing, and later costs of performing increased O&M and upgrades to the plant. These upgrades include not only hardware, but software.

PAM requires far greater focus on implementation of upfront specification and reliability and maintainability practices. This method calls for far greater accuracy in **SE/R**epowering™ effort, timing, more reliable component specification and installation, industrywide training, application practices, and operations skills, among others. It does not use or apply "marketing assumptions"! It is based on detailed data and the factual analysis of existing systems. These are derived from science and engineering processes requiring well-substantiated data to ensure flexibility and deliver effective choices.

In short, effective plant **SE/R**epowering™ requires the additional steps of assessing all stakeholder's long-term[1] wants, needs, and commitments, to address key elements of system delivery. At the same time, we acknowledge that some industry professionals have begun to make real though limited progress. Nevertheless, the industry as a whole has only begun to realize much less address these critical issues.

The challenge for the industry is to determine which is more effective and valuable:

- Bowing to fiscally destructive and counterproductive first cost pressure, or
- Lowering actual lifecycle costs of energy with greater infrastructure life and stability.

We propose focusing on the actual cost of PV energy over time and its potential inherent stability. We note that failing to address the former has resulted in significant financial losses and frustration, while the latter is based on the establishing of the actual cost of energy over time for cost stability.

The long-term outcome will actually be a lower cost for energy at the wholesale and retail levels.

4.2 What Is the SE/Repowering™ Process?

SE/Repowering™ is a systemic mission critical path, planning, monitoring, and implementation process initiated, preferably beginning at system concept. It includes technology investigation, specification *prior to EPC bidding and contract*, enhanced data collection, analysis, engineering, and decision-making. It then evaluates all possible essential changes

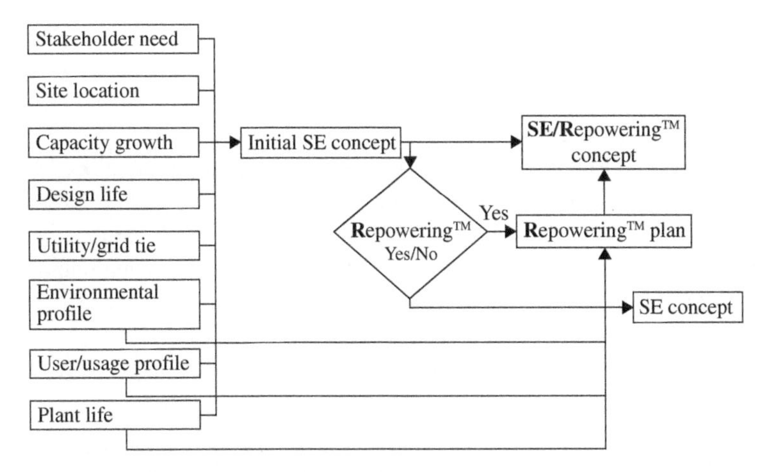

Figure 4.1 Overview of **SE/R**epowering™.

to the plants during their lifecycles. **SE/R**epowering™ is a causal process. It addresses reliability and serviceability as a functional **SE/R**epowering™ set of tools providing optimal design supported extended lifecycles. It goes beyond today's project planning, Table 4.1, it is more thorough while addressing all potential and plausible issues that impact system life and its ability to meet its required mission. However, it is important to remember **SE/R**epowering™ requires accurate information based on serious, precise reliability, and systems engineering study (Figure 4.1).

Simple or massive replacement of parts without addressing a functional and long-term, planned improvement program to correct lifecycle plant health and condition, is not "SE/Repowering™" and should not be considered as such. Doing so, otherwise, is the result of critically poor ownership, planning of financial, engineering, management skills, practices, and oversight.

SE/Repowering™, when effectively implemented, results in extending the productive lifecycle of the system, reducing LCOE, thereby improving the economics of the PV plant. It is accomplished by focusing on the following at minimum:

- The reduction and/or elimination of weaknesses in existing PV project delivery stratagems resulting in unanticipated challenges.
- A defined and agreed upon systemic process to collect, secure, curate,[2] and analyze essential data to clearly delineate system long-term lifecycle needs based on better more reliable component selection.
- Improved component specification and design to deliver essential criteria-based needs. (As an Example: For a given combination of PV Inverters and Modules, understand degradation and derating thresholds for both as they delivery negative energy (i.e. losses) and fiscal impacts. This includes module capability and inverter operating voltage/current changes, and at which points they require replacement, service, or supplementation to achieve original owner, PPA, and future owner wants, needs and expectations. Or vice versa, given expected module or inverter degradation and defects,

what inverter specifications should be in place to handle production as expected, plus some margin of error.)
- Complete all project lifecycle planning, with operational integration, financial viability, analysis, and profitability projections.
- Address the component and plant availability impact of continuing often excessive changes in technology and that of the operational plant (IEC TS 63265: 2022). Photovoltaic power systems – Reliability practices for operation.
- Ensure proper specification of components, by identifying weaknesses in components selected primarily on least cost only, while adequately assessing reliability, availability, and future labor replacement lifecycle costs.
- Strategically shift from a 25- to 30-year deliverable operational life to a 50-plus years of lifecycle.

(Note: At this point in PV industry history, we do not believe that many if any existing plants will actually operate "as expected or projected," for even 20 years much less 25–50 as present approaches are applied to infrastructure.)

- Consider consequences of not looking at a broader horizon for an integrated PV approach with or without storage as energy infrastructure for at least a 50-year lifecycle.
- Assure management does not allow the shifting risk and cost while choosing to leave essentially critical details or decisions to later stakeholders. For example, the common phrase, "Just leave it (i.e. unresolved issues) for the O&M guys/experts to solve!"

SE/Repowering™ planning and system delivery is a considerable undertaking. As such, there are 10 initial questions that should be addressed prior to submitting the plant specification and **SE/R**epowering™ plan to EPC bidding. Today, the primary focus is on "PV project delivery" as compared to "PV system delivery."

That approach is nominally based on "marketing" information, with a plant delivery targeted on a 25- to 30-year life that does not account for all necessary systems attributes. For effective PV **SE/R**epowering™, the reliable energy infrastructure approach is to address ongoing inclusion with specific planning for maintainability and serviceability. It includes the ability to adjust the timetable to rebuild specific sections or subsections of the plant and eventual site restoration. The **SE/R**epowering™ process develops clearly defined "triggers" or "hooks" based on curated, shared data and analysis that may result in planned changes earlier or later than initially projected. These triggers and hooks are delineated by data and RAMS as a well-defined range of condition/s not only specific set time periods. This gives the stakeholders a *template* with options that are not available today. This vision is accomplished at system specification to address the entire lifecycle along with issues that will arise during each phase of the plant life.

System long-term considerations include:

1. What are the defined asset health, condition, performance, and revenue challenges, considerations, and metrics?
2. If the value is in long-term higher revenues, are there provisions in the planning process to accommodate one or more major **SE/R**epowering™ phased efforts to improve long-term profits and reduce risks, which also improves total commercial asset value?
3. What are the long-term financial realities, accounting rules, elements, changes, and how will they be addressed? These include O&M, financing, or O&M contracts.

4. What are the landowner and property issues including extended leases, taxes and tax benefits or liabilities, property ownership, potential for change, and how will they be resolved?
5. Have land use plans for adjacent properties been accounted for the SE lifecycle plan?
6. What are the code and standards trends, specifications, and requirements for the project independent engineer (IE) and/or project engineer (PE) to address and sign off on, especially for component and design choices that are not designed for 20–30 years?
7. What is the potential and process for modifications to PPAs that may be determined, negotiated, and/or modified over time?
8. How does a buyer or seller determine what the true asset value is for the plant at any specific point in its life? The timeframe includes stages during planning, financing, specification, and commissioning or COD, at years 1, 5, 10, 20, 30, and beyond, or at any time the plant is sold, refinanced, or reinsured?
9. Have the stakeholders "properly budgeted" for and secured an experienced IE or PE that is fully capable of delivering a functionally accurate evaluation of the completed system at its benchmarking and commissioning? … and are they funding a full evaluation? (Unfortunately, many planned evaluations are carried out with limits on time and budget which result in incomplete inspections, analysis, and results.)
10. How are long-term revenues and return on investment defined, identified, and maximized?

Considerations for **SE/R**epowering™ lifecycle financials include but are not limited to the following:

1. Mean time between unscheduled maintenance (MTBUSM)
2. Mean time between scheduled maintenance (MTBSM)
3. Mean time to repair (MTTR)
4. Initial development cost
5. Maintenance repair cost
6. Internal rate of return
7. Preemptive and preventive maintenance costs
8. Spares and storage costs
9. Overhead charges known and potentially inherent in the system design life
10. Depreciation
11. Salvage value and risks
12. Life of asset and asset phases
13. Cost of parasitic power to repair (for non-daytime maintenance and operation)
14. "All" other organizational costs that are generally not defined or considered.

It is critical to consider that when adding energy storage to a project (see Chapter 9), addressing these finance and cost issues become far more urgent. Whether storage is included in the initial first-phase planning or not, it is generally best to evaluate and consider a range of options for the addition of storage at a later date during the initial system planning. This is in part because masking (Balfour, et al. 2021) issues will be far more visible early on in plant life and operations when the option details are considered and evaluated. This is especially critical when potential energy production outside of the clipping range is harvested. It will result in potentially greater levels of energy collected, while exposing many weaknesses within the system when including later storage additions. Yet, practicing with today's project delivery model does not fully address the details that must be resolved at specification. As a result, numerous storage-based systems may include many energy losing and wasting elements.

Stakeholders should begin to clearly understand the difference between scheduled and unscheduled maintenance,[3] major repairs, and large or small product failures as being different from "Preemptive" maintenance. The idea is to illuminate and eliminate any challenges by better planning prior to EPC bidding and finalized long before the plant is completed. This assessment starts by defining a lifecycle plan for the continuing maintenance and rebuilding of the system. This knowledge and evolution in long-term thinking has begun to take hold in the industry. Yet, for substantive improvement, institutionalization takes time!

Today, most O&M plans, programs, and contracts do not adequately deal with, much less resolve, "all of the basic issues." They usually produce budgets that do not address the reality of what is to come for so many underaddressed unknowns, until they become crises.

Going into a plant at 6, 8, or 10 years of operations and doing a major reconstruction or replacement of the majority of modules or inverters and calling it repowering "Is Not SE/Repowering™"! SE/Repowering™ is a complete PV systems lifecycle planning process."

There are several complex, perspective-based questions that must be resolved. The short-term methods of simply replacing parts to temporarily boost performance outputs often do not address the core issues. Also, by performing additional maintenance or refurbishing, without a structured planning process, results in the plant being set up for short-, medium-, and long-term failures.

4.3 There Is a Continuous and Contentious Complaint about Lifecycle Performance

Yet, it has always been a myopic view of the system based on incomplete annual estimates of the plant performance without fully addressing IEC 61724-3, "Part 3: Energy Evaluation Method," IEC TS 63019:2019, and IEC TS 63265:2022.

The need is for understanding the source of all lifecycle costs, then reducing or optimizing those cost for the reduction of risk through improved performance and service life. For **SE/R**epowering™, it is important to ensure that the system specification/design has been reviewed to establish a complete risk–benefit profile. The details and benefits include lowered risks and unbudgeted expenses, greater system resilience and sustainability, and more predictable revenues over many more years without starting from scratch. All this helps to retain and enhance greater commercial energy production and asset value!

For new systems a critical question is: "What are the long-term engineering, operations, and maintenance challenges that must be addressed?"

For existing plants, it is the same question/s plus a special review of the current state of the hardware and software: "Have all the areas of improvement been clearly identified and noted?" "Do we know what is causing suboptimal performance or has led to failure to meet contract?"

There is a growing understanding in the industry that the requisite set of usage profiles, for historical and current environmental, including other considerations are truly not being addressed accurately or completely. The present project delivery model, concept, specification, design, build, and O&M phases tend to be underaddressed and incomplete. This lack of understanding has led to significant plant failures and additional unplanned costs. A major instance is the selection of components that may be appropriate and functional for a specific design in one environment or location, yet wholly inadequate and inappropriate for a plant located in a different region or country that has different climatic and environmental variations.

The reliability of a new or repowered versus a **R**epowering™ plant is based on the level and quality of the reliability engineering considerations addressed initially. Today's repowering does not adequately include SE and RAMS as project Keystones. However, the **SE/R**epowering™ process and approach includes the details to look at lifecycle solutions based upon real data and its application. These will have a major impact on O&M cost throughout the project life. Incompletely addressed at specification with a corresponding failure to perform validation and verification, i.e. "the missing data," will result in overwhelming replacements of components and the need for additional spares, which may no longer be available.

Without this system engineering effort of looking at the entire temporal picture will completely undermine attempts to identify the true problems (costs) resulting in an inaccurate LCOE. Confirming the details of component and subsystem reliability requires that all stakeholders have a basic understanding of reliability attribute values and the data sets from which they are derived.

When we talk about reliability, we were talking about measurable reliability and not vague or misleading reliability marketing language or guarantees.

One of the main reasons why this is not clearly seen in the industry is because plants that have major failures can result in the partial or total removal of, or complete rebuilding of systems. This is the result of not including those essential yet underaddressed costs, which go into the global costs to finance those systems. In essence, by removing all of the most serious cost cases, the rest of the industry continues to see skewed and faulty results financially. Although discussed within the industry during IEC, IECRE, IEEE, ASTM, and other standards discussions, making a more complete case to report operational costs including IEC TS 61724-3, IEC TS 63019:2019, and IEC TS 63265:2022 for operation completed has yet to gain sufficient traction. As the need for better energy production, lower documentable LCOEs, and a move toward delivering PV plus energy storage as infrastructure, these standards will be strengthened and better adopted. Many passive and active lobbying efforts are still focusing on the short term while not addressing long-term industry challenges and fiscal waste. At some point in the near future, this must be resolved; however, the industry must begin to understand the consequences of not doing so.

Carefully addressing the above requirements will affect the economics of the project. It is also important to understand there will only be a "best option[4]" while there will never be a "perfect option," when looking at the future expectations from the perspective of the initial, prior to EPC bidding specification. This "system" optimization, as an alternative to first cost optimization, results in the best and most cost-effective solution for the LCC of the plant.

Unfortunately, in many instances this same need is misinterpreted and used to buy the cheapest but often not the best (or least optimal) components. The focus strictly on cost overlooks component reliability and its impact on the system performance. There is some hope. Since about 2015, field experience is beginning to illustrate how least-cost price practices can be counterproductive to project success for both the owner and customer.

It is always easier to use or do what you are comfortable with as compared to specifying for the actual site conditions! It is an unfortunate and dangerous habit to get too attached to existing processes and procedures without continually addressing process improvement.

Many of the PV myths and assumptions today that go into financing, building, and operating plants do not provide sufficiently detailed or accurate enough financial information to resolve long-term issues. Addressing this gap early will have a major impact on the flow of revenues and expenses at the time that **SE/R**epowering™ phases and activities are most advantageous.

One source of information that can be very useful for improving the lifecycle of a PV plant can be found in how utilities had previously "historically" improved lifecycle performance of existing and future power generation plants. There is, however, a caveat to keep in mind when researching historical utility power generation data.

Even mature industries at times begin to neglect their lessons learned from decades of successful SE/Repowering™. It may take the utility industry decades to get back on the RAMS, resilience, and O&M cost control track. That is no reason for the PV industry to wait!

An unfortunate outcome of historical analyses is the tendency to document and then "forget" about essential lessons learned. Poor specifications as a result of budget slashing can lead to a high-speed crash of profitability in the out years of the equipment operation.

This practice shifts the assignment of essential fiscal assets and practices. It blinds many if not most stakeholders to the negative value of reducing RAMS. We have seen this repeatedly in the industry. We are not immune to or alone as we continue to see these destructive organizational practices. Unfortunately, even today, these poor business practices continue, hence the need to move the industry to use more practical, financially responsible, and effective methods.

The negative impact on revenue from poor initial business decisions drives organizations and industries back to the time-honored practices and applications as they begin to reel from the negative impact on revenue. Indiscriminately slashing budgets always seems to end in loss of significant revenue, market share, and project value. The practice is often one of buying/controlling assets, making the numbers look good with a little "chefery" by simmering or cooking the books, then dumping the company or plant on an unsuspecting buyer. It is exacerbated by tying organizational staff bonuses to "not spending money" on critical items, which were budgeted, for specific sets of conditions. This creates a major temptation to make budgetary actions look good making management happy in the short term, while deferring appropriate choices and decisions.

Some of these "flipped PV power plants" end up as publicity disasters, ending in lawsuits that result in personal damage to people's credibility and potentially outright loss of the property. It is less costly and stressful to solve the problems properly before one gets hauled into a legal situation that was created by an organization's greed, ignorance, stupidity, or negligence.

A very public example in aerospace is the Boeing 737 MAX 8 and the bottom-line dollar-driven practices[5] and poor design practices that led to two fatal crashes and loss of tens of billions of dollars. An appropriate PV example is the Sun Edison bankruptcy,[6] which resulted in multiple lawsuits and significant financial and power generation capability loss in the billions of dollars.

What processes allow for the development of cost-effective specifications? These are needed to provide a sound basis for planning and accommodating **SE/R**epowering™ as per a "**R**epowering™ Plan" developed in concert with the design of the plant.

If you want effective and meaningful change, you must want it enough to commit your whole organization until it becomes second nature as truly functional business practices!

There is a growing awareness – as evidenced by an increasing number of papers and presentations at conferences, in industry magazines, white papers, trainings, and webinars – that *least-cost systems do not provide a stable and long-term financial viability or even operational capability*. The evidence that is available over the last 10 years or so has become a topic like the weather. Even though there are mitigations to resolve many weather and climate issues, people tend to talk about solutions that are viewed as academic, hypothetical, or the result of paralysis of analysis. There continues to be a lack of industry vision and action on the obvious long-term costs of taking those risky illogical and unresponsive positions. Yet, the most effective PV mitigations are simply good planning, specification, and systems engineering!

In most instances, there is insufficient reliability data, due to a failure to gather it or share it, to make effective, long-term choices. This well-known continuing problem in developing and maintaining good data for RAMS prevents the decision makers from using it to aid them in determining the type and extent of **SE/R**epowering™ or later growth phases to be implemented. As a result, the usual fixes are short term, costly, and often drive up the cost of energy.

What does the flow of capital look like to achieve the maximum PV plant asset value?

Many, if not most, of the critical issues of asset value are not adequately addressed today. In better planned and built plants, what does the flow of capital look like after 5, 10, 20, 30, or 40 years, and what are the existing examples that give us a reasonable view of what takes place over time? By not extrapolating those details over time, numerous opportunities for improvement are overlooked.

4.3.1 Risk Reduction

SE/Repowering™ directly reduces risks while providing far better control of initial and lifecycle costs. This results in improved system performance, health, condition, and revenues. **SE/R**epowering™ planning must be flexible with well-documented, fact-based data decision-making! Failing to address or complete the following items increases risk of cost, schedule, fiscal loss, and liability:

(a) It is critical to understand that the **SE/R**epowering™ process is not a prescription to make specific changes at specific times without having the flexibility to be adjusted to meet real conditions and technology changes. For example, energy demand may grow faster than anticipated resulting in the need to **SE/R**epowering™ plant segments or phases earlier than originally planned.

(b) It is not simply prescriptive! Effective **SE/R**epowering™ requires a detailed review of cause and effect while addressing "if then" questions, which are seldomly performed during project concept through commissioning. It addresses risk shedding and data sharing issues. As stakeholders make determinations of how they will replace all the modules in X number of years, all the inverters in Y number of years, or make other plant improvements in Z number of years, they are in all probability making decisions based on unsupported assumptions using inadequate information or data.

(c) Expanding the awareness of what is actually taking place in PV plants will allow owners and operators the ability to refine component selections through better data collection and analysis. Doing so requires visualizing component lifecycle as a ***time- and condition-based range***, not a specifically defined number of years. This perspective brings clarity to real, not assumed, costs.

(d) It is important to keep in mind that the existing tools and processes generally in place today are not sufficient to apply the most effective and actionable **SE/R**epowering™ strategies. The underpinning need is to collect and curate far more accurate and higher quality data. It includes additional data, which is not being collected or lacking in quality, but can provide a substantively more comprehensive analysis. Such data collection and curation are addressed elsewhere in the book (Chapter 7, Maintainability, Chapter 10 Data Collection and Analysis). A topic outside this book is the existing and new sensor technologies and their ability to replace time-consuming labor with rapid and more accurate analysis of modules, inverters, and other components.

(e) It is not unusual to hear a knee-jerk reaction in response to the insufficiency of existing tools by proclaiming it is "too expensive" to develop and/or apply those tools. Tool development is an investment. If your team demonstrates clearly why it is "too expensive," there must be a counterargument or definitively better set of solutions. Yet, the facts are clear and simple, those capabilities have not yet been applied across the PV industry. Digressing to the "too expensive" excuse is not fact or data based without addressing those requirements, and the unplanned costs or results that can arise. Figure 5.11 illustrates a simplified optimization of the cost of O&M versus reliability versus O&M and acquisition costs.

(f) These tools have been developed in other industries and are beginning to be explored, developed, and applied to the PV industry. Reducing modest systemic losses on existing plants will more than pay for the application of these new tools, which will be essential for effective root cause analysis (Chapter 6) while reducing labor costs.

4.3.2 SE/Repowering™

SE/Repowering™ [7,8,9,10,11,12,13] of PV plants has four basic types: **CONCEPT, EXISITING, DISTRESSED, AND RELOCATION** discussed further in Section 4.6. These are based on the particular point in the life of the plant that the **SE/R**epowering™ planning and process is applied.

The industry is beginning to grapple with the challenge of what an **SE/R**epowering™ process really is versus current "repowering" or "refurbishment" or "revamping" has been discussed and presented in the industry today. To date, "repowering" is often a marketing term and little more based on limited or incomplete studies.

Industry confusion will continue to stimulate further discussion. To gain the potential billions of dollars in savings and profitability, the industry must address a myriad of questions. Those include the broader issues of what **SE/R**epowering™ is "in reality," and not defined in marketing terms. It requires a set of effective **SE/R**epowering™ processes and procedures that look at PV systems as PV infrastructure as compared to just jumping on the "repowering bandwagon."

Figure 4.2[14] projects a potential market view to 2030. It highlights explosive growth internationally, of repowering which is in good part the result of inadequate PV plant specification, design, and delivery processes (not our **SE/R**epowering™ planning). Unfortunately, much of the predicted repowering may be a little more than a series of emergency unscheduled patches or rebuilds that may just be stopgap solutions and make many more situations worse in the longer term. The actual results from today's systems are generally not discovered until later in the plant's life, often beginning in about 7–10 years. It is our position based on this information, that in all likelihood system growth challenges will continue to be problematic until effective **SE/R**epowering™ planning practices are

Figure 4.2 Global Repowering Market Projected Growth. (Source: Wood Mackenzie US Solar Market Insight Q3 2021).

in place throughout the industry. At that point, the market will begin to level off and be more consistent due to more effective industry operations and practices. Present growth creates a worldwide potential of tens, if not hundreds of thousands of plants that will basically need to be rebuilt almost from scratch. Failure of the industry to take advantage of established SE planning approaches can add billions to system costs. Yet, an effective planning process can dramatically extend the life of those plants while reducing costs to the owners, utilities, and their customers throughout the plant's lifecycles.

4.4 Cannibalization

Definition for Sustainable PV Cannibalization (PV Component Testing[15] & Reuse Process)

Cannibalization in the PV/Storage operations and maintenance universe is the act of removing components from an operational or nonoperational plant, testing/rerating and installing that component as the repair or replacement for another like and/or similar PV system application.

Please note: At the time of publication, there were no international IEC, IECRE, ASTM, IEEE, or other PV or related component recertification standards. Therefore, any cannibalized components are reused not recycled and must be tested, verified, and certified to meet a defined and documented component specification even if it is only an inhouse process, preferable, by an independent third party.

In this context, repair may not be accomplished using new inventory spares. Components may come from currently installed equipment at the same plant, or one in a different location. In addition, one could use cannibalization to perform **SE/R**epowering™ of one plant using parts from a different plant if accomplished by meeting our definition above and the listing below. This latter method of **SE/R**epowering™ has a number of significant issues and must not be applied haphazardly, including:

1. Inadequate accounting of problems with component life issues being unknown,
2. High probability of inadequate failure tracking data history,
3. High potential for induced damage (multiple handling scenarios),
4. An existing business model based on short-term fixes,
5. Requires documented processes and procedures for extra care in dismantling, handling, and transportation,
6. Requires clearly define storage processes, procedures, and capability, (inventory, environmental control, proper handling during movement, proper storage packaging),
7. Clearly defined removal, cleaning testing (standard certification rerating) package, store and/or ship for installation, retest.

4.4.1 Additional Requirements for Success for Commercial & Industrial (C&I)

For items that are removed from one plant to be used in another, there is often a need to handle and store this equipment for some period creating inventory challenges (lost, unaccounted for, misplaced, damaged by handling or uncontrolled storage environment, part number (P/N) or serialization tracking problems if not properly accounted for during removal). This requires installing recycled/reused components in a new site without documentation, prefabricated wiring, connectors (unless retested to meet mfg. requirements demands recycling versus simple reuse). It is not acceptable to just remove components from a plant to be shipped to another country for reuse without completing sufficient testing, grading, and packaging. Modules for example should undergo Module Retest and certification according to IEC TS 92915: 2018. *This also indicates that there is still a gap in anything done with removing modules as any part of* **R**epowering™.

4.5 Impacts of SE/Repowering™

4.5.1 Improved and Higher Asset Valuation

The ability to establish comparable metrics and asset values for the PV, residential, commercial, and utility systems benefits the whole industry and normalizes the long-term cost for energy. **SE/R**epowering™ elements create a substantively improved comparative and competitive market. As an implemented standardized approach, **SE/R**epowering™ allows for determination of both an accurate health status and overall system condition that defines the commercial value of a plant or product. The approach is central to the development of an entire PV industry infrastructure expansion and its technologies.

4.5.2 Long-Term Energy Production

In addressing the accurate economics of building or buying a plant, the **SE/R**epowering™ planning process allows for a more precise capability to predict what long-term energy production and component condition will look like. The ability to build on and improve RAMS attributes are key factors in the financial package, which are often underassessed today. Identifiable and measurable data drives better modeling, delivering improved decision-making and profitability.

4.5.3 Plant Viability

The emphasis on clearly establishing health, condition, and status is the baseline for defining system long-term viability. The process of reducing faults, failures, common errors, and anomalies dramatically reduce not only the O&M costs; they reduce a variety of soft costs that result from challenges, which can make or break a project over time. These soft costs include but are not limited to additional administration, operations, maintenance, organizational fire drills, legal, insurance, and organizational angst, along with a broad range of involvement and disruption for a number of organizational members, usually uninvolved with O&M activities. All of these soft cost elements are contributors to organization and corporate angst, consuming organizational energy, finances, and productivity. The earlier these are addressed and resolved, the greater the opportunity for project success throughout and while extending its life.

4.5.4 Revenue

Revenue, which continues beyond the existing (25–30 year) lifecycle to 50 years or more, produces far greater energy production, profits, and incentives for buy and hold investors. *PV plants degrade more slowly, while production and revenue curves continue longer with greater consistency. This is what moves the industry from a high-risk speculative effort to a far more levelized mature industry.*

Effective financial modeling includes revenue projections. However, when those revenues are eaten up by avoidable unplanned costs, they tend to undermine fiscal viability of the project. When plants are built using SE processes with substantially better planning and implementation, not only do the revenues for the first phase of life (5–10 years) become more accurate. They also provide improved infrastructure of the plant to reduce costs for the next temporal and operational phase. For those institutional fleet owners, all of this provides a much higher investment grade opportunity at a lower risk, which results in greater longer-term success.

4.5.5 Operations Analysis and Decisions

With the details of an effectively designed **SE/R**epowering™ plan, ownership operations analysis and resulting decisions can be more precise, realistic, and simplified. There is an increase of useful information to be analyzed automatically, providing a clearer picture of the entire system and its operations. Although the plan will not have addressed everything that can potentially be an issue, it addresses the vast majority of problems that owners will now be able to resolve before benchmarking and COD. This visibility can result in a template for effectively improving the discovery of issues with a set of basic plans to deal with them before, during, and after the issues have been resolved.

Better, more complete processes provide greater stability for plants over time, thereby reducing costs. This results in additional unasked questions, some of which may never be completely answered, substantially identified, measured, planned for, or resolved.

4.5.6 Commercial & Industrial (C&I) and Utility Economics

The facts clearly indicate that any improvement in the industry is driven by money spent and assumed benefits gained, which is loosely referred to in our industry as a "cost–benefit analysis." We say loosely, because until the actual details and factors on both sides of the equation are sufficiently identified and accurate, the analysis is incomplete and thus flawed. In essence, if you do not have a good plan, how would you know what the details are? Over the last few decades, we have repeatedly seen that numbers are massaged to meet the expectations, requirements, and arguments, or satisfy the case being made to assure financing. Yet, the vast majority of our more experienced PV colleagues have come to the conclusion that stakeholders are beginning to see a bigger picture. It is still not a clear industrywide picture, yet it indicates that there exists a myriad of problems that are solvable to make a project more financially viable.

Historically, the authors of this book have yet to find anybody who can actually and factually tell why it is "too expensive" to take the longest-term cost-effective steps. We suspect that as long as there are no penalties other than a potential loss or bankruptcy, corner cutting will remain "King." We suggest you discuss this among your colleagues because there are serious cost reduction answers that can easily be drawn if you have the right information. However, getting your teams attention will probably be your greatest challenge.

During his ongoing development of PerfectPower (1998–2011), as president and CTO, Dr. Balfour began listing the items, actions, and considerations that his solar company did, that his competitors did not. His presumption was that his company provided at least 12 specific actions to achieve more resilient and higher performing systems beyond his competitors. At the top of the list in the early 2000s, a five-year warranty with annual inspection was included in the purchase of all plants. The list included items that were fairly well accepted by his organization. However, at the first review, approximately 36 additional items were identified, later referred to as identified "limiting factors" or systemic constraints.

That began a series of meetings with Joe Cunningham, an employee of PerfectPower at that time (later with Sunny Energy LLC), where an enhanced list was developed. It included many of the lessons learned from mistakes made related to issues that needed to be addressed at specification and design. He saw and documented issues continually throughout the process, including numerous conversations with colleagues which added to the list. Much of this list was based on unsupported industry assumptions, mistakes made, lessons learned, and corrections introduced resulting in a growing number of policies and procedures for all phases of PV system specification, sales, design, construction, operations, and maintenance. These developed into a growing list of "limiting factors" that must be addressed for essential PV improvement.

4.5.7 Financial Viability

As stakeholders look at the overall financial viability and reliability of projects, as they impact plant value, they will demand greater verified reliability in the products that are being purchased. This request can also be encouraged through the advancement and development of new and improved reliability standards like IEC 62093 (PVQAT TG 11/discussions of IEC 62093) and EC TS 63265 ED1, Reliability Practices for the Operation of Photovoltaic Power Systems. These standards strengthen manufacturing validation testing, which is a step forward in improving the quality of the product. However, as pointed out in Chapter 6, quality and reliability are not the same thing.

4.5.8 Insurability

Because we are looking at system specification and design service life (Chapter 5) and lifecycles beyond todays 25- or 30-year claims, more robust system components will need to be specified, especially as it relates to the insurability of the project and insurance costs. Present component and system generally do not meet expectations especially under extreme weather and other environmental conditions. This need for robustness is true to minimize additional O&M costs while improving plant lifecycle. The historic utility model clearly indicates how a minor increase in system robustness can result in additional systemic performance and operational capability.

This begs the questions:

- Has your organization understood the desired service life and lifetime system requirements and derived appropriate specifications for lifecycle system effectiveness?
- Have these requirements been allocated appropriately to individual components including component subsystem grouping and do each of those have service goals identified?
- Has your organization carefully reviewed more robust products and identified those few that meet these goals?
- Has your organization asked the manufacturers what they have done to address climate extreme specification issues and durations to improve robustness and reliability?
- If not, why have these steps not been considered, ignored, or never adopted?

If your team does not specify "real" requirements and required improvements, they will not see them. They may assume that the product(s) they are looking for would be the best product for a more intense and accurate set of environmental conditions than is often the case.

The act of avoiding by "preemption," the actual organizational, hard, soft, and corrective action costs, with or without using existing insurance in place, will more than pay for themselves in the first few years. The cost of avoiding one claim and its impact on raising insurance premiums, in many instances, pays for itself alone, because one incidence could raise premiums for an entire fleet or region. It is important to keep in mind that many of these extreme weather issues may no longer be insurable. Thus, it becomes a primary business risk reduction strategy to consider and address these issues at specification as they impact insurability. In most instances, by avoiding one moderate or serious climate-related set of damages, owners and operators may recoup some or all of the additional expenses in that first event.

The reality is that when the steps are taken and documented properly for all the limiting factors, many marginal or potentially peripheral decisions are more quickly quantified, mitigated, or eliminated.

4.6 Types of SE/Repowering™

SE/Repowering™ a plant has a foundation in the specific phase of life. As mentioned above, there are four main types, **SE/R**epowering™ at concept, existing, distressed, and relocation. The most effective approach begins with a detailed and documented "**SE/R**epowering™ Process Plan" at the new plant concept phase.

Our view is beyond discussions where the focus is primarily on ad hoc repowering equipment choices, selections, and actions taken. The PAM process is to see and apply **SE/R**epowering™ flexibility as a set of system specification, design and delivery, financial investment, management tools, and decision-making processes. It is not a rote process and requires substantial initial effort. It is a complete detailed systems plan, development, delivery, and O&M support process program model. For the greatest value, it begins at concept and ends at site restoration. As a result, plant service life and lifecycles can be extended from 10 or 15 or 20 to 30 years to 50 or 75 years or more as we will show.

Changing component technology plays a central role in the **R**epowering™ process. This includes everything from the PV cells through and including the inverters, where efficiencies, new chemistries, and IGBTs etc., have been changing at the rate of about every 18–24 months.

These technology changes create havoc with specifiers, designers, and maintainers. Today, it may take 1–3 years to develop and deliver a plant. The components that were originally designed for use might change, might no longer be available, might be almost through their market cycle due to the rapid technology development, or worse, might be a from a specified single source that has gone out of business.

The inclusion of all of the stakeholders results in far more accurate and detailed systems information to be fed into current as well as future projects. This also provides learning what is unique to each stakeholder and puts the project team on a common ground of process, approach, and delivery understanding. It is not uncommon for a minor technical or even philosophical oversight to become a major expense later! An example of this can be found in a utility project that did not include soil chemistry testing to reduce the upfront costs due to an unsupported assumption, i.e. "it was too expensive!" That choice of not spending a few thousands of dollars cost over a million. It resulted in complete removal of the system after mounting structures began to dissolve (oxidize) in a relatively short period of time. Bypassing a critical and necessary soil chemistry expense cost an entire system! This was such an embarrassment to the owners, that today, they do not wish to even acknowledge that that system was ever built.

4.6.1 SE/Repowering™ at Concept

Planning for **SE/R**epowering™ during a new project phase concept must be in concert with plant baseline stakeholder specification. It must include information and/or an assessment of the specification[16] for existing plants for tracking and comparison. The focus is on addressing many of the functionalities that are often underaddressed or asked "what if" questions that impact the plant over time. To answer these questions, a different set of analytical approaches and options are considered that will reduce and optimize lifecycle system delivery budgets.

Note: In the PV plant aging process, performance metrics tend to look at the rear view mirror, historic (past), results of plant's health and condition in terms of energy production. A comprehensive and effective Repowering™ process, focuses on past, present, and future plant health, condition (failure) data (Chapter 10) and status to identify, resolve, and improve plant longevity (Chapters 5 and 6).

It does so while improving the physical capability, health, and condition of the plant throughout its life. By following a variation of the previous historic utility model, mirrored in many other mature industries, the fiscal viability and risk reduction for plants are dramatically improved. The results are major improvement in the lifecycle economics and profitability.

Because the specification has been very carefully determined, refined, and modeled prior to EPC bidding, plants are designed, built, operated, and maintained far more effectively. We prefer the term *effectively* instead of *efficiently!*

This distinction is needed because we can build a marginal plant quickly, cheaply, and profitably (efficiently) without delivering the long-term performance, reliability, production, and revenue essential to a mature PV as infrastructure industry. Thus, delivering plants requires determination of what are the essential elements needed to support a long-term profitable plant.

It is all a matter of priorities and the choice between short-term least-cost thinking and lifecycle benefits, which deliver a variety of additional RAMS and fiscal value. That effectiveness is heavily impacted by the amount of data, data quality, and its analysis to continually improve current and future specifications and changes to the plant during its lifecycle. **To take it out of the context of what some people might refer to as being academic or too scientific, the reality is the better quality information results in far better choices/results**.

Yet more to the point, effective and focused processes determine the financial stability and viability of the project to its owners. This stability dramatically improves the capability of knowing what to do, when to do it (scheduling), and how to do it for the best "most effective" outcomes. As a result, it continuously gives PV stakeholders the ongoing checks and balance information to address alternatives analysis while updating and comparing real LCOE values.

Improved, advanced manufacturing and testing methodologies allow for many faults, failures, and defects to be identified at specification and prior to commissioning. Examples include modules and inverters that may have been damaged during shipping or installation yet may not show substantive and warrantable negative impact until much later. As a result of better PAM processes, many of these problems will be resolved as training, education, and reduction of warranty issues are improved along with O&M practices addressed early. With sufficient data, meaningful changes to discover actual plant health and condition provide a better time-based approach to all forms of maintenance and replacement. This method keeps a substantial amount of budget creep within the warranty period and out of the O&M budget.

As a result of an incredible lack of consistency in data collection and analysis, there are often unanticipated serious systematic issues that eventually become visible and consequential. When there are significant problems, the owners may flip the plant, develop a financial restructuring package, execute a fire sale, or declare bankruptcy. As a result, the plant may not have had a clear and in-depth problem or risk analysis, resulting in numerous challenges and questions for the new owners. This can result in purchasing a substantive number of unidentified costs.

SE/Repowering™ value begins at plant concept through delivery on to site restoration. It is critically important, valuable and requires building in plant serviceability (maintainability), reliability, durability, and resilience. Doing so dramatically reduces the number of unexpected and expensive repairs, replacements, and unplanned reconstruction. Today, it is common for PV professionals, during concept and design, to focus primarily on component costs or current performance attributes and not adequately consider the lifecycle labor, management, and replacement costs and future entitlements. A least-cost focus on development and installation results in higher post warranty O&M costs as cheaper, less reliable parts begin to fail earlier than "planned" or "expected." This perspective also requires more spares and their associated cost (Chapter 7). Effective, well-defined **SE/R**epowering™ is the only industry paradigm we are aware of that addresses the industrywide challenges that effectively respond to the comment and concern, "It's too expensive." The real facts are that "It's all about the money!" This truth is where the PAM **SE/R**epowering™ Process is the most effective approach, as it focuses on addressing all the lifecycle financial issues!

With effective PV **SE/R**epowering™ initiatives, it is essential to include all stakeholders, their insights, perspectives, wants and needs, while keeping them updated throughout the whole plant lifecycle. The stakeholders must also understand how their wants and needs may affect other stakeholders. They must be prepared to accept compromises that other stakeholders require.

4.6.2 Existing Plant SE/Repowering™

An existing plant may initially have the capability to deliver its estimated contractual performance numbers, thereby meeting *initial* expectations. However, as a result of "masking" (Balfour et al. 2021), a data blindness to the actual physical health and condition of the plant exists that needs to be addressed early (at specification) in the plant life. The plant owners may begin to notice that expenses for O&M are creeping, then moving up more rapidly than expected while performance is not as consistent or as positive as it was previously. This is often further confused by complex weather variability.

Existing best practices continue to focus on contract "performance" as presented in IEC 61724-3, yet bypassing seriously underaddressing plant health, condition, and status, unanswered questions and, due to a lack of documentation, a variety of hidden issue and costs. This unbalanced approach is consistent with, compounded by, and supported by existing industry myths and assumptions!

Though PV O&M practices are still emerging and evolving, a number of utilities and third-party service providers are working on these practices and developing a better sense of what level is required for system and site-specific costs and benefits (Balfour et al. 2021, EPRI 2010).

Whether the plant lifecycle is at concept or is 1, 10, or more years old, an **SE/R**epowering™ plan and analysis should be developed and implemented, especially if the plant is going to be bought, sold, refinanced, or reinsured. The reality is that, at the minimum, that plant will need to be reinsured during that period of time. However, the commonly referred to repowering usually tends to take place because substantial numbers of faults and failures have occurred and threaten the projects financially. The performance numbers begin to plunge, and the plant can no longer meet its contractual requirements without substantial unbudgeted dollars (Chapter 6).

The downside of this form of historic repowering is twofold:

1. First, it lacks consistent and coherent data acquisition input for analysis. Owners making project changes may lack critical information to make the most effective decisions. In some instances, those incomplete decisions and details can ripple through resulting actions thus creating additional problems that are actually manifested by this "pseudo" or "faux" repowering approach.
2. The second is where this form of repowering may only be a partial effort that is good for a few years (a patch), at which time the trend toward serious plant degraded performance continues to occur. Once serious issues occur and are identified, there may

be little or no budget available to make critical improvements. This can result in project and organization failure, and/or project or possible plant derating or organizational bankruptcy. This situation is a reason why many systems are flipped around year five, seven, or earlier.

Many experienced developers, EPCs, and owners are aware that this approximately five to seven year period is when least-cost budgeting begins to see ugly flaws arise. Therefore, the risk-shedding strategy is to dump the plant before plant overbuild (which "masks" these conditions due to high DC-to-AC ratios or overbuild), exposing performance that is no longer protected by the excess DC STC capacity. More detailed information can be found in Chapter 3, "Current PV Component Technologies." The PV industry continues experiencing a growing number of these situations, which are quite avoidable. However, existing PV myths and assumptions have not yet been effectively replaced by fact-based metrics to address the range of system issues.

For example, the automotive industry has been around for about 130 or so years. Beginning that time, the automotive companies responded to their customers to improve their products, if they wanted to continue selling them. This consumer pressure forced improvement in cars and trucks by simplifying, improving, adding, and modifying the automobile capability. And, as expected, the automotive industry passed along the costs for these items directly to the customer.

What was not seen is that as cars became more complex, they started having more failures to which the consumer responded by no longer buying cars that broke down too often. The industries' response to this was to implement warranties based on a set of user/usage (Chapters 5 and 6) profiles. The illustrated example is that the automotive producers would charge a "little" more for each product but give the consumer the "warm fuzzies" that if something broke it would be replaced by the dealer or manufacturer under warranty during a specified time period. This "warranty" approach was used to build product confidence.

Automotive users are happier since they are getting all the extra features they want, and are willing to pay for, and as a consequence cars sales keep climbing. Today's drivers are also more inclined to drive cars for only 5–6 years or so and buy a "newer, better" model with more features. In the past, it was not uncommon for owners to drive their cars for 15 or more years. The hidden reality here is that based on this approach by the buying public, the majority of cars never really reached the point of significant failures for first owners, as the expected life for many electronic devices and some of the more complicated mechanical devices is on the order of 5–10 years.

Technology complexity being balanced by cost–benefit is an attribute illustrated by today's cars having, in a single vehicle, greater computing power than National Aeronautics and Space Administration (NASA) had for the Moon missions. Along with this increased capability came issues with software (Toyota), hardware no longer repairable by the owner, and safety, Tesla, and other battery fires.

PV has been taking a similar path but is way behind the curve based on the continuing forcing function of producing the PV plant for the lowest cost possible. NASA found this to be unsustainable when they tried to use and practiced **"faster, better, cheaper"** for their projects only to find out that they could only achieve two of these, "faster/cheaper." This included the Hubble telescope and space shuttle *Challenger* explosion examples that resulted in losses exceeding $1B. In the PV universe, the faster, better, cheaper has failed as evidenced by plants that are not built as **R**epowering™ energy infrastructure.

One additional element that has developed in cars that has not reached PV yet is more effective standards, codes, government rules and regulations. Europe, using the IEC, ANSI, IEEE, ASTM and other standards, has reached the point of requiring compliance to many standards related to PV. This is to assure the owners and public that relying on that these products can and will produce power reliably. While standards and certification are improving, there are still major weaknesses. Overcoming those insufficiencies is still hampered due to insufficient requirements for SE and RAMS as critical elements essential to achieve project success. In the United States, there is little real oversight and regulations other than those to meet basic safety needs (NEC) and provide compatible utility grade power to the interface. However, once enough users become annoyed enough to push for regulation, it is possible that the US government may generate code/regulations/laws to establish relevant levels of consistency. This would be in line with the acceptance of tax incentives like the 2022 IRA, the 2022 Canadian investment tax credits (ITC), and a growing number of other international incentives programs to deliver better more productive plants.

While philosophically, many in the business community abhor rules and regulations, this often results in many instances, where plants do not meet their marketed service life capabilities. There is of course the issue of over regulation. However, when a buyer or group of investors are buying a PV plant for multiple hundreds of millions of dollars, there should be a foundational level of comparable information to make sure that they are receiving what they believe they are buying.

The need for regulation is because the market will demand to know and understand what they are buying and how it will function over time as presented. These must be tied to government incentives, through clearer direction and lifecycle effectiveness for environmental impact realities and needs. Stakeholders are beginning to see existing system challenges as critical to sustainable business growth and national security. They are beginning to focus on current plant and project designs that are often incapable of operating in the wide range of site climates and conditions. This is even before we begin to address growing awareness and realities of climate change.

By simply reviewing performance data and having a quick and/or insufficiently structured inspection does not, nor can it, provide the level of information necessary to make the critical effective financial decisions about an existing plant.

Is the plant worth the stated or negotiated price based on current condition, or will it require substantive and often expensive changes to reach that value?

This question quantifies a new owner's deepest concerns regarding value. Yet, that information is presently not fully or effectively available or applied with existing PV plant standards, analysis, practices, and techniques.

4.6.2.1 New Ownership

Much of what will drive all of these issues are related to bankability and insurability. As the risk goes up, banks, financiers, and insurance companies will either decline funding or require much higher interest rates, deductibles, rate of return, or premiums, if at all. In essence, the banking and insurance institutions are becoming aware of discrepancies and inconsistency of accurate plant health, condition, and status value metrics provided in the industry today. Much of the movement toward far greater bankability and insurability was launched in the US market by Jon Previtali, at that time Vice President Environmental Finance, Wells Fargo Bank. Jon masterfully began to link greater levels of critical PV metrics and analysis with improving bankability qualification guidelines and choices. This initiated trend has driven the issue of whether a project financier or insurer will fund a project and determine the cost of money and insurability as well. PV projects will not be funded if they cannot meet those evolving and improving core, due diligence requirements. If you cannot finance or insure a plant, you will most likely not be able to buy or sell it.

4.6.2.2 Major Financial and Legal Issues with Current Plant

This loss of capital and insurability, if not addressed early, could become epidemic, thereby reducing industry bankability and the number of bankable PV power plants. When stakeholders can no longer determine what the actual value and risks are involved in the project or plant, there may be no funding or insurance available. This push is one of the reasons that the authors became actively involved in the development of the IECRE PV Plant Rating and Certification Standard, which began around 2019. A plant that cannot clearly be valued, contributes to a potential loss of overall bankability for the industry, not just for that specific owner!

Plant rating and certification will drive the cost of money and availability of insurance. Yet, for well-designed, well-built, and operated plants, those costs will be substantially less as associated project development and delivery will progress more rapidly. This route is the quickest, safest way toward long-term qualitative power production and industry stability. In a world where we have the Kelly Bluebook for automobiles, professional appraisals of all kinds of properties, and a variety of different types of inspections for other technologies, banking and insurance institutions will expect actual values to be more accurately meaningful and determined. These methods for buying or selling are not perfect; nevertheless, they do provide a substantial amount of additional and more accurate information that allows buyers and sellers to make better decisions.

4.6.3 Distressed Plant SE/Repowering™

Distressed plants are more common than many believe. Distressed plants are plants that have not been specified, designed, built, and operated as PV or energy infrastructure. They have been conceived and delivered as projects or plants to meet the needs primarily of the initial stakeholders with incomplete/ineffective processes and procedures that do not or cannot meet the needs of later stakeholders.

Buyers may or may not have a clear idea of the actual health, condition, and status of the plant they are buying even when the plant appears to be meeting its initial contractual requirements.

All too often, insufficient plant inspection, data collection and analyses have not taken place to determine the real health, condition, and status of the plant. Their assumptions primarily based on their due diligence are insufficient for an effective and thorough technical or financial analysis and inspection especially during the first few years. As we often hear, "The plant is meeting its numbers, therefore it must be operating and producing energy as per design, construction and contract." Or, a buyer assumes that in spite of the issues they have discovered, they believe they have effectively determined most if not all its value, health, condition, status, and future capability.

Owners of distressed plants will often move to flip the system before the projected costs become too great to create a significant financial impact. This decision may return greater profits for them, while continuing the cycle of selling plants that in all probability may not be worth the money spent by the new buyer. Still, the original owners may be marketing their minimal or marginal upgrades as a "thorough repowering." Yet today, results tend to be strong on marketing and short on real and effective medium- or long-term results. With this growing number of distressed plants, many aggressive buyers are evaluating the cost of bringing them into long-term revenue-producing assets. However, many of those same buyers are not addressing the real issues that are preventing long-term viability.

To confuse things further, in a bankruptcy or a fire sale, there is seldom if ever enough time, effort, or investment made to do an adequate PV plant condition analysis. Because an overbuilt plant is still primarily gauged on performance metrics, the missing elements are in the form of plant health and condition – the true value metrics of a plant's commercial value. It is the physical plant health, condition, and status, that defines "real" asset value, not short-term, ephemeral, or assumed performance metrics. The health, condition, and status, data trends and associated information are far more valuable. For the more experienced buyers building a fleet to hold as a long-term investment, they may be able to afford some underperforming plants with questionable health, condition, and status, as long as they purchase a sufficient number of healthy plants to make up for it. This investment strategy is high risk, which may show its fair share of winners, yet, most likely far more losers.

A growing uncomfortable issue that is continuing with an insufficient level of detail to physically qualify for incentives is a lack of health, condition, and status requirements for qualification. This can lead to investors buying and flipping underperforming plants by continuing with least-cost practices while claiming a "repowered plant," yet still not addressing SE and RAMS practices.

Stepping into a **SE/R**epowering™ planning process requires a much deeper dive. It identifies and uncovers more of the challenges in conceptual or existing plants while allowing for a more accurate health, condition, status, and fiscal analysis. A deep dive requires an effective set of processes and procedures focused on discovery of the issues. Or, as we indicated earlier from our friend and colleague Bob Holsinger, "If you don't look for those problems, you're not going to find them!" The greater the accuracy of the evaluation, the more negotiating power the buyer has in the quest for a reliable energy production and revenue asset. For a seller, if their documentation shows added value in the plant health, condition, and status they will earn more money in the sale no matter what the circumstances are for selling the plant. These issues are of greater interest to financiers and insurers in addressing risk, based on more clearly defined bankability data.

A properly defined and executed **SE/R**epowering™ plan following COD is an extremely valuable choice of action to take! However, if a distressed plant repowering does not include a well-defined plan or processes, it may not improve the plant, its capabilities, or its value over time significantly or in a meaningful manner. Doing so effectively has many powerful and positive results as addressed throughout this chapter. Nevertheless, without a complete and effective planning **SE/R**epowering™ process, it is generally an expensive exercise, raising plant costs over the medium and/or long term. Success requiring the right component selections, engineering, and installation is seldom completely implemented today.

4.6.4 Relocation Plant SE/Repowering™ or More Likely Common Repowering

A growing number of plants are becoming inconvenient in their initial or current location. They may need to be moved for a variety of reasons. Those reasons may include redevelopment of an area or property, environmental concerns affecting the plant (like flooding or some other disruptive event), relocation of a roadway/highway, the construction of additional structures that now or will shade the plant, or an eminent domain challenge. Indeed, there may be others.

Based on information available today, unless a plant is specifically planned and designed for relocation/portability, it likely will be a very expensive process and most likely to fail while not being cost effective. Its components may have some cannibalization value.

Addressing plant relocations may have a series of challenges, risks, and legal matters that need to be overcome *prior to the owner making "any" final decisions to proceed.* Those could include items like:

(a) If there is any potential for a move of the system, it must be addressed in the original PPA language during specification before signing. This may be of value in all PPAs.
(b) If there is a need to move the system with no contractual provision for such a move, the PPA will require renegotiation possibly in its entirety. There is also no guarantee that all of the PPA stakeholders will agree to such a change.
(c) If there is a PPA, has this been adequately addressed in the contractual details? If not, then some form of solution must be found between all the project stakeholders.
(d) What is the length of the PPA along with all the terms and conditions for relocation?
(e) Are there requirements regarding equipment/component types, sizes, or other limitations that will be a challenge after the PPA has been under way?
(f) The issues of ownership need to be included and require a clear analysis of type of ownership of the plant, property ownership, and any legal or technical issues that would result in the transfer to another site. It can be complicated by leases, options, and other site and/or project contractual issues over time.
(g) What is the condition of the equipment? That includes all structural, electronic (including modules), inverters, racking/tracking, transformers, system controls, data acquisition and analysis (DAS) or SCADA, any unmentioned Balance of System (BOS) components, etc. After five years or so, the original equipment may be of some, negligible, or no value.

(h) Will those components be usable, or will they have to be disposed of properly and in an appropriate, environmentally sensitive manner? That means this equipment will have to be identified, recycled, or disposed of as solid or hazardous waste.
(i) What can actually be reused, recycled, and what will be disposed of?
(j) Consider that much of the equipment, including electronics, software, modules, and even wiring and cabling, may no longer meet or be capable of meeting updated local or national codes or standards.
(k) How will the team address the value of what is usable or recyclable, and the costs to remove and transport the existing system to the new site or a proper disposal site? This estimate must also include the cost of packing, shipping, unpacking, and rebuilding that system, and the difference in meeting all of those Authority Having Jurisdiction (AHJ) and PPA requirements that may define newer technology and newer solutions.
(l) As the team resolves all of these issues, they will have to uncover all of the real costs and legal challenges that will include negotiations for the changes. How will all of the finances be addressed and resolved, and what are the new requirements after the PPA has been modified or replaced for the new site?
(m) What are the unanswered questions? There are always issues that are not identified. After all of the regular **SE/R**epowering™ requirements are identified, there should be very few unanswered questions. Expect there to still be several surprises. The remaining unknowns will be covered if the steps are taken, documented, and clearly shared with the project stakeholders. However, each of these projects will be incredibly different from each other, and each will be an ongoing set of learning opportunities for all involved.

4.7 Preemptive Analytical Maintenance SE/Repowering™ System Planning

4.7.1 Planning Element Requirements

The following is a minimum set of **SE/R**epowering™ planning elements that can be used for any PV project. All of these are elements of system engineering effort (Chapter 5) that need to be identified beginning at specification. Each should be treated as living plans as time, new technologies, and new guidelines and codes may require updates for execution.

The above plans and efforts require long-term scope, goals, and objectives while including alternatives analysis and flexibility for unexpected situations and challenges. Flexibility is not "carte blanche." As information, component values, and needs change, there must be flexibility to make the necessary adjustments and clearly document them. While there is a learning curve, once an organization develops its **SE/R**epowering™ planning processes, procedures, and actions, benefits rapidly outweigh costs issues, while reducing time. This benefits all lifecycle stakeholders financially, profitability wise, with a reduction of missed opportunities and organizational angst.

All of these planning elements need to be addressed when considering project "Success and Failure" for the broad range of stakeholders.

Please note that many of our colleagues presently believe they are addressing most if not all of the above planning points in their existing processes and procedures. At this time, we believe that many organizations address some well, while overlooking an effective coherent process. However, their goals and objectives are substantively different from the PAM **SE/R**epowering™ process. PAM goals and objectives strive toward defining ***real*** "long-term" costs, actions, capabilities, and results, while looking well beyond the marketing of plants for quick profits and shorter lifecycles at a higher actual lifetime cost. The real and greater profits and customer satisfaction are found over longer lifecycles. The results of diluting or skipping **SE/R**epowering™ and preemptive planning processes lead to an abundance of unexpected faults, failures, defects, and errors made during the EPC (design-build) and O&M phases. This lack of attention and failure to address results in the underassessment of actual costs to build, operate, and maintain the plant, compromising project delivery. The outcomes may or will result in the risk of project derating, underperformance, potential loss of PPA, or possible project or organization bankruptcy and liquidated damages.

The PV system delivery challenge is exacerbated by the fact that many plant owners and manufacturers have been reticent to provide or share the defect, fault, failure, and test data necessary to improve systems and components for next-generation projects. It is often held as proprietary or intellectual property (IP). This fact and the lack of useful failure or operational data collection, transmission, and security can make existing data, whether available or not, of questionable quality.

Important requirements that are often underaddressed or not addressed at all become points of significant cost and schedule risk. Requirements, such as temperature operating range, voltage, current and/or power temperature coefficients and component derating, are necessary to ensure plant operation over all of the local site conditions. This is where addressing potential climate change impacts should be taken seriously as components within the entire plant experience greater extremes. The site environmental data for example must be appropriately utilized to specify conditions to which the components will be exposed, and that the equipment must survive that exposure throughout its operational life. This part is often where historical environmental conditions are evaluated, with the appropriate risk (confidence level [CL]) noted and the operating conditions within the equipment controlled and evaluated prior to final design decisions. Further evaluation should be done as *issues presently occur, after the equipment fails early, or one or two years or so after installation.*

4.8 RAMS for SE/Repowering™

A minimum recommendation for any system (residential/commercial/utility) is to perform the basics of the RAMS/SE tasks with the inclusion of:

- Systems analysis,
- Revisiting the specific site environmental data,
- Assessing and developing the system and reliability models,
- Updating the maintenance assessments and actions,

- Upgrading the DAS or SCADA systems to provide the needed data acquisition for system performance and energy delivery over time,
- In addition, there are new and borrowed technologies that increase a variety of data collection needs which dramatically improve issue identification and actionability.

Certain tasks as shown in Chapters 5 and 6 have cautions that must be integrated into the RAMS/SE tasks. When performing the failure mode effects analysis (FMEA), the process is to examine a single failure mode and its effect on the system. It does not address the effects that would include prior existing failures, cascading failures (unless a known effect), or simultaneous failures. For this work, the fault tree analysis (FTA) is the tool of choice. **SE/R**epowering™ issues to address for operating ambient conditions for non-airconditioned sites include:

- Inverters are not designed to operate at the maximum/minimum temperatures found at many sites. Inverter derating does not necessarily protect them adequately from reduced output or stress and damage over the medium/long term.
- To offset, the developers and/or manufacturing will accept a derating of the inverter for lower output at higher operating temperatures to attempt to reach a reasonable level of reliability. This route effectively derates the plant capability and its energy production while potentially stressing and shortening high-power component life.
- Failure to design for ambient air temperature plus the induced temperature in the inverter interior is referred to as the Sol-Air Temperature[17,18], which may result in significant overstress on the power components as well as the rest of the electronics within the cabinet. Although primarily use for building energy analysis, the inverter is treated as a small metal or plastic building. Thus, if exposed to sunlight, there may be a need to install sun shades to moderate excessive out-of-specification temperatures, both inside the inverter and on the surface of its container.

Table 4.2 is a summary using data for Phoenix Sky Harbor Airport. The caution with this data is that it is based on hourly measurements, so it is probable that there are higher or lower extremes that have not been recorded or considered. In addition, the temperature

Table 4.1 PV Project Planning Elements.

1. Timing and scheduling plan	2. Finance plan
3. Insurance plan	4. Force majeure plan
5. Community action plan	6. Regulatory preparation plan
7. Data acquisition and analysis plan	8. Training and education plan
9. Site and system environmental impact plan	10. O&M plan
11. Project management plan	12. Energy storage plan
13. SMART specification	14. Systems engineering plan
15. Integrated RAMS plan	16. Design plan
17. Commissioning and benchmarking plan	18. Site restoration and recycling plan
19. Construction plan	20. Procurement plan
21. Grid interface (utility) plan	22. Other site/local dependent plans

and other environmental vary substantially between the airport and downtown Phoenix. This would hold true for any two data sets separated by more than a couple of miles apart and with altitude differences of several hundred feet or more. For this reason, the 95% CL is applied to the minimum or maximum as appropriate. Rain average is only 8.34 inches per year. Adding the standard deviation times a z-value of 1.96 results in 11.80 inches per year at 95% confidence, while there is already a demonstrated maximum event of 19.6 inches for 1995. There are also 3 other events that exceed 15 inches per year. Consideration should also be made for rain events to exceed 1 inch from a single thunderstorm event.

Most stations currently do environmental samples once an hour. At lower sampling time intervals of 5–15 minutes, the irradiance and temperatures can and will spike above/below the hourly values. One needs to understand the response time of the sensor when using this information. More frequently sampled data is better to understand the actual stress on the equipment, Phoenix, Arizona, for example, has had these recordings:

- 122 °F (50 °C), Phoenix AZ, 26 June 1990 (Ambient),
- 128 °F (52.3 °C), Lake Havasu, AZ, 29 June 1994, and 5 July 2007 (Ambient),
- 133 days of 100 °F (37.8 °C) (Ambient), Phoenix AZ, or more in the year 2023,
- 54 days of 110 °F or more 2023, and 53 days of 110 °F or more 2020,
- June 26, 1990, 120 °F (49 °C) 1995 & 2023,
- Sixteen (16) of the last 27 years have had 56 days or more of temperatures greater than or equal to 105 °F (41 °C).

The data in Table 4.2 covers only 28 years between 1995 and 2023. The value in performing a statistical analysis and allows for a higher CL in setting the design requirements that can be seen from the additional, detailed information. The use of the CL for temperature, for example, is that the estimate of 95% CL results in a design point upper limit of 125 °F (52 °C). This does not completely address the out-of-range stress on components, especially inverters where internal heating, when combined with the ambient temperature and the thermal effects of insolation on the cabinet, can cause the internal temperature to rapidly reach extremes of 130 °C or more with the loss of cooling.

What is often overlooked in this Phoenix, Arizona, case and other desert and high-altitude locations is the shift in substantive daily maximums and minimums. The effects of climate change and global warming are driving ambient temperatures higher more often and for additional daily hours. PV plants are also beginning energy production with higher morning and afternoon temperatures resulting in higher daily operational temperatures. This means that while these are ambient temperatures that are being recorded, they do not reflect the internal temperatures of modules, combiners, and inverters. We suggest that the industry have substantively more thorough and intense discussion regarding these. Then, it is highly recommended for the industry to update a variety of standards that will be impacted by these conditions.

For example, the above bullet list reflects some of the past history for Phoenix:

- In Phoenix Arizona, in 2020, the record number of high summer temperatures shifted from 33 days (2011/2019) over 110 °F (43.3 °C), dating back to 1896 when that record was first set, to 54 days in 2023.

Table 4.2 Example Environmental 27-Year Summary Data.

Environmental metric		Min	Max	Confidence level Avg	Std Dev	0.95 95% CL
Temperature (°F/°C)	**Annual max**		120/49	116/47	3/2	125/52
	Annual min	29/−1.7		34/1.1	3/2	25/−3
	Annual avg			76/25	2/1	80/26
Annual temperature exceedences	**Days ≥ 100 °F/38 °C**	75	140	105	14	129
	Days ≥ 77 °F/25 °C	210	272	245	17	273
	Days ≥ 105 °F/41 °C	20	80	54	14	77
	Days ≤ 32 °F/0 °C	0	4	0	1	6
Dew point (°F/°C)	**Max**		81/27		2.2	
	Avg			39/4	2.5	
Humidity (%)	**Min**	0	100.0	3.5	1.8	100
Wind speed (mph/kph)		31/50	54/87	43/69	8/12	66/106
Pressure (Hg)	**Max**	29.10	30.59	30.33	0.34	31.15
	Min	28.30	29.53	29.32	0.34	28.97
Rain (in/cm)	**Cumulative**	2.67/7.62	19.62/50.8	8.34/20.32	3.46/7.62	26.17/66.47
	Rain days	8	58	21	10	75

- The previous record for 115 °F (46.1 °C) was 7 days, while the 2020 record was for 14 days.
- The number of days over 100 °F (37.7 °C went from 143 (1989) to 145 days (2020)

This last item should be seen as a warning sign of how the changing environment will impact system health, condition, and performance. This kind of additional temperature impact is seldom if ever considered and addressed in specification and design.

As a result, inverters and modules will experience more stress over time as compared to what average data is telling us, which increases the risk of unanticipated failures.

As you look at the numbers above, you may wish to consider the following:

- These are ambient air temperatures, which should be compared to inverter operational temperature ranges before you consider that the temperature inside the inverter will be substantially higher. This is not so much because of the heat generated within the inverter itself. As it sits in the sunlight, the internal temperatures are substantially higher. In our perspective of systems in a growing number of locations around the world, those inverters will operate outside of their design temperatures and will be derated earlier in the day and later in the afternoon.
- As for modules, they can reach temperatures as high as 212 °F (100 °C). This is above their design parameters and excessive touch temperature. While the module and its components are seeing greater stress with the higher temperatures, the results are lowering power output.

- As for many microinverters, they may have the opportunity to experience the worst of both conditions. When mounted behind the module, they are artificially hotter in temperature as long as the solar radiation is falling on the module. This issue has been on our minds since the late 1980s and early 1990s, and has not been sufficiently addressed within the industry.

While we addressed Phoenix locations, other international locations have higher seasonal temperatures, while the extreme trends (highs and lows) appear to increase over time. Therefore, we would suggest that you consider design temperatures and the actual capabilities of the equipment more closely than has been the common practice in the past.

PV plants and many of their components are not designed to operate in these types of conditions.

All too often these extremes are referred to as aberrations. Yet, temperatures are trending up in many locations internationally, while these so-called aberrations are becoming more commonplace. However, historically over the last few decades, most specifiers and designers have not addressed these environmental issues adequately in their models, decisions, and component making choices. The reliance on averages tends to wash out and minimize the impact to system components and their ability to meet their mission.

Careful consideration of the extremes can make a difference between a system that can weather them versus one that is taken out by the first or second big or unusual heatwave, storm, or another event. This suggestion does not indicate, nor are we encouraging, that somebody is attempting to build a "bullet proof" project. Our position is clearly that most system specifications and designs do not adequately address the reality of the local site conditions. However, we encourage you to consider and then address the issue of "At what point does insurance and the potential of **SE/R**epowering™ per PAM or rebuilding come in to play." With what has been learned in the initial era of climate change, averages and the percentages that go into determining probabilities are changing and must be considered for the entire lifecycle of the system.

Has the whole team been effectively informed about what the probabilities of damage from an environmental issue can have on the short-, medium-, and long-term life of the project and system?

One attribute of a site is the amount of rain, hail,[19] ice, or snow received in each storm. Data from 1, 5, or 10 years may not cover the actual potential for erosion damage due to excessive rain, flash flooding, flooding, or wind/hail destruction from storms. Near Houston, Texas, for example, during Hurricane Harvey (2017) the "total rainfall hit 60.5 inches in Nederland, Texas, a record for a single storm in the continental United States that created an unprecedented 1000-year flood event…" with "two feet of rain [falling] in the first 24 hours." More importantly, it took several weeks for many areas to again see dry ground. The growing frequency, strength, and volume of hail damage is becoming of sufficient scope as to severely impact local energy production and the cost and availability of insurance coverage. Examples include:

1. "In May 2019, the solar industry was faced with a disaster unlike anything ever seen before, when a massive hailstorm passed through West Texas. In the path of the storm sat 174 Power Global's 178 MW Midway Solar Project, bolted to the ground, and pointed to the sky on 1500 acres near Midland....Once the storm had passed, the industry learned that it had left behind the largest weather-related single-project loss in its history. More than 400,000 of the plant's 685,000 Hanwha Q cell modules were damaged or destroyed; insurance losses totaled $70 million, and most everyone involved endured at least a few sleepless nights."[1]
2. "**Urban land and aerosols could accelerate hail storm formation,** Research from the US Department of Energy has found that urban landscapes and man-made aerosols have the potential to accelerate hail storm formation, make their winds harsher and direct these storms toward cities." *PV Magazine*, 6 January, 2021
3. "**Insuring against hail is all about risk management for solar project developers**," PV Magazine, 7 April 2021
4. "**PVEL adds hail test to its module certification program,**" PV Magazine 12 October 2021

The industry issues and consequences at hand are growing, costly, and disruptive. They will continue to impact module and other component design along with the ability to finance and insure PV projects internationally. Many of us in the industry have been concerned about these and a broad range of issues for some length of time (decades). Without making a concerted industrywide shift, PV system delivery as reliable energy infrastructure will continue to underaddress reducible and avoidable additional lost time, energy, and cost challenges.

Therefore, it is not sufficient to just collect data for documentation's sake; it must be properly validated, analyzed for the site, and cover more than a brief span of time. More importantly, stakeholders must be educated to the realities prior to decision-making. Anything less than 25 years does not provide the potential to detect and account for historic environmental extremes, much less the growing extremes of climate change!

While many PV professionals around the world pretend that climate change is not real or taking place, we counsel our friends and colleagues to consider carefully and address this trending series of extreme temperature and weather curves and their consequences. While doing so, consider that as PV plants are built today, in many instances they are not prepared for the weather trends of the past much less the future.

Previous standard elements that went into hail testing were basically a statement and testing methodology, which was in need of updating for some time. All stakeholders have an interest in them providing these types of standards especially when it comes to asset value and insurability.

[1] "Storm season has the US solar industry looking to protect assets from costly hail damage," PV Magazine 3/29/2021.

In our working group for IEC TS 63397, "Guidelines for qualifying PV modules for increased hail resistance," a substantial amount of discussion of what the standard should entail including what forms of testing requirements was for size and type of hail, angle of contact, positions of testing glass, frame, position, and velocity, among other issues. As we have mentioned previously about PV being a series of physics challenges, the hail issues have major O&M, asset value, and project viability issues. Hail damage affects LCOE values and plant viability.

There are a broad range of climate change considerations, some of which are addressed below:

- For climates that have cold nights but hot/humid days, the inverters can condense water from the air resulting in potential for electrical shorts.
- For inverter and module operation in extreme cold climates, there is an enhanced effect of thermal transient at "power on" in the morning/first light with the albedo reflection from snow/ice, which can spike or significantly increase the irradiance seen by the modules. This effect can present a potential overvoltage/current condition.

Note: In the early 2000s. Dr. Balfour's company sold a package for a complete set of system PV components for a university project that was being set up in Antarctica. When he spoke to his contact he asked, "Why are there so few inverters for this size of the array? If you size the system as you are, you will suffer overvoltage damage to the inverters and as a result will shut down your entire system while damaging the inverters." "I'd be happy to provide you with a set of specifications to address the appropriate ratios for modules to inverters for that environment if you would like," he offered. (Please note: At that time, there were no available charts or graphs of module and inverter sizing data aside from the data on cut sheets, that gave us the details to graph voltages and current at low temperature to establish a usable set of guidelines.

His contact replied confidently, "We have a professional licensed electrical engineering PhD on our team and we're sure that everything will be all right."

Because the equipment had to be shipped to Antarctica, it was going to take about six months to get it there and about three months to set up the system, test it, and make it operational. About nine or ten months after the equipment was shipped, the contact called him one afternoon.

After some polite discussion, the contact indicated that he would need additional inverters. Dr Balfour then asked, "How long did the plant operate before it shut down after the sunlight hit the modules?"

The contact was silence for a few moments and answered, "From the time the modules had full light, to the time the first inverter burned out was about 45 seconds or so."

While the cold temperatures and albedo (reflected energy from the ice and snow) were not addressed in the specification and the design, the results on the inverters were catastrophic. Dr. Balfour provided the correct number of inverters and two spares in case of a component failure.

- As per the example above, when designing a system for cold temperature extremes, the system will work as a system and not just as a bunch of components. The conditions that will impact the system were greater than the temperature and/or wind chill. In areas of snow and ice, the reflection (albedo) can dramatically raise the amount of energy going into an inverter and it can happen so fast the inverter may not be able to protect itself.
- Although inverters may derate or shut down during extremes, the actual operation temperatures within the inverter can reach 195 °F or 91 °C above ambient (Chapter 6). This measurement does not account for the thermal radiation from the backside of the material or the reflectivity of surface finish. However, this temperature is indicative of a potentially higher internal temperature for the inverter. An ambient temperature of 122 °F (50 °C) could result in a surface temperature of more than 265 °F (130 °C) – direct irradiance. The radiative component in the airspace behind the wall could aid in lowering the surface temperature.

It is equally important to illustrate the time being spent at each of these ranges. This is important as it impacts useful life as time at temperature extremes are also quite important. Derating will not recover environmental site lifetime, unless time and actual temperatures are factored.

A chain is only as strong as its weakest link – that is true in PV also. In addressing relevant limiting factors, we begin by reviewing each system element against the detailed site environmental factors and other design elements. This review requires establishing which components, subsystems, and systems are required for plant specification, measuring them against reliability documentation and data, while determining if all of the quantifiable metrics from environmental impacts have been adequately addressed, documented, and adjusted for.

Environmental issues that impact the PV plant lifecycle are shown in Figure 5.8.

For example:

- Have 25 or more years of historic climate data been established for temperature, humidity, wind speed, rain, hail, snow, etc.? Note: *More environmental data is better!*
- How much of the data is factual and how much is based on estimates or extrapolation (synthetic data)? (Chapter 10)
- Is the data complete or does it have a series of known gaps that would impact minimums, maximums, or averages?
- In some cases, data from a nearby standard weather station can be used. This data may be obtained from National Oceanic and Atmospheric Administration (NOAA), National Centers for Environmental Information (NCEI), airports, and other similar international organizations and sites.

Best practice: To understand the stresses on the equipment, a statistically significant set of data must be gathered and analyzed to set the specification, design, and operation limits.

For larger systems, setting up a portable weather station at the site, when it is being considered, has been selected and/or approved, allows a comparative assessment to be made on the data differences. It can clarify the accuracy of data from the local weather station site, satellite generated, or modeled data set to provide a better design baseline. Then if a site is a few degrees hotter or colder than the primary source data, adjustments can be made in the base data to accommodate the true site variance. If there is little consistency between the data researched and the site data, then additional measures must be taken to ensure that the plant equipment will function properly. Special care must also be taken to not "cherry pick" the local data to provide the best operating conditions that would result in a higher point estimate of good system performance. This type of method is not an acceptable substitute for a serious, comparative engineering climate study. Potential data sources to be used for historical comparison include:

- Local television or newspaper weather reporting,
- State or regional weather services,
- Local and regional airport data,
- Weather Underground, https://www.wunderground.com/,
- National Weather Service, https://www.weather.gov/gis/NWS_Shapefile,
- NOAA, https://www.climate.gov/maps-data/dataset/past-weather-zip-code-data-table,
- Available satellite and related data sources,
- Historic newspaper and magazine data which may not be digital in nature, and
- Local certified and uncertified weather stations, only if there are no other data sources.

A comparison may result in improved and valuable technical climate design assumptions. Care also must be taken to ensure that the data collection is somewhere close to the proposed/existing site, as a variance of more than one or two miles (1.6–3.2 km) or less can potentially have significant microclimate different values.

4.9 SE/Repowering™ Considerations

The four **R**epowering™ types previously mentioned are concept, existing (operating), distressed, and relocation.

4.9.1 Plant Design at Concept

SE/Repowering™ is a complete flexible modular system delivery process from cradle to grave that begins at project concept and ends at system site restoration. Decisions are data driven and not marketing language driven.

4.9.2 Existing Operable Plant SE/Repowering™

These are plants that will stay under control of an existing or new owner and/or owner operator. **SE/R**epowering™ practices are instituted after the plant goes into operations, where, at some point, the plant owners institute the **SE/R**epowering™ planning process while addressing present condition and operations. The plant may be underperforming, have serious issues and challenges, and must be brought up to a RAMS state of operability, performance, revenue generation, and profit. Or the owners come to the conclusion that if they wish to get the most out of a plant, they will redefine how they look at plants and operate them for the long term.

4.9.3 Photovoltaic Module Recycling: A Survey of US Policies and Incentives Distressed Plant

Distressed plants, including existing, abandoned, under or nonperforming, are usually flipped, sold in a fire sale, or bought in a bankruptcy due to a variety of plant energy production or O&M problems. The most acute is that of failing to meet contractual energy delivery requirements. These plants tend to exhibit numerous component and design challenges and have lost substantive asset value.

4.9.4 Plant Relocation

SE/Repowering™ that takes place if a PV system is going to be moved to a new location. We see this type of **SE/R**epowering™ to be risky and do not recommend its use.

The difference between the four types is that primarily, the initial new plant lifecycle stakeholders reap the greatest benefits. Nevertheless, all these variations, except plant relocation when initially planned at the start of the concept, have some level of benefit for the owners, operators, and O&M.

4.9.5 Stakeholder Need – Requirements

The stakeholders have needs, wants, and desires that must be accounted for, dispositioned, and documented. There are other points of consideration beginning with entire plant history, such as asset value, viability, energy production, and revenue to name a few. The following addresses some of the more common points between most stakeholders.

4.9.5.1 Substantively Better and More Accurate Asset Valuation

The ability to establish asset values for PV residential, commercial/utility systems benefits the entire industry and normalizes the long-term cost for energy. It creates a comparative market. This standardized approach allows for determination of both an accurate health, status and overall system condition that defines the value of a plant or product. The approach is central to the development of an entire PV industry infrastructure expansion and its technology.

4.9.5.2 Long-Term PV System Energy Production

In addressing the accurate economics of building or buying a plant, the **SE/R**epowering™ planning process allows for a more precise capability to predict what long-term energy production and component condition will look like. The ability to improve reliability, system availability, its maintainability and safety are key factors in the financial package which are often underassessed today. Identifiable and measurable facts begin to drive better modeling, driving profitability.

4.9.5.3 Plant Viability

The emphasis on clearly establishing health and condition is the baseline for defining system long-term viability. The process of reducing faults, failures, common errors, and anomalies dramatically reduce not only the O&M costs; they reduce a variety of soft costs that result from challenges, which can make or break a project over time. These soft costs include additional administration, operations, maintenance, organizational fire drills, legal costs, and organizational angst, along with a broad range of involvement and disruption for a number of organizational members, usually uninvolved with O&M activities. All of these soft cost elements are contributors to organization corporate angst, consuming organizational energy, finances, and productivity. The earlier these are addressed and resolved, the greater the opportunity for project success throughout and while extending its life.

4.9.5.4 Revenue

Revenue, which continues beyond the existing (20–30 years) lifecycle to 50 years or more, produces far greater profits and incentives for buy and hold investors. Plants degrade more slowly, while production and revenue curves continue longer with greater consistency. This is what moves the industry from a high-risk speculative effort to a levelized mature industry.

All financial modeling includes revenue projections. However, when those revenues are eaten up by avoidable unplanned costs, they tend to undermine fiscal viability of the project. When plants are built with better planning and implementation, not only do the revenues for the first phase of life (5–10 years) become more accurate, it also provides improved infrastructure of the plant to reduce costs for the next phase. For those institutional fleet owners, all of this provides a much higher investment grade opportunity at a lower risk, which results in greater longer-term success.

4.9.5.5 Ownership Operations Analysis and Decisions

With the details of an effectively designed **SE/R**epowering™ planning process, ownership operations analysis and resulting decisions can be more precise, realistic, and simplified. There is an increase of useful information to be analyzed automatically, providing a clearer picture of the entire system and its operations. Although the plan will not have addressed everything that can potentially be an issue, it addresses the vast majority of problems that owners will now be able to address before COD. This visibility can result in a template for effectively improving the discovery of issues with a set of basic plans to deal with them before, during, and after the issues have been resolved.

Better, more complete processes provide greater stability for plants over time, thereby reducing costs. Failing to implement these processes will result in incomplete questions that may be never completely answered, substantially identified, measured, planned for, and/or resolved.

4.9.5.6 Bill of Materials (BOM)

With an approximately 18- to 24-month product cycle time for major changes for products and the BOM, the short production time range tends to result in inconsistent product models from batch to batch (Chapter 3). The impacts to the industry are "baked in" during that period. The same product model number does not necessarily indicate or include the same product attributes or reliability. The inconsistencies create replacement and integration challenges affected by module rating, voltage, current outputs, chemistry, and form factor, among others. This variation in the BOM may have and has resulted in a specified product no longer being appropriate for use as initially required and intended. For inverters, newer modules may become more difficult to match by warranty, through defects, damage, failure, age, form factor, or other replacement issues.

The reality today for an order of say, 100 inverters, may take a manufacturer 6–12 months to build and deliver for larger power units, especially as PV growth continues to expand. Some parts have longer lead times even under "Just in Time" delivery. (This was the case even before 2021 and 2022 from Covid supply change challenges. There is a growing probability that following Covid, there may be an entirely different emphasis on "Just In Time" versus additional warehousing to offset supply disruption.) A simple substitution of a constituent material, component, or subsystem arising from supply shortage, alternate supplier planning, or value engineering can wreak havoc with the field reliability experience. The component for the BOM will most likely change over relatively short periods of time – not only during design and manufacturing but also after certification, even for the same model. Although not always the case, a small change in the design can significantly and dramatically change product RAMS attributes.

These changes can make it almost impossible to know what has actually been purchased. Without proper verification, there is no guarantee those changes will provide a better or similar product since there may be a negative impact on the actual product and PV plant reliability. Therefore, it is generally well worth paying for the services that do manufacturing inspection and reporting.

However, what might improve the ability to know essential product data would be an industrywide push to accept, apply, and if necessary, update the application of IEC TS 62915: 2018, "Photovoltaic (PV) modules – Type approval, design and safety qualification – Retesting"[20] and IEC TS 62093, "Power conversion equipment for photovoltaic systems – Design qualification testing."

These are critical IEC Technical Standards and steps toward improving product selection, reliability, and appropriateness for a specific environment, component function, and system viability.

We believe the industry and all of its stakeholders are not adequately addressing these standards, additional tests, and recertifications, which could result in dramatically better field results and useful product life. In project/system contracts, this and a number of other standards should be required and confirmed at installation and beyond. Yes, it would increase the amount of testing required by an OEM; however, it would have two beneficial impacts to those OEMs:

- The first is it would eventually result in fewer modifications based on what has been learned previously and the fact that the results would include potentially better selection of product components.
- Secondly, this would result in a better quality of product reaching the market, which has benefits to all stakeholders and may dramatically reduce the number of warranty claims, organizational fiscal challenges, and as we have seen from the past, manufacturer bankruptcies.

Table 4.3 Impact of Improved Reliability for Inverter Spares Reduction.

Average Down Time	36 Hours	Average Cost per Spare	$5,000	Average Cost of Maintenance Call		$1,000		
	Hours	**1/MTBF**	**Units**	**E(x)**	**Number of Spares**	**Average Annual Cost**	**Operating Down Time**	**Lost Power Generation**
MTBF 1	15000	0.000067	100	23	29	$138,000	828	23,000,000
MTBF 2	20000	0.000050	100	18	23	$108,000	648	18,000,000
MTBF 3	30000	0.000033	100	12	16	$72,000	432	12,000,000
Op Hours	3500							
Confidence	90%							

4.9.5.7 Spares

Better component selection and installation practices have a primary fiscal impact on product and/or component replacement. This step is the first toward reducing the number of components to be replaced throughout the plant life, along with the labor otherwise needed to have them identified, diagnosed, removed, and reinstalled. Fewer replacements over time will result in fewer needed spares, lower spares storage costs, and lower maintenance costs.

Table 4.3 (Chapter 6) is a simple example of assessing the impact of improving the MTBF of an item with an initial value of 15,000 hours (approximately five years of nominal operation). For **SE/R**epowering™ to improve power production over time as well as profitability, it is necessary to specify, verify, and validate the reliability of the system. However, many manufacturers include the cost of change for those components, while not actually improving the reliability of the item or product. This method is simply not weighing the actual "cost," which includes burdened labor, loss of production, and/or the impact to plant performance. The result will drive up the lifecycle costs. These are considered cost of ownership (COO) issues that are often poorly or not effectively addressed during the early project stages of concept, planning, and finance.

Shortened product development cycles further push the effect of putting products in the field that may have the same model number but not necessarily the same revision numbers. It is not simply a matter of improving RAMS attributes. It requires greater organizational commitment, especially from management, to drive long-term reliability affecting profitability. It zeros in on more robust design and test standards, reliability testing, and certification to improve the development of higher system quality that lead to a longer product life. To reduce risk requires additional, thorough investigation of what the RAMS specifier is recommending and what is actually built.

4.9.5.8 Education and Training

Education and training of not only the technical staff but engineering and project management, and field service technicians (FSTs) is crucial to success. Personnel only trained to perform robotic repairs can miss many potential points of plant health and condition that, if found and corrected early, provide for exceptional long-term savings.

Education: In the industry today, the vast majority of project or system stakeholders who make critical decisions are often specialized. They may not have a basic or broader base of cross-training of education in PV components, systems, subsystems and how they work in the field over time as a complete system. This results in a limited level of understanding of the impact of sunlight hitting on a surface and how that eventually turns into electricity then into usable energy. They may be proficient in law, accounting, business development, marketing/sales, or finance, among many others in the use of common terms within the industry, or while being aware of common RAMS, yet without having been trained to understand the relationships.

By default, we have many PV professionals who make critical decisions that affect the system RAMS throughout its lifecycle. This further creates confusion as the industry rapidly changes, while what might have been accepted previously whether myth or reality may no longer be valid.

There is an ongoing refrain of referring to "the experts" in those decisions. What we see throughout the industry is, with a lack of fundamental knowledge of PV, with components and how systems deliver energy, the reference to experts often is a default based on the "hopes" that somebody else will get the important details right. Therefore, major decisions default to assumption.

In our experience, we have interfaced with all of these types of professionals who may be exceptional in what they have been trained in. We have worked for developers who have asked for solutions to resolve issues whether it be system design, specification, or operations and maintenance. Often, they fall into three groups.

1. The first love to hear the language and what could happen as long as they do not have to provide resources or do anything about it.
2. The second are evolving with the industry, understand there are a growing series of challenges, and do not have the language or sufficient information to effectively address the cause and effect of decisions made.
3. The third are actively moving in the SE-RAMS directions, while having to overcome a broad range of internal conditions and decisions which are piecemeal. They are making progress. However, they do not have a clearly defined well laid out PAM/**SE/R**epowering™ set of guidelines, planning processes, activities, well-defined capabilities, and cross-organizational program, yet they are working on it. In essence, they are fighting the least-cost juggernaut inch by inch!

Industry results are that, all too often, we tend to do the same things over and over again as an industry, which at times delivers a form of insanity, because we are not speaking the same language of what critical wants and needs are. If we do not adequately educate the broader industry to understand that better decisions make more money even if you have to spend a little, we are stuck in an endless loop of substandard decision-making which on the surface seems to make sense.

The industry must come to terms with the overwhelming need for a clearly defined set of educational requirements and common language for all stakeholders in order to make better decisions. We might call it PV system delivery 101–102 with updates that explain the technologies and their most effective application. Doing so addresses how they work, how they fit into systems, and how the radiance from the sun passes through the system to create that electricity, and how it is stored and eventually used. It must be balanced with a basic understanding of the existing and newly discovered challenges found throughout the System Delivery process, what they are, their causes, how to preempt or mitigate them, and the necessary solutions.

Training As an EPC, it was not unusual to continually reinforce basic training on what needed to be done from selecting functional specification and design, components, handling, and installation practices in the field, including O&M. As our teams would go to trainings and speak to other PV professionals, we often had to clarify, what they had learned that was a value, and what was not. The almost universal comment for poorly determined action was: "That's what everybody else is doing," and/or "It's easier this way!"

Whether with PV participants from the office, site managers, or the PV on-site workforce in the field, they require a broader level of training and on-site management knowledge. This assists those stakeholders to ship, unpack, install, construct, test, or maintain while understanding what they are doing, why they are doing it, and provide them with enough information, to indicate when something has been done correctly or incorrectly. This means that many issues that are introduced during any phase of system delivery must be understood by the workforce and management to identify and reduce those challenges.

Unfortunately, just having one or two well-trained individuals on a project or crew is grossly insufficient. All too often, workforce participants are trained on specific activities or actions and are seen only as "labor."

- They may be taught to do the same thing repeatedly without fully being trained as how it and some variation impact the system.
- They may be assigned a variety of tasks without adequate cross-training.
- They may see issues that they feel are not adequately handled, while not knowing with certainty and as a result are silent.
- There are some installation staff that see and understand problems yet are supervised by someone who does not have the training or personal self-confidence to accept input from individuals who are viewed as his or her inferiors.
- Finally, there is the issue of "assumption" without supporting training or capability. Dr Balfour spent years in training his crews and employees to **"never assume anything"!** He commonly told them that when you assume anything it seems to cost us $50,000 in lost time and in corrections. In many instances, the costs and losses were much higher when asking for additional information could have avoided a number of issues, challenges, and errors.

Consistent high-quality training must be as regular and reinforced as with safety training and continually broadened and updated. Effective and valuable training is a substantive cost-cutting measure.

4.9.5.9 Codes and Standards

A number of system requirements are based on health codes, safety codes, and standards for design and construction, which are normally enforced by the Authorities Having

Jurisdiction (AHJs). Many other standards are the responsibility of the stakeholders to apply and confirm. The interpretation and application of appropriate standards and codes, which are applied to the design and/or operation, provide for a generally safer system, but not necessarily a more functional or reliable one. The assumption that following codes and standards is sufficient for improving system performance, reliability, and output has been a soap box topic the authors have focused on for years. It never ceases to amaze us how many engineers know the NEC code, yet have never read the introduction, which we have cited above. **The majority of codes are strictly for safety and nothing more!!!**

Some codes, for example, are mandatory and are an absolute minimum for health and safety. Currently, however, many of these codes and standards included in the contract are not included as part of the specification; or if they are, there is often no effective verification and/or validation process on their impact on reliability, quality, or performance attributes. Codes and standards are generally the result of a significant event(s) that later dictate that certain design or components are required to avoid health and/or safety problems from arising.

In our specifications, for example, especially in hot climates, we always oversized conductors and conduit as compared to what was required in code; this and many other minor changes helped us improve performance in the system on an annual basis. On the maintenance end of things, slightly oversized conduit made it easier to pull, while reducing insulation nicks or cuts and allowed us to make replacements in some systems that had been modified.

The development of standards (local, regional, state, national, and international) is a public, technical, and political process. Except where there is a clear immediate/vital health and safety issue, it is often a complex consensus-based process, which takes time and effort to issue an improvement or change. Consensus is a complex time-consuming process, as all the stakeholders involved have some aspect of knowledge, need, goals, self-interest, or cost involved. There are also times when this approach is hindered because supporting organizations tend to believe that their approach is best, that change is "too expensive," or that they will lose market share because the proposed approach is not currently performed by their organization. Indeed, even organizational internal stakeholders have conflicting goals and objectives that result in conflicting internal organizational interests. It is only later that they may discover that this conflict may not be accurate, true, or factual.

This juncture is when the regulatory agencies (e.g. NEC, NESC, NFPA, FAA, NERC, FERC, FCC, etc.) can and do force requirements on the industry. Once the regulation is in place, the affected industries must ensure and be able to provide proof of compliance. Then there is the issue of nonuniform interpretation. Not all labs or professionals interpret the standard requirements the same way, while not all certificates are from accredited institutions and not all AHJs impose or enforce similar requirements. Nevertheless, with all the codes, standards, and regulatory requirements, it is not unusual for organizations to do their best to influence, water down, gloss over, or ignore these requirements. This lack of effort can be seen in acquisition when the supplier claims that the parts were built in accordance with a standard, but do not have the proper in-house documentation to support compliance. This phenomenon is a cultural issue seen not only in PV but most other industries as well. Failure to provide verification of compliance leads to negative impacts on the system reliability, performance, system revenue, and profitability.

Financiers and insurance companies have increased premiums and decreased coverage or payouts as the result of a significant number of claims under force majeure and other events. They are recognizing the fact that many of the plants that they have insured are not meeting the basic capability to operate in the known site environmental extremes that should have been accounted for in the concept, specification, and design. In short, this means insurers are spending more money on these projects after the fact than anyone anticipated. (This is a data acquisition/analysis issue and challenge.) Because of the excessive claim payouts, many insures and financiers are no longer covering, or abandoning industry segments due to excessive losses or are not funding projects because of perceived or real risk. This has a negative impact on project viability. Once identified, this lack of attention is quickly noted to be unacceptable. This situation is all simply the result of NOT addressing the real-world issues of lifecycle bankability.

These identified issues and challenges are forcing new and existing projects to rethink or modify their approaches. To accommodate the new realities of the business, the industry is moving toward a PV system delivery process based on applying system engineering and advanced reliability approaches to improve specification, design, and construction.

How do these factors impact plant availability?

During the development of the IEC Availability Technical Standard, IEC TS 63019:2019, there was a long consensus process necessary to agree on the basic fact that availability was impacted by plant health, condition, and status (RAMS), which drove performance. Some participants believed that plant availability and performance are basically the same thing and can be measured by the energy output each day or other measurement interval. The authors' and other experienced professionals' perspective was that availability and performance must be viewed separately. Availability refers to whether the plant is *capable* of producing the power/energy when demanded, while performance is the active measurement of the power/energy being delivered. The former is based on the reliability and maintainability attributes of the plant, while the latter can be a direct measurement at a single point at the output of the plant. As shown in Chapter 8, there is a distinct relationship for availability between reliability and maintainability.

The final, standard consensus was based in part on a previous and long-term set of views from the NASA on PV reliability and availability being linked, Lalli, 1980. The link was direct because of the challenges in making a NASA "truck roll" (to dispatch a vehicle for equipment maintenance) to an extraterrestrial site – whether it be a satellite or potentially a different planet – is extremely expensive. The Hubble Space Telescope repair cost totaled over $2 billion, $500 million for upgraded optics and $1.2 billion for the shuttle flight,[21] were the perfect example of addressing this set of points of view. "If detected prior to launch, the repair would have cost $2 million, an investigation into the cause of the flawed mirror found."[22]

Another case in point is NASA's development of PV to support satellite power, which goes back to the 1950s. Reliability and availability were crucial to performance over time. Between the 1950s and 1980s, NASA developed requirements, measurement methods, and testing as an approach to address their needs for satellites to have more power and to last 20 or more years in extreme environments. The PV lessons learned from NASA have been available for 70 years, and yet the PV industry has generally ignored many if not most of these lessons, which has resulted in the repeat of the same problems decade after decade.

If it were not for NASA's (and other space agencies) work over the last 70 years along with the efforts in the automobile, telecom, computer, medical, aerospace, and other industries, we would not have the quality of many everyday products that we do today. While the needs of terrestrial PV are not as extreme, the example proves to highlight the link between reliability, serviceability, and availability. It also clearly allows us to understand that these improvements are achievable.

Once PV embraces RAMS needs and requirements as an essential facet of the **SE/R**epowering™ Process, the industry will improve dramatically while becoming far more financially stable. These points we are addressing and presenting in this book will have success and acceptance in PV applications once the value of that information is applied.

This perspective was clearly presented decades ago as documented in the following NASA report excerpts:

> "NASA Lewis senior management requested that the Office of Reliability and Quality Assurance assist the Photovoltaic Project Office to deliver hardware that would be reliable and safe. Figure 2 shows the relationship of R&QA with the Photovoltaic Project Office.
>
> These offices worked together and identified, evaluated, and either eliminated or controlled undesired system events with the potential to:
>
> - Damage system or support equipment and facilities,
> - Injure personnel, and
> - Render system unavailable.
>
>   These reliability and safety issues were accomplished by: Identifying the equipment functions and operations that may result in undesired events; assessing those events for impact and probability; by instituting methods to eliminate those events by reducing the event to an acceptable risk; and by verifying implementation of control measures in design, operating controls and procedures for installation, test, and maintenance."

The Concluding Remarks began with:

> "The modern solar cell power system is a product of the procedures, practices, and technology developed and used by the utility, construction, and aerospace industries. Development of solar energy as an acceptable, low-cost energy source requires solar cell power systems to demonstrate reliable, safe operation. The reliability assurance program developed for these systems makes use of the lessons learned from these industries and the safety, reliability, and quality assurance tools developed by them."

And:

> "The operational reliability of solar cell power systems is directly related to the prevention of single point failure modes. These failure modes are aggravated by the variable environment to which the power system is subjected and the need to operate and maintain the system for maximum availability.

To meet these challenges, an engineering reliability program was developed and utilized. This program involves a definition of the solar cell power system natural and operating environments, use of design criteria and analysis techniques, an awareness of potential problems via the inherent reliability and FMEA methods, and the use of a fail-safe and planned spare parts engineering philosophy. It is expected that this program, when coupled to an effective quality assurance and system checkout program, will demonstrate that solar energy systems, such as the village power system in Africa, will meet the reliability and safety objectives of the Federal Solar Energy Program."

Much of the previous work over the last 60 plus years in PV lay the foundations for PAM, the papers we published and this book. We included historic information that indicates that somehow, somewhere along the way, the industry lost sight of the very issues of system/plant capability, availability, reliability and how they are directly related to revenue and profit. If one is willing to focus strictly on first cost revenue and profit, that mindset alone will provide the foundation for why we have so many problems that have repeatedly been resolved decade after decade.

Thus, it is important to keep in mind we are not presenting unproven approaches that are new! With six decades of PV behind us, it would behoove the PV industry and all stakeholders to take stock of the progress made to date and use it to propel the industry toward true maturity with all the resulting benefits.

Figure 4.3 is the IEC committee modified Revenue Stream model, while Figure 4.4 illustrates the initial basic flow of information that the PAM approach advocates. It is based on the need for information that leads to an understanding of how specific decisions and actions can result in improved revenue (energy production).

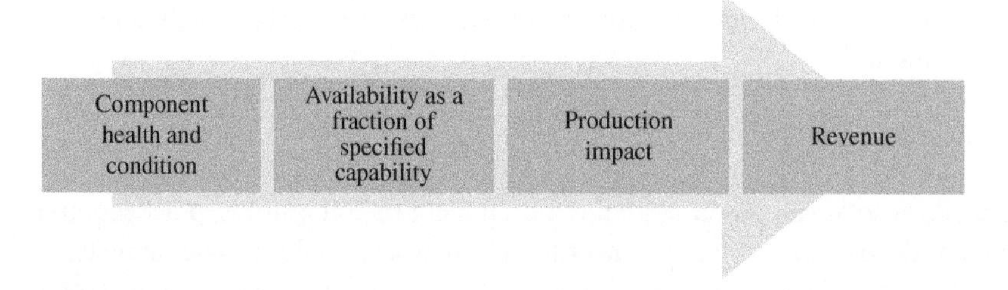

Figure 4.3 IEC 63019 Revenue Stream Model with standard Discussion Based on Initial PAM Publication.

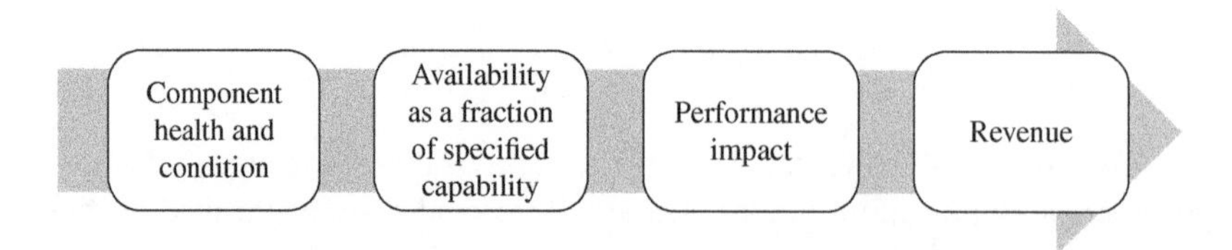

Figure 4.4 Flow of Information (Source from PAM paper).

Within the working group, there was some differentiation in the discussion centered on performance and production. In IEC TS 63019:2019, the term "Performance Impact" was changed to "Production Impact" by focusing on actual energy production. Doing so separated the calculations of energy produced from the process of measured actual output for saleable energy delivered. While we were initially communicating using the accepted industry language of "system performance (IEC 61724-3)," the discussion evolved to the critically related reality of production and associated revenues. The focus on our initial terminology preceded the IEC standard for availability, while the PAM process is driven by focusing on actual system performance and production revenue that results in energy production including short-term measurement of energy production (Chapter 6). As a result, we think the evolution of the terminology as codified in the IEC TS 61836:2016, and **Orange Button**[23] is an improvement on our work and support it completely.

However, what might improve the ability to know essential product data and health would be an industrywide push to accept and apply the application of IEC TS 92915: 2018,[24] PV modules – Type approval, design, and safety qualification – Retesting. This is a critical step toward improving product selection, appropriateness for a specific environment and component and system viability. It also requires many BOM changes to be retested which are not today.

Our position is that many product BOM changes for modules and inverters that are not retested should take place while many retests for certification have been ignored. By not addressing this as an industry, many products may not be able to meet published specification, thereby resulting in reduced performance and energy production revenue. This lack of application drives up component and labor costs unnecessarily. It is an issue for specified products and eventual results.

We believe the industry and all of its stakeholders are not adequately addressing the standard, additional tests, and recertifications, which could result in dramatically better field results and useful product life. In project/system contracts, this and a number of other standards should be required and confirmed at installation and beyond. Yes, it would increase the amount of testing required by an OEM; however, it would have two beneficial impacts to those OEMs and buyers:

- The first is it would eventually result in fewer modifications based on what has been learned previously and the fact that the results would include potentially better selection of product components.
- Secondly, this would result in a better quality of product reaching the market which has benefits to all stakeholders and may dramatically reduce the number of warranty claims, organizational fiscal challenges, and as we have seen from the past, manufacturer bankruptcies.

The Availability Index is a metric that establishes the relationship between the component or subsystem probability that shows how much the component will meet its specified and designed mission requirements (Chapter 8.). In other words, if a component or system element is not sufficiently specified to meet that mission, the associated reliability metric will indicate the probabilities of that element failing earlier than anticipated or expected. Variations in probability are primarily based on environmental and usage conditions over time. This means it is important to understand those changes on the component at subsystem levels.

4.9.5.10 What Does Conformance/Compliance Verification Entail? (Benchmarking, QA/QC)

Without clearly documented compliance verification, the result may be fiscal loss, potential shutdown, legal actions, contractual termination, and/or bankruptcy. These pitfalls are easily avoided by establishing an SE verification and validation process (Chapter 3).

Conformance/compliance can be verified in at least five different ways:

1. Design specifications that address requirements and limitations of financial and insurance industries.
2. Requirements developed and validated for and by the owner/s (first and future owners).
3. Certification, for example, established in an emerging IECRE PV Plant Rating and Certification Standard.
4. Validation and verification of compliance to local, regional, state, and country codes and standards.
5. Providing all the documentation for referral and use by later stakeholder review and use.

All these industry requirements are the result of a growing number of stakeholders attempting to offset the damage, losses, and unanticipated costs created through a first cost/least-cost project development model. The growing trend over the last few years is where additional stakeholders are beginning to address and discuss the weaknesses in the current process. Nevertheless, there is a difference between concern and discussion, and the commitment to make the necessary changes to offset the cost spiraling out of control.

4.10 Technology Fatigue

"How do Owners and Operators (O&O's) and asset managers address the rapid and short duration of a particular technology, which changes on average every 18–24 months or less, when delivering new product development and obsolescence that are seemingly endless?" A part of this consideration is that obsolete components may have a limited manufacturer support life and/or limited production requiring consideration of lifetime buys for spares.

Please note that a minor change in BOM can have a serious impact on component capability.

One of the most insidious issues today is continual "technology fatigue." It raises its ugly head in two ways that complicates specification and **SE/R**epowering, requiring additional consideration and planning. Generally, this issue is not adequately understood, discussed, and addressed during specification, design, and/or resolved early. This lack of attention and confusion drives up LCOE and O&M costs. They include the ongoing rate of the existing industry technology change model. This model forces rapid change and industry introduction without adequately addressing the reliability and long-term costs or performance of products. Although many or most products are improving, such improvement is inconsistent and may not have a long-term benefit, due to the lack of reliability information on which to base predictability or to plan the O&M and spares. Often, an older product is more reliable than newer variants or new models for the reason that the issues are largely already addressed, and field fixes are operationalized.

In a PV magazine article titled, “Bigger Is Not Necessarily Better” by Emiliano Bellini, 28 August 2020, Hongbin Fang, Longi’s director of product marketing, explained that the decision is in line with a clear strategy. “Our development logic is to look at every aspect of PV project deployment with holistic view and find best overall solution: not only ingot, wafer, cell and module manufacturing processes, but also module deployment processes, such as transportation, installation and system integrations.”

In our follow-up discussions with Hongbin, he agreed that the industry needed to take a deeper look at product change, not just for the specific item, but how those changes impacted the rest of the PV system delivery process. Because there is no format or forum at this time to discuss or address these issues, change accelerates geometrically while the inherent challenges created by that change seem to accelerate exponentially. Those changes are physical to the components while impacting the whole organization, all of the stakeholders, and the economics of systems as delivered today. Addressing these issues must bring a substantial segment of the PV industry together to come to a common perspective as to how to address this for the future.

An article in *PV Magazine* (2020/0828) quoted Mr. Fang (Longi Director of Product Marketing) as stating that “...increasing the bifacial module width beyond 1.3 m would make a crossbeam on the backside necessary to maintain sufficient mechanical strength. Crossbeams create shading on module backsides and reduce the power generation of bifacial modules. This will also increase the cost of module bringing limited value to customers.”

This is an awakening in the industry’s awareness that the industry has to discuss and resolve these issues if we want to address all of the RAMS issues and impacts. What may appear to be good decisions by a manufacturer to deliver an improved product may come back around to haunt them as a result of how the user must change their strategies and technologies. This growing underaddressed physical and financial industry issue is still flying below the radar.

4.11 Data Collection

One critical aspect of a **SE/R**epowering™ plan is the inclusion of Digital Acquisition System (DAS) and or Supervisory Control and Data Acquisition (SCADA) historical and current data for the plant (Chapter 10). The data process requires answering the following questions:

(a) How do we establish an effective data collection and curation process to determine what data is necessary and how it will be protected?
(b) What do we do with the information collected and analyzed?
(c) What does conformance/compliance verification entail?
(d) What are the data essentials?

It is considerably easier to gather as much data as possible and then cut back to the essentials. It is not unusual for an organization to identify essentials while falling short on necessary data. The most critical issue is based on what types of data are collected. For example, RAMS-related data regarding systems and subsystems should be a high priority and collected on a very granular basis. This is where you get the most out of data collection analysis to resolve existing and future problems. The more information collected and shared with the project stakeholders will lead to better designs that translate into demonstrably better and more financially stable and rewarding PV systems. In addition, this information is critical for addressing the reliability and maintainability aspects of the plant (Chapters 5–11).

When possible, it is imperative to implement methods of anonymizing data (Chapter 10) that can be shared between projects and PV professionals across the industry in general. The sharing of information in anonymized form should include the prime suppliers and their third-tier suppliers (components or assemblies). As discussed in Chapter 10, it is incumbent upon the PV industry to provide a storehouse of detailed PV plant fault, failure, and defects on specification, components, design, construction, operations, maintenance, performance, health, condition, and status and other data. This information will greatly improve new and **SE/R**epowering™ designs as the data will reduce the uncertainty of the various component attributes as well as reducing risk by identifying weak or poor parts and practices.

For example, does the inverter manufacturer provide all the status and failure data that the SCADA can process and store? To be specific, it is incumbent on the owner and O&M to contractually request, require, demand, and have access to copies of all failure reports and test results. If the contract calls for such data to be delivered and it is weak or unusable, there may be a need to perform a "deep dive" with the manufacturer to assess the problem(s). If the units have high failure rates and if the manufacturers do not provide the owner with all the relevant data, there may seem to be good reasons not to share that information from their perspective, but not to or for the benefit of their client or industry.

The lack of detailed failure data holds the industry back. It results in the inability to make the best decisions possible. Sharing of detailed data is critical to current and future improvements in the industry. When stakeholders of projects demand a better accounting of the RAMS attributes of the components or end items, this crucial foundation is fed back to specification. The results allow them to make far more informed, cost-effective, and qualitative decisions.

Example: The need for more detailed data is illustrated by the substantial module browning problems that were resolved in the late 1990s. The information led the industry to address the problems with the glass and ethylene-vinyl acetate (EVA) chemistry mismatch. Most industry professionals who dealt with that issue then believed that the root causes had been eradicated and were no longer a concern for the industry.

In the 2000–2012 time frame, some manufacturers opted to use module component chemistry that was incompatible, thereby bringing back the problem again just to shave pennies off the cost of modules. This shortened module life. This case shows another area where addressing the limiting factors is especially critical as they relate to the module bill of materials (BOMS). Using PAM methods provides for an approach to address these relapses making them much less likely to take place and with significantly reduced impact.

Assessing the two options based on improved component reliability data and substantially better information, the comparison can be made to indicate a probable value in both time and money as to the time and physical benefits of a specific selection. This finding can be balanced with a cost summary, which may indicate that the truly best product is more expensive than improved budgets can allow. However, when including all of the hard and soft costs, not just first costs of product replacement, i.e. components and labor, the view of what is too expensive shifts dramatically. We would expect, in time, that a better analysis of the true value of the more or most expensive components will be factored in based on the broadest range of potential benefit as compared to the limited least-cost approach today.

Questions about the long-term viability of a manufacturer are critical. However, that data is not going to be published in OEM marketing data, brochures, or online. The issue of product quality for components again requires digging deeper into the RAMS, SMART engineering, and a broad range of manufacturing and testing metrics. While standards such as the IEC, IEEE, ASTM, and others are actively being established and improved over time, they still do not provide sufficient information and guidance for risk reduction. To confidently accept many standards as solving long-term technical issues, just because the manufacturer, developer, EPC, or others are meeting existing standards, is a first step. However, success requires doing more. Our experience has been, when discussing the details of standards as claimed in bidding documents, that it requires delving into the following issues:

1. Are those stakeholders actually applying those standards or just giving them lip service?
2. Are they using current standards (latest updates)?
3. Are those standards appropriate for the claims being made?
4. Do the stakeholders actually have an available "inhouse" copy of applicable standards?
5. Do they understand the standards as evidenced by a verification process?
6. Do the standards as applied meet the requirements of the project goals and objectives in addressing limiting factors and the project specification?
7. Do the standards and specifications address the entire set of project risks?
8. How does the project plan address standards that may be inadequate?

It does little good to put standard requirements in documents and contracts that are not actually being read or understood, current, accurately applicable or implemented, verified and validated. Never the less, lip service is common. Just hiring an engineer or designer who has a ***copy*** of those standards and has read them does not necessarily translate into those standards being part of the specified, designed, and constructed plant. We have found disconnects between the understanding of contractual and AHJ requirements in the processes from specification and design through commissioning. In addition, when possible, through experience and data, it is advisable, when possible, to exceed those standards.

4.11.1.1 How Are the Metrics Being Measured Today? (Data Collection & Analysis)

A full analysis of the components and the companies that manufacture them elicits another critical set of questions, generally neither raised nor addressed. An example is the Bloomberg PV module manufacturer rating (previously mentioned) that refers to the companies as being either Tier 1, 2, or 3. Although the classification is supposed to address the financial viability of the manufacturer, or company involved, it is often mistakenly applied to the quality and reliability of the products, which makes it little more than a marketing term. Thus, this misused rating system can create avoidable, serious, and expensive consequences. People often make the assumption that the Tier 1 group may be of the greatest quality and reliability, whereas in many instances it is not! (See 2021 PV

Module Quality Report, PI Berlin: Trusted Solar Advisors, data and graphics from the PI Berlin Chapter 3 for greater clarification on module defects and challenges).

This situation is rather insidious. In a number of conference presentations, Dr. Balfour asked, "How many of you use Tier 3 Modules?" Although everyone looked around, nobody raised their hands. He next asked, "How many of you use Tier 2 modules?" Again, the same results. Finally, he asked, "How many of you only use Tier 1 modules?" Everyone raised their hands. And they did so with the level of pride that indicated that they knew what they were doing and that they had always made the superior finance selection. Though the poll was rather unscientific, it illustrated a broad-based lack of understanding regarding the actual use of the Bloomberg rating.

He looked around at the audience and asked, "If all of you are using Tier 1 modules, who is using the other ones, the Tier 2 and 3?"

If this Tier assumption is true, then why are Tier 1 companies going out of business? Other informal polls indicated that some to many developers, EPCs, and other solar professionals understood the intent of the rating system as being financial only; however, those individuals are far from being the majority. But as Dr. Balfour noted at that time, "So far, I have not been able to find anyone who admits to using anything but Tier 1 modules, and if none of you are, who represent a good section of the industry, then who?"

He wrapped up that issue by stating, "The fact is that there appears to be few Tier 2 or Tier 3 modules purchased because there is primarily a preference on reliance on a Tier 1 list of manufactures, and unfortunately this has become strictly a marketing term and claim. The reality is that all tiers are being specified in projects today and what may actually be installed may not be what the stakeholders assumed. ***The Bloomberg tier process does not guarantee the viability or bankability of the manufacturers or their products.*** That requires separate services. The rating system is only a 'snapshot' of the current bankability of the listed companies, which represent a segment of all PV module manufacturers. In short, 'Who is buying from the remaining modules from the other companies?' As recent history has shown, even previously rated Tier 1 companies have either quit the business or gone out of business, in part because often, the quality and/or reliability of their product did not hold up in the field. As a result, if you are relying heavily on this one metric, you may end up with some ugly surprises by not having a clear understanding of what is being produced, how it is being produced and how it is changing as far as the BOM is concerned following certification."

4.11.1.2 "Who Should Gather the Data?"

These steps will require both a learning curve and additional research. It will soon become second nature where most of the required data is automatically gathered in the process. In a relatively short period of time, those additional research elements will more than offset O&M costs in the first few months or years compared to not anticipating the actual site conditions and specifying the appropriate components.

Most serious losses related to cost in the process result from not having a clear, focused, and effective data acquisition and analysis plan and system that works. By determining not only what information needs to be gathered and protected, it must include selecting the right equipment and software to make the process work. Too many companies seem to pay for new equipment and software in short replacement cycles. This is cost prohibitive yet essential due to many AHJ, regulatory, grid access, other requirements, or *an unclear vision of their real needs.*

However, success requires the additional data from visual and nontraditional site data and testing that may be acquired through service inspections, and an array of multi-spectral and wavelength tools.

4.11.1.3 What Metrics Are Being Taken, and How Are They Being Measured Today?

When an organization begins the **SE/R**epowering™ process, it is critical that they look at what metrics are being collected for projects compared to what metrics are needed to understand the *current and future* condition of the system. Likely, they will be substantively different. In all instances, they will need to be more comprehensive because each metric has a clearly defined value and use. While developing the evolving list of limiting factors, we quickly discovered that most of the issues that need to be addressed do not have any or consistent metrics related to components or subsystems, or other potential issues. In the real world, some of these metrics are more in the form of incomplete explanations of process and procedure that must be adhered to.

The importance of collecting the relevant data is to clearly see the differences between what the condition of the current system is and to generate the guidelines and requirements for meeting all existing and future systems requirements. It is also important to remember that the foundations for buying better equipment require greater product information to deliver greater lifecycle value over time.

We began the development of the **SE/R**epowering™ process and approach when we delved into the concept of system 'limiting factors.' These 'limiting factors' include all the thousands of elements and attributes of PV System Delivery that define the capabilities of a system to meet its intended function of generating and delivering power/energy to customers as planned, specified, operated, and maintained throughout its fiscal/energy production lifecycle mission. Using **SE/R**epowering™ methods, processes, approaches, and milestones events, such as upgrades, refurbishment and replacement must be planned and accommodated as part of the initial design process. The attributes necessary for the successful application of this process include environmental site related conditions, equipment capability organizational philosophical approaches, goals, objectives, decisions, and actions to be taken or not taken. All of these, in total, define the PV plus ESS system lifecycle, that affects the asset value, how much energy can be produced and at what cost.

The view we developed over time is that a flawed business model delivers a set of flawed assumptions, choices, and outcomes, that has resulted in the poor economic performance of many of today's PV plants. This is supported not only by the experience of many actively engaged professionals, it is supported by over four decades of public and private industry research, development, and documentation.

We also understand that many PV professionals are aware of many of these flaws, which are avoidable or that can be improved upon when the system plans and specifications are detailed and complete. However, the present business model seldom empowers these professionals to make better, more fiscally sound lifecycle decisions. This includes, in many instances, the appropriateness of and real environmental capability of many components, items, and products. The specification, application and use of these parts greatly affects design and system choices. It requires, for example, that a more in-depth and detailed statistical analysis of actual site environmental conditions is essential to understand the stresses on components derived from shared field failure data.

Today's repowering is often little more than an O&M Band-Aid™ to slow the haemorrhaging or bleeding of money as a result of lost energy due to failures and masking. When taking a serious view of lifecycle "All In Costs," the most ***credible*** lowest cost or LCOE, favors a PV systems lifecycle planning and delivery approach, which we define as **SE/R**epowering™.

Bibliography

Balfour J.R., Hill R.R., Walker, A. et al. (2021). *Masking of photovoltaic system performance problems by inverter clipping and other design and operational practices*. United States. https://doi.org/10.1016/j.rser.2021.111067.

EPRI (2010). Addressing Solar Photovoltaic Operations and Maintenance Challenges, A Survey of Current Knowledge and Practices", EPRI #1021496 July 2010.

Hill R.R., Klise G., and Balfour J.R. (2015). Precursor Report of Data Needs and Recommended Practices for PV Plant Availability Operations and Maintenance Reporting. United States. doi: https://doi.org/10.2172/1169447.

IEC TS 61836:2016. Solar photovoltaic energy systems – Terms, definitions and symbols.

IEC TS 63019:2019. Photovoltaic power systems (PVPS) – Information model for availability.

IEC TS 63265:2022. Photovoltaic power systems – Reliability practices for operation.

Lalli V.R. (1980). *Photovoltaic power system reliability considerations*. NASA Lewis Research Center, Annual Reliability and Maintainability Symposium, San Francisco, California, January 22–24, 1980.

PVQAT TG 11/discussions (2021) of IEC 62093-2022. Photovoltaic system power conversion equipment – Design qualification and type approval.

TamizhMani, G. (2014). *Reliability Evaluation of PV Power Plants: Input Data for Warranty, Bankability and Energy Estimation Models*. Workshop. http://www.nrel.gov/pv/pdfs/2014_pvmrw_04_tamizhmani.pdf.

Notes

1 Long-term stakeholders include those that may not be present at the start of the project or even present during the first 10 years of operation.

2 Data curation is the organization and integration of data collected from various sources. It involves annotation, publication, and presentation of the data such that the value of the data is maintained over time, and the data remains available for reuse and preservation. Data curation includes "all the processes needed for principled and controlled data creation, maintenance, and management, together with the capacity to add value to data" (Miller, R. J. (2014). "Big Data Curation." In: 20th International Conference on Management of Data (COMAD) 2014, Hyderabad, India, December 17–19, 2014), https://en.wikipedia.org/wiki/Data_curation.

3 The various traditional types of maintenance include: Preventative (or sometimes Preventative) maintenance (PM), Corrective or reactive maintenance, Condition-based maintenance (CBM).
4 Best Option – This case is where an FMEA and reliability predictions has been performed, and the best combination of cost and performance is used to select the component or product.
5 "Engineers were pushed to submit technical drawings and designs at roughly double the normal pace, former employees said. Facing tight deadlines and strict budgets, managers quickly pulled workers from other departments when someone left the Max project. Although the project had been hectic, current, and former employees said they had finished it feeling confident in the safety of the plane." https://www.nytimes.com/2019/03/23/business/boeing-737-max-crash.html.
6 "SunEdison went bonkers on the acquisition front," said one former SunEdison employee. "There was no management around the table to say this does or doesn't make sense financially." https://www.latimes.com/business/la-fi-sunedison-collapse-20160504-story.html.
7 "Repowering" What It Is & How Does It Impacts the Bottom Line!", New Energy Update Conference, San Diego California, November 7-8, 2018, John R. Balfour.
8 "Repowering: The Impact of Components Across the Lifecycle of the Plant", Session 7Aii, Solar Asset Management North America, March 14, 2018, John R. Balfour.
9 "Repowering" What It Is and How Does It Impact the Bottom Line," Session 7A, Solar Asset Management North America, March 14, 2018, John R. Balfour.
10 "PV Plant Repowering, the Utility 50-Year Systems Model," http://www.renewableenergyworld.com/articles/2017/04/pv-plant-repowering-the-utility-50-year-systems-model.html, John R. Balfour 2017.
11 "Preemptive Analytical Maintenance (PAM), Introducing a More Functional and Reliable PV System Delivery Model," John R. Balfour and Russell Morris 2018.
12 "The Solar PV Life Cycle Dilemma," http://www.renewableenergyworld.com/articles/2017/05/the-solar-pv-life-cycle-dilemma.html, John R. Balfour 2017.
13 "PV SYSTEM REPOWERING, FEASIBILITY, VALUE AND PRODUCTIVITY" https://sam-northamerica.solarplaza.com/news-source/2018/2/22/pv-system-repowering-feasibility-value-and-productivity.
14 https://www.woodmac.com/research/products/power-and-renewables/us-solar-market-insight/.
15 Provide Testing requirement from PI-Berlin paper.
16 An assessment specification is performed to determine where the existing plant is and what the **R**epowered plant needs to be.
17 Sol-Air Temperatures, Chapter 25, 25.4 ASHRE Fundamental 1977.
18 "Standard Method of Test for the Evaluation of Building Energy Analysis Computer Programs," ANSI/ASHRAE Addendum a to ANSI/ASHRAE Standard 140-2017, 31 July 2020.
19 IEC TS 63397 ED1 Guidelines for qualifying PV modules for increased hail resistance.
20 EC TS 62915:2018(E) sets forth a uniform approach to maintain type approval, design, and safety qualification of terrestrial PV modules that have undergone or will undergo modification from their originally assessed design. Changes in material selection,

components, and manufacturing process can impact electrical performance, reliability, and safety of the modified product. This document lists typical modifications and the resulting requirements for retesting based on the different test standards. This document is closely related to the IEC 61215 and IEC 61730 series of standards.

21 https://www.space.com/11358-nasa-space-shuttle-program-cost-30-years.html.

22 https://www.baltimoresun.com/news/bs-xpm-1993-11-28-1993332073-story.html.

23 https://www.energy.gov/eere/solar/orange-button-solar-bankability-data-advance-transactions-and-access-sb-data.

24 EC TS 62915:2018(E) sets forth a uniform approach to maintain type approval, design, and safety qualification of terrestrial PV modules that have undergone or will undergo modification from their originally assessed design. Changes in material selection, components, and manufacturing process can impact electrical performance, reliability, and safety of the modified product. This document lists typical modifications and the resulting requirements for retesting based on the different test standards. This document is closely related to the IEC 61215 and IEC 61730 series of standards.

5

System Engineering

5.1 Introduction

"A system is an assemblage or combination of functionally related elements or parts forming a unitary whole…" (Blanchard and Fabrycky 2011). The Photovoltaic Power System (PVPS) (Krauter 2006) is comprised of a relatively few different component types, but in extremely large numbers. These include cells, modules, panels, combiners, inverters, other balance of system (BOS) components, and transformers. As mentioned in Chapter 1, our focus is on establishing a basis for a paradigm shift from the current project models based on the lowest cost to an optimized system model that meets all stakeholder needs and requirements. These efforts use well-known established but often ignored processes that are available today to accomplish this goal. Among these are Systems Engineering (SE), **R**epowering™ planning, reliability (R), availability (A), maintainability (M), and safety (S) or RAMS.

This chapter does not present a single case study but uses various examples while pointing out some of the tools that can be used on projects to be successful over the entire lifecycle of the plant. The intent is to provide the reader with a good foundation of information to be able to conceive, specify, design, develop, install, and operate profitable PVPS. To that end, we use a 100 MW PV plant as a general example. The consideration of the **SE/R**epowered™ process is to plan for the life of the plant, say 50–100 years, which requires some level of overbuild to compensate for the variables associated with PV technology. This is a significantly different approach than the current PV project delivery method.

SE has many definitions, but they all essentially state that SE covers all aspects of a project. A good definition and description of SE are given by INCOSE-TP-2003-002-04 "Systems Engineering is a transdisciplinary and integrative approach to enable the successful realization, use, and retirement of engineered systems, using systems principles and concepts, and scientific, technological, and management methods." See also Woolf (2012), Kerzner (2003), Ventre and Messenger (2010). In other words, SE is based on a world view of each project by understanding the functional and performance expectations of the system to meet the needs of the stakeholders over the life of the system including PV, site restoration. Using this information, the SE engineer models the attributes of the system to support the initial and detailed system design (software and hardware) including hardware/software specifications, constraints, and other specialty engineering functions such as RAMS, logistics, and supplier quality. Verification of performance has been outlined in IEC 61724-1:2017.

Photovoltaic (PV) System Delivery as Reliable Energy Infrastructure, First Edition.
John R. Balfour and Russell W. Morris.

This chapter will cover the SE function and what processes are doing at each phase while providing the reader with enough information to:

- Understand the definition of systems and SE.
- Understand the need for applying SE processes to PV projects.
- Provide tools to develop a plant concept, system design, and allocation of requirements for sending out to potential contractors or EPC companies to bid on through the request for quote (RFQ) process.
- Development of Measures of Effectiveness (MOE)[1] for comparison of competing designs.

One limitation of this book is that we cannot hope to cover all potential configurations of PV plants from an SE/RAMS perspective. There are string and central inverter, module/panel inverter configurations, as well as fixed tilt, one-axis, and two-axis tracking systems. Therefore, we use fixed tilt module/panel installation and string-inverter configuration for the working example. All of the points are, however, applicable to other configurations.

5.2 Why Systems Engineering

The answer is simple: SE provides the means for the project team and stakeholders to not only understand their requirements but the impact those have on the other stakeholders and the specification of the plant. It also integrates the design and specialty engineering into a team that can trade cost and benefit that optimizes the concept, specification, design, and O&M of the plant.

One specific study, using data from programs, was performed by NASA. Figure 5.1 illustrates the impact of employing SE methods on a project along with the relative cost impact of doing so. The more effort that went into SE, the lower the overall lifecycle cost and schedule impacts on various projects. And as noted in the graph, there is an optimum point at which the percent of budget spent is no longer providing a marked improvement in the product. This is in the range of 5–15% for the NASA projects. One could argue that NASA has a narrow focus and "needs" or must use SE for their cost, human, and schedule risks. But personal experience of one of us (Morris), with over 50 years working in a wide variety of industries and programs, has found that failure to use SE principles and practices leads to not only significant cost overruns and schedule slips but failure to meet the customer's needs and expectations. When comparing the NASA 5–15% SE front-end cost investment to the needs of PV, it should be noted that the PV plus energy storage system (PV+ESS) technologies and products have very different risk extremes. PV + storage risks tend to be far lower on the risk scale. The message from the NASA study and industry experience is that a reasonable PV investment of 5–6% allows for improvement in the lifecycle cost reduction while improving the overall performance of the system.

The SE team consists of the systems and specialty engineers, the lead design engineer or EPC equivalent, representatives of the stakeholders, and the owner. The SE team acts as the "glue" by bringing the stakeholders together, to identify unique needs and requirements,

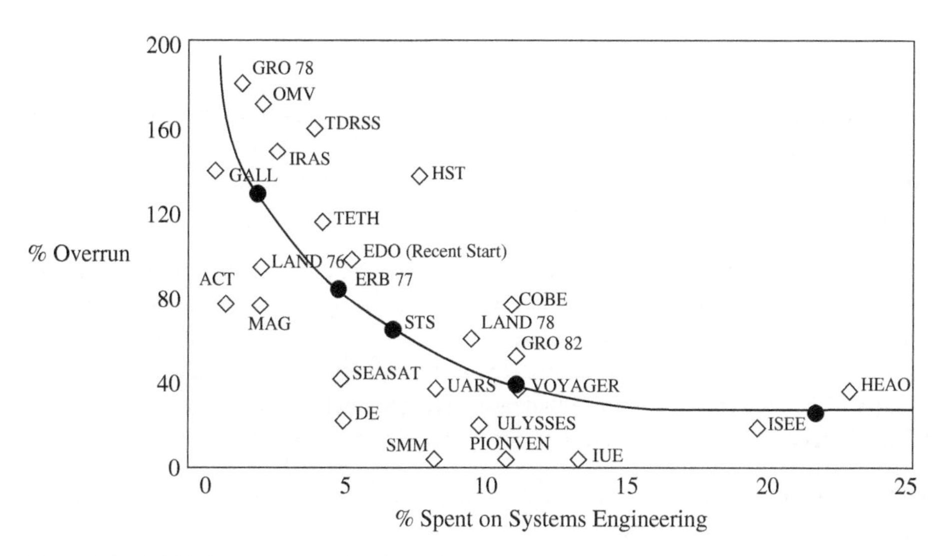

Figure 5.1 NASA Assessment of Cost Overruns versus Percent of Budget for Systems Engineering.

defines success and failure, and then builds an optimized system specification that reflects all those requirements.

NASA has developed a number of SE processes as documented in the NASA Systems Engineering Handbook. This emphasizes the need for understanding requirements, costs, and schedule.[2]

Examples (with little or no public data, but public exposure) include the 787, which was well over three years late, the F-22[3] (contract award 1991), due in 2000, was not operational until 2005 and did not achieve full capability until 2011. Many of these delays and cost overruns are related to changing requirements, management errors, and design and development problems. As a generalization, projects that provide no or truly little funding to perform SE generally end up with cost overruns, schedule slippages, loss of name value, loss of income, and other financial impacts to the project. These are in addition to changing requirements, a major problem with Department of Defense (DOD) contracts, which for PVPSs should be avoidable.

Without proper scheduling, the stakeholders cannot accurately assess the impact of a lack of detailed or missing changes to requirements on the cost and schedule of a project. Project delays, cost overruns, and poor performance problems often show up initially as schedule delays. Stakeholders must understand a project's schedule as a requirement for an accurate budget and for tracking performance and progress.

The prior history of PV has been fraught with well-intentioned efforts to design low-cost plants believing that PVPS is simple, easily maintained, and inherently cost-effective. As noted in Chapter 1, this is a significant myth that must be addressed to achieve effective renewable energy from the sun. General issues with PV plants have included many technology and engineering deficiencies. These include (but are not limited to):

- Lack of understanding the technologies and their limitations,
- Excessive cost for O&M,
- Failure to address all stakeholder needs in detailed specifications,

- Bidding assumes that the technologies used will meet all the manufacturer's specifications without validation or verification (such as reliability testing),
- SE is missing from many existing plants given plant failures, bankruptcies, fire sales, or abandonment,
- Reliability and qualification testing are done on a limited basis, and
- Field data are missing, or critical elements of the data are limited.

Over the last 30 years, there have been an increasing number of PV plants built and added to the local, regional, and national power grid. Unfortunately, and within a relatively short period of time (7–10 years), many of these plants began to fail to meet contracted energy delivery, or were delivering energy at an excessively high cost, or were just too expensive to maintain. This trend can be circumvented by applying SE methods and processes beginning with the concept. The fundamental steps for PV requirements development are:

- Obtain and analyze stakeholder needs and requirements on functionality at the beginning of the development cycle,
- Develop system requirements from stakeholder needs wants and desires,
- Design synthesis and system validation while considering the complete problem,
- Define the system lifecycle, i.e. concept to site restoration,
- Fully understanding all the stakeholders involved,
- Performing trade studies to reach a compromise that all the stakeholders can agree on, and
- Environment and community concerns.

From Chapter 2, Tables 2.1 and 2.2 provide a summary of the internal and external stakeholders. These people or organizations should be contacted directly and have a representative in person at project inception.

SE is a set of tools and processes that provide for an optimized design to ensure that a project's stakeholder's needs and requirements have been identified, documented, accounted for in the design, and met over the lifecycle of the PV plant. The SE process and tools are applied during all phases to integrate all stakeholder needs and requirements for them to be successful.

The SE function, when integrated with the rest of the project team, is expected to perform the following tasks:

- At concept, the SE is part of the system delivery leadership team that addresses the full range of SE tasks as presented in this book,
- Monitoring and oversight of all installed PV systems and infrastructure,
- Establish, configure, test, and maintain the configuration of PV plant operating systems, application software, and system management tools,
- Evaluate the existing PV systems control elements and diagnostic or fault isolation capability and provide the technical direction to software engineering staff,
- Oversee the development of PV plant-specific software and hardware requirements and testing,
- Plan and implement systems automation as required for improved effectiveness and efficiency,

- Formulate and design the security system in place to provide physical security to maintain plant and data safety,
- Oversee the availability of technical resources,
- Maintain and supervise the inventory, and
- Timely reporting on the performance/trouble log sheet for the rapid response to any problems that arise in performance, safety, and RAMS.

Although there are many books, such as references Blanchard and Fabrycky (2011), Kalogirou (2013), Gevorkian (2011), Messenger and Abtahi (2020), Kim 2015, INCOSE-TP-2003-002-04, that support implementing SE over the life of a project, the INCOSE SEBoK (SE Book of Knowledge) manual provides extensive detailed coverage of SE principles and practices. The reader is encouraged to obtain a copy for reference. Within the SEBoK, SE has a further description: "SE is a transdisciplinary approach and means to enable the realization of successful systems. Successful systems must satisfy the needs of their customers, users, and other stakeholders."

To address the transdisciplinary characterization, the SE team generally includes the following specialty engineering skill sets:

1. **Reliability, availability, and maintainability**: perform analysis needed to understand the state of the physical plant and predict future performance, *DOD Guide* 2005,
2. **Human engineering**: performs analyses of maintenance, tools, handling, and safety,
3. **Security engineering**: provides plans to control plant accessibility for disruptive action, theft, and maintenance access,
4. **Electromagnetic interference/electromagnetic compatibility**: provides data acquisition and reports for EMI/EMC interference in the surrounding community,
5. **System resilience**: works in conjunction with RAMS to determine the capability of the plant to absorb and control failures in addition to understanding operations, performance, and life of equipment for specific site environments,
6. **Manufacturability and producibility**: answers questions about technologies being planned for implementation through the RFQs and if there are any cost, schedule, technology, and operation risks,
7. **Affordability**: contracts, subcontractor monitoring, and control,
8. **Environmental engineering**: consider environmental impact issues like site, climate, soil conditions, pollution sources (air, noise, light), and proximity to sensitive installations,
9. **Safety**: this topic covers not only regulatory safety, but also general operational safety of the system.

The reliability, maintainability, human engineering, systems resilience, and affordability topics are covered in Chapters 6–7. Safety touches many disciplines, is covered as a topic within systems, reliability, and maintainability, and impacts the **SE/R**epowered™ process and energy storage (Chapters 4–8 and 10). Other topics in the list above are not covered in detail in this book but are critical to success.

One additional note is the need for an SE team is regardless of the size of the company. A small company may only have one or more systems engineers while larger companies will (or should) have a large number of systems engineers as a project resource. Smaller companies may hire consulting solar power SE engineers as they have the requisite extensive knowledge and broad backgrounds in concept, specification, design, and the operation of PV plants. Regardless, the SE engineers must be well-versed in PV.

5.3 SE Process

SE is an integrated multiple specialty discipline that relies on establishing the best understanding of project requirements while recognizing that even for the best of practices and intents there are unknowns that can and do crop up and result in poor performance and/or early failure. The goal of expending effort in SE up front and throughout the lifecycle is to reduce project/stakeholder risk by identifying and resolving problems before they have large technical, financial, and schedule consequences.

A step in ensuring that the system and detailed design are consistent and meet specification is to establish a PV plant functional diagram that increases in detail as the project progresses. Figure 5.2 illustrates the general form and function in an SE context. The PV plant input is, of course, solar irradiance, and the output is AC electricity delivered to a customer through an interface, generally a utility. Control has two elements, local and remote. The local control is most often performed through a SCADA subsystem. The SCADA remote monitoring and control are performed by an offsite agent that monitors the status of the plant, generates work orders, tracks spares usage, and tracks energy delivery.

Note: This is NOT a least-cost system's design. It is one that trades investment cost for the long-term optimized system output and longer-term maintainability, with the lowest operational cost resulting in a lower risk to all the stakeholders.

Figure 5.3 provides the next step in the system concept development of a PV plant. The "I1" and "O1" are simply the "input" or solar irradiance, and the "output" is the electrical energy to the customer/end users. Using the INCOSE SE Handbook, INCOSE-TP-2003-002-04, the "control" and data acquisition function are performed by the SCADA system. The "Enablers" for the PV plant design include equations, Computer Aided Design

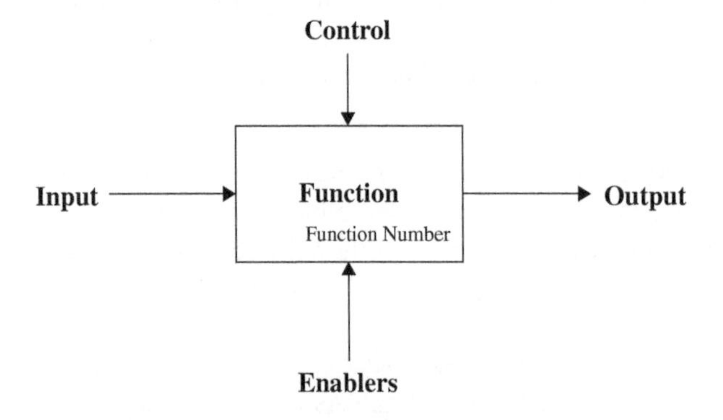

Figure 5.2 Basic SE Function Block.

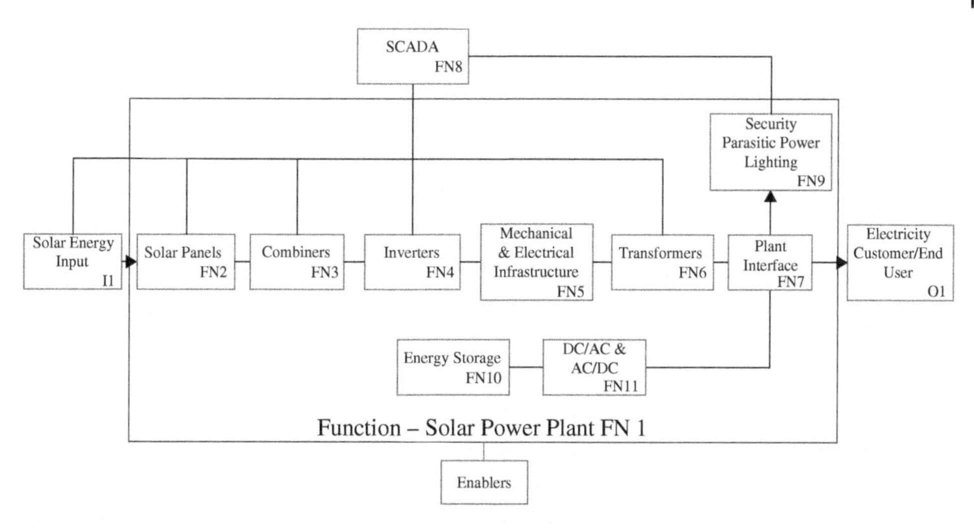

Figure 5.3 Example of a PVPS System with Energy Storage.

(CAD)/Computer Aided Modeling (CAM) drawings, diagrams, operational attributes and acceptable margins, software algorithms for control, and heuristics or lessons learned and testing. Enablers can also include UV, IR, and other remote energy sensing. The remaining functions are the top-level subsystems that perform the function of energy conversion. The systems engineer/s provides a description of each function to begin the process of developing the concept but also to develop the specifications. As is covered later in this chapter, the specification proceeds from the system function level to the detailed function.

The initial functional descriptions are:

- **FN1**: Convert solar irradiance to AC electrical energy to supply to their customers. This is the global function of the plant.
- **FN2**: Convert solar irradiance to DC electrical energy consistent with the electrical requirements of the inverter.
- **FN3**: Aggregate the outputs of multiple solar panel strings into a single input to each inverter.
- **FN4**: Inverter converts DC energy to AC energy.
- **FN5**: Provide mechanical structural support for the solar panels, wiring interconnect (wires, connectors), and pads for inverters.
- **FN6**: Buffers between the inverter electrical output to the grid measure the input/output energy and sum the energy output from every inverter.
- **FN7**: This is generally one or more utility-grade transformers and associated switch gear, breakers, and point of interconnection (POI) measurement of energy delivered to the customer/end users.
- **FN8**: Provide the monitoring capability for the inverters, combiners, strings, as well as monitoring the temperature and energy quality of the output transformers and perform emergency shutdown of an inverter or the plant when a fault is detected. This is the control function from Figure 5.2.

- **FN9**: Provides for the site fencing, nighttime lighting, data acquisition, tracking and storage, security monitoring, and other plant functions within the plant boundaries.
- **FN10**: Provides surplus energy storage for smoothing energy output when clouds reduce the energy input and to aid in generating a stable ramp up and down of the energy that the plant provides to the customer.
- **FN11**: Provides the conversion of AC to DC for energy storage. The stored energy is then converted from DC to AC to supply smoothing energy during power drop-off due to shading conditions, such as clouds, structures, or vegetation, and as an additional energy reserve during a transient power demand. Provides surplus energy storage for smoothing energy output when clouds reduce the energy input and aid in generating a stable ramp up and down of the energy that the plant provides to the customer.

A point of clarification. Although functional diagrams and reliability diagrams appear alike, functional diagrams do not always convey the information needed to perform a reliability analysis. An example would be using a block to indicate the strings of modules or panels that feed into a combiner. To adequately describe the reliability configuration, each string needs to be broken out such that reliability predictions, failure mode and effects analysis (FMEA)/FMECA, Sarr, Omar. (2017), and fault tree analysis (FTAs) can be performed to reflect the capability of the physical plant down to the lowest level of repair. The systems functional block generally aggregates the components at the string level.

An additional requirement is that the PV plant program or project should provide for the "sustainable development" Carson (2016), over the lifecycle of the system. "Sustainable development seeks to meet the needs and aspirations of the present without compromising the ability to meet those of the future." In performing SE for PV plant stakeholders, it is necessary for the systems engineer to meet the stakeholder needs and ensure that the plant is designed not only for the first year but the whole of the service life included in **R**epowering™ planning. For stakeholders to understand if the system is meeting expectations, it is necessary to define both success and failure.

5.3.1 Success and Failure

Success is common to all stakeholders, but failure is not the absence/lack of success. Every stakeholder view failure differently. More specifically success is obvious to most stakeholders. Success, such as the plant providing power to the grid per specification, there is a positive profit, or contractual requirements are being met. Failure on the other hand does not necessarily mean a lack of success as partial success may be acceptable to some stakeholders. It is important for each stakeholder to define failure based on the needs of each of them.

Failure has or can have multiple meanings for each stakeholder. A financial institution may declare failure if they only achieve 80% of the expected return on investment (ROI). Insurance on the other hand tends to be more cut and dried with payouts exceeding expectations. Excluding force majeure, the owner may declare plant failure if the profit drops below some set percentage of gross profit. The suppliers may fail due to excessive warranty costs, and the O&M may fail due to excessive maintenance events within the contractual limits.

Technology plays a factor in success for example. The commercially available Si-based PV cell efficiencies have been improving from approximately 15% efficiency in 2000 to as much as 27% in 2020.[4] The latter number is a development lab value with manufacturing values of cell efficiency running between 20% and 22%. That said, these seemingly significant improvements have also been tempered by the large number of product defects, degradation (Makrides et al. 2010), shipping and installation damage, and plant failures (IECRE 2017; IEC 2017; Flicker 2014; Engle 2010). Plant failures have been manifested as bankruptcies (insufficient understanding of costs), defaults (energy income versus cost variance), or fire sales. The inability to generate the contractual power/energy demanded due to poor specification and design reliability and/or SE practices. This conflicting progress and results can generally be found due to poor reliability, contracted O&M services with excessive exceptions, and the common fallacy that solar is simple, reliable, and requires little maintenance.

5.3.2 Stakeholder Requirements

The stakeholders must not develop their requirements in a vacuum, i.e. independent of the other stakeholders.

The owner and operator (if separate), financial, insurance, regulatory, and other stakeholders that do not have equipment or component type requirements, need to establish the budgets, schedules, ROI, assurance of compliance, and other elements unique to their specialties.

The SE function, working with potential suppliers, defines the specification, performance, and other attributes of equipment that are planned for the plant. If these are left to the EPC, then during the bidding process, the SE team must monitor and assess potential suppliers for performance, reliability, quality, and other attributes to ensure proper design specification for the request for proposal (RFP)/RFQ.

Carson (1998) and Ventre and Messenger (2010) provide templates for establishing the specification for a system. With the requirements developed, the systems engineer can then work with contracts to provide potential bidders, such as EPC companies, with an RFP or RFQ based on those specifications. What specific metrics (requirements) are needed to ensure successful operation and support?

The SE process mandates an understanding and agreement of the top-level requirements and constraints. This leads to being able to develop the plant systems model. The stakeholders must communicate their requirements and needs across the project, and it is the function of the SE concept to address and resolve conflicts in these requirements.

In addition to the initial SE plan for a new or existing plant, a preliminary **SE/R**epowering™ plan is needed to address how the plant is to be **R**epowered™ over the service life of the plant. This latter item is preliminary as it will change over the course of the life of the plant. An **SE/R**epowered™ plan looks at the expected plant's service life and defines the "hooks" and "triggers" needed in the initial concept, and design to facilitate plan execution. This considers known life limitations and rate of technological advancement and considers when the plant should be taken partially or completely offline to allow for updating/upgrading or commonly discussed repowering. The traditional use of repowering does not focus on longevity or infrastructure maintenance practices of the plant. If the project is an existing plant that is being **R**epowered™, the SE plan utilizes the requirements of the **SE/R**epowered™ plan to establish new system plan and coordination with the stakeholders. Chapter 4 goes in depth on the design considerations for the **SE/R**epowered™ process in the new plant concept as well as considerations for an existing plant.

The SE concept and plan, and **SE/R**epowered™ plans are brought together with the insurance, regulatory, and financing stakeholders. These stakeholders provide additional limitations on the project that may have already been considered with the initial effort but should be revisited to ensure compliance. This review of limitations ensures that all costs of the PV plant as energy infrastructure, instead of just a project, are accounted for and finalized for the payback period. This aids in defining an accurate levelized cost of energy (LCOE) and expected annual profit projections from commissioning to the first major upgrade/**R**epowering™, i.e. major module and inverter replacement point.

5.3.3 The Initial Concept Development and Feasibility

The development of the concept for the plant *Photovoltaic Array Performance Model, 2004* may seem easy, i.e. produce electrical power from solar energy. There are however tasks, as easy as they appear, that if not done properly will result in plant failures. The concept needs to identify the specific PV cell technologies, the type of plant (fixed tilt, one, or two axes), energy storage, security needs, the plan for the design, development, installation, and operation of the plant, e.g. the **SE/R**epowered™ plan. The initial concept is identified by potential owners/operators working with the customers to determine the size (available land area), power (W), energy (GWH), Volts AC (Vac) (utility required), Current AC (Iac), load factors (resistive or reactive), and other requirements needed for the plant. This is the start of the SE process for consideration of stakeholder requirements, limiting environments, accessibility, and other plant attributes.

The initial concept is also an aggregation of customer, regulatory, financial, and insurance requirements to address the plant viability, critical attributes, and constraints. A list of stakeholders is identified, and their inputs are solicited to fill in their requirements for concept development using Figure 5.4.

- **Service Life** is the planned duration over which the plant is expected to provide power to the grid, customer, and end user. Hydroelectric, e.g. is planned around 50–100-year life with periodic upgrades and refurbishments that can extend the life further.
- **Environmental Profile** includes elements such as plant physical location, size (acreage), terrain, ground composition, accessibility, and based on the latitude the variation in solar resource/irradiance or AM index. Based on location, the natural environment (temperature, humidity, rain, snow, etc.) is defined. The solar data for candidate sites (insolation, latitude, longitude, altitude, and hemisphere) are gathered, analyzed, and needs are developed to cover the planned service life.
- **Regulatory** (local, regional, national) requirements include all the safety, operation, and other limitations that are or could be imposed at each level.
- **Financing/Insurance** consideration typically go together with the project schedule. The time-budget profile allows the financing institution to establish critical review points on the project expenditures and ensures that contingency budgets have been set aside to handle the inevitable unknown unknowns. Insurance requirements can stop a project from even getting started. If the insurer is concerned about aspects of the project that

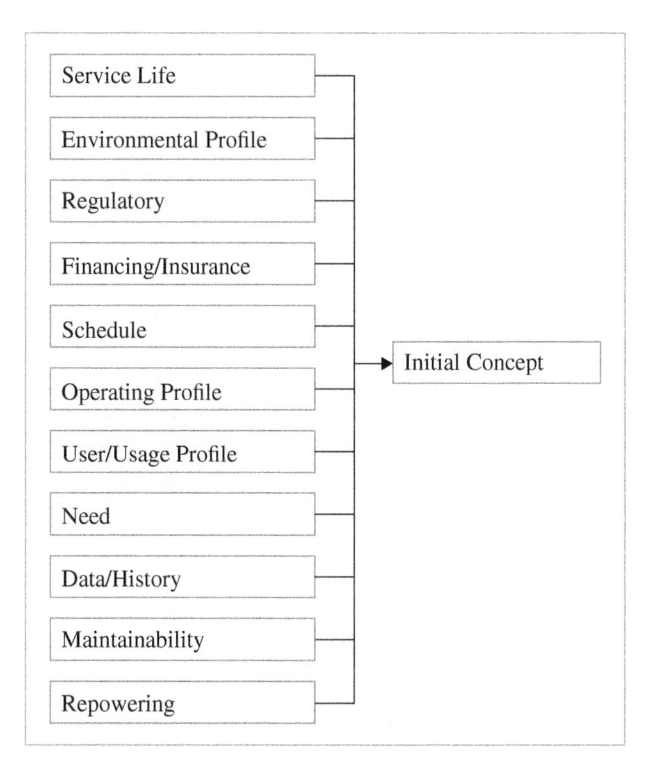

Figure 5.4 Development of the Initial Concept.

lead to excessive risk, they can put a hold on further work until risks are refined and mitigated or ameliorated.

- A **Schedule** is needed to define major milestones for the project/program management to monitor in order to control costs and understand the impact of schedule delays and project cost overruns. There is a myriad of schedule software tools such as GANTT, Program Evaluation and Review Technique (PERT), Project Scheduling Software, etc.
- **Operating Profile** is the planned operating attributes such as days of operation, hours of operation over the year, planned shutdowns, planned maintenance intervals, and time of day for scheduled or various types of maintenance. This profile assumes that the plant will produce power if the sun is shining, i.e. the plant is available to deliver energy to the user(s).
- **User and Usage Profiles** need to be established (Chapter 6). User profile is the how, when, and where of the project. This information generally comes from the stakeholders including users such as utility, commercial buildings, or other customers. The usage profile defines the operating times, expected power demand (including curtailments), and critical operating periods. This also assumes that there is sufficient solar irradiance to provide the demanded power.
- **Need**, as in what power in Megawatts (MW), energy in Megawatt Hours (MWh), and other requirements do the various stakeholders require from the proposed project. Need establishes the size and potential expansion of the plant as well as the type of infrastructure needed to support production and transmission of energy. Need also identifies

the type of plant (fixed tilt, one-axis, or two-axis) based in part on the geophysical and terrain location and space available for the plant, type, and load as defined by the customer demand. A consideration for the plant layout would be the need to consider future expansion of the plant.

- **Data/History** is the information gathered from previous plants of the proposed or similar types. This includes lessons learned as well as the failure reporting and repair history of the components. This latter item should come from the suppliers, experience, and shared data.
- The **Utility** provides the requirements for the connections to the grid or substation feeding the grid as well as the specifications for frequency, phase, and amplitude requirements and control attributes.
- **Maintainability and Human Engineering** (Chapter 7) input defines the way the plant and its components will be laid out, what the support functions are, safety attributes, data acquisition requirements, and how the plant will be monitored and managed.

5.3.4 Management of PV System Delivery

Solar power projects, or any project for that matter, have distinct phases that they progress through, whether well documented, addressed or not. Although there are exceptions, a project needs to have the following information, and potentiality more, available for project management and stakeholders at the beginning of the project:

- What are the stakeholder needs, specifications, and requirements?
- What work products are produced, by whom and to whom is it delivered to?
- What tasks need to be done?
- What are the task start and end dates?
- What dependencies does each task have that may prevent a succeeding task from starting?
- What is the project's overall calendar?
- Development of contractual work packages for each contractor/supplier?
- What are the project budget and milestones for progress to fill in the calendar above?
- What resources are needed and when are they *available*?
- Establish the milestone and schedule risk analysis.

The above illustrates the need for visibility and clarity for project management and all stakeholders to track their milestones and provide the earliest possible flags that something is potentially going wrong.

One of the first efforts needed for any complex, large scale, or large-value project is to establish the lifecycle of concern and the major phases of the project. Figure 5.5 illustrates a basic flow of a project using a modified Vee diagram. This modified Vee diagram illustrates when various phases of the system engineering process occur. The validation and verification are shown as a large arrow is intended to place importance on the effort to verify or validate the plant design and operation over the life of the project.

The concept of operation for example results in the system requirements and specifications and architecture. The architecture may require that the concept of operation be modified due to technical problems or excessive costs to achieve the planned operation. During commissioning, the testing may find problems with the EPC design (design error,

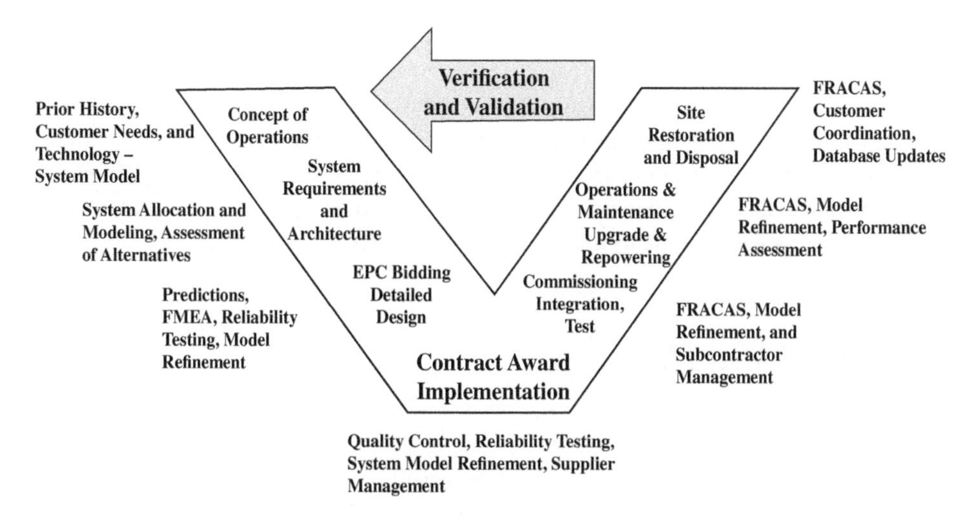

Figure 5.5 Principle Management Project Phases. (Reference INCOSE SE Handbook)

poor supplier performance, etc.), and this will result in changes affecting the EPC. Failure Reporting and Corrective Action System (FRACAS) provides data to validate the design and verify compliance with the system requirements and even the architecture.

The first four phases are pretty much the same for every project, regardless of size and the major elements of the investment costs. The cost after installation and commissioning is the variable annual cost of operation. One must start with the initial concept and establish the cost, how much power and energy are required, alternative system designs, what is the sustainability of the system, and how to achieve the required life. Then the concept must be turned into a set of specifications and system design that the stakeholders can see will meet their needs.

Once the system specifications are completed and signed off by the stakeholders, they are submitted to the EPCs in an RFQ process that the EPCs can use to bid on the project. Note that the original SE team can develop the system specification and design, or the EPCs can be tasked as part of the RFQ based on the constraints of the system specification. Following a best and final proposed design selection by the owners, SE, and required stakeholders, the contract is let, and the detailed design phase begins where the top-level requirements to provide a certain amount of power/energy are allocated out to the various suppliers for design, built, test, and compliance checking. Upon completion of the design, acceptance, and manufacturing, the site contractor then preps the site and installs all the components called out in the design and as delivered by the suppliers.

During the entire project, the SE team must perform risk management by identification of potential problems, assessing their likelihood of occurrence, and their consequences to the stakeholders and the project. At the beginning of the project, the stakeholders are the source of information input to define their risks. During operation, the field data gathered through FRACAS or other data acquisition tools is assessed for risks, and the appropriate stakeholders are then notified of these results.

After installation, as the system begins power generation for its customers, the SE team tracks performance and plant operating parameters (such as environmental conditions) to verify the initial prediction models as well as to monitor infant mortality issues or failure trends indicating or: any plant design weaknesses. The ongoing effort is to take the information provided from the field data (data acquisition system (DAS), supervisory control and data acquisition (SCADA), original equipment manufacturer (OEM), root cause analysis, field addendum reports, etc.) and as analyzed by the specialty engineering, to assess trends in performance, reliability, and maintenance. These assessments provide the foundation for future plants as well as provide feedback to suppliers about their products.

5.4 Project Phases Overview

5.4.1 Concept

One must start with a concept that includes assessments of the cost, how much power and energy are required over time, alternative system designs, and how to achieve the required performance over the life of the plant. The system concept is the result of first acquiring the stakeholders' requirements and developing an SE plan to implement those requirements into a fiscally sound project. It is at this point that the stakeholders must understand that the lowest bid is not something that will produce the best service life of the plant unless it has been carefully specified.

Operation, usage, user, and environmental profiles provide the basis for the development of the system concept and specifications as they set the boundaries for equipment performance. Examples include the reliability (North American Electric Reliable Corporation (NERC) Reliability Guideline BPS, 2018) calculations for large systems (>20MW), the design to support redundancy, the design of the energy storage based on rapid changes in demand, and the impact on O&M planning and execution.

5.4.2 Detailed Design

Detailed design services are generally provided by an EPC. This covers the spectrum of tasks to develop the final system specification, incorporate all regulatory elements, establish maintenance and safety requirements, find, and contract suppliers, and determine the plant investment cost in capital equipment.

The specification and site information sent out to EPCs or contractors to bid on is also the basis of the bid evaluation by the SE team. Variance to specification should only be allowed with well-documented rationale and the plan for correction. Emphasis should be placed on design margins (such as volts rated versus applied voltage) and accounting for the effects of temperature and irradiance (Jordan and Kurtz 2010), etc.

5.4.3 Manufacturing, Build, and Test

The manufacturing processes are required to provide the most reliable product that meets the system lifecycle performance requirements, based on using quality processes. Typically, this is one of the ISO 9000 series certifications, not just being compliant with the standards. It is important to note that a design that is reliable may not be when it is fielded because of poor manufacturing processes and/or poor quality of parts. Systems and Quality Assurance/Control (QA/QC) engineers play the role of monitoring both the building and testing as well as the certifications that suppliers and manufacturers need to do.

The new technologies in PV cells, power transistors, magnetics, and software require that the latest designs in modules, motor controllers, SCADAs, power controllers, sensors, and inverter systems require that prototypes not only be tested for performance but also their reliability. Testing, quality, or reliability is based on having statistically significant samples. Chapters 6 and 7 cover various testing requirements needed to assure that the equipment will meet performance specification over time and that the system can be maintained within a reasonable budget.

5.4.4 Installation and Commissioning

Installation and commissioning involve SE and quality functions that are used to ensure that the plant is built and installed per the engineering specification and drawings. The SE team monitors the progress of the installation, and as various sections are completed, are part of the evaluation team assessing that equipment during commissioning and benchmarking. Commissioning is based on the plant passing the regulatory and customer requirements and certifications as per the contract/s.

5.4.5 Operation, Upgrade, and SE/Repowering™

Half or more of a plant's life is spent in operation, with short periods of repair, upgrade, and **SE/R**epowered™ activities. Upgrading is the process of replacing product(s) with a newer or more current versions of the same product(s)., usually due to discontinuing an existing technology or just the natural improvements in an existing technology. **SE/R**epowered™ of existing plants should consider and be evaluated on current and proposed or desired capability and how the modules, trackers, inverters, and other BOS components are performing and the cost and timing of replacement.

> To guarantee a successful fiscal return over the lifetime of the PV project, it is best to prepare for a catastrophic failure by having, at minimum, a rough plan in place and adequate financial reserves to replace primary components like an inverter or transformer. Having a repowering plan as part of an overall operations and maintenance strategy can save a huge amount of time, money, and anxiety if a problem happens. The best time to develop this strategic approach is during the plant conceptualization.[5]

5.4.6 Site Restoration

Site restoration has already been defined as usually returning the site to the same state or condition that existed *before* the PV plant was built and installed. SE planning tasks for this phase provide for the removal, recycling, disposal, and restoration of the plant site. This is done with the initial concept and updated as the plant's useful lifecycle is ending.

Among the costs to be absorbed are disposal and reclamation of solar materials (i.e. silicon, arsenic, phosphorous, gallium, etc.), construction material (aluminum, steel, and copper), and materials of use or reuse such as concrete chain link fencing.

The planning for this activity should begin with the initial concept and include an estimate of the money (future USD/Euro) needed to perform this task. The project managers should make sure that there is an appropriate set aside to fund site restoration.

5.5 Systems Engineering Tools

SE, aside from using standard engineering CAD tools and processes also uses process tools that greatly aid in obtaining consistent and uniform results from project to project. Two key tools include SIMILAR (state, investigate, model, integrate, launch, assess, and reevaluate) and SMART (specific, measurable, assignable/achievable, realistic and/or relevant, time-related, and traceable).[6] These ensure that the attributes of the system needs are clearly defined and that the progress from concept to end of life allows for correcting for problems in the field, learning and documenting lessons, and using those to improve the current and future product. Another resource for the specialty engineering aspects of the SE process can be found in the DOD guide to achieving reliability, maintainability, and availability 2005.

5.5.1 SIMILAR

The SIMILAR Engineering process provides a basis for the development of a successful product or plant. This process is usually comprised of the following seven tasks:

- **State** the problem,
- **Investigate** alternatives,
- **Model** the system,
- **Integrate**,
- **Launch** the system,
- **Assess** performance, and
- **Reevaluate**.

The core attributes of the SIMILAR process are illustrated in Figure 5.6. Note that the process starts by gathering all stakeholder needs and requirements and proceeds through to generating the possible solutions and developing the design and system models and on to manufacturing, installation, and operation.

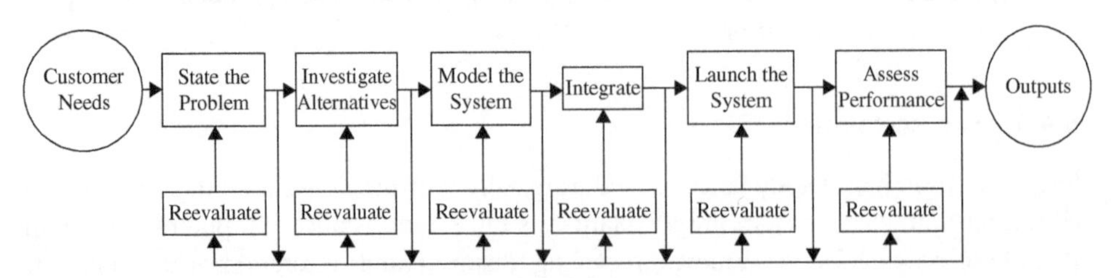

Figure 5.6 Basic SIMILAR Process. (Reference INCOSE SE Handbook).

These principles are product-oriented and include all the contractual metric requirements demanded by the stakeholders. In this context, stakeholders include suppliers and their sub-tiers or third-party providers. In addition, there is a specific need to understand the total operating environment of the product over time. The approach is to establish the project pillars (requirements) and domains (specialties) and then to iteratively work each, in how they impact or affect the overall project from concept to disposal.

SIMILAR data feedback is taken at each point in the process and *used to refine and update the prior established requirements* and needs from each of the stakeholders.

5.5.2 SMART

Beginning in the early 1950s, systems became more complex, and the problems associated with integration, operation, and maintenance became increasingly more complicated. With the electronic revolution of semiconductors, the system problem was further compounded as a result. Beginning in the 1980s, there was a movement to develop an integrated process that started with the initial customer requirements and finished with the delivery to the customer of the final product. This included the delivery of an integrated support system to cover the warranty or guarantee period (Doran 1981).

Doran (1981) provided the first enumeration of a management control process that he called SMART and expanded on by Bogue (2018).

SMART was originally conceived as a management tool to track or trace management goals and objectives by ensuring that a specific goal has been identified with a way to measure progress toward meeting the goal and ensuring that someone has the responsibility to ensure completion. All of this assumes that the objectives are realistic and that the schedule supports completion within the assigned resources.

Over the next 20 years, those dealing with SE noted that this process was directly applicable to the project or program process from concept to delivery of a product to the customer. The application to SE has resulted in the translation of the project process to include the SMART approach, resulting in specifications to be:

- **Specific**: the requirement is specific to the system, subsystem, module, or component,
- **Measurable**: the requirement is measurable using tools currently available,
- **Assignable**: the responsibility for the requirement has been assigned to a supplier, or specific engineer, or other entity,
- **Realistic**: the requirement is within the bounds of current technology and will perform over the plant's operational and usage profiles, and
- **Time-Related and Traceable**: the requirement is for a defined period of performance and measurement and has traceability to the top-level system requirements.

The SMART/SIMILAR premise is based on fundamental actions and attributes:

- Traceability of requirements from stakeholder needs/requirements to solutions,
- Improved designs by ensuring an ongoing validation and verification process,
- Provides for hierarchical models to visualize the architecture of the plant,
- Identification and mitigation or reduction of ambiguity in requirements and stakeholder conflicting needs, and
- Establishes the ability to understand the lifecycle of the plant and the elements of knowledge and data capture needed to monitor the plant.

SMART attributes of time-related or traceable relate to differences in hardware and software. Software and hardware specifications should be capable of showing traceability to the top-level systems requirements. Software must also show that its execution can be performed on the platform and does not create any hazardous safety conditions. In SE, this process is called "requirements validation" based on the "SMART" approach. The key characteristics are verifiable (how it is written so that it is measurable), necessary (cannot satisfy a stakeholder need without it), sufficient to satisfy the need, and feasible (can be implemented with acceptable risk). Time-related is of importance for performance metrics since the reliability, hence the availability of the PV system is performance measured over time.

Development tools such as PVSyst,[7] Western Electricity Coordinating Council document generic photovoltaic system dynamic simulation model specification[8] to predict PV plant performance over a specified time period, PV Watts,[9] System Advisor Model (SAM),[10] NSRDB,[11] and RDTools[12] are among the many that address PV-specific analyses and data. All of these models do have certain limitations or use average data (such as Typical Meteorological Year (TMY3)) that can leave holes in the system analysis and thus should be used *as a solution, not the solution*!

5.5.3 Risk Management

SE can also be thought of as the technical foundation of risk management. The SE engineer or team needs to spend significant time and effort understanding the problems and defining the requirements from the lifecycle perspective. This diminishes the risk of errors in the design process due to misunderstandings between stakeholders and errors introduced by improper or incomplete specification. There is a direct relationship between applying SE or risk management where the effective results are lowering the identified and other costs. By properly identifying the risk attributes for the various stakeholders, it is possible to improve the initial design and system concepts and reduce future costs while also reducing the unknowns in the system.

Risk is an event or condition that, if it occurs, adversely affects the ability of a project to achieve its outcome objectives. Each stakeholder has both a common and different set of risks that need to be addressed and coordinated with other stakeholders, through risk management.

A tool often used and abused is the risk priority number. It is based on an assessment of severity, consequence, and ability to detect. It is open to interpretation between specialists, even with documentation. Just having a number of 800 (assuming that each of three elements has a weight from 1 to 10) does not demonstrate what the underlying driving factor is without considerable effort. For that reason, we stick to the probability of occurrence (likelihood) and consequence. The methods presented here are based on performing a direct risk assessment using tools such as FMEA, reliability predictions and assessments, FTA, cost-benefit analysis, and hazards analyses. Risks, especially if ignored, are the costliest elements of the project. As shown above, there are many risk areas possible in the project and from different sources. Since the risks of financial, performance, cost, etc. are historical and prevalent throughout the PV industry for decades, it requires the SE and project teams and stakeholders to specifically identify the improvements needed or performance required to reduce or mitigate the total number of risks. More specific risks include module degradation rates Makrides et al. (2010), Jordan et al. (2010), Jordan and Kurtz (2010), and Sample (2011), and module and inverter reliability as two of the primary drivers of annual maintenance calls and events. Dirt, dust, and other residues collecting on the surface of the modules/panel soiling will further reduce the power generated by the plant. NERC[13,14,15] found after several fires in California that there needed to be additional design controls put in place to reduce the problems found as a result. Inverters have been analyzed by a number of authors, for example, Atcitty et al. (2011); Formica et al. (2017); and Zhou et al. (2015) (http://www1.eere.energy.gov/solar/international_reliability_2009_workshop.html; Vazques and Rey-Stolle 2008; Ristow et al. 2008; https://www.nrel.gov/pv/pvrw.html; Fu et al. 2019; Doran 1981; Bogue 2018; IEEE/ISO/IEC 29148-2018). Chapters 6–8 cover the RAM attributes that need to be developed as part of the SE effort as well as illustrate some of the different problem areas that result in risk to the performance of the plant.

Consequences of failure are the direct and often immediate effects of components, parts, and modules (DeGraaff et al. 2011; Collins et al. 2009; Vazques and Rey-Stolle 2008), cards, assemblies, software (IEEE/ISO/IEC 29148-2018), or end items failing to perform to specification. Failures that occur after delivery and acceptance may or may not be covered by warranties or contractual language. This is a requirement burden on each stakeholder. The general categorization shown in Table 5.1 has different meanings depending upon the level at which the FMEA is performed and the specific type (process, functional, and detailed) of FMEA. Tables 5.1 and 5.2 provide a general summary of the different types of consequences associated with failures and their likelihood or probability of occurrence.

For Table 5.2, the typical measurement interval is per year (approximately 2,000–3,000 hours for solar components and 8,760 hours for devices operating 24 hours a day).

The combination of Tables 5.1 and 5.2 provides a means of quantizing risk in the form of a stop light chart as shown in Figure 5.7. A high likelihood of occurrence with the consequence of significant or major impact requires that the design be corrected and/or manufacturing processes corrected and/or corrective maintenance action taken at the plant. This is assigned the color red. A combination of not probable or unlikely with a consequence of negligible or marginal would be assigned a green.

The risk management elements are based on experience or formulary from the stakeholders. Figure 5.7 illustrates that each stakeholder has identified their specific risks, the likelihood of those risks occurring, and the consequence of that risk for them and the team. For stakeholders that cannot identify the likelihood, the SE team can perform an assessment with the stakeholders(s) and reach an agreement on what is an acceptable value.

A financial institution risk, for example: Is there some chance that the project will not even start after the loan is complete or fail before the loan balance and interest have been paid? This can include all or part of their capital investment, loss of income via the ROI, and risk that they may be part of any litigation that occurs because of the failed plant. Financial stakeholders may be satisfied with risk levels at 10^{-3} to 10^{-4}. The O&M may need risks

Table 5.1 Failure Consequence Category Definitions.

Rating	Severity	Process	Functional	Detailed
V	Catastrophic	Plant production shutdown with loss of revenue and potential death to plant personnel and potential lawsuits	Loss of function with potential loss of one or more lives or major damage with cost exceeding \$10,000,000 and/or customer damage claims	Loss of specific equipment with loss of one or more lives or major damage with cost exceeding \$10,000,000
IV	Significant	Plant production slowdown with significantly reduced or loss of revenue potential injury to plant personnel	Loss of capability or capacity with potential major injuries or loss of limb or damage exceeding \$1,000,000 and/or customer damage claims	Loss of specific capability or capacity with potential major injuries or loss of limb or damage exceeding \$1,000,000. Backup or alternative may exist
III	Major	Loss of several production days with delays in shipment possible minor injury and cost incurred to replace lost items	Reduction in capability with potential injury to one or more persons or damage not exceeding \$500,000 and/or customer damage claims	Reduction in capability with potential injury to one or more persons or damage not exceeding \$500,000. Redundancy exists with minimal loss of function
II	Marginal	Loss of part of shipment with resulting cost incurred to reproduce lost items	Reduction in capability with potential minor injury or damage exceeding \$100,000	Reduction in capability with potential minor injury or damage exceeding \$100,000
I	Negligible	Minor cost impacts and/or delays in production – no lost revenue	Reduction in capability with damage less than \$100,000	Reduction in capability with damage less than \$100,000

as low as 10^{-4} or even 10^{-5}. Certainly, the likelihood of a plant catastrophic event from a design or component flaw needs to be low, 10^{-5} or lower as determined by FTAs.

Part of the SE team job is to solicit the information from all of the stakeholders, ensure that the risk and consequence are adequately described, and then coordinate and work with the EPC, owner and Operator (O&O), and O&M stakeholders and take actions to mitigate or eliminate each risk to the satisfaction of the stakeholders. The EPC, for example, needs to perform a risk assessment of the likelihood of suppliers not meeting their specification or performance metrics and what are the cost and other consequences of that failure. For

Table 5.2 Likelihood of Occurrence.

	Probability of Occurrence	Action
Certain	10^{-1} to 0.9999-	Requires immediate design or other corrective maintenance action
Likely	10^{-3} to 10^{-1}	Requires design changes or other corrective action
Possible	10^{-5} to 10^{-3}	Requires design or process changes, potential for alternative part
Unlikely	10^{-7} to 10^{-5}	To be corrected if the failure manifests itself or a planned upgrade is to occur
Not Probable	10^{-9} to 10^{-7}	No action needed

a well-established O&O with an EPC under contract that has an extensive background in designing and building these plants, the likelihood may be in the 10^{-5} range while a new EPC or one with a questionable track record may have a risk that is in the 10^{-2} or 10^{-3} range.

The application of SE to the project lowers the overall risk. However, the SE process should include a risk called "unknown unknowns" with a non-zero likelihood.

Why not a zero likelihood of occurrence?

Quite simply, there are always external and internal factors not properly addressed or missed, which can occur resulting in project failure or other consequences. One such example of a near zero likelihood is the loss of power generation due to surrounding grass fire. ***Assuming*** that the plant owner works to keep the grass/bush cut short under the plant modules during the year, then nearby fires have a lower likelihood of propagating to the plant and damaging or destroying parts of the plant. This risk has a generally very small likelihood, but it is not zero.

Risk consequence in every day experience can be viewed as an assessment of the damage that can be done if an event or condition occurs. A car driving on surface streets in a town runs over a nail. The driver pulls over and proceeds to change the tire. The risk of something bad happening to the driver during the tire change is relatively low. That same event occurring on a freeway with narrow shoulders results in the risks being exceptionally high. For PV project financiers, they can mitigate some of their risk by requiring collateral or

Consequence						
	V		F1, I1	I2	O&M1	F2
	IV	E3	E2			E1
	III				O&M2	
	II			F3		
	I					
		10^{-7}–10^{-9}	10^{-7}–10^{-5}	10^{-5}–10^{-3}	10^{-3}–10^{-1}	10^{-1}–1.0
	Likelihood of Occurrence					

Figure 5.7 Risk Matrix Example.

charging higher interest rates. The plant on the other hand has hundreds and thousands of comparable items. So, events that occur in one part of the system have a potentially higher chance of occurring in another part of the system. This is the basis of requiring increased reliability through robustness, redundancy, and easily and quickly repaired equipment.

As previously noted, the consequences of failure will also vary from stakeholder to stakeholder. These can range from failure to produce enough energy, loss of investment, high customer LCOE, to cancellation of contracts. This illustrates one method of plotting the various risks for analysis.

The elements of risk are dependent upon the specific stakeholders. For example:

Financial (F) –

a. **F1**: Natural disaster. Consequence: loss of invested funds
b. **F2**: O&O bankruptcy. Consequence: loss of invested funds, loss of profit
c. **F3**: Inadequate plant performance. Consequence: Customer brownouts or blackouts, failure to meet payment schedule, Bankruptcy

Insurance (I) –

a. **I1**: Inadequate risk evaluation. Consequence: Excessive insurance payments of damage or default to the project
b. **I2**: Internal Fire Hazard. Consequence: Loss of all or major portion of plant, resulting in excessive payout

EPC (E) –

a. **E1**: Supplier(s) fail to meet quality, performance, or reliability requirements. Consequence: Poor plant performance excessive warranty costs.
b. **E2**: Supplier delays in shipments not caused by natural disaster. Consequence: schedule slide.

O&M (O&M) –

a. **O&M1**: Excessive failures (before and after) warranty. Consequence: Excessive maintenance and spares costs.
b. **O&M2**: Excessive maintenance-induced damage. Consequence: Unplanned expenses that occur but meets the specification limits.

Figure 5.7 is an illustration for the systems engineers and stakeholders to identify the areas of risk that affect them and the consequence, provide an assessment of likelihood and the consequence of failing. The SE team job is to understand each stakeholder's risks and aid in assigning the likelihood and consequence of those events. Working with the design teams and suppliers, the project SE seeks to either eliminate the risk or mitigate it such that the likelihood and/or consequences are reduced to acceptable levels. The consequence of an event may be different for each stakeholder, however, the risks are clearly identified! The likelihood of occurrence is typically in the range of 10^{-7} or lower probability. The need for such a small probability is for safety type events. These are the serious risk to the health and safety of maintenance and operating personnel or the potential for significant loss of plant generation capability for an extended period.

In this part of the matrix, the financial group and other stakeholders need to mitigate or ameliorate the consequence to less than 10^{-3}. Note: a 10^{-3} is for only one risk. The aggregation of risks to the financial or insurance providers can easily exceed a 0.05 likelihood, which may force an additional reduction in likelihood for specific identified risks.

More specific physical plant risks include module degradation rates (Sample 2011; IEC 2019; IECRE 2017; IEC 2017; IEC 2016) and module and inverter reliability (Ristow et al. 2008; https://www.nrel.gov/pv/pvrw.html) as two of the primary drivers of annual maintenance calls and events. Dirt and dust collecting on the surface of the modules/panels (soiling) will further reduce the power generated by the plant. NERC[16,17,18] found after several fires in California that there needed to be additional design controls put in place to reduce the problems found as a result. Chapters 6–8 cover the RAM attributes that need to be developed as part of the SE effort as well as illustrating some of the different problem areas that result in risk to the performance of the plant.

5.5.4 Project Management Tools

There are a variety of management tools and processes to facilitate a smooth execution of the project. One reference for project management is the body of knowledge published by Project Management Institute (2013). Scheduling, supplier monitoring, materials costs, project cost are among the major concerns of the project. A few of these tools needed are as follows:

- GANTT charts is a project management tool that aids in the planning and scheduling of PV projects from residential to full scale utility and assists in structuring complex project, Geraldi and Lechter (2012), Burkhard et al. (2005) and Kumar (2005).
- PERT reference Woolf (2012); Kerzner (2003) identifies the time it takes to complete project tasks or activities.
- Critical Path Method (CPM) is an equation of form that graphically shows the longest possible time that a project should take Woolf (2012).

In addition to tracking the work packages via an integrated schedule and to ensure completeness Carson (1998), the project should develop a work breakdown structure that identifies how many tasks and deliverables are needed to achieve a final product. It is beyond the scope of this book to address the management of PV projects as this is a specialty itself. Cost and schedule impacts of projects decisions are addressed where data can support it. The reader can find any number of project management books and articles that cover this topic in depth.[19]

5.6 System Versus Project Delivery Method

The project delivery process generally focuses on the lowest cost and shortest schedule to design and build, a plant. System delivery uses SE methods and processes presented here to achieve a longer service life, optimized investment, and operational costs for the stakeholders. As previously noted here, each PV project has essentially the same end point, the delivery of converted solar energy as electrical energy for consumption by its customers. In this section we start adding the details for delivering a cost-effective PVPS.

5.6.1 Concept

A PVPS is both simple and complicated. As shown in Figure 5.8, the lowest item in the power chain is a solar cell with a cell voltage of 0.6–0.7 V and current depending on size of the cell, 100's of milliamps to 10 Amps or so. It weights fractions of an ounce (grams) and yet provides the foundation of making use of "free" energy.

The complication for this simple idea is that there are 60–72 cells per 250-Watt, 290-Watt, 300-Watt, or 305-Watt modules. In the market today there are modules rated at greater than 450 WDC. Thus, for a 100 MW (120 MW when sized for degradation and redundancy) plant using 300 WDC modules there are some 400,000+ modules or 24+ million cells. There may be as many as 120 individual 1 MW inverters (containing thousands of circuit cards and assemblies), a thousand combiners and hundreds of thousands of connections. The concept for a PV plant must consider this complexity and the associated probability of single, cascading, and multiple failures (Chapter 6). The concept phase stakeholders generally include the owner or owner/operator, financing institution, insurance, utility or customer, site authorities having jurisdiction (AHJ), and the project systems engineer or team. During the early concept phase, the principal stakeholders must arrive at critical decisions including:

- How much power is needed?
- What is the expected energy demand from the customer on a standard sunny day?

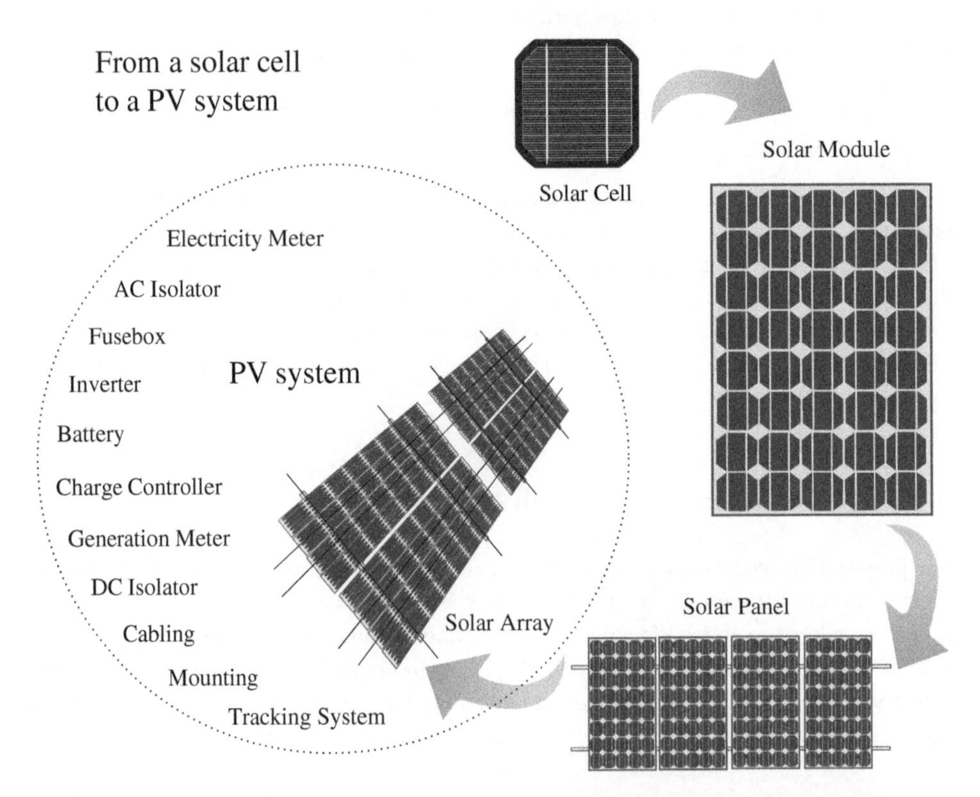

Figure 5.8 Solar Cell to PV System. Source: Rfassbind/Wikimedia Commons/ Public Domain CC BY 3.0.

- How much over capacity should be designed into the plant?
- What is the site environmental, regulatory, owner, and other limitations?
- What are the required physical and software security measures needed?

This last item, above, was demonstrated when the Colonial Pipeline[20,21] was hacked and there was a loss of several days pumping of gasoline products, resulting in financial loss both for the company with the loss of profit and the cost of bringing in personnel to fix the software.

Although the stakeholder requirements vary from project to project, the underlying plant subsystems generally are consistent. Figure 5.9 provides an initial functional diagram, an expansion of Figure 5.3, to illustrate the various elements and their relationship to each other.

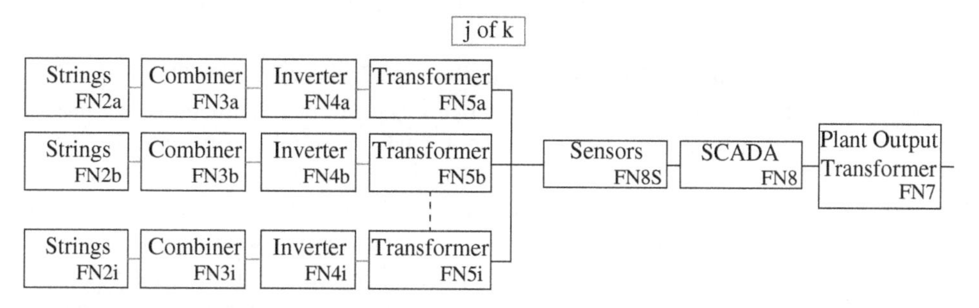

Figure 5.9 Simplified PVPS Functional Diagram.

Note that the duplication of functional elements (FN2a, FN2b, …) allows for a general simplification of the analysis. The "j of k" designation refers to the functional capability of the system to absorb certain failures such as reduced power from several strings. So, a string may be designed to have 600 VDC at 100 A (nominal at standard test conditions [STC]) or 600 kW, but only generating 490 kW due to failures. To improve the system capability for a 100 MW plant to absorb a variety of system faults, the design should include at least three additional 1-MW generation units over those needed to allow for plant degradation.

5.6.1.1 Determine Stakeholder Success and Failure Definitions

Defined success and failure criteria are needed by the entire team, so that all the stakeholders can see how their requirements match up or conflict with the other stakeholders needs and requirements. For example, for the owner or O&O, failure may be not meeting the annual power generation requirements at the specified cost. This may result in contractual default and or having to pay liquidated damages or, if the power generation drops low enough, the plant will be shut down by direction of the customer. For the O&M, it may be an excessive number of failures of end items, that result in excessive cost to spare and repair act budget either as an out-of-pocket cost or by having to increase the billing to the O&O.

Financial Institutions or investors would view failure as not being able to collect on the principle or the interest due, or not achieving the contracted ROI. In addition, there is the problem with the initial investment if the plant has cost and schedule overruns. These generally delay the time when the project should be producing energy which also affects the LCOE.

Insurance companies are concerned with excessive payouts. Utilities may experience conditions requiring them to cut back power distribution (power) to brown-out conditions because it is unable to deliver the power its customers are demanding. An example of this problem is in California where customers experience different forms of scheduled or unscheduled rolling blackouts.

Second- and third-tier suppliers can be caught in several ways by failures in their equipment. Typically for protection, suppliers set up a warranty reserve (90 days to 1 year is typical) to cover initial failures due to manufacturing errors or failures due to random events. Warranty periods run from 90 days to one year, but caution should be taken for exclusionary clauses. Excessive failures due to design or poor quality can lead to loss of reputation, which greatly affects future sales. This latter point has been demonstrated in the PV module and inverter manufacturing industry as many have simply gone bankrupt.

5.6.1.2 Establish Concept of Operation, Usage, User, and Environmental Profiles

Concept of Operation Time is a critical element in the calculation or estimation of the energy output of the plant, the reliability of a component, module, assembly, subsystem, or system as well as the time to perform maintenance and other site conditions. For PV systems, time has an additional element of variability associated with the PVPS location, or more accurately its site location latitude, longitude, altitude, and local annual weather patterns. All of these are part of the concept of operation of the plant. "The Concept of Operation (ConOps) or the System Operational Concept (OpsCon), depending on your practices, are useful in eliciting requirements from various stakeholders of a project and as a practical means to communicate and share the organization's intentions."

- Stakeholders of the system are identified.
- Required characteristics and context of the use of capabilities and concepts in the lifecycle stages, including operational concepts, are defined.
- Constraints on a system are identified.
- Stakeholder needs are defined.
- Stakeholder needs are prioritized and transformed into clearly defined stakeholder requirements.
- Critical performance measures are defined.
- Stakeholder agreement that their needs and expectations are reflected adequately in the requirements are achieved.
- Any enabling systems or services needed for stakeholder needs and requirements are available.

Establish Operating and Maintenance Concept An initial maintenance concept might be to perform all except emergency maintenance during the evening or nighttime hours. Emergency maintenance would be undertaken during the day but carries additional safety concerns since the modules/panels can produce high voltage DC and currents that could injure or kill even on cloudy days. At this point the owner needs to identify the planned maintenance responsibility, i.e. will the O&M be done by the owner, or an O&M specialty house, or even the owner performing the operation management while a third party does the maintenance. Each has unique benefits and limitations. The project team needs to understand these and adjust their expectation accordingly. Night operations, for example, will require the use of parasitic power drawn from the grid for the plant while with daytime maintenance the 24-hour parasitic power is needed only for lighting, SCADA, security, and sensors.

Usage Profile A usage profile is based on what the customer demand is for the power that the plant can produce – on average. If the user tends to require curtailment, or reduced demands during the spring and fall months, e.g. the owner and/or O&M can plan other actions to be performed during these reduced demand times. When the customer provides a schedule of demand it is possible for the O&M to schedule major maintenance for portions of the plant in need of it.

Although it would be preferential to have this information ahead of time for planning purposes, it may take two to three years of monitoring to establish the typical actions of the user over time. Clouds alone can cause significant variation in how much power is delivered at a 5–15-minute interval. Figure 5.10 illustrates the potential variation in power generated against the potential demand. It should be noted that there is a high degree of daily performance and demand from the end users. Daily performance is tied to the plant weather (Clouds for example) as is demand. Very hot or cold days drive demand more so than workdays versus the weekend might.

Expected Number of Days of Operation This can be developed from the environmental data gathered to support the reliability analyses (Chapter 6). One source of local area hours

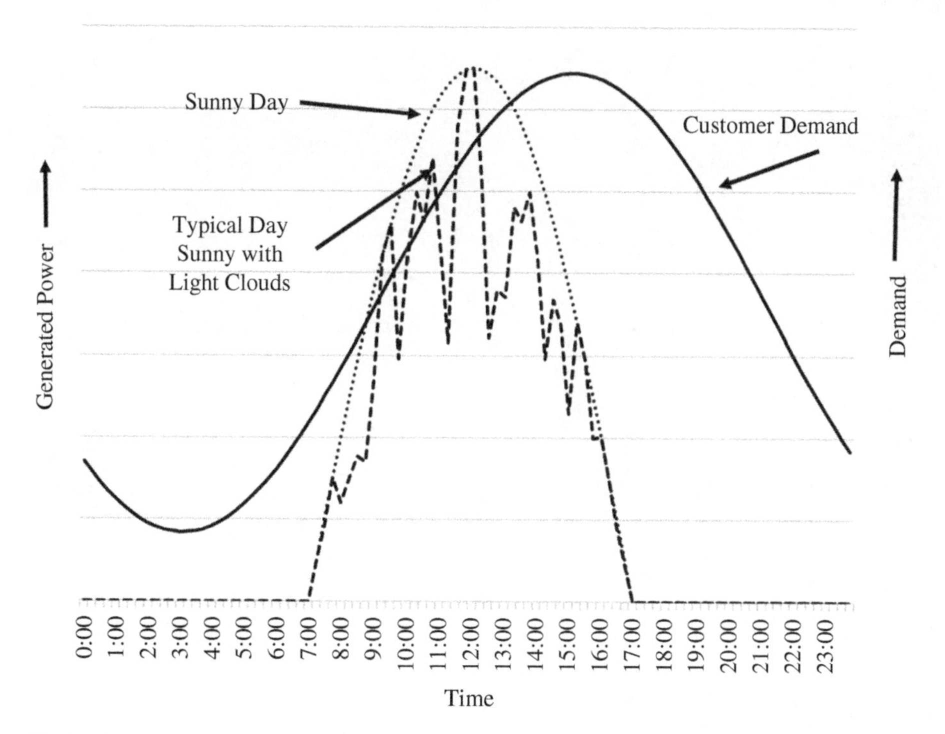

Figure 5.10 One-Day Example of Consumer Energy Demand and PV Power Generation.

Table 5.3 Hours of Sunshine by Month for Various Cities.

Country	City	Ja	Fe	Ma	Ap	Ma	Ju	Ju	Au	Se	Oc	No	De	Year
Afghanistan	Kabul	177	179	205	233	310	353	357	340	304	283	253	182	3,175
Argentina	Buenos Aires	279	241	229	220	174	132	143	174	189	227	252	267	2,525
Australia	Brisbane	264	223	233	234	236	198	239	267	270	276	270	265	2,968
Brazil	Belo Horizonte	176	191	195	211	221	230	241	242	203	197	167	153	2,425
Canada	Calgary	120	145	177	220	249	270	314	284	207	175	121	114	2,396
China	Qingdao	186	181	220	222	245	219	183	223	219	220	189	183	2,490
Colombia	Barranquilla	283	245	241	211	187	194	217	208	166	167	191	252	2,561
Congo	Pointe-Noire	157	156	164	160	150	132	119	106	70	93	123	144	1,574
Egypt	Dakhla Oasis	295	280	316	315	357	366	384	375	336	329	300	291	3,943
Falkland Islands	Stanley	198	161	169	115	77	57	69	90	128	189	200	198	1,651
France	Marseille	150	156	215	245	293	326	366	327	254	205	156	143	2,836
Georgia	Tbilisi	99	102	142	171	213	249	256	248	206	164	103	93	2,046
Greece	Athens	130	134	183	231	291	336	363	341	276	208	153	127	2,773
Hungary	Budapest	62	93	137	177	234	250	271	255	187	141	69	52	1,988
India	Chennai	233	240	291	294	301	234	143	189	195	257	261	211	2,849
Iraq	Baghdad	192	203	245	255	301	348	347	353	315	273	213	195	3,241
Ireland	Dublin	59	75	109	160	195	179	164	157	129	103	71	53	1,453
Israel	Tel Aviv	192	200	236	270	329	357	369	357	300	279	234	189	3,311
Italy	Cagliari	150	163	209	218	270	311	342	321	243	209	150	127	2,726
Ivory Coast	Ferké	279	249	253	229	251	221	183	151	173	245	261	262	2,757
Japan	Nagoya	170	170	189	197	198	150	146	200	151	169	163	172	2,092
Madagascar	Antsiranana	189	170	215	256	285	257	273	284	293	307	282	229	3,039
Mali	Timbuktu	264	250	270	255	275	235	249	255	249	273	274	259	3,107
Mauritania	Nouakchott	267	250	302	311	320	284	271	265	266	263	268	267	3,333
Mexico	La Paz	200	234	271	292	332	322	287	258	257	272	233	190	3,148

New Zealand	Auckland	229	195	189	157	140	110	128	143	149	178	188	197	2,003
Pakistan	Quetta	229	210	233	273	335	327	313	313	294	307	279	239	3,341
Philippines	Manila	177	198	226	258	223	162	133	133	132	158	153	152	2,103
Poland	Warsaw	43	591	115	150	211	237	226	214	153	99	39	25	1,571
Puerto Rico	San Juan	237	231	282	268	255	259	281	268	235	227	202	217	2,964
Russia	Omsk	68	125	184	235	284	319	321	248	180	105	71	61	2,201
Saudi Arabia	Abha	267	266	295	282	288	276	233	239	273	291	273	267	3,248
South Africa	Upington	353	299	298	284	291	270	290	307	300	329	345	367	3,732
South Korea	Busan	199	183	193	210	222	180	166	201	167	209	194	204	2,327
Spain	Cádiz	184	197	228	255	307	331	354	335	252	228	187	166	3,024
Sweden	Stockholm	40	72	139	185	254	292	260	221	154	99	54	33	1,803
Switzerland	Geneva	61	96	161	187	212	246	269	242	184	116	65	48	1,887
Taiwan	Kaohsiung	175	166	187	189	199	200	221	194	176	182	162	162	2,212
Turkey	Ankara	78	99	161	189	260	306	350	329	276	198	132	71	2,450
Ukraine	Kyiv	31	57	124	180	279	270	310	248	210	155	60	31	1,955
United Arab	Dubai	254	230	254	294	344	342	322	316	309	304	285	254	3,509
United Kingdom	Edinburgh	54	79	115	145	188	166	172	162	129	101	71	46	1,427
United States	Seattle	70	109	178	207	254	268	312	281	222	143	73	53	2,170
United States	Washington,	145	152	204	228	261	283	281	263	225	204	150	133	2,528
United States	Wichita	191	186	230	258	290	305	342	309	246	226	170	169	2,922
United States	Yuma	268	271	336	366	407	415	393	376	342	320	270	253	4,015
Uzbekistan	Tashkent	118	127	164	216	304	363	384	366	300	226	150	105	2,824
Vietnam	Ho Chi Minh Ci	245	246	272	239	195	171	180	172	162	182	200	226	2,489

Adapted from Wikipedia.org.

of sunlight can be found on Wikipedia https://en.wikipedia.org/wiki/List_of_cities_by_sunshine_duration. A sample of a couple of cities is shown in Table 5.3. These should not be taken as the numbers one can expect at a specific site as those number may be for a specific year or averaged over an unknown number of years. It does provide a basis for technical assumptions in the initial concept modeling of potential annual operation and a rough revenue flow.

Determine a Service Life of the Plant A major element of any power plant design, from paddle wheel to nuclear, requires establishing the expected life of the plant. This includes the need to establish the schedule of planned updates and/or **SE/R**epowering™ hooks and triggers for specific actions for the plant. For a 50–100-year service life, the plant may need to be designed for as much as 115% plus an additional 3–5% of the name plate power to extend power generation between **R**epowered™ events. Much of this depends on the quality of products selected and their capabilities over time. If the plant will be expected to operate on the order of historical utility power plants, then the plan must include the service life with four to six the appropriate **SE/R**epowering™ points and planned technology upgrades for some items (e.g. inverters). These **R**epowering™ events are initiated by carefully defined sets of hooks and triggers to deliver an improved lifecycle costs and more consistent energy deliver. These do not exclude the need for the design to have the hooks and triggers to catch other potential events.

5.6.1.3 Plant Stakeholder Requirements

Plant requirements are what the stakeholders have identified to meet their obligations. These requirements detail the functions that the plant (system) must perform to achieve the required power/energy production. Requirements also should identify the metrics needed to verify plant performance at commissioning and benchmarking.

The regulatory requirements associated with each region in North America must be met. North America has five major regions: Eastern Interconnection, Western Interconnection, Texas Interconnection, Quebec Interconnection, and the Alaska Interconnection. With these requirements, there are additional questions to answer. They are (i) What specific metrics (requirements) are used to ensure successful operation and support? (ii) Who has responsibility for each metric?

The plant's electrical output distortion, noise, and transients must meet the criteria specified by the customer (e.g., grid operator) and the plant power frequency must meet basic time standards as millions of clocks use the line frequency to establish the time for the public. The plant may be required to have a minimum 100 MW hour output that is typically measured on a sunny day at solar noon, for 25 years or a similar contractual time frame.

One example of a general requirement for all forms of power generation is frequency control and accuracy. The frequency accuracy requirements are based on the use of motor-driven clocks or clocks that derive time based on using the powerline 60/50 Hz that are driven by the powerline. Table 5.4 illustrates the frequency of error corrections based on time errors in the United States (NAESB WEQ 2005).

These numbers affect the inverter requirements. What this indicates is that a single inverter design may not be acceptable to a plant if it cannot meet these requirements.

The financial, insurance, and bonding stakeholders will be viewing the project by its cost, schedule, and the establishment of the LCOE. LCOE varies by completeness and accuracy of the "all in" costs and region depending on what the primary electrical supply sources

Table 5.4 Frequency to Time Correction Metric.

	Initiation			Termination		
Time (seconds)	**East**	**West**	**ERCOT**	**East**	**West**	**ERCOT**
Slow	−10	−2	−3	−6	±0.5	±0.5
Fast	+10	+2	+3	+6	±0.5	±0.5

are and which of five major regions the plant is sited in. There is some overlap between US and Canadian utilities for the West and East regions and to a lesser extent with Mexico for the West.

There should also be a concerted effort to understand the lifecycle costs (LCC). This involves understanding the cost of items, reasonable projections of cost growth and adjustments for cost of living, cost of labor and repairs, and the cost of operating the plant whether local or remote. LCC is the cost of investment and includes O&M, spares, maintenance overhead, insurance, cost of money over the life of the project, and upgrades as per **SE/R**epowering™ planning and/or actions which need to take place. Failure to do so will result in inaccurate LCOE.

It is during the concept phase that the design factors (technology dependent) should be identified and monitored by the SE group. Testing requirements, including power quality, reliability, and maintainability, and specific performance characteristics provide the basis for tracking the progress of the plant design from the concept through the design and installation phases.

The planned O&M approach, i.e. in-house by the owner, third-party O&M contractor, split with operation done by the owner or other party and maintenance or done by a contracted (bid) party should be established during concept and refined through the SE design and contracting efforts.

Table 5.5 is an example of an overall plant system requirements. This table is not all-encompassing for a 100 MW plant but provides an example of the type of information needed. The operational profile provides the day-to-day environmental changes over an average year. The extremes are found by statistical analysis of the daily data over a minimum of 25 years. Not listed, for example, is the plant output voltage and current and the reactive element and tolerances as these are unique to each plant based on the utility requirements for transmission.

5.6.1.4 System Lifecycle Optimization

As mentioned previously, an optimized system is not, generally, the lowest cost alternative. True optimization, or optimal balance of energy generation is between cost and profit. In the SE context, it is taking the long view of the total LCC. These include all system costs (e.g. development, installation, O&M, and expected site restoration) and ROI (Fu et al. 2019). These determine what the systems cost-driving attributes are that feed the LCOE calculations. Optimization is controlling not just the solar modules and inverters design and use attributes but also the overall plant infrastructure. The infrastructure includes the plant layout for accessibility, minimizing wiring lengths, and adding warehouse capability where local spares can be easily accessed. One of the most critical LCC considerations is that funding for emergencies and **R**epowering™ events must be documented in the *original concept and specification.*

Table 5.5 Minimal Example Summary of PVPS System Technical Requirements.

Metric	Value SI	Notes English	
System size power	100,000,000		Watts – Min Power required at 20 Years
Cost per Watt	$2.00		
System life	>50		Years
Repowering™ interval	20		Years – Staged over one year
Standard hours/year	8760		Hours – Continuous Operation Equipment
Annual sunshine days	294		Average – based on 25-year analysis
Solar operating hours/year	2352		Hours
Module average efficiency – STC	21% ± 1.5%		
High temperature (95% confidence)	56 °C	124 °F	
Low temperature (95% confidence)	1 °C	33 °F	
Max wind (95% confidence)	87 mps	54 mph	
Max gust (95% confidence)	106 kph	65.8 mph	3 s gust
Cum rainfall (95% confidence)	61.5 cm	24.2 in.	
No. rain days (95% confidence)	71		
DC/AC ratio	1.425		DC_AC
Total DC power	142,500,000 W		Watts
Cells per module	60		
Volts per module – minimum – STC	36		VDC
Power per module – STC	305		WDC, average at installation
Modules per inverter	4679		Nominal 305 W/module
Inverter power	1,000,000		Output AC Watts – Max @ 45 °C
Inverters per System	103		Redundancy, Inverter outage for Scheduled/ Preventive Maintenance
Contract availability warranty	0.99		Annual required availability
Degradation/year average – STC	.0012		Module degradation
First-year degradation	0.02		Max
Power conn/module	2		
String voltage	1000 VDC ± 50 V		
Duty cycle	0.342		

Table 5.5 (Continued)

Metric	Value SI	Notes English
Module cleaning cycle – minimum	2	Per year minimum + performance cleaning as required (5% threshold)
Worst case days lost (99% CL)	15	
Plant MTBF(95%)	48 h	Per operating hours
Plant Mmax(95%)	4.7 h	Hours
Maintenance efficiency	0.8	48 min per logged hour
Site MTTR	2.6	Hours
• Travel To	0.7	Hours
• Documentation (*E*)	0.2	Hours
• Travel from	0.7	Hours
Labor cost/hr	$65.00	Average
Min crew – standard maintenance dispatch	2 + 1	Must meet 1472/OSHA, +1 = safety monitor for HV work
Maintenance shifts per day	2	
Energy, Power metering		String, Combiner, Inverter, POI
Energy storage	5	MWH
Battery tech	Li-ion	
Minimum battery life	5000	Cycles – deep discharge (20%)
Energy storage operating voltage	1000	VDC
Battery capacity degradation	10% +5%, −3%	Specified by supplier
Average annual irradiance Variability	5–11%	

Figure 5.11 illustrates a simplified optimization of the cost of O&M versus reliability and acquisition costs. In this instance, specifically, the mean time between failure (MTBF) can for example be used to balance investment and operational cost. As shown, the early design and concept decisions drive the acquisition costs, while these same decisions also drive the O&M costs. The lowest acquisition cost results in the highest operational costs with the attending lowest reliability.

There has been a vexing assumption that solar power plants are simple, easy to design, install, and even simpler and easier to maintain. The reality is that this approach results in attempts to keep the initial investment costs down while expecting the plant to last at least 25–30 years. System specification and design optimization are the primary methods for controlling the system lifecycle costs of the plant over its service life. For preemptive analytic maintenance (PAM), system lifecycle is the planning, process, and implementation of SE methodologies to optimize the concept, design, build, operate, and restoration of a photovoltaic system.

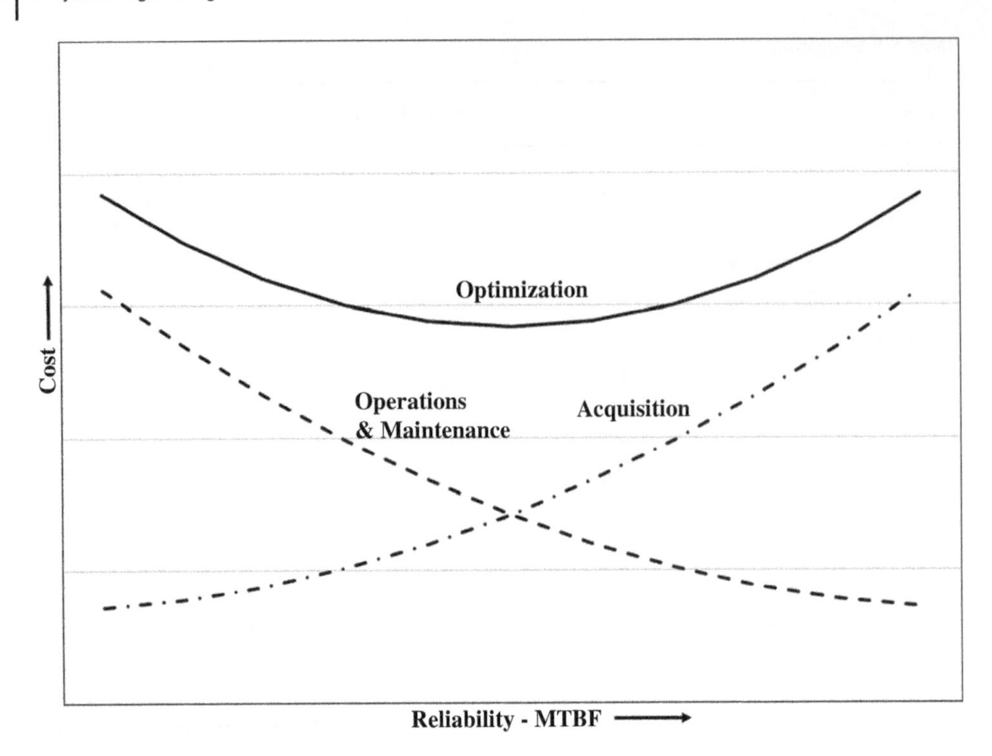

Figure 5.11 Lifecycle Cost Optimization.

Chapter 6 provides a cost impact example based on the cost of improving reliability, which results in a reduction in truck rolls due to improvement in O&M costs due to fewer man-hours expended per truck roll as well as the capital costs of the support equipment. Doing so slightly increases acquisition cost while improving energy production and system revenue. A simple example would be improving the reliability of the power semiconductors by 50%. This improvement might increase the acquisition cost of those components by 150%. The actual impact on the inverter cost may be on the order of 2–3%. The inverter reliability may improve by 20–30% since power components account for a large portion of failures in an inverter. The reduction in truck rolls reduces O&M costs for the inverter by approximately 30% and easily pays for the manufacturing cost increase.

It is important that the SE team addresses the need for optimization of the overall project by balancing the cost of equipment at manufacturing against the cost of added maintenance events due to poor reliability during O&M.

5.6.1.5 Daily Variance in Power Generation and Load

The simplified PV power generation model of Figure 5.10 for consumer demand and capability for the typical PV plant output is not usable for utility-grade projects. It does not account for potential energy storage (Chapter 9) that may smooth some of the variations or transients as well as the significantly different power demands for high-energy users.

The non-utility *customer* demand for extra energy, as with many things concerned with PVPS, is subject to location and seasonal variation. During the winter months, the demand for power may be greater due to morning and evening lighting and heating, while during the summer the demands may be driven by late afternoon and evening air conditioning.

5.6.2 System Specification and Architecture

Valid specifications are the foundation for the development and execution of a reliable, dependable, maintainable, and profitable PVPS. Invalid, poor, incomplete, or bad specifications will yield higher maintenance costs, delays in repairs, reduced availability, underperformance, and/or failed business. The operational penalties generally come in the form of liquidated damages for failing to meet the contract for energy delivery.

One common approach in a wide variety of industries is to organizationally silo the specification, design, and manufacturing functions. Over the last 50+ years, there has been an increasing tendency for management to toss requirements over the wall to the engineers, who in turn toss the design over the wall to manufacturing. This would result in the final product often proving to be something unrelated to what the customer really wanted or needed. In addition, with the potential for having to meet revised specifications, there are often hardware design changes and block changes to software (IEEE/ISO/IEC 29148-2018) or wholesale redesign of the system. This results in cost overruns and schedule delays. It also results in unhappy customers, and EPCs, owners, and/or O&M companies having significant financial difficulties until the process, causal events, and product selection is corrected.

5.6.2.1 Specification

Application of SE processes allows for the incorporation of all the stakeholder's needs and requirements for the success of the PV plant. It allows sharing information, validation of requirements, and a common basis of analysis and assessment of potential risks based on the reviews of external and internal variables that can affect system operation, energy production, and availability. Several specific points of value for the use of SE include:

- Traceability of requirements from needs/requirements to solutions,
- Improved designs by ensuring an ongoing validation and verification process,
- Providing for hierarchical models to visualize the architecture (physical and energy delivery) of the plant,
- Identification and mitigation or reduction of ambiguity in requirements and stakeholder conflicting needs, and
- Establishing the ability to understand the lifecycle of the plant and the elements of knowledge and data capture needed to monitor the plant system design.

Using specification, determine the overall top-level system design to include the plant-level hardware availability requirements and the level of redundancy needed to ensure operation over the service life. Service life is the duration of operation of the plant from commissioning to plant shut down and site restoration. The initial concept has the customer requirement for power, planned service life, and development of the system specification.

5.6.2.2 Maintenance Hours

Daily nonoperating maintenance hours are easily established based on the specific site location, and latitude. It varies throughout the year from 18 hours or so in Seattle in December to 6 hours a day in June. On the equator, on the other hand, it is 12 hours plus or minus since the sun is essentially overhead throughout the year.

Seattle, Washington, has as much as 15 hours of sunshine at summer solstice, with approximately 8 hours being good for power generation.[22] During the winter solstice, there are about eight hours of sunshine, but good power generation may be for a total of only two to three hours (not counting the general cloud coverage). The sun also only reaches a solar noon[23] zenith of about 20° at the winter solstice which often would make the plant susceptible to shadowing from plants or buildings. Phoenix, Arizona, has more consistent hours varying only from 12 hours during the winter and 14 hours during the summer and a solar zenith of 33° at winter solstice.

5.6.2.3 Variation in Solar Irradiance

There is also the annual variation in solar irradiance. This is weather, latitude, geographic region, and altitude-dependent. Figures 5.12 and 5.13 illustrate the differences in the time that the sun is above the horizon for Seattle and Phoenix assuming that there are no shading effects on mountains, buildings, trees, etc. While Phoenix has some 3,800 hours of sunshine a year and about 210 clear days, Seattle (West of the Cascades) often has around 2,100 hours of sunshine but only 60 clear days. Thus, the same-size plant design will only deliver about 50%, or potentially less, of the energy (kWh) in Seattle.

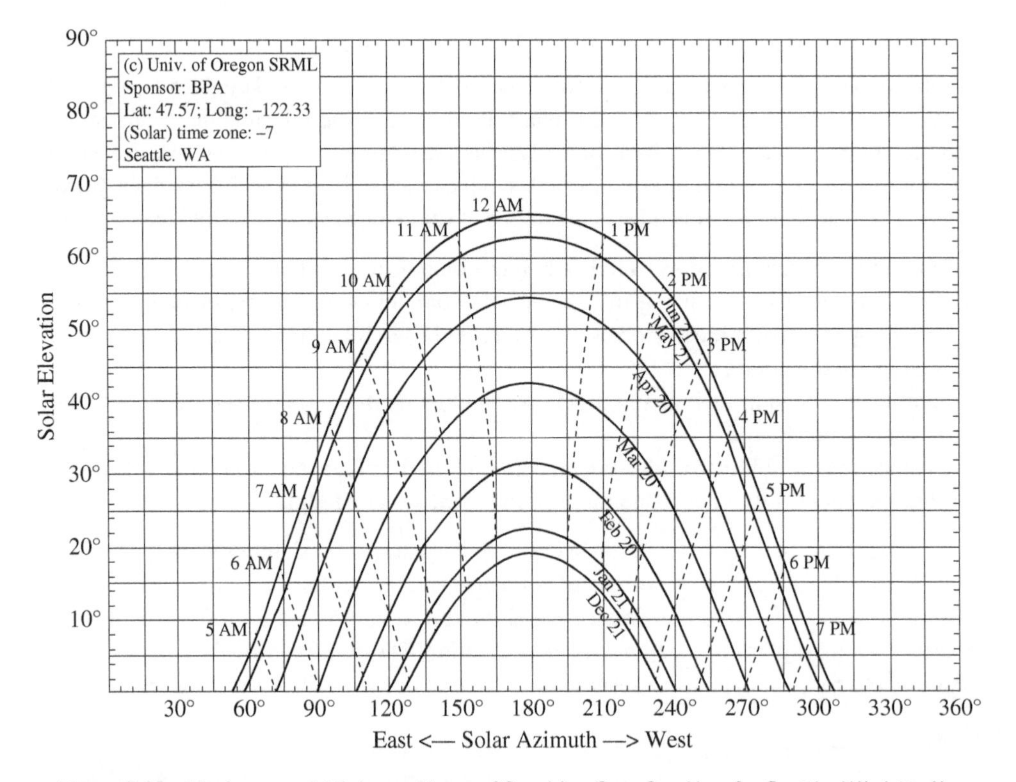

Figure 5.12 Maximum and Minimum Hours of Sunshine Over One Year for Seattle, WA. http://solardata.uoregon.edu/SunChartProgram.php.

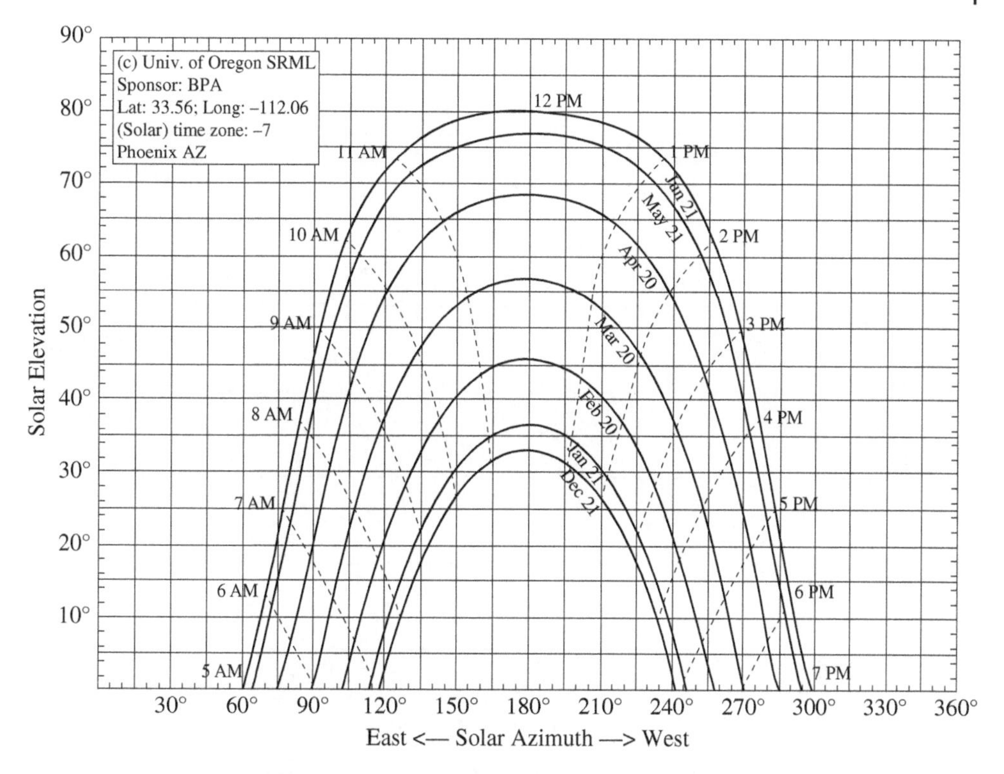

Figure 5.13 Maximum and Minimum Hours of Sunshine over One Year for Phoenix AZ. http://solardata.uoregon.edu/SunChartProgram.php.

Although the sun may be shining it is still possible that there can be a utility curtailment or grid event that limits the total energy delivered on a day-to-day basis. If terrain or structural shadows can be held to less than 10° above the horizon, then there is minimal loss of energy production for a fixed panel system or plant.

Figure 5.14 shows the variation in irradiance that can occur from full sunshine to intermittent clouds or heavy overcast, including clipping. Again, the location of the plant and its associated weather must be considered to ensure that the plant can provide the required power when needed. Figure 5.14 represents only the first year. With time and associated degradation, failures, and the effects of maintenance by year 10, the 143 MWdc plant may be down to a peak capability of only 115–120 MW or lower in 15 years or less.

The Duck Curve, Figure 5.15, illustrates the changing utility annual power demand for California from 2012 to 2020. Compared to the residential demand it shows that commercial demand dominates the overnight (midnight to 0700 power demand) while the residential demand decreases significantly from 0700 till 1800. Arriving home in the evenings results in greater demand for electricity to power cooking, heating/air conditioning, and entertainment (TV, computers, radio, etc.).

5.6.2.4 Acquisition Costs

The Pareto diagram in Figure 5.16 for example illustrates the cost improvements that have been occurring within the PV industry from 2016 to 2020 (source: EnergySage).

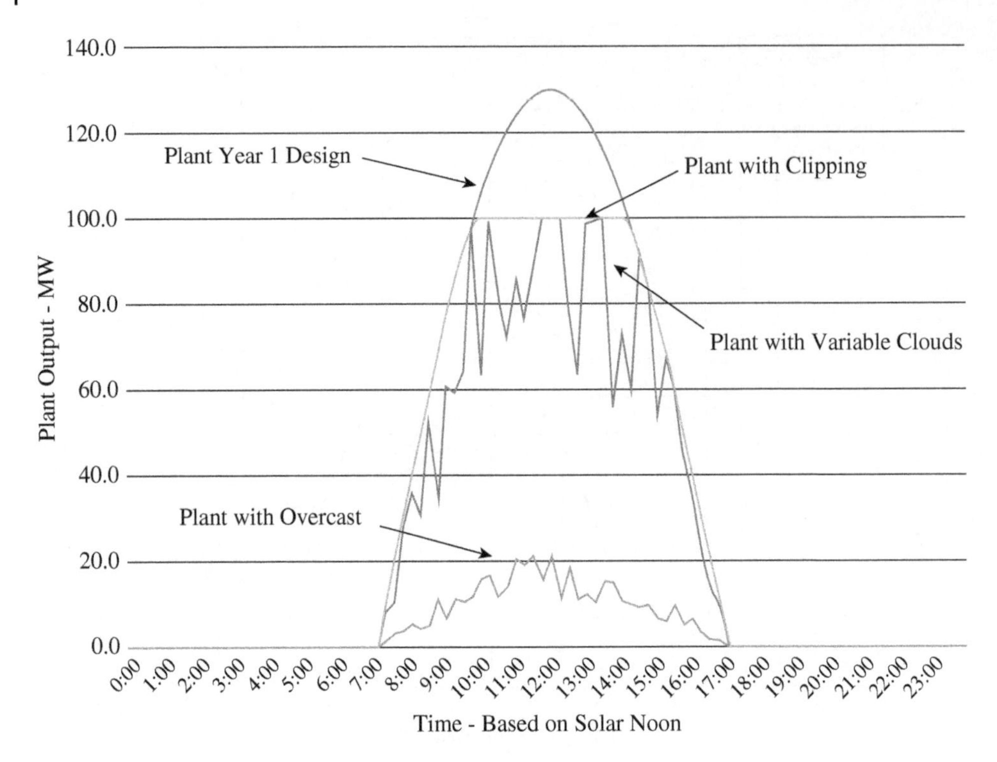

Figure 5.14 Typical Variation in Daily Irradiance. Source: Adapted from [24].

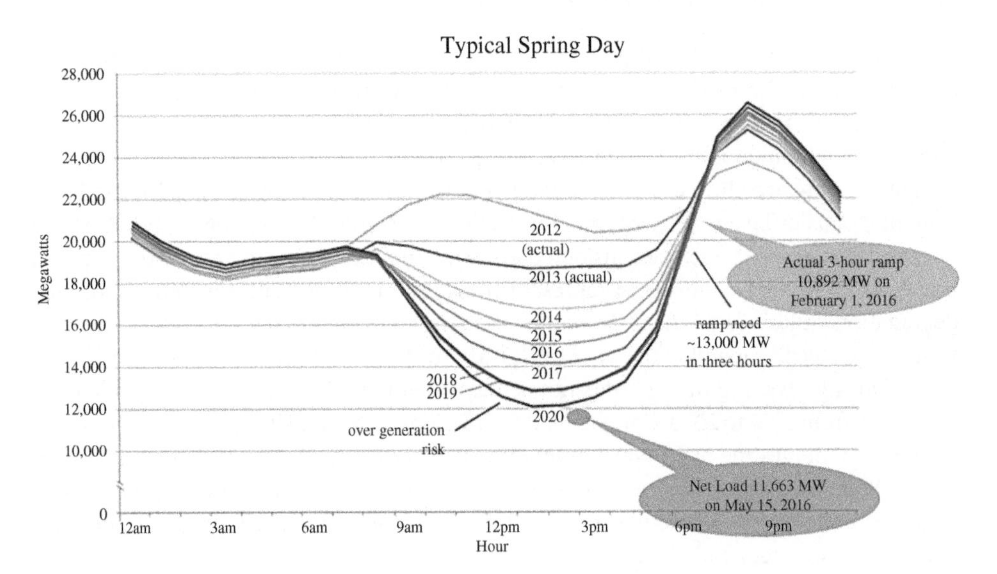

Figure 5.15 Duck Curve. Source: [25] California Independent System Operator.

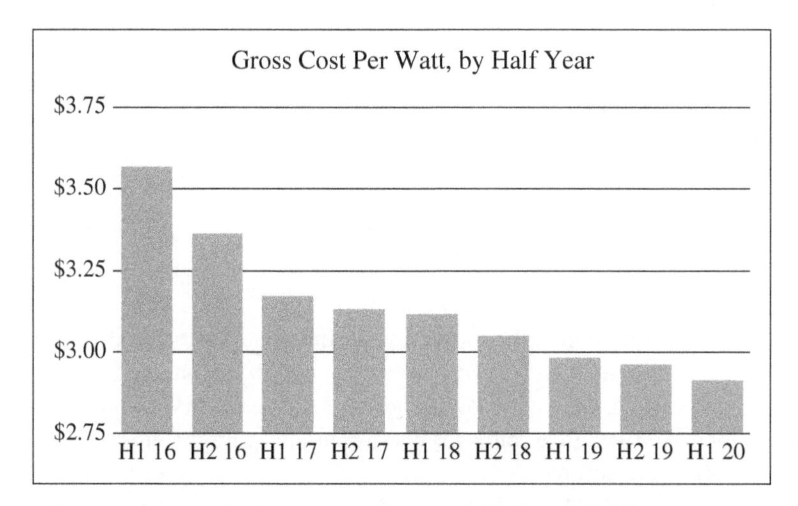

Figure 5.16 Solar Panel Cost Reduction from 2016 to 2020. Source: Adapted from[26].

Of importance is that the system costs have been dropping, in part driven by improved efficiencies of PV cells and inverter power generation capability improving from 100 kWac to 250 kWac to 1 MWac or more. This reduces the number of inverters needed for equivalent power generation, while the price per unit (cost per megawatt) has seen only minimal increases or even decreases. One drawback is that the larger the inverter, the greater the percentage loss for each inverter or generator unit failure.

General acquisition costs include but are not limited to:

- EPC contract specification
- Parts
- Development costs
- Project management
- Research
- Testing
- Installation and commissioning
- Contract monitoring and control
- Systems engineering costs
- Design engineering costs
- Manufacturing costs
- Non-recurring overhead
- Quality engineering
- Experimental tooling
- Reliability testing/verification

Support costs consist of the day-to-day cost of operation and include but are not limited to:

- Spares
- Repairman labor costs
- Engineering (SE, RAMS, and support)

- Inventory entry and supply management
- Support equipment
- Technical data and documentation
- Training and training equipment
- Logistics management
- Operators utilities
- O&M overhead

5.6.2.5 Energy Loss Budget

The energy loss budget is the variation in the conversion of sunlight to electricity delivered and includes elements as shown in Table 5.7. This variation would be accounted for in the design of the plant. For example, with a 5–11% energy variance, a 100-MW plant would need to be designed for 143-MWdc to account for energy variance and the effects of cell degradation over time and temperature.

To mitigate the initial effects of 5–11% energy loss or variance, the plant infrastructure approach is *designed using design margins for manufacturing defects, degradation, shipping/handling damage, failures, and adding redundancy to compensate for the loss of large blocks of power production.* Cabling, transformer, and shading losses may be fixed while soiling, deviation from STC, and module and inverter degradation can vary somewhat from year to year. STC conditions are 1000 W/m^2 solar irradiance, 25 °C cell temperature, air mass equal to 1.5 (AM 1.5), and ASTM G173-03 standard spectrum (standard tables for reference solar spectral irradiances: direct normal and hemispherical on 37° from horizonal tilted surface).

Degradation variance is also the result of temperature variation that the site may experience. For years when the temperature is excessively hot, the lifetimes of the modules decrease more rapidly.

Solar cells have median degradation between 0.5% and 1.0% or more and an average value of 0.7–1.5%. The actual range of annual degradation is in the range of 0.3–2.2%.[27] Of importance here is the term average. This is different than the efficiency of solar cells which also must be accounted for in the sizing of the plant. The theoretical limitation of crystalline silicon cells is approximately 29%, but most commercial cells *average* around 20–21%[28] with no statistical data given to understand the distribution. Having the manufacturer perform lot acceptance testing and require periodic third-party (e.g. PI Berlin, PV Evolution Labs) testing to ensure that the lot-to-lot variation is minimal.

On top of the losses of Table 5.6, there is an annual degradation of the inverter outputs. This is also an effect of elevated temperature, but electrical transients can also affect the inverter output by an episodic increase in temperature.

The net result is that the design of the plant should account for the initial uncertainty of 5–11% and an annual degradation rate of 1.2% or more (allowing for a higher confidence level in the design). Systems today generally fail to address detailed specific planning for a **R**epowered™ event as the variables of Table 5.6 and a conservative degradation are not completely accounted for.

For a 50–100-year plant with a planned approximately 15–25-year **SE/R**epowered™ event interval, the design for a 100-MW plant would require something on the order of a 25% overbuild to compensate for those variables (as compared to existing plant specification and delivery). Note, as stated in Chapter 4, the planned 15- to 25-year repowering event

Table 5.6 Summary of Sources of Initial Uncertainty in Plant Attributes.

Source of uncertainty	Uncertainty range [%] (one standard deviation)	Individual	Combined
A: Irradiation	Satellite GHor	3…5	4…6
Direct/diffuse-ratio and GPOA transposition	2…4		
B: Shading	Horizon	0…0.5	1…4
Inter-row	1…4		
External	0…3	n/a	
C: Soiling	1…4	n/a	
D: Deviation from STC	Reflection	< 1	< 2
Spectral losses	~1		
Irradiance intensity	1…2		
Temperature	1…2		
E: Actual PV capacity and DC-losses	Module power and mismatch	2…5	< 5
Cabling	0.5		
F: Inverter losses	Model related	< 1.5	< 2
Power limitation	< 0.5		
G: AC-losses	Cabling	0.5	< 1
Transformer	0.5		
Combined total initial uncertainty		Typically 5…11%	

Table 5.7 Summary of General Losses in a PVPS Plant.

Inverter losses	(4–10%)
Temperature losses	(5–20%)
DC cables losses	(1–3%)
AC cables losses	(1–3%)
Shadings 0–80% !!!	(specific to each site)
Losses due to dust, snow, etc.	(2%)
Other losses	(?)

interval is only one of several triggers to actually perform the activity. Excessive degradation and failures can trigger a **Repowered**™ event earlier than expected. An additional design requirement is to handle some number of multiple failures to reduce the chances of the plant not meeting customer contractual expectation or demand. It should be noted that a report from IEEE[29] indicated average degradation rates of 1.3% to 2.73% per year. As noted in the report, these high degradation rates indicate the need for an extremely high level of quality control from the manufacturers as well as a sound design.

One final variation that must be accounted for is the individual cell failures that can cause the loss of 1–30 or more cells with the attendant reduction or loss of the module output power. The variance in loss of cells depends on the use of bypass diodes. Some modules have one bypass diode for every cell while others have one for every 3–10. Again, there is quite a variance among manufacturers and products which must be considered. With module failure rates[30] of 0.05%, the plant can expect to experience the loss of 250 or more modules a year and with a degradation rate of 1.2% or more per year while the annual loss or reduction of available DC power generation can exceed 1.5% per year. This does not include the potential for induced losses (such as latent failures or damage due to maintenance or misapplication) as well as failures in connectors or fuses in the combiners. Latent failures manifest themselves after the installation and initial operation, and the source of these problems can be from packaging and handling, shipping, and/or installation damage.

$$P_{\text{dc}}(t) = P_{\text{peak}} * \frac{I(t)}{I_{\text{stc}}} * (1 - DP * \Delta\theta) \tag{5.1}$$

Where:

- $P_{\text{dc}}(t)$ = the available DC power at time t, in kW
- P_{peak} = the installed peak power of all the PV modules under STC, in kW
- $I(t)$ = the solar irradiance at time t, in W/m^2
- I_{STC} = 1000 W/m^2 is the solar irradiance under STC
- DP = a coefficient that expresses the power reduction due to the temperature rise of the cells. A typical value, for crystalline silicon cells, is DP ≈ −0.5%/°C (IEC 2019; IECRE 2017; Collins et al. 2009)
- $\Delta\theta$ = is the difference in temperature from the STC conditions
- $\frac{I(t)}{I_{\text{STC}}}$ is the fraction of solar irradiance at a given point in time (day and time of year)
- $1 - DP * \Delta\theta$ is the fraction reduction in PV cell power due to temperature

A typical value for the junction temperature rises above ambient, $\Delta\theta$ is approximately 30 °C due to solar heating, and the temperature rise due to current–voltage (V–I) drop across the diode.

Forced Outages Quoting from IEC 63019: "The category FORCED OUTAGE is obtained when damage, fault, failure, or alarm has disabled components or systems." This can be detected manually or automatically. FORCED OUTAGE is active when such events occur simultaneously on inverters or on other components/elements of the PVPS, which prevents all or parts of the PVPS from performing the service or functions.

Examples:

- General component failures
- Inverter failure

- Cable fault
- Control (software) failure
- Circuit breaker trip
- Tracker failure
- Outage time during response, diagnostics, coordination, and repair."

SE RAMS specialists utilize this as the foundation for performing FTAs and FMEA.

Masking and Clipping Masking obscures or hides defects, degradation, random faults, and failures, by virtue of having a overbuild and design margin and not having the capability to detect and identify that loss of power/energy. This results from failing to include sufficient fault detection (FD), fault isolation (FI), and reporting capability throughout the plant. Since adding this capability does increase front-end costs, it is often one of the first things dropped in the project delivery method.

Lifecycle design margin specifically accounts for defects, degradation, random faults, and failures of the system for which attribute compensation can be designed in. Note: this does not address designing over and above best design practices. This includes not only the required performance output and revenue over the system lifecycle but also the margin to protect out of specification performance and system failures. This is part of the RAMS attributes needed to ensure a cost-effective system. Unsupported design margin will result in lowering the reliability of the system and increasing the investment costs leading to higher LCOE.

Masking, along with erroneous assumptions about the reliability of the system, will result in insufficient O&M funding for the time and technology needed to identify and perform the needed maintenance. FD, FI and reporting can be part of the plant design down to the lowest level of repair or by employing such tools as infrared/ultraviolet (IR/UV) of remote sensing by monitoring DC power input quality to inverters. Masking only hides these issues for a few years after installation and commissioning. When enough defects, faults, and failures have aggregated, the plant will experience a sharp increase in truck rolls, demand for spares, and large portions of the plant being down for maintenance.

Clipping is a control process by the inverter to hold the power output at its maximum rated value when the DC input power available exceeds the power needed to reach the inverter AC rated value (Figure 5.17). A slightly refined definition of clipping is either an inverter derating its output to meet its maximum power rating "or" an O&O decreasing the maximum power point tracking (MPPT) value for the inverter at the request of the customer.[31]

In addition to the losses that occur for which maintenance has some effect, the losses in Table 5.7 are the systemic losses that must be accounted for in the design of the system.

User Profile The user profile is a compilation of the user's (customer's) energy demand that the PVPS is designed to deliver. The end user/customer may be a utility, heavy industrial, critical infrastructure (hospital), commercial business, airport, or residence. Each will have specific energy needs with variations in the amount of power and when it is needed. The utility, commercial business, residence/microgrid, and airport for example have power demands that vary by season and typically have a reduced energy demand depending on

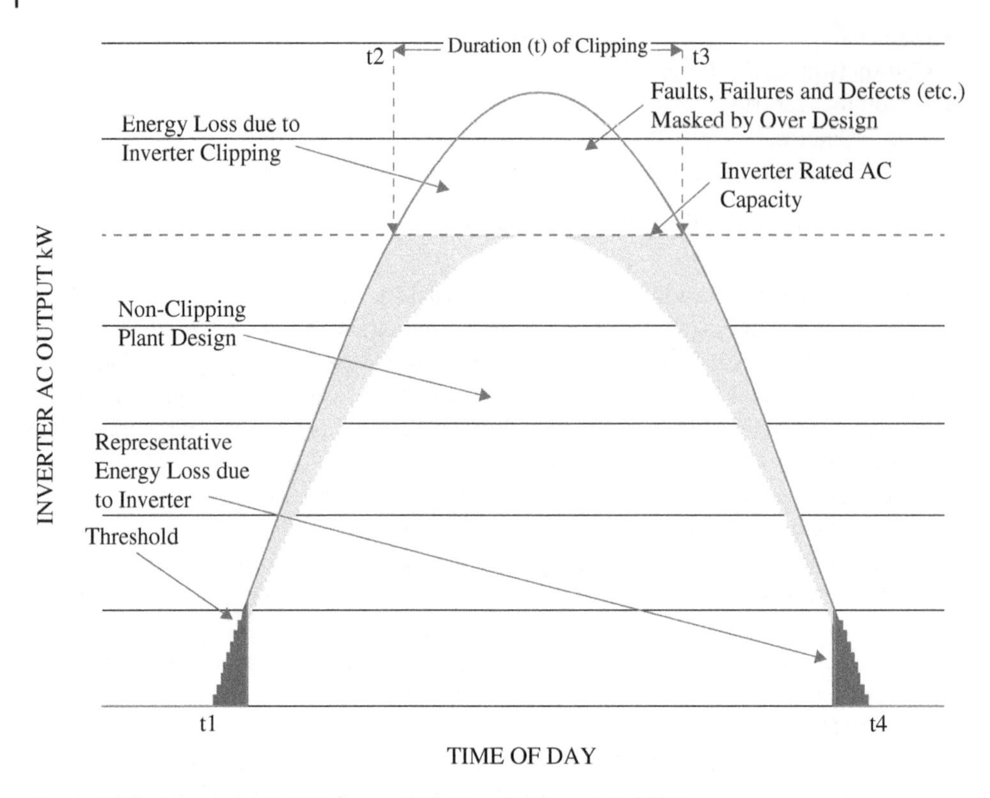

Figure 5.17 Losses due to Clipping and Masking (Balfour et al. 2021).

local time and environments/temperature. Hospitals and critical infrastructure will, however, need a steady supply of power over the entire 24-hour day.

Power quality varies among large energy users as they generally have multiple distribution points and can provide reactive (capacitor, inductor) filtering. (Electric power quality is the degree to which the voltage, frequency, and waveform of a power supply system conform to established specifications.)[32] This is especially true of PVPS as there are large power swings due to variations in irradiance or cloud cover for example. Demand will also vary by season with summers generally demanding the greatest amount of power and winters less electrical power when the use of heating oil and gas increases significantly.

Environmental Profile Table 5.7 is a summary of the environmental data needed to define the limits the PV system must operate in. A special point is that each site will be unique, and it is possible that sites even within a 5-mi (8 km) radius may have an elevation difference of 1000 or more feet (300 m) and accordingly the daytime temperatures, winds, and other weather elements can differ dramatically.

The environmental profile is developed based on the site's natural and local environments such as new structures or trees as an example, over a period of at least 25 or more years. This data is critical to understanding not only the average environments but the extremes that are possible for the site that must be accounted for in the design. Table 5.8 is an example of a Phoenix AZ, US site's environment over 25 years with the statistical analysis based on a

Table 5.8 Environmental Dataset Requirements.

Metric	Attribute	Value	Worse Case
Temperature		max/min °C	Max/Min +/−2 or 3 sigma
Relative humidity	Non-condensing	High/Low %RH	
Visibility		mi/km	e.g. Less than 5 mi
Winds	Max sustained	mph/kph	Max +/2 or 3 sigma
Wind (Gust)	Max	mph/kph	Max +2 or 3 sigma
Precipitation	Rain	Daily inches	Max +2 or 3 sigma
Pressure	Pa/ in Hg	High/Low	Max +2 or 3 sigma
Events	Thunderstorms, hail, lightning	Count high	Max +2 or 3 sigma
	Ice Storms	Cumulative thickness	Max +2 or 3 sigma
	Snow	Cumulative thickness	Max +2 or 3 sigma
	Microburst	Windspeed	Max +2 or 3 sigma
	Hurricane/Tornado	Windspeed	Max +2 or 3 sigma
	Earthquake	Displacement	Max +2 or 3 sigma
	Flooding	Depth of water – Flow rate	Max +2 or 3 sigma

90% confidence level. Note: As will be discussed in depth in Chapter 6, reliability designing to a confidence level does not cover all the potential 1 in 100 or 1 in 1000-year events that can and do occur. As has been evidenced in the last 15 years, the 1 in 100 and the 1 in 1000 year event trends are occurring more frequently and have been more consistently exceeding expected worse-case maximums.

Figure 5.18 illustrates the increasing trend of the number of hot days (greater than 104 °F (40 °C)) for Phoenix. There is year-to-year variability, but the impact on the plant is that solar cell power decreases with increased temperature, while more days of excessive heat will result in significant energy loss, module stress, and reduced reliability. To understand the impact of weather variability including daily, monthly, seasonal, and annual on power production, we use statistical analysis (assuming normal distribution) of daily weather data over 25 or more years. Using average weather based on data from noncontiguous site conditions grossly underestimates the impact on the power generation, reliability, and availability of the system, as well as the efficiency of the system. To understand, for example, the impact of clouds on power production, we first need to estimate the number of cloudy days each year. Or, if we wish to understand the impact of snow on power production, we need to look not only at the days it is snowing but also how many days are lost due to solar panels being covered by snow or ice after the storm has passed.

Why worry about the number of days of high temperature over 104 °F (40 °C)? Reliability is impacted by the ***cumulative total time*** that semiconductors such as photovoltaic cells

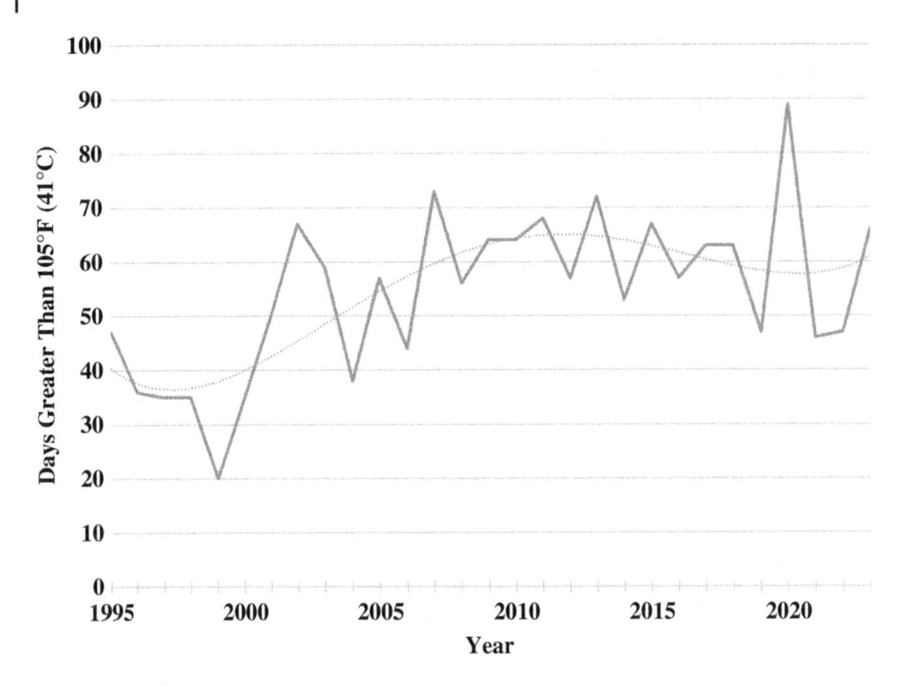

Figure 5.18 Phoenix – Days Per Year Greater than 104 °F (40 °C).

and power devices (integrated gate bipolar transistor (IGBT), HEXFET, MOSFET, Thyristor) spend at high temperatures. This trend for a higher number of days exceeding 104 °F (40 °C) may continue for the foreseeable future and may continue for the next 50–100 years. This is a worldwide issue, not just a regional or national. Chapter 6 covers this in more detail. The system specification should not only address the high temperature, but it should also address the number of hours/days that the plant can expect (with confidence) to spend at those high temperatures. Based on the information in Figure 5.18 and the environmental profile shown in Table 5.8, it would result in a plan for the plant to be capable of handling at least 86 days a year or more, where the daytime temperatures are greater than 104 °F (40 °C). Also. In the Phoenix case, it is good to know that the record high is 122 °F (50 °C).

Temperature has a negative impact on power production through a reduction in the efficiency of photovoltaic cells. Most cells have a power reduction of –0.3% to –0.5% per degree centigrade over the reference test temperature, 25 °C. If the operating temperature of a cell is 55 °C and the cell has a temperature coefficient of –0.4%/°C and an efficiency of 20% then the actual power is down approximately 24 W for a typical 400-Watts panel. However, if the operating cell temperature is 90 °C and the temperature coefficient is the same, then the actual power is down approximately 104 Watts for the same panel. If the O&O and/or a utility needs 90% confidence of having at least 30,000 MWh/year of solar energy, then the statistical weather information and irradiance are needed to estimate the number of hot and cloudy days that degrade performance below expected or projected need, and then oversize the system to compensate. This needs to be added to the systematic variations shown in Table 5.6. A tool to understand the potential variations is a Monte-Carlo simulation using

Table 5.9 25-Year Analysis Operating Environments for Phoenix, Sky Harbor Airport.

				Confidence Level	0.95	
Environmental Metric		**Min**	**Max**	**Avg**	**Std Dev**	**95% CL**
Temperature (°F/°C)	Annual Max		120/49	116/47	3/2	125/52
	Annual Min	29/−1.7		34/1	3/2	30/−1.3
	Annual Avg			77/25	2/1	79/26
Annual temperature Exceedences	Days>100 °F/38 °C	75	140	106	14	129
	Days>77 °F/25 °C	210	272	245	17	273
	Days>104 °F/40 °C	37	89	63	14	86
	Days<32 °F/0 °C	0	4	0	1	6
Dew Point (°F/ °C)	Max		81/49	76/24	3/3	85/30
	Avg			39.1	2.6	43
Humidity (%)	Min	1.0	6.0	3.6	1.6	3
Wind Speed (mph/kph)		31/50	54/87	43/69	8/12	66/106
Pressure (Hg)	Max	30.31	30.59	30.43	0.06	30.69
	Min	28.58	29.53	29.40	0.19	29.22
Rain (in/cm)	Cumulative	2.67	19.62	7.68	2.75	24
	Rain Days	8	58	19	8	71

randomizing ambient temperature and weather conditions that match the historical information statistically.

One source of daily/monthly weather data is WUNDERGROUND®. It provides a relatively uniform basis for weather data essentially worldwide. It is, however, data that is not certified or calibrated and thus must be handled with some expectation of variance or error. The other major consideration for the data is that it is often from airports and/or from commercial and private weather stations connected through an Application Programming Interface (API) to the WUNDERGROUND® site. For this example, it is data from the Sky Harbor Airport at Phoenix not near a potential solar power plant site north, south, east, or west of the airport.

Best Practice – For the critical environmental conditions, gather the data from the plant site or the nearest National Weather Service weather station and allow for additional variation.

5.6.2.6 Additional Environmental Effects for Concept and System Design

Wind Temperature is not the only variable to be considered. For the site construction of module/panels structure, for example, wind speed needs to be carefully evaluated. For a standard case, if the surface area of the array (horizontal) is 10 ft^2 (~3 m^2), then the pressure on the surface can be calculated as:

$$FW = \frac{1}{2}\rho\, v^2 A \tag{5.2}$$

where:

- Fw = Force applied to a panel
- ρ = air density in kg/m^3 or lb/ft^3 the STP air density is 1.2 kg/m^3 or 0.075 lb/ft^3 at sea level
- υ = air velocity in m/s or ft/s
- A = area m^2 or ft^2

Additional information is needed for the coefficient of drag (c_d) to assess the impact of the modules/panels being mounted at an angle α to ensure that the array is pointed at the sun at noon for fixed tilt and one-axis installations. (Some design considerations may result in the actual rotation off of true geographic south.)

The major effect of high winds is that it acts as an airfoil. As such the maximum winds (at high confidence) must be accounted for. Statistically, therefore, a realistic design point would be to have 95–97% confidence for the wind speeds and loads on the modules/panels.

One aspect of wind speed often overlooked is microbursts from thunderstorms. These down drafts can have a velocity exceeding 60 mph (100 kmph). One additional aspect is that the peak wind speeds or gusts can easily exceed 110–140% of the sustained winds. The Phoenix airport data indicates that there appears to be no significant wind gusts, but there are sandstorms as well as thunderstorms that can generate high-speed microbursts. Again, data from the airport is not a complete assessment of a PV plant site data. In addition, for a plant in the Central United States, there is a finite possibility of being hit with a tornado/thunder storm and its associated wind speeds from 60 to 300 mph winds. These values are well outside the expectation of survival assuming even a low 5% risk. Mitigation could include insurance, and *setup a protected reserve fund* set aside to have an O&M pool of ready monies to begin affecting repairs until insurance compensation is available.

On large panels or arrays, the mounting must also consider the potential for wind/gust-induced flutter.[33] The center of large modules or panels tends to not be structurally attached and may begin to oscillate resulting in excessive stress on the glass, frame, or the mountings resulting in failure.

Winds also bring the potential for additional significant wind-induced metal and paint erosion and glass pitting due to impact from airborne sand and dust. Any wind speeds over 10 mph will pick up fine dust particulate and winds over 20 mph will begin moving larger grains of sand. Higher wind speeds can destroy the modules and their mounting hardware.

Rain From the data set in Table 5.8, the two extremes in rain and dew point, 19.6 inches and 31 F both fall within the 95% confidence interval. The consideration for dew point is that it is possible to condense moisture from the air inside inverters on interior cool or colder metal surfaces as well as enclosed non-hermetically sealed equipment such as modules, motor controllers, gear boxes, and so on. Rain in areas where soil conditions combined with significant rainfall can lead to flooding or flash flooding leading to excessive erosion and loss of structural integrity. Large one-day rainfall amounts will result in flash floods in the open land as well as flooding of low points in the site terrain until it runs off, absorbed, or evaporated.

If the conditions are above specification, this translates to a force majeure, or superior force or in contract language extraordinary and unforeseen event, for which there cannot be a 100% certainty that they have been accounted for. These can constitute contract conditions that in common law, is "unforeseeable circumstances that prevent someone from fulfilling a contract." Such circumstances often exist when several environmental events occur at the same time, such as severe thunderstorms with lightning, significant rainfall, or hail in a short period of time, and excessive winds that create additional stresses on the plant equipment.

Soil Conditions There should also be an accounting for the type of ground the site is located on. Soil loading accounts for installation equipment such as heavy cranes, dump trucks, and other heavy equipment. Just performing a compression (load) test on the ground may not indicate that the ground is susceptible to erosion, soil flow, or collapse once the panels are installed, and the ground cover dies out due to lack of direct sun or reduced watering. The type of soil, slope, and local weather effects that could lead to erosion and loss of ground structural support must be addressed.

Additional surveys are needed to account for types of ground or soil outside the site. Sandy soils, for example, can be picked up by winds and impact the structure and panels resulting in erosion or pitting reducing the life of both. Soft soils for farming, brush, or local fire soot or other airborne material may result in excessive dust in the air several times a year that will coat the panels reducing efficiency in solar conversion. Also, salty (NaCl), oxidizers, basic, or acidic soils will cause corrosion of most metallic surfaces resulting in a weakening of the structure. Application of SE processes allows for the incorporation of all the stakeholder's needs and requirements in the success of the PV plant. It allows sharing of information, validation of requirements, and a common basis of analysis and assessment of potential risks based on the reviews of external and internal variables that can affect system operation. All of this must be based on relevant and extensive data and its appropriate curation.

5.6.3 Safety

Safety is a broad topic that covers everything from chemicals, electrical potentials, working conditions, working hours, and a variety of other topics. It is beyond the scope of this book to address all of these. In the prior paragraph, we addressed the need for planning and execution of policies and procedures to handle material classified as hazardous to prevent direct or indirect injury to maintenance personnel.

However, there are several system and design-level requirements that need to be met to qualify as compliant to national and international safety standards. In addition, there are standards that are applicable to handling waste products.

The SE team must identify and ensure compliance or adherence to safety standards that have been identified including those from AHJs. Safety considerations also are required for the O&M plan as well as ensuring that third parties (outside the fence) cannot gain ready access to the site and be exposed to the various hazards associated with solar energy production.

5.6.4 Common System Engineering Problems

During project concept development and system design efforts to define the details of the plant, there are six major issues that tend to result in excessive costs, schedule slides, either upfront during the design and development phase or during the installation and operation phases. These are:

- **Scope creep**: The introduction of new, "little" requirements here and there because they seem "easy" or to make the customer happy. Not only do they soon add up, but there is also an impact on all the project activities including test, documentation, and training.
- **Gold plating:** Doing more than is required to solve a problem or adding code when someone thinks, "While I'm doing this I may as well do...." This can put the actual requirements at risk. Like scope creep, it creates a lot more work. A random number-generating algorithm does not have to incorporate the latest thinking of mathematics Nobel laureates.
- **Lack of communication** between stakeholders – While it is unhelpful to have the users constantly scrutinizing the team's work, it is worse to keep them out. The earlier that user feedback is received, the less costly it is to incorporate.
- **Unidentified stakeholders:** It is essential that all stakeholders have been identified and consulted. Look for those who are unusually or unexpectedly impacted by the product(s) being built. For example, a paper-saving system may affect the job of an unseen archivist who works in the basement.
- **Management performing requirements engineering:** When there is resistance to change, the solution often employed for managers is for them to enter information into a reliability, availability, and maintainability systems design tool and perform the analysis by pushing a button. With few exceptions, management knowledge base is often either too limited or too far out of date for them to provide meaningful input to the systems and RAMS data. Not only does this introduce a useless burden on management, the project, and stakeholders, but it also creates opportunities for human error and often results in poor design and poor reliability.
- **Assuming a tool will solve the problems:** *Throwing a tool or money at the problem is like giving a Maserati to a pizza delivery person who cannot drive.* It is important to provide adequate training, not only in the tool but also in the basic technical attributes and underlying theory, and the process improvements designed to take advantage of the new tool as well as understand its limitations.

Of note from the above for PV power plants, there are often strong indications of a lack of communication between stakeholders and SE, such as finding unidentified stakeholders late in the process. When the EPC/(O&O) are at odds because of poor communications, the plant when commissioned will have problems in the first year or so with spares and repairs due to poor performance related to low reliability, excessive maintainability, or lack of funding.

5.6.5 System Design

The previous section covers a number of the critical elements of establishing the system specification. The complete package will be done once all of the stakeholders have provided their requirements for each of their areas and an agreement is reached on any adjustments or compromises. After the stakeholders signoff on to the project and SE plan, they can begin the process of selecting a company that will perform the detailed design of the plant. For the most part this is done by EPCs, but the project team should not necessarily limit the bidding to EPCs. The project team needs to put out the system specifications in a formal RFQ/RFP. The RFQ/RFP will contain the plant specification, the historical environmental data, and any other items that can be set out contractually. Remember, "If it is not in the contract, it should not be expected." This includes all attributes of concern such as performance, reliability, maintainability, and safety.

With the proposals in hand, the final system design work can begin through a review of the proposals and the selection of a winning bid. Any refinements needed for the proposed plant due to bidding exceptions or alternate design proposals are then worked on with the team and an agreed-upon consensus of the stakeholders is reached.

Once the RFQs have been returned, the team then solicits via RFP and reviews each proposal to verify through the established MOE that:

- The proposed designs will meet the system specification,
- The proposed designs fall into the range of cost expected,
- That there are no imposed limitations that had not previously been discussed and agreed upon, and
- The proposed suppliers have a demonstrated history or unique capability that will ensure that the equipment will work as needed.

5.6.6 Detailed Design

Detailed design is spilt between the company that will design, install, test, and commission the plant and the manufacturers of the modules, inverters, combiners, and other balance of plant equipment. Although contractually the EPC may be responsible for all the equipment and the detailed design, the project team should make it a point to also monitor the suppliers and sub-tier suppliers. Many programs and projects have run into cost and schedule impacts by not keeping track of the entire project, not just the EPC/Prime Contractor (PC).

5.6.7 Build and Test

This phase of the project is where typically the EPC/PC may be building up discrete elements of the system to verify and test the design. In addition, the manufacturers should be performing final acceptance testing as well as lot testing to ensure that the product being built meets the contract specification and that their manufacturing processes are under control, using ISO9000/9001 or other appropriate QA/QC efforts. Reliability testing should be considered where new technologies and/or much newer components are being used for which there is no history.

A word of caution: Being compliant to and performing the ISO 9000 series, QA/QC are two different things. From experience, some manufacturers state that they are compliant to ISO but have not been certified and do not have the third-party reviews needed to ensure compliance.

For quality manufacturing processes, there should be documented testing and sample testing throughout the manufacturing cycle. Both the EPC and project team should request status reports from and monitor all the primary manufacturers and even the sub-tier suppliers. The rationale for this is not only to assure that the plant will operate per specification but that the project team owner/operators know and understand the products being installed and the limitations of the technology being used.

One common point for all projects as noted in Chapter 3 is that new technology cycles tend to last only for 18–24 months with product support for those technologies often lasting only 48–60 months. The succeeding technologies come out at about a two-year interval. With a production life of only two years, one major concern for the owner/owner operators is whether to buy lifetime supplies of critical parts. This issue also affects the planning for upgrades, modifications, replacements, spares, and **SE/R**epowered™ planed and unplanned events. It is where this planning resolves many of the planning elements to assure stakeholder system expectations are met.

5.6.8 Installation and Commissioning

Installation can have a significant negative effect on the ability of the system to perform if not properly performed. The initial installation specification elements should cover the effects of shading by buildings, trees, system equipment structures, and other obstacles located on or offsite.

Installation should account for spacing for module/panel maintenance as well as access to the structure itself. An assumption that no maintenance is required is not valid, as not only does the structural PV support for the modules need a periodic inspection, repair, and/or replacement.

Upon completion of the installation and commissioning, the PV plant must be benchmarked. This should also be performed on an annual basis.[34] A PV benchmark is a defined process to acquire data documented in the PV system specification, that can be compared to technical expectations, e.g., how does performance compare to specification at year 0, 1, 2, …, 25, etc.

> Commissioning is the process of assuring that all systems and components of a PV plant are designed, installed, tested, operated, and maintained according to the operational requirements of the project's owner or final client.[35]

Note that the commissioning process includes a defined and implemented O&M. Commissioning is generally defined by regulations of the local government or AHJs, and the customer and owner(s).

5.6.9 Benchmarking and Certification

Benchmarking is a process of testing a system after the installation has been completed to document its capability against the system specification prior to the beginning of operation, the Commercial Operation Date (COD). This is not a statement that the plant meets the specification but what the actual capability and production numbers are for the plant (Sarr 2017; Atcitty et al. 2011).

A limitation of benchmarking is that it often is not a detailed assessment of the entire plant but an evaluation or assessment of just the major component functions, such as inverters and transformers. Without benchmarking is seldom possible to know at COD what the individual strings are providing to the inverters or what if any variation is occurring at the various levels of combiners used in utility-level plants.

5.6.10 Operation and Maintenance

This is the longest phase of the project. It begins with the final certification of the commissioning process and ends on the last day of operation of the plant. This is also the highest cost item of the program as it accounts for all day-to-day operations over the life of the system. This includes maintenance (preemptive, preventive, scheduled, and unscheduled), warehousing, and acquisition of spares, maintenance of support equipment, monitoring the daily power parameters, and performing data acquisition for the various stakeholders. This latter activity may be automated but should be monitored by the O&M organization and owners to document any trends indicating plant degradation outside of specification.

Much of the field reliability (Chapter 6) and maintainability (Chapter 7) data acquisition and assessment is collected during this phase. It includes the acquisition of field operating and failure data, tracking failures back through the supplier for root cause analysis, and implementing repairs or updates as dictated by the root cause analyses.

Design for maintainability should allow for the lowest level of technical support, requires using standard off-the-shelf tools, extensive diagnostics with electronic sensors allowing for fault isolation to the failed unit, and ease of access for maintenance. The project also needs to decide how it is going to include active switching among modules/panels and inverters (and perhaps batteries) so as to maximize availability and capacity and ensure there is not a loss of system output for a single, isolatable failure. Again, this needs to be considered during the architecture/system design phase, based on the MOEs[1] (power, energy, availability, accessibility, reliability, etc.) of the selection process.

There are several ways to set up the O&M for a PV plant (Chapter 11). The O&O can perform the maintenance themselves; a third-party O&M company can be contracted, or the maintenance can be contracted with the suppliers as part of the warranty or warranty extension. The "maintenance concept" is critical to assessing availability. It may be location-dependent because of local environments (precipitation, snow, sand/dust, etc.), and there will be a trade between system availability/power output and maintenance costs. Regardless, to ensure that the plant will meet specification over its life it is necessary to document operation and failures to the degree where root cause can be determined, and corrective action is taken. Failing to do so leaves a potentially expensive and catastrophic failure in the field.

5.6.10.1 Maintenance

Maintenance is the major cost item once the sunk (purchase and installation of the plant equipment) is completed. Maintenance itself is a combination of unscheduled (failure)

Table 5.10 Power Loss Associated with PV Plant Example.

Cause of Energy Loss	Percentage Loss (%)	Design or Maintenance
Shading	7	Both
Dust and Dirt	2	Maintenance
Reflection	2.5	Design
Spectral Losses	1	Design
Irradiation	1.5	Design
Thermal Losses	4.6	Design
Array Mismatch	0.7	Design
DC Cable Losses	1	Design
Inverter Losses	3	Design
AC Cable Losses	0.5	Design
Total Loss	23.8	

Adapted from SolarEmpower.

maintenance and scheduled and preventive maintenance and the spares or disposable materials used to perform maintenance. From a SE standpoint, tradeoffs must be performed during system design to optimize the cost-benefit of maintaining the equipment and at what point do upgrades and/or **R**epowering™ events need to occur. LCOE contains the "all in" cost elements of the plant and can serve as a basis for trend analysis that can indicate when there are increasing problems that need to be addressed by the owner or O&O.

The maintenance schedule needs to allow for potential variables such as weather-related reduction in power, transportation delays for critical maintenance, source (spares and tools) material delays, and shipping delays of critical parts due to manufacturing delays.

Consideration should be to the detail of what causes loss of power production in the system. Table 5.10 is one summary of those losses. As can be seen, without proper design, reliability, and maintenance, the amount of energy that can be lost can be substantial.

5.6.10.2 Levelized Cost of Energy

LCOE is a calculation that is done over a defined period of time. As the plant ages, there is the potential for companies to go out of business, changes in technology, and running out of warehoused spares. All these result in significant demands on finding spares and addressing "form factor" or component physical geometry changes and/or changes in component electrical attributes. This makes form, fit, and function replacements extremely difficult to accomplish. In addition, there exists the potential for "hidden" costs when considering maintenance on equipment that is no longer manufactured or for which form, fit, and function replacements do not exist. This latter item can result in a full replacement of the component/s versus a repair.

To achieve the lowest possible LCOE requires that the stakeholders provide all their metrics to the systems engineer to ensure that the requirements for the plant are properly identified, documented, coordinated, allocated, and tracked.

$$\text{LCOE} = \frac{I_o + \sum_{t=1}^{n} \frac{I_t + M_t}{(1+r)^t}}{\sum_{t=1}^{n} \frac{E_{t,el}}{(1+r)^t}} \tag{5.3}$$

where:

- LCOE = Average levelized electricity generation cost
- I_t = Investment expenditures in the time period "t"
- M_t = O&M expenditures in the time period "t"
- $E_{t,\text{el}}$ = Electricity generation in the time period "t"
- r = Discount rate
- n = Number of time periods

A word of caution, today, the LCOE is often presented in a limited form and does not include *all aspects of the total annual cost (A_t) over time or the site restorations costs.* For example, many plants also have significant government subsidies that are not truly accounted for in the LCOE. These subsidies vary year by year and can be stopped with limited notice. The LCOE must include all the benefits and costs of the O&M of the plant to include specifically the preventive, scheduled, and unscheduled maintenance cost, the cost of spares, and overhead. It requires including the real cost of recycling modules and hardware during plant operation after component removal, replacement, site maintenance, and its return to original or contractual condition during the time interval "t" (Equation 5.3).

From Equation 5.3, the LCOE is driven by two principal cost elements. The cost to produce the power/energy and the investment costs and the cost to produce energy include the costs of operations, maintenance, and spares as well as all the associated overhead costs. The desire to minimize or to exclude certain costs (e.g., site restoration) from being classified as investment results in inaccurate LCOE calculations. This minimization of the investment cost is the primary source for the poor reliability and system performance (Jahn and Niasse 2003) of some plants as well as the increased O&M costs associated with lower reliability in general. In addition, there are soft costs such as software maintenance, updates, and with additional management and labor-related though not always considered costs that can add substantially to the maintenance element.

LCOE variability is influenced by the LCC of the project. LCC includes all the investment and operating costs associated with a project including for example the elements shown in Table 5.11.

As a side note, current renewable power plants, as are other forms of energy generation, are heavily subsidized by national governments and their agencies. At some point in the future, the cost of those subsidies will become too great to maintain, and the true LCOE costs will appear.

5.6.11 Upgrade and SE/Repowered™

Upgrade, refurbish, repower, and **R**epowered™ have often been used interchangeably. In this book, an upgrade is an ongoing process of installing new technology as part of a planned replacement cycle. Other industry examples include upgrades to military and commercial subsystems, such as radio, radar, and engine technologies, and hydroelectric

Table 5.11 Example Summary of Cost Elements.

Fixed Costs	Ongoing Costs
Concept	Incentives
Specification	Discount rates
Development	Insurance costs
Design	Interest on financing
Build	Cost of living adjustment (COLA)
Install	Forced outage
Commissioning	Scheduled/preventive maintenance
Operation	Partial outage
O&M, including spares	Unscheduled maintenance
SE/Repowered™	Spares and warehousing
Decommissioning	Overhead (management)
Site restoration	Taxes
	Benchmarking
	Insurance

power plants that are generally planned to have technical updates every 5–10 years depending on the technology. These are subsystems that are replaced as improvements have been made and integrated into a form of fit and function capability for the system. In aviation, this does not include the cost of a new airframe. Airframes are only replaced when their economics dictate, generally around 20–25 years for commercial or 10–15 years for *most* solar-based systems today. The B-52 is the most notable exception, which is going on 60 years of operation of the airframe. The internal electrical, electronic, and other parts have been replaced roughly every 7–10 years to keep its capability current. The service life of the B-52 is expected to extend to 90–100 years, a testament to the 1950s design. Hydroelectric power plants pull generators and replace brushes, power couplings, and other shelf items and then reinstall the generating unit.

SE/Repowered™ on the other hand is a process that applies to the plant in its entirety, which is still improving the service life of the plant as a system. Although, and probably without exception, **R**epowered™ events will include new technologies, it also may require modification to the base module/panel supporting infrastructure to accommodate different-sized modules. **SE/R**epowered™ events will also include the redesign of the plant software and firmware.

5.6.11.1 Upgrade

Upgrading a system for the purposes of this discussion is a lifecycle process where components, assemblies, and end items are periodically (scheduled) replaced as new technologies arrive without performing a **R**epowered™ or redesign of the plant. Historically in aviation, marine, and energy production, these take place on an established timeline based on defined triggers and hooks, or as the result of manufactures no longer producing the current components but providing a form, fit, and function capable part that performs at lower power consumption, higher efficiency, or at higher speed for the same power.

For inverters, new semiconductor processes provide for transistors/printed circuits (e.g. IGBT) that have the capability to handle higher voltage and current with lower series resistance, resulting in lower temperature rise (https://www.nrel.gov/pv/pvrw.html; Fu et al. 2019; Doran 1981). It may also be possible to upgrade the inverter controller processor with a newer technology while having generally better attributes for sine wave generation, reduction of harmonics, and reduced switching noise.

A guiding principle for planning upgrades is that there are supplier contractual notifications in place that allow the performing O&M organization to do a cost-benefit/performance analysis and determine if it is better to buy an expected lifetime supply of spares of the current technology or accept the new parts as spares or replacement items as needed.

Software and firmware can be new or upgraded as well as sensors, cameras, and computing systems. Many of these can be classed as off-the-shelf and commercially available.

5.6.11.2 PAM SE/Repowering™

SE/Repowered™ (Chapter 4) is the plan and process whereby the plant power components (mainly inverters and modules) are replaced as they are at, or near, their end of life. This is normally a planned event, but it is possible that after a significant force majeure, or significant reliability problems, it may be necessary to perform **R**epowered™ tasks due to excessive damage to the plant components. If it is being considered because of poor performance (most likely in a new plant purchase), this implies that the initial design was poor and/or the parts specified or used were of poor quality. The plan covers the broad scope of what is entailed in the process of specifying, designing, building, testing, manufacturing, installation, and recommissioning of the plant whether new or existing. For the PV modules, the combination of degradation, failures, technology changes, and the need for achieving the original module power specification provides the basis for returning the plant to the original level of contracted energy. If greater power output is desired, then **SE/R**epowered™ plans and processes provide for the redesign of major portions of the plant to achieve greater energy generation. The initial plan, developed during the concept phase, should focus on the planned date and/or based on trigger or triggers criteria for executing the plan.

SE/Repowered™ is also a fact of life for existing plants. It is highly recommended that the project team systems engineers perform a cost-benefit analysis as well as estimate the total cost of replacement of the plant's major or minor components in need of upgrading or replacement. Again, this plan is based on poor performance and should not be a gross replacement of parts trying to get to where the contract requires the plant to be. The language in a variety of new and modified existing contracts must clearly address the **SE/R**epowered™ realities. Not employing SE methods is a recipe for early failure or "more of the same."

5.6.12 Site Restoration, Equipment Removal, Disposal, and Recycling

Retirement of the plant is defined as when the plant can no longer operate profitably and/or functionally (end of the service life) and when upgrading **SE/R**epowered™ is not fiscally possible. Of course, the original contract and local AHJ expectations may be for the plant to be taken down and the site restored to its original usage.

In the event of expected plant retirement, clearly defined plans need to be made for removing the infrastructure and components and site restoration to original or contracted state.

5.6.12.1 Site Restoration

Site restoration is the process of returning the site to its prior use state. This includes and is where the removal of all conduits, wiring, concrete pads, fencing, lighting, waste, pollution, chemical spills, and other facility items are removed (Curtis et al. 2021). To be done correctly it may mean that employment of local biologists, horticulturists, and naturalists to collaborate on the necessary flora to be planted versus those that may naturally fill in the site within a short period of time. It is critical to determine to what extent reserve funding would be essential to cover the costs of site restoration. This requires developing a protected management reserve fund gathered during the life of the plant to return the site to its intended condition. This is in part driven by to what extent the O&O is going to be liable both financially and legally. Or will they simply abandon the site and leave it to the local authorities to fix? This latter approach may still leave them legally responsible, which can be more expensive than having a plan for handling the restoration and disposal.

5.6.12.2 PV Removal through Disposal

PV equipment (all-inclusive) has some degree of hazardous materials contained within them. This includes silicon (glass) that has been doped by phosphorus, or boron, or other precious metals, gallium, indium, etc. It is recommended that the project identify the appropriate disposal requirements and implement them at the beginning of the project. There are several IEC and other standards for the handling of these materials if they are to be disposed of after the plant is dismantled. In addition, there are international standards that must also be considered.

5.6.12.3 Recycling

The plant infrastructure includes metals (Al, Cu, Ag, Au, steels, etc.), all of which have some value and should be reclaimed versus being disposed of in a landfill. These make up the macro scale parts and are generally easily separated. The semiconductor material that makes up the photocells and computing/power circuitry consists of silicon (Si), with phosphorus (P) or boron (B) and other elements. Unfortunately, these materials have been hard to extract. Silicon or glass can make up more than 70% of a solar module (monocrystalline or polycrystalline). The other ~30% is copper, substrate, and connections. Glass, even ultra-pure silicon, however, has little intrinsic value and extracting the dopants such as B, P, Be, Ag, etc., or less than 5% of the wafer requires extensive use of chemical and electrical processes. None of these are environmentally friendly in particular: lead (Pb), arsenic (As), and beryllium (Be), for example. Even so, they can be more environmentally safe than other forms of non-hydro energy generation such as coal, oil, and for disposal of radioactive wastes.

A basic process is to disassemble the modules to recover the materials, then heat is needed to burn off any adhesives or non-metal materials. Depending on materials used in the construction of the PV modules, for example, they may give off toxic fumes that must be collected and disposed of. The cells are then exposed to acids to extract the glass, leaving the doping materials behind. Further processing with acids is used to recover the different precious metals.

In the United States, hazardous materials are controlled by the Federal Resource Conservation and Recovery Act (RCRA, in 40 CFR[36] 261) and by state-level laws and regulations. In addition, there are state, regional, and local policies that govern waste disposal and disposition. The United States currently has little consistency in regulatory requirements, which makes cookie-cutter plant designs nearly impossible. Washington state, for example, requires that the PV manufacturers provide for the removal and proper disposal of PV products.

The Solar Energy Industries Association's (SEIA) PV Recycling Working Group has been developing recycling partners across the United States since 2016. In 2019, it published a common checklist for handling the disposal of PV modules.[37]

5.6.12.4 Restriction on Hazardous Substances (RoHS)

The RoHS is a requirement in the European Union or for countries that subscribe to the EU standards on what materials cannot be used in manufacturing unless there is a specific waver for a special condition.

The IEC/EN 62321 covers methods for testing material and components for the seven commonly restricted substances, which are:

1. Lead, Pb
2. Mercury, Hg
3. Cadmium, CD
4. Hexavalent Chromium, Cr VI
5. Polybrominated Biphenyls, PBB
6. Polybrominated Diphenyl Ethers, PBDE
7. Bis(2-Ethylhexyl) phthalate, DEHP

Although many of these restricted substances are no longer available, the project safety person should require all suppliers to submit materials lists. A review of the list ensures that no prohibited items are being used in the project. Note that some states, provinces, and countries have additional items that have been banned. Exceptions can be made, but they must have an engineering rationale that explains why the hazardous material is the only approach to make a certain design work.

It is recommended that all projects implement elements of EN 50581.[38] This is a management process-based standard for controlling the supply chain materials. This allows manufacturers with a large number of components/materials in their products and who may often change materials in the production lines a more efficient way to ensure that their products are RoHS compliant.

RoHS restricted substance examples are of materials that, for example, contain lead and are required to handled according to the regulations:

1. Paints and pigments
2. PVC (vinyl) cables as a stabilizer (e.g. power cords, USB cables)
3. Solders
4. Printed circuit board finishes, leads, internal and external interconnects
5. Glass in television and photographic products (e.g. CRT television screens and camera lenses)

6. Metal parts
7. Lamps and bulbs
8. Batteries
9. Integrated circuits or microchips

5.6.12.5 Manufacturing Compliance to Waste/Recycling

For PV modules and panels, there are a number of IEC standards addressing their design and manufacture. IEC 62548, for example, Photovoltaic (PV) arrays define a set of design requirements and describe design phase, IEC 62941:2019 Standard | Terrestrial PV modules – Quality system for PV module manufacturing, OD-405-1 3.1 IECRE Quality System Requirements for PV Module Manufacturers – Part 1: Requirements for certification and OD-405-2 2 IECRE Quality System Requirements for PV Module Manufacturers – Part 2: Audit Checklist. The project should apply these standards for their suppliers and the EPC, in particular, for plants being designed for sites outside the United States.

Waste Directive 2008/98/EC of the European Parliament and of the Council of November 19, 2008, on waste and repealing certain directives covers how hazardous materials are to be handled when they are waste to be disposed of. Additional rules for the calculation, verification, and reporting of data on waste in accordance with the amended Waste Framework Directive can be found in Commission Decision (EU) 2019/1004.

Within the United States, the National Pollutant Discharge Elimination System and elements of 40 CFR cover the effluent guidelines of the Environmental Protection Agency and provide regulations regarding hazardous waste. Each state can have additional restrictions on handling and disposal of hazardous materials.

Solders Lead-free solders have been shown to have an impact on safety and reliability. Many of the lead-free solders have, in military and commercial applications, demonstrated lower reliability due to increased brittleness and susceptibility failure from thermal cycling, vibration, and shock. This does have direct applicability to PVPS in operation, transportation, and handling. There are areas in the panels/modules and inverters where there can be large temperature swings over the course of a day, resulting in significant numbers of thermal cycles over 10 years.

5.7 Conclusion

This chapter does not attempt to cover all the SE tools and processes but points to some that will provide immediate benefit to a project. We have provided the definitions, explanation, and usage for that set of SE tools and processes that if implemented should provide for greatly improved systems. The reader should be able to explain these and understand the need for and results of applying the SE processes to their project. The reader should also be able to define when the various tools are used and how they are applied.

Our specific point is that the application of SE processes will significantly improve the probability of a successful long-lived energy production PVPS. We also have pointed out that failing to apply these tools (and others) will result in significant risk and unplanned costs for all stakeholders.

Using these processes with SE personnel who are well-trained and experienced in PV power plants will yield a well-defined set of requirements and a system design that will meet the owner and utility energy demands over the expected life of the system.

Bibliography

Atcitty S, Granata J.E., Quintana M.A., and Tasca C.A. (2011a). *Utility-Scale Grid-Tied PV Inverter*, Reliability Workshop Summary Report, Sandia Report SAND2011-4778, Unlimited Release, Prepared by Sandia National Laboratories Albuquerque, New Mexico 87185 and Livermore, California 94550.

Balfour, J.R., Hill, R.R., Walker, A., et al. (2021). *Masking of photovoltaic system performance problems by inverter clipping and other design and operational practices.*. United States. https://doi.org/10.1016/j.rser.2021.111067.

Blanchard, B.S. and Fabrycky, W.J. (2011). *Systems Engineering and Analysis*, 5e. Prentice Hall.

Bogue, R. (2018). *Use S.M.A.R.T. goals to launch management by objectives plan.* TechRepublic.

Burkhard, R.A., Meier, M., Rodgers, P. et al. (2005). Knowledge visualization: a comparative study between Project Tube Maps and Gantt Charts. In: *5th International Conference on Knowledge Management*. Graz, Austria: University of Kent Retrieved 17 September 2017.

Carson, R. (1998). Requirements completeness; a deterministic approach. *Eighth Annual International Symposium of the International Council on Systems Engineering*, Vancouver, British Columbia, Canada.

Carson, R. (2016). Quantifying sustainability in system design. *Conference: International Symposium of the International Council on Systems Engineering At: Edinburgh, Scotland, UK*.

Collins, E., Dvorack, M., Mahn, J. et al. (2009). Reliability and availability analysis of a fielded photovoltaic system. In: *34th Photovoltaic Specialists Conference*, vol. 1, 002316–002321. https://doi.org/10.1109/PVSC.2009.5411343.

Curtis, T., Heath, G., Walker, A., et al. (2021). Best practices at the end of the photovoltaic system performance period technical report NREL/TP-5C00-78678 February 2021.

DeGraaff D, Lacerda R, Campeau Z, Xie Z, 2011, *How do qualified modules fail – What is the root cause?*, 15 July 2011 NREL International PV Module Quality Assurance Forum. San Francisco, California, https://www.pvqat.org/pdfs/10-ipvmqaf_degraaff_sunpower.pdf (accessed 2020).

Doran, G.T. (1981). *There's a S.M.A.R.T. way to write management's goals and objectives. Management Review.* 70 (11): 35–36.

Engle, P. (2010). *MTBF and reliability—a misunderstood relationship in solar photovoltaics*, Electronics Design, https://www.electronicdesign.com/energy/mtbf-and-reliability-misunderstood-relationship-solar-photovoltaics (accessed 2021).

Fife, J.M. (2010). *Solar Power Reliability and Balance-of-System Designs*. Renewable Energy World.

Fife, J.M., Scharf M., Hummel S.G., and Morris, R.W. (2010). Field reliability analysis methods for photovoltaic inverters. *IEEE: 35th IEEE Photovoltaic Specialists Conference.*

Flicker, J. (2014). *PV inverter performance and component level reliability*. https://www.nrel.gov/pv/assets/pdfs/2014_pvmrw_35_flicker.pdf (accessed 2021).

Formica, T.J., Khan, H.A., and Pecht, M.G. (2017). The effect of inverter failures on the return on investment of solar photovoltaic systems. *IEEE Access* 5: 21336–21343. https://doi.org/10.1109/ACCESS.2017.2753246.

Fu, R., Feldman, D., and Margolis, R. (2019). *U.S. solar photovoltaic system cost benchmark Q1 2018. National Renewable Energy Laboratory* https://doi.org/10.7799/1503848.

Geraldi, J. and Lechter, T. (2012). Gantt charts revisited: a critical analysis of its roots and implications to the management of projects today. *International Journal of Managing Projects in Business* 5 (4): 578–594. https://doi.org/10.1108/17538371211268889.

Gevorkian, P. (2011). *Large-Scale Solar Power System Design, An Engineering Guide for Grid-Connected Solar Power Generation*, 1e. McGraw-Hill's (GreenSource Books) (May 2, 2011).

IEC:2003 60300-1 and 60030-3, https://webstore.iec.ch/home.

IEC TS 61724-2:2016 Photovoltaic system performance – Part 2: Capacity evaluation method, https://webstore.iec.ch/home.

IEC TS 61724-3:2016 Photovoltaic system performance – Part 3: Energy evaluation method. https://webstore.iec.ch/home.

IEC 61724-1:2017 Photovoltaic system performance – Part 1: Monitoring. https://webstore.iec.ch/home.

IEC 63019 (2019). Technical specification information model for availability of photovoltaic power systems (PVPS), https://webstore.iec.ch/home.

IECRE OD 501 (2017). IEC System for Certification to standards relating to equipment for use in Renewable Energy applications (IECRE System). https://webstore.iec.ch/home.

IEEE/ISO/IEC 29148-2018 Systems and software engineering — Life cycle processes — Requirements engineering.

Jahn, U. and Niasse, W. (2003). Analysis of long-term performance and reliability of PV systems, IEA-PVPS Task 2 Report, https://iea-pvps.org/key-topics/analysis-of-long-term-performance-of-pv-systems-2015-2/ (accessed 2020).

Jordan, D.C. and Kurtz, S.R. (2010). Analytical improvements in PV degradation rate determination. In: *Proceedings of the 35th IEEE PV Specialist Conference*, 2688–2693. Honolulu, HI, USA.

Jordan, D.C., Smith, R.M., Osterwald, C.R. et al. (2010). Outdoor PV degradation comparison. In: *Proceedings of the 35th IEEE PV Specialist Conference*, 2694–2697. Honolulu, HI.

Kalogirou, S.A. (2013). *Solar Energy Engineering: Processes and Systems*, 2e, Kindle Edition. Academic Press; 2e (October 25, 2013).

Kerzner, H. (2003). *Project Management: A Systems Approach to Planning, Scheduling, and Controlling*, 8e. ISBN 0-471-22577-0.

Kim, H.-J. (2015). *Solar Power and Energy Storage Systems, 2019*. Jenny Stanford Publishing.

Klastorin, T. (2003). *Project Management: Tools and Trade-offs*, 3e. Wiley ISBN 978-0-471-41384-4.

Krauter, S. (2006). Photovoltaics. In: *Solar Electric Power Generation-Photovoltaic Energy Systems*, 28–37. Berlin: Springer-Verlag.

Kumar, P.P. (2005). Effective use of Gantt chart for managing large scale projects. *Cost Engineering* 47 (7): 14–21. ISSN 0274-9696.

Makrides, G., Zinsser, B., Georghiou, G.E. et al. (2010). Degradation of different photovoltaic technologies under field conditions. In: *Proceedings of the 35th IEEE PV Specialist Conference*, 2332–2337. HI, USA: Honolulu.

Messenger, R.A. and Abtahi, A. (2020). *Photovoltaic Systems Engineering*, 4e. CRC Press.

Morris, R.W. and Fife, J.M. (2009). Using probabilistic methods to define reliability requirements for high power inverters *Proceedings of the SPIE 7412*, 20 August.

NAESB WEQ 2005, *Manual Time Error Correction Standards* – WEQBPS – 004-000 NAESB WEQ Standards 211 January 15, 2005 Copyright © 2005 North American Energy Standards Board, Inc. All Rights Reserved. http://www.naesb.org/pdf2/weq_bklet_011505_tec_mc.pdf (accessed 2019).

NERC Reliability Guideline BPS (2018). *Connected Inverter-Based Resource Performance* September 2018, NERC. https://www.nerc.com/comm/OC_Reliability_Guidelines_DL/Inverter-Based_Resource_Performance_Guideline.pdf (accessed 2020).

Project Management Institute (2013). *A Guide to the Project Management Body of Knowledge*, 5e. Project Management Institute. ISBN 978-1-935589-67-9.

Ristow, A., Begovic, M., Pregelj, A., and Rohatgi, A. (2008). *Development of a methodology for improving photovoltaic inverter reliability. IEEE Transactions on Industrial Electronics* 55: 2581–2592.

Sample, T. (2011). Failure modes and degradation rates from field-aged crystalline silicon modules, NREL PV Module Reliability Workshop, Golden CO, USA, Feb 2011,

Sarr, O. (2017). Analysis of failure modes effect and criticality analysis (FMECA): a stand-alone photovoltaic system. *Science Journal of Energy Engineering* 5: 40. https://doi.org/10.11648/j.sjee.20170502.11.

Stein, J.S., and Klise, G.T., (2009). *Models used to assess the performance of photovoltaic systems.* United States: N. p., 2009. Web. doi:10.2172/974415.

Systems Engineering Handbook, 4e, 2003, INCOSE-TP-2003-002-04.

Vazques, M. and Rey-Stolle, I. (2008). *Photovoltaic module reliability model based on field degradation studies, Progress In Photovoltaics: Research and Applications*. Published online in Wiley InterScience (www.interscience.wiley.com) https://doi.org/10.1002/pip.825.

Ventre, J. and Messenger, R.A. (2010). *Photovoltaic Systems Engineering*, 3e. CRC Press.

Woolf, M.B. (2012). *CPM Mechanics: The Critical Path Method of Modeling Project Execution Strategy*. ICS-Publications. ISBN 978-0-9854091-0-4.

XXX DOD Guide for Achieving Reliability, Availability, and Maintainability, Active, August 3, 2005

Zhou, Q., Xun, C., Dan, Q., and Liu, S., (2015). Grid-connected PV inverter reliability considerations: a review, *2015 16th International Conference on Electronic Packaging Technology*, 978-1-4673-7999-1115/$31.00 ©2015 IEEE.

Notes

1 "A Measure of Effectiveness (MOE) is a measure of the ability of a system to meet its specified needs (or requirements) from a particular viewpoint. This measure may be quantitative or qualitative and it allows comparable systems to be ranked. These effectiveness measures are defined in the problem-space. Implicit in the meeting of

problem requirements is that threshold values must be exceeded." N Smith, T Clark – 11th International Command and Control Research and Technology Symposium (ICCRTS), 2006.

2 https://www.nasa.gov/sites/default/files/atoms/files/nasa_systems_engineering_handbook_0.pdf

3 Lt Col Christopher J. Niemi, The F-22 Acquisition Program, Consequences for the US Air Force's Fighter Fleet, November–December 2012 Air & Space Power Journal, p. 53.

4 https://www.energy.gov/eere/solar/sunshot-initiative

5 Thorsten Hoefer, "When to repower aging utility-scale solar projects," SPW, November 9, 2020.

6 SMART has several variations. Assignable can also be Attainable, Realistic can also be Relevant, and Traceable can also be, for reliability requirements, Timely or Time related. https://corporatefinanceinstitute.com/resources/knowledge/other/smart-goal/.

7 https://photovoltaic-software.com/pv-softwares-calculators/pro-photovoltaic-softwares-download/pvsyst

8 https://www.wecc.org/Reliability/Solar%20PV%20Plant%20Modeling%20and%20Validation%20Guidline.pdf

9 https://pvwatts.nrel.gov

10 https://www.nrel.gov/state-local-tribal/assets/pdfs/sam-for-cities.pdf

11 National Solar Radiation Database, https://www.nrel.gov/docs/fy17osti/67722.pdf

12 Accurate Degradation Rate Calculation with RdTools, https://www.nrel.gov/pv/rdtools.html.

13 900 MW Fault Induced Solar Photovoltaic Resource Interruption Disturbance Report Southern California Event: October 9, 2017, Joint NERC and WECC Staff Report February 2018.

14 https://nerc.com/comm/RSTC/Documents/Wildfire%20Mitigation%20Reference%20Guide_January_2021.pdf

15 https://www.nerc.com/pa/rrm/ea/Documents/San_Fernando_Disturbance_Report.pdf

16 900 MW Fault Induced Solar Photovoltaic Resource Interruption Disturbance Report Southern California Event: October 9, 2017, Joint NERC and WECC Staff Report February 2018.

17 https://nerc.com/comm/RSTC/Documents/Wildfire%20Mitigation%20Reference%20Guide_January_2021.pdf

18 https://www.nerc.com/pa/rrm/ea/Documents/San_Fernando_Disturbance_Report.pdf

19 Kerzner, H. *Project Management: A Systems Approach to Planning, Scheduling, and Controlling*, 12e, ISBN: 978-1-119-16535-4 April 2017.

20 On May 7, Colonial Pipeline shut down its 5,550-mile gasoline pipeline following a cyberattack on the company's computer systems. https://www.nytimes.com/2021/05/09/business/energy-environment/colonial-pipeline-shutdown-gasoline.html.

21 https://www.whitehouse.gov/briefing-room/presidential-actions/2021/05/12/executive-order-on-improving-the-nations-cybersecurity/

22 This depends on the percent fraction that the customer wants for ramp up.

23 At solar noon, by definition, the sun is exactly on the meridian, which contains the north–south line, and consequently, the solar azimuth is 0°. Therefore, the noon altitude α_n is: $\alpha_n = 90° - L + \delta$

24 A good example of annual variation can be found at https://www.njweather.org/content/exploring-njwxnet-solar-radiation-observations.

25 http://www.caiso.com/Documents/FlexibleResourcesHelpRenewables_FastFacts.pdf#search=What%20the%20duck%20curve%20tells%20us%20about%20managing%20a%20green%20grid

26 https://news.energysage.com/solar-panel-efficiency-cost-over-time/

27 Jordan, D. and Kurtz, S. (2013). Photovoltaic degradation rates—an analytical review. *Progress in Photovoltaics: Research and Applications*. 21. https://doi.org/10.1002/pip.1182.

28 Lucio Claudio Andreani, Angelo Bozzola, Piotr Kowalczewski, Marco Liscidini, Lisa Redorici, Silicon solar cells: toward the efficiency limits, Article: 1548305, Advances in Physics, Received 16 Oct 2017, Accepted 08 Nov 2018, Published online: 05 Dec 2018.

29 Bora, B., Sastry, O.S., Kumar, R., et al. (2021). Failure mode analysis of PV modules in different climatic conditions, *IEEE Journal of Photovoltaics*, 10.1109/JPHOTOV.2020.3043847, 11, 2, (453–460).

30 Researchers at NREL Find Fewer Failures of PV Panels and Different Degradation Modes in Systems Installed after 2000 April 10, 2017, https://www.nrel.gov/news/program/2017/failures-pv-panels-degradation.html.

31 Balfour, John; Hill, Roger, Walker, Andy et al., Masking of photovoltaic system performance problems by inverter clipping and other design and operational practices, 2021-04-21, Sandia National Lab. (SNL-NM), Albuquerque, NM (United States); National Renewable Energy Lab. (NREL), Golden, CO (United States).

32 https://en.wikipedia.org/wiki/Electric_power_quality

33 A classic example of induced flutter is the Tacoma Narrow Bridge in 1933, which was destroyed by high winds (35 mph (56 km/h)) inducing an unbounded oscillation of the bridge deck that resulted in its destruction.

34 Benchmarking is defined as the evaluation or checking (something) by comparison with a standard.

35 https://www.pv-tech.org/inside-commissioning-the-latest-trends-in-getting-solar-projects-operationa/ "Inside commissioning: The latest trends in getting solar projects operational," Sara Verbruggen, July 28, 2020.

36 Code of Federal Regulations.

37 https://www.seia.org/initiatives/recycling-end-life-considerations-photovoltaics

38 European Standards. Technical documentation for the assessment of electrical and electronic products with respect to the restriction of hazardous substances.

6

Reliability

6.1 Introduction

This chapter has two principal sections. The first addresses reliability as a general topic for any reader. The second contains the more detailed technical and mathematical treatment of reliability as it applies to systems engineering (SE) and reliability. The high-level material in Sections 6.3–6.4 covers reliability from the perspective of the nontechnical stakeholders, who have a vested interest but not necessarily the technical background to understand and/or apply SE, the nuances of reliability, physics of failure (PoF), and the mathematics of statistical analysis of field data for status and trends. Section 6.9 provides the foundation mathematics and statistics to allow the Operators, Operations and Maintenance (O&M), Engineering, Procurement and Construction (EPC), and others to apply the theories and procedures to improve the delivered system/s. This provides for reducing and stabilizing annual and lifecycle O&M costs and improving power/energy availability. This is not a complete treatise on the speciality of Reliability engineering (RE). The reader is referred to references DOD GUIDE (2005), IEEE Std 493™ (2007), Kececioglu Vol 1 (2002), Kececioglu Vol 2 (2002), Elsayed (1996), Kailash and Pecht (2014), System Reliability Toolkit (2005), Mil-Hdbk-338B (1998) for more in-depth coverage.

Reliability is a term that is used and abused throughout many industries. Terms such as "The most Reliable blank (fill in your favorite commercial)," or the blank (fill in your favorite commercial), "has the Highest Reliability in the Industry," and our favorite: blank "has the Highest or Best in Class Reliability," What all of these have in common, and fail to convey, is that *Reliability is a measurable attribute of a product* and as such when one is talking about reliability, it should be discussed using the *appropriate language and metrics*. As such to achieve more reliable infrastructure, one must address the reliability of inverters, combiners, PV modules, trackers, and so on. Because of the complexity of systems and the different attributes that customers and stakeholders use to define success, reliability has a large number of direct and derived metrics that allow for a complete prediction and/or analysis of a specification, design, or plant from each stakeholder's perspective.

Many proponents of PV as "green" energy appear to believe that PV itself is quite simple, requires little or no maintenance, and will easily last for 20–25 years. This, however, is far from the established reality. Statements such as "having a high reliability" are meaningless without context. Because reliability is an umbrella term with many metrics, there is the attendant problem that when two or more parties, such as financial, insurance, and EPCs,

Photovoltaic (PV) System Delivery as Reliable Energy Infrastructure, First Edition.
John R. Balfour and Russell W. Morris.

sit down to discuss a program, they all have different interests and the "reliability" metrics that they are concerned about. This means that unless there is an agreement between the stakeholders as to terminology and definitions (taxonomies and ontologies), the stakeholders can and will be talking past each other since their interpretations of the terms may be entirely different.

The most important definitions to begin with are success and failure for each stakeholder and then selecting from the available reliability metrics, the attributes to be reported to them.

6.2 Why Reliability

To explain why it helps to get back into the reason, i.e. start with the grid. In recent years there has been a general realization that the U.S. power grid is becoming less reliable (Amin 2011; North American Electric Reliability Corporation (NERC) 2016). Grid availability is directly tied to the sources of energy/power, e.g. hydroelectric, nuclear, coal-fired, turbine engine, photovoltaic, hydrodynamic, and wind, as well as the extremely high voltage transmission lines, switch gear, and power stations associated with them.

To address some of the issues with grid availability and reliability, the NERC published "Reliability Considerations for Clean Power Development,"[1] in Canada and United States as well as Baja California, Mexico. This is a general agreement on the reliable *transmission* of electrical power. In time this may include Mexico as their grid development evolves and becomes an extended element of an overall North/Central American grid. The NERC effort is in part due to an aging infrastructure with increasingly poorly-addressed and funded problems/failures. Nevertheless, it also relates to the need to ensure reliable power generation and delivery to the customers. To support the need for a more reliable system, NERC has implemented reliability reporting requirements on power generation of 20 MW[2] or greater that connect to the grid. These reports can be used to identify weak areas of the power system and initiate action to correct the problems. Each geopolitical sector around the world tends to experience the same types of challenges with different reporting agencies, guidelines, and approaches. At the low end of the scale, home or small-scale power systems, there is generally little to no reliability data available. This creates additional issues.

California for example, in 2018, passed legislation requiring solar power on every house beginning 1 January 2020. This then becomes an exceptionally large, distributed network with sub-nets being tracts or blocks of homes and businesses connected to bring on additive power for the load during the day. This could lead to significant power generation challenges with large distributed systems due to installation, sub-net connections, monitoring, energy storage technology, and connection to a grid/utility station. Also, there is a strong possibility that there will be inconsistent application of SE and reliability, availability, maintainability, and safety (RAMS), if they are even applied, or used to begin with. The following are simple examples of the effect the PV module has on plant operation:

- Module reliability in aggregate is a driver for cost.
- Module reliability does not dominate system reliability.
- Module reliability enables/improves affordable maintenance for PV power.
- Key module stress factors must be addressed to improve module reliability.

In the solar industry, it should be noted that there are several key reliability issues that can drive the lifecycle costs (LCCs) and levelized cost of energy (LCOE) of the system. These have impacts resulting in reduced or no output power when the system should be producing the expected level of specified power. For photovoltaic modules/panels,[3] there are five major reliability problems (failure modes) that affect more than 95% of the units in the field. These are but are not limited to:

1. Hot spots on the panels
2. Cracks and micro-cracks
3. Snail trail contamination
4. Potential Induced Degradation (PID) effect
5. Internal corrosion, delamination

A number of failure mechanisms applicable to modules/panels include:

- Condensation
- Differential thermal stress
- Vibration, wind-induced, mechanical rotation (motor)
- Chemical aging (degradation)
- Browning
- Front surface cracking
- Seal water leaks
- Fire
- Corrosion
- Manufacturing
- Shipping
- Handling/installation

The above list is the information for performing an initial failure modes and effects analysis (FMEA). What detailed information is needed to develop the failure modes and effects?

- A clear understanding of each technology, what it does, how it does it, the conditions within which it works, and the impact of operating outside of those parameters
- A fundamental understanding of degradation mechanisms and what stress drives those factors
- Knowledge of the products as manufactured and the application of products in the field
- Correlation of final test and reliability testing data to real outdoor performance data

We use the 100 MW plant throughout this book, although many of the rules and design guidance will be applicable to systems as small as a single residential unit. For large plants, the EPCs must ensure that the system specification developed in the concept phase by the system engineering team and with input from the stakeholders follows the SMART specification requirements (Chapter 5). These allow the Owner and/or Operator (O&O) to provide a request for proposal (RFP) or request for quote (RFQ) to bidding companies and to then perform a comparative assessment of several proposed plant(s) by competing EPCs. In addition, the RAMS and performance specifications must be properly allocated to the third-tier suppliers. As part of this effort, the requirements must be appropriately verified and validated for all the stakeholders. By using this method, the owner compares the proposed plants against a specific set of measurable requirements and can then pick the optimum design (Chapter 5).

Residential, commercial, industrial, and utility PV systems have historically had consistent challenges and unresolved problems meeting contractual requirements within the first five to seven years of system delivery. For PV systems, the problem has been equipment not meeting expectations of reliability (specified, contractual, expected, or implied). This is directly attributable to the lack of awareness and incorporation of RAMS into the specifications and the appropriate acquisition of field and test data to support verification and validation.

Some of the confusion is that reliability metrics have a variety of forms and definitions (see Glossary), while in general applications, reliability terminology is seldom properly defined and just thrown around rather loosely creating more confusion.

> The underlying attribute of reliability is that an item continues to meet its specification or to perform its intended function over time. This requires that stakeholders define success and failure for the plant. This criterion must apply over the life of the plant and be incorporated into any contractual agreements.

6.3 Success/Failure

As has been stated in Chapters 4 and 5, for stakeholders to understand if the system is meeting expectations, it is necessary for them to define both success and failure. Success is obvious as it means that the system provides power to the grid per specification and contractual requirements based on stakeholder wants, needs, and expectations. One disruptive assumption is that if the plant is providing power at the contracted level, the plant components are operating to spec. Failure on the other hand has multiple meanings for each stakeholder and for different parts of the plant. For example, the plant may be meeting contractual requirements, but lower-level items may be failing (strings, for example).

For the O&O, failure is not meeting the contracted annual power generation requirements at the specified cost due to excessive number of failed components. This may result in contractual default and/or having to pay liquidated damages or, if the power generation drops low enough, the plant will be shut down as being unprofitable. For the O&M, it is the number of failures of end items. For an excessive number of failures, the O&M may exceed their annual contract budget either as an out-of-pocket cost or by having to increase the billing to the O&O impacting their profitability. This disregards contractual exclusions that may be in place (Klise and Balfour 2015).

Financial institutions would view failure as not meeting the return on investment (ROI), monthly energy production, or even loss of the initial investment due to bankruptcy. Utilities may experience conditions requiring them to cut back power distribution (power) to brown-out conditions or address the problems with rolling blackouts because it is unable to deliver the energy its customers are demanding. Suppliers can be caught in several ways by failures in their equipment. There are also the losses due to warranty repair that may exceed their warranty reserve and potential loss of reputation for being a quality supplier. This will result in loss of future sales and, as has been demonstrated in the PV module and inverter manufacturing, the company can go bankrupt.

Therefore, reliability is based on the success/failure definition, see Chapters 4 and 5. This also requires that all stakeholders define the metrics that apply to their specific performance, wants, needs, and requirements.

Reliability probability of success as a function of time (R(t)), probability of failure (F(t)), a mean time to failure (MTTF), mean time between failure (MTBF), mean cycles to failure (MCTF), mean time between maintenance (MTBM), and a variety of other specific terms are covered below. Energy delivery is covered by an availability metric based on both reliability and maintainability. This often results in confusion among those who are not trained or understand the wide variety of terms used to define and express reliability for different stakeholders.

6.3.1 Stakeholder Metrics

For simplification, reliability metrics assume they are following a Poisson process. That is, a failure of any one specific item is assumed to follow a Poisson process, which can be found in DOD GUIDE (2005), Kececioglu (2002), Kececioglu (2002), Elsayed (1996), Formica et al. (2017), etc., if the following conditions are met:

1. All failures are assumed to be independent and identically distributed random processes for like and similar items.
   a. Independent, i.e. that the part fails by itself, and the failure is not the consequence of another different failure.
   b. Identically distributed implies that the underlying failure distribution is the same for all the failures for the same items.
   c. The rate of occurrence is assumed constant over the measurement time period, generally one year for PVPS.
2. The probability of an event occurring is related to the length of the time period.

In the event that the failure events begin to appear as pattern failures, i.e. failures are occurring under a set of common usage, environmental or operational profiles, RE looks at the common cause attributes of the failures and attempts to address the problem or condition.

Reliability terminology has general parameters which can be reduced to three basic forms that can be applied to stakeholders. These are probability of success (R(t)) as a function of time, or failure (F(t)) as a function of time, mean time to (or between) failure (MTTF or MTBF), and failure rates (FR) or failure per unit time.

$$\mathrm{R}(t) = 1 - \mathrm{F}(t) \tag{6.1}$$

$$\mathrm{MTTF} = \frac{Nt}{n} \tag{6.2}$$

$$\mathrm{FR} = \frac{n}{Nt} \tag{6.3}$$

where n = number of failures in time t; N = number of units that are operating over time t; t = the operating time for a specified interval.

As shown in Eqs. (6.2) and (6.3), a generalization can be made for MTTF and failure rate (FR) as they are the inverse of each other. However, caution should be applied to these because they are assuming an exponential distribution, i.e. constant FR. Section 6.9 has the more detailed development of the mathematics of reliability.

R(t) is the probability of successfully reaching some defined time "t" without failure and F(t) is the probability of failing as a function of time "t." The function R(t) is not realistically a probability of operation without failure unless it is accompanied by a stated confidence (50%, 60%, 80%, 95%, etc.). Again, having a high confidence in a mean value does not mean that it will in fact be that exact value or even within a small distance of the mean. Confidence levels and intervals are covered in Section 6.9.4.

Although there are some that believe that there are things that cannot fail, however, without an explanation of the limitations, requirements, conditions, and/or restrictions specific to that statement the reality is: *Everything fails; it is a matter of why, where, when, and how badly*.

So, what then are the definition(s) of reliability? The following eight terms provide a general description of reliability. The general definition that is used for the term "reliability" is:

RELIABILITY: The probability that an item can perform its intended function for a specified interval under stated conditions. This is also known as Mission Reliability (as a function of time) where the time element is specifically identified (Elsayed 1996; Kailash and Pecht 2014; System Reliability Toolkit 2005; Pecht et al. 2006).

Using statistical analysis of field or historical data allows deriving "mean time" calculations for a number of parameters such as:

- MTTF (for nonrepairable items)
- Mean time between scheduled maintenance (MTBSM)
- MTBF for repairable items
- Mean cycles to failure (MCTF)
- Mean time between critical failure (MTBCF)
- Mean time between maintenance (MTBM)
- FR (generally written out as failures per million hours or in the case of semiconductors it is listed as failures per billion hours (Failures In Time [FIT]).

Note that reliability is generally defined in terms that can be translated to time, more specifically operating, non-operating, and/or total time. Even the term cycles will be converted to time to be able to estimate how often one can expect to perform maintenance on relays, for example.

For the FIT metric stating failures in billions of hours would imply that nothing would ever fail. But when one considers that a circuit card may contain 100 or more parts with FIT values of 5–100 or greater, then the circuit card itself has an FR measured in the 10^{-4} to 10^{-6} range or 10,000 to 1,000,000 hours (about 114 years). Failures can occur at any time during operating and non-operating periods. Capturing operating failures is obvious as this is when stress or overstress is applied to the part and cumulative damage (stress) or instantaneous (overstress) results in failure. Non-operating failures are more difficult as they can occur when unpowered, power is removed or applied, for example making a "root cause" assignment problematic. For example, corrosion can occur on many metal pieces whether they are powered or under operational stress (such as flexing) but will manifest themselves when equipment is on and vibrating. A general rule of thumb is that for most electronic parts, the non-operational FR will be on the order of 1/100th that of the operating FR. For long operating periods, much of the PV plant tends to spend more time off than on. Mechanical non-operating failures are much harder to estimate as these generally do not manifest themselves immediately with operation. These failures can often be avoided when an effective preventive maintenance program is employed.

Critical failures represent the highest cost impact on the plant's power generation as well as the cost of O&M and reduce the profitability of the plant. In this case, it is assumed that critical failures are a subset of the failures making up the MTBFs and that there are repairable and non-repairable (repair by replacement) items within the unit. Single Point Failures (SPFs) are classed as critical failures since a single point is a piece of equipment or component, that if it fails, the functionality of that item is lost.

MTBM is one such useful Measures of Effectiveness (MOE) as it projects the total number of maintenance events that the project can expect from all sources, i.e. unscheduled, scheduled, false alarm, and preventive maintenance over a specified period. For example:

A standard assumption for solar power is that there is a maximum of 4380 hours of sunshine per year regardless of where the plant is located. Since this is assumed to be the number of operating hours, then one would expect that with an MTBF of 16.000 hours it would operate for almost four years before failing. The reality is that there is a finite probability that the inverter will fail in any given year over its life.

Although the general term applied to equipment or components is MTBF or MTTF, the probability of success or failure applies to systems where the design compensates for failures. Table 6.1 contains a sample of metrics that might be used for the components of the PV plant. Raheja (2019) notes that this is an important distinction. Things will fail, but the important question is "Can the system (plant) complete its mission?" To improve the probability of successfully providing contracted power to the utility can be done through robust design or redundancy either functional or physical.

Functional redundancy is having several power sources that can potentially compensate for failures in the PV plant. For example, a system could consist of solar, wind, diesel generator, and energy storage to compensate for the reduced energy output of the PV plant over varying periods of time. Physical redundancy is done through the implementation of identical capabilities such as having 103 inverters, when only 100 are needed to meet demand. An additional redundancy can be built into the arrays or strings by adding extra modules/panels to provide additional power to compensate for the loss of any number of modules/panels or an entire string.

Table 6.1 MTXX Table for Major Components of a PV Plant.

Subsystem	FR	MTBF	MTTF	MTBSM MTBPM	MTBR	MTBCF	MTTR	Mmax
Plant		X		X		X	X	X
SCADA		X		X		X	X	X
Lighting			X	X	X		X	
Inverters		X		X		X	X	X
Combiners	X			X			X	X
Solar IRR Sensors			X	X	X		X	
Modules/Panels			X	X		X	X	
Cells	X						X	
Connectors			X	X			X	
Wiring			X				X	

In historical power plants, there are, for example, four generators required to meet power demand, but a fifth generator is on standby in the event that one online unit fails for some reason. In addition, because the maintenance of large mechanical systems takes time to perform unscheduled, scheduled, and/or preventive maintenance, a sixth unit is available but offline for that maintenance.

Solar power plants have distinct differences from historical power generation. First and foremost is the sheer quantity of items (individual components) that make up a plant. An example of a 100 MW plant using fixed tilt installation is shown in Table 6.2.

6.3.2 Lowest Level of Repair

Defining failure requires listing the components, parts, assemblies, or equipment that will be tracked over time. We use the *lowest level of repair* definition to provide that list

Table 6.2 Item Quantity for 100 MW Solar Power Plant.

	Quantity	Failure Distribution
Cells	>30,000,000	Exponential
Modules	500.000	Exponential
Panels	50,000	Exponential
Combiners	10,000	Normal
Inverters	50–500	Mixed-normal
Interface Transformers	50–500	Weibull
Circuit Breakers	500–1500	Weibull
Fuses	5000	Weibull

Table 6.3 Examples of Lowest Level of Repair.

PV Module/Panel[a]	SCADA	Inverter Components
Combiner(s)	Sensors	IGBTs
Wiring	Software	PCBs/cards (control, communications, accessory, sensor, driver)
Wires	Computers	Capacitors
Relays	Uninterruptible power supply (UPS)	Contactors
Contactors	Routers	Heat Mgmt. systems (fans, motors, pumps, liquid, filters)
Switches	Network interconnect	Sensors
Circuit Protection	Firewalls interfaces – physical	Fuses
Fuses	Lights/lighting	Switches
Over Voltage Protection	Lightning protection	Power supply
Grid Tie Transformer	Security monitoring	Reactor/inductor
Over Current Protection	Monitors	Breakers
Circuit Breakers	Cameras	AC output
	Movement detection	Enclosures
		Transducers

a) The combination of module/panel is used to reflect that some installations are done with single modules with attendant support and attachment while other prefab panels of two or more modules. Thus, the lowest level of repairable item may be a module or a panel dependent on the system design.

(Table 6.3). In brief, the lowest level of repair is the item level at which maintenance will be performed and for which failure data will be tracked. This provides for the identification of spares (Chapter 7) and maintenance tasking (Chapters 7 and 11).

6.3.3 Failure-Free Operation

A common myth is that an MTBF or probability of success (as a function of time) implies that the item should not fail before this time. A mean or average time to failure (TTF) or FR is commonly based on a simple ratio. What is required is statistical analyses or in some cases a Physics of Failure (PoF) (Pecht 1995) analysis of test or field data. These analyses result in a number that covers a *population* and so there are some items that will fail before the mean and some that will fail after the mean. Applied to a single item there is a requirement to define the confidence level (CL) attached to the value. For example, an item with an MTBF of 30,000 hours with no data to support the value has, at best, a 50% confidence level. A large population of items with data taken from the date of initial operation, has not only the average or mean but the ability to assess what the confidence level MTBF would be based on the statistics of that population.

6.4 Overview

This section addresses reliability from a high-level perspective. This is for those that need to understand what reliability is and why it is important but may not need a detailed understanding of mathematics or the ability to apply the rules of the PoF. Math is kept to a minimum but is still needed to illustrate and understand reliability impact for all the stakeholders, not just the O&M and/or EPC.

Reliability has two strongly linked major divisions:

- Design or predictive reliability and
- Analytical or statistical reliability

Reliability as an engineering discipline requires both as one cannot perform predictive reliability without historical data from the field. The former is related to the analysis of a design(s) to determine the principal attributes applicable to the stakeholders by comparing owner or EPC specification, concept, and design proposals. The predictions are based on the part or component history as documented by field reliability data. Design specifics, such as design margin or derating (GEIA STD 0008 2011), redundancy, worse case analysis, environmental controls, and quality of the components, provide the basis for predicting the reliability of a paper design. Along with the assessment of the design, the predictive process accounts for the usage profile, environmental profile, and operational profile all of which need to be known about the specific plant site.

These analyses also require usage profile documents, based on-site and user data, and what the demand for energy will be to assess the electrical stresses on the design. Design-dependent attributes that can drive the usage profile are, for example, what type of tracking system, if any, is used. The addition of a PV tracking system (one or two axes) decreases the reliability of the system and will also increase the cost of the system if additional arrays are added to compensate for drive failures. The user demand will vary by time of day, time of year, and the ambient environments that the plant is exposed to during the year.

The environmental profile used to perform many reliability tasks (i.e. predictions, FMEA, Fault-Tree Analysis [FTA], etc.) is based on the ambient environments including but not limited to the examples in Table 6.4.

Reliability has, as shown above, a variety of metrics that are used and must be understood to apply the information gathered and analyzed from the field. The more generally used ones include:

1. FR is expressed in failures per unit hours. Most systems and reliability engineers use failures per 10^6 hours. Semiconductor manufacturers, however, use FITs (failures in time) that are expressed as failures per 10^9 hours.
2. MTTF is the average or MTTF of a non-repairable item. By this, we mean that the item is replaced with an identical unit.
3. MTBF is the average or MTTF of a repairable item. Thus, a driver card in an inverter may have an MTTF of 28,000 operating hours while the inverter itself can have an MTBF of 5000 operating hours.

Table 6.4 Example Environmental Data Needed for System and Detailed Design Analysis.

Temperature	Humidity
Wind (Steady and Gusts)	Sand and dust
Salt	Rain
Hail	Snow
Ice	Power transients (On/Off) related to forced shutdowns such as high winds
Lightning Strikes	Shipping (vibration and shock), handling, and installation
Fungus	Manufacturing assembly environment

4. MTBCFs are those failures that result in the loss of a system-level performance metric/s such as meeting energy demand due to the failure of too many units in a redundant function.
5. Probability of failure or probability of success, as a function of time.

Management wants to know the probability of success, while the SE/RE is looking at the probability of failure. Using this as a starting point, i.e. definitions of success and failure, reliability has a statistical property most often represented as:

$$R(t) = e^{\frac{-t}{\eta}} \tag{6.4}$$

where

- R(t) is the cumulative probability of success as a function of time,
- e is the exponential,
- t is the operating interval, and
- η is the characteristic life of the distribution.

Conversely, the probability of failure F(t) is the cumulative probability that the item will have failed by time "t" as given by Eq. (6.1).

In other words, this is what the probability is for an item surviving over a given time (t) as a function of the characteristic life (η) or MTBF. F(t) is also known as a survival function in many references. Graphically the reliability plot in Figure 6.1 provides an easy way of viewing the probability of success over the life of the item. For the case of $t = \eta$ the result is 0.368. Thus, the probability of failing by the time the item reaches its characteristic life is 0.632. As will be discussed later, this oversimplification has a significant impact as the plant ages.

What drives R(t) (and F(t)) is that different components and technologies have different FRs and failure distributions over time (Kececioglu 2002; Elsayed 1996; Kailash and Pecht 2014). These failure distributions are generally exponential, normal, and Weibull.

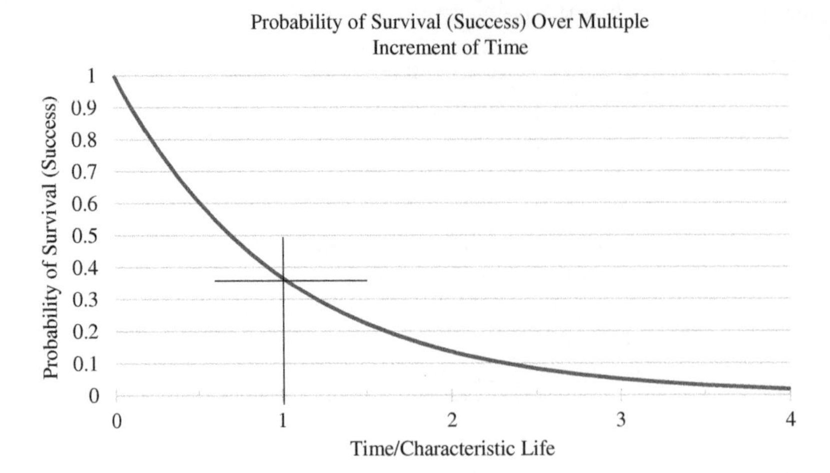

Figure 6.1 Exponential Probability Density Function.

6.5 Reliability

Although a Solar Power Plant operates during the daylight hours the SCADA, lighting, security, and maintenance systems must operate 24/365. The operational profile, then, differs from the usage profile in that it accounts for all the operating time as well as standby/maintenance time. It also provides the profile for any storage system and other parasitic power demands the plant may impose on the grid to maintain operation of the plant.

Design reliability is the application of PoF to a design with the intent of looking for weaknesses in the design. Also, if the equipment is already fielded, reliability is documenting and tracking failures to understand *why* and how something failed. As such, reliability spans the lifecycle from concept to end of life and is an integral part of SE. It is also one of those engineering disciplines that you either do or do not perform on a project. There are no real appropriate halfway measures to achieve a reliable design or reliable plant. It requires a budget and schedule to perform the analyses and gather the data, and management must act on the results.

The above form a core set of MOE, but there are other MOE metrics needed to understand the system and identify what is driving failures, maintenance, and/or failure to meet contractual obligations.

6.5.1 Synthetic Reliability Data – Bayes Theorem

Synthetic reliability data is data developed using techniques such as Bayes Theorem, Interpolation, and Engineering Estimate. Interpolation and Engineering estimates will be left to the reader. The former is a simple math technique taking the initial value and next interval value and finding an average value between them. The latter *requires* a systems engineer or reliability engineer experienced in the reliability discipline and the underlying engineering and the data from like and similar equipment or item to provide an estimated value.

Bayesian prediction techniques[4] use apriori data that has some basis on like and similar equipment for which actual data or an inference can be made about an expectation for the equipment under analysis. There is however a major drawback in the RAM domain.

Bayes Theorem, stated in Eq. (6.5), is the probability of occurrence based on apriori knowledge. Although this does work for some equipment, it does not work well for new technologies or for changing technologies.

$$P(A|B) = \frac{P(B|A)P(A)}{P(B)} \tag{6.5}$$

The rational for using Bayes method is that the *data must be representative* of the equipment being assessed. Caution should, however, be taken since the underlying assumption is that $P(B|A)$ (the probability of B given A) and $P(A)$ (probability of A) and $P(B)$ (probability of B) are known with some degree of accuracy.

6.5.2 Interpolation

Interpolation is taking data sets and parsing the numbers into sets in which values between valid data points are calculated by assuming a linear or other function between those two points. Provided that the data are not significantly different in value, interpolation will lead to a smoothing of the "curve." Caution should be used in interpolation in temperature, for example. If the temperature was 90 °F on Monday and the next data point is on Thursday at 90 °F, it is too easy to assume that it was 90° on Tuesday and Wednesday when it may have varied significantly on those days.

High temperature readings vary from minute to minute and by as much as 5 °F within an hour.

6.5.3 Engineering Estimate

An engineering estimate begins based on *a reliability engineer's experience* with a specific part type or class of like and similar equipment and operating conditions. Although there are many managers that believe their engineers produce reliable designs, history in multiple industries indicates otherwise. For example, a design engineer may accept as fact that a roller bearing of a certain size and load might have a demonstrated life of 15,000 hours (about five years of operation) under a stated set of conditions. But applying this to a new larger design may not scale as one would think. With the same bearing and race material a larger bearing with a higher load might not work as reliably well as the strength of the material is a reliability attribute.

In the 1970s, TTL (Transistor-Transistor-Logic, made up of bipolar transistors) circuits had geometries of 1–2 microns and very long-life expectancies, about 50 years or more. Today's CMOS (complementary metal oxide semiconductor), made up of field effect transistor devices with geometries of 10–20 nm have mean lives of 7–10 years or less because of the physics involved at these extremely small dimensions. The physics and chemistry of extremely small geometries have put a significant burden on long-lived exceptionally reliable systems.

6.6 Stakeholder Needs

An important aspect of any system's design is how the various stakeholders get their reliability information as well as establishing the interfaces to ensure good technical communication, understanding, and expectations for data. Figure 6.2 provides an example of a RAMS-centric view of stakeholders and the products that each could use to status their portion of the project.

Stakeholder reliability is a systems-specialty engineering skill that requires personnel trained in SE, software, hardware design, and safety (Figure 6.2). It requires a working knowledge in the PoF, statistical analysis, and specialty tools to support the model and analyze proposed designs to enable optimizing the system design and setting the reliability requirements for each piece of repairable equipment in the system. One additional point is that:

- **Human factors**: Human factors are that aspect of the plant that specifically deals with the human interface to the system which includes performing maintenance and operations. From circuit Breakers to computer input and including specifying and designing the system to be repairable with the easiest accessibility that can be allowed within cost constraints. Identifies special tools and test equipment and any specialized support equipment to be stored on-site or leased regularly. Affects the MTTR, Mct, Mmax, and MLDT (see Chapter 7) for the item and system.
- **Systems engineering**: SE is the group responsible for the overall architecture, system specification covering the plant lifecycle, allocation of requirements to the main

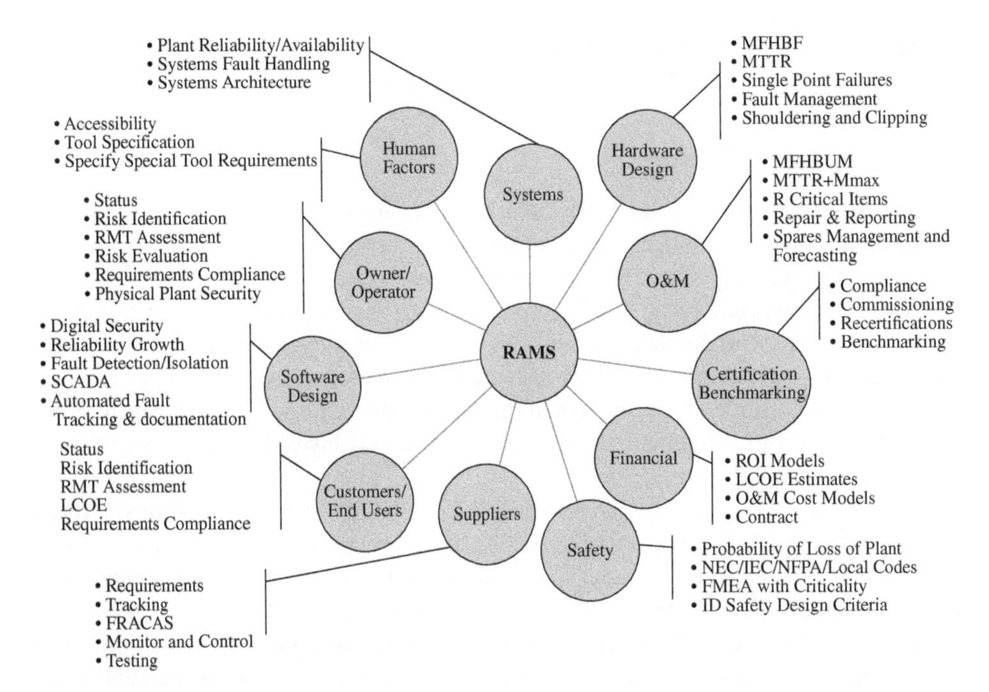

Figure 6.2 Typical Stakeholders Requiring RAM Information (Source: Morris).

suppliers, and coordination of the stakeholder needs as defined and documented in the system specification. This group includes the system engineers, reliability, maintainability, and safety engineers, logistics specialists, and additional stakeholders, including potential EPCs. This group generates the requirements and specifications for the plant and equipment to achieve a clearly defined plant availability and energy generation capability. Part of this process includes the identification and invitation to EPCs to review the draft lifecycle system specification. EPCs are requested to make comments, suggestions, and recommendations to improve the quality and completeness of that draft specification. (Note: all communications [written or oral] between the SE team and individual EPCs must include Non-Disclosure Agreements [NDA].) SE is also looking for public and/or EPC data including root cause and corrective action (RCCA) history that tracks MTBF, MTBM, MTBCF, A_i(inverter), A_i(strings), A_i and A_0 (Plant), and system-level MTTR and Mean Corrective Maintenance time (Mct) of their prior experience and projects.

- **Hardware design**: This is the selected EPC hardware or physical plant design and layout as well as including the design teams for everything from the modules/panels up to the interface to the Point of Interconnection. It is a concurrent effort with the software engineering and SE efforts to design the system to meet the system design and architecture. This affects the MTBF, MTTF, MTBM, MTBSM, MTBPM, and MTBUSM.
- **Software (SW) design**: Software design includes the SCADA[5] system as well as any inverters, motor controllers, site pyrometric monitoring, site security, and fault detection (FD) and fault isolation of plant failures down to the repairable item level. Software is unique in that it does not have a physical form and exists only as 1s and 0s in memory.

Software does, however, have defects and faults that manifest as failures (bugs and improper or incomplete function specification) through the hardware it controls.

As a consequence, it is important to look at the history of failures associated with the inverters, motor controllers, and SCADA to identify those failures associated with SW. Software can be assessed with an increasing failure rate early in the project and decreasing after final integration and test. At release, the metric is generally shown and tracked as an MTTF. Ramasamy and Govindasamy (2008) provide a model and evaluation that address questions about how to model software reliability.

Software has a basic MTTF attribute that can be modeled using the Weibull distribution with a $\beta \leq 1$ based on data gathered from the initial development testing. In theory and practice, SW failures will decrease over time with one caveat, that is that the core SW remains the same. As defects in the code are found and fixed or as updates are made to the SW that:

(1) full regression testing is made and avoids introducing new defects,
(2) defects, when found, are corrected within a reasonable period of time, and
(3) updates are much less than 5% of the total lines of code.

This latter limit is arbitrary but assumes that the larger the number of lines of code changes that occur, the higher the probability of introducing new or uncovering previously hidden or unused code defects.

Software defects have another attribute that is unique to SW, which is that the conditions under which the defect manifests itself as a failure, it will *always* occur for those same conditions in other equipment, or simply "if the defect exists in one unit, it exists in all identical units." For example, since all the inverters in a field will have (or should have) identical SW, there exists what is called a common cause failure. This is where if one unit fails then there is a high probability that other units will fail at the same time given the same conditions. The MTTR (Chapter 7) for software can range from weeks to months depending on the defect and its effect on the system. Depending on the fault that is manifested, this can have a major effect on the operation of the plant and its capability to deliver power to the customer.

- **Certification/Commissioning**: During the installation, certification, and commissioning phase, the plant will undergo many tests, especially if the plant is built in phases where each phase is successively added to the grid or customer load (IEC 62446-1 ED 2). During this testing and operation phase, many early failures will occur for weak components, transportation damaged components, installation errors or abuse, and SW failures during integration of the different blocks of equipment in the plant. These problems must be documented and tracked and root cause analyses (RCA) performed to determine the sources of failures and properly classify them to perform corrective actions for current and future installation work. Metrics tracked should include MTBF, MTTF, MTBM, and scheduled and preventive maintenance times.
- **Financial/Insurance**: Financial/Insurance requirements for reliability are the aggregate need for protecting the funding for the project. Higher reliability systems always require improved specification and cost some fraction more than the common model approach for existing project delivery model today. These systems do, however, reduce the risk for the Return on Investment (ROI) and decrease the cost of insurance as the probability of having to do a payout due to poor equipment and system design is significantly reduced. The financial stakeholders have a vested interest in insuring that the project SE addresses, tracks, and corrects any identifiable weaknesses early in the project and during the design and testing of system equipment. Financial and insurance institutions should primarily be tracking, MTBM, A_i and A_0, MTBCF, and the results of supplier final acceptance testing.
- **Safety**: Safety, particularly electrical safety, is part of the code certification process with periodic reviews of the plant to ensure it continues to meet or exceed the national, state, regional, and local safety regulation requirements. It is also an ongoing process for the O&M organization and therefore important to track safety-related incidents, determine cause if related to human error, equipment failure, or process error. The human and process errors need to be corrected immediately and verified, documented, and validated as soon as practical. Major equipment electrical faults (blown capacitors, overheating inductors and transformers, and any form of arcing) and failures need to be listed as critical and documented as MTBCF. If determining the root cause is a frequent practice used in the plant construction, the root cause should be determined and documented (lessons learned), corrective actions performed, and polling using all the stakeholders with an interest notified. Note: NEC Code is not a certification standard.
- **Suppliers**: Suppliers form a major part of the RAMS foundation work. Those that have RAMS functions as a part of their design and SE teams should provide a better performing, higher quality, and more reliable product. Suppliers should provide the

MTBF, MTBCF, MTBM, scheduled maintenance and preventive maintenance intervals, MTBSM, and criteria for inspections.

- **Project management (PM)**: Project management, aside from being charged with program performance, cost, and schedule success, needs to track or monitor how the project is doing by monitoring all the stakeholder's critical metrics. This includes the RAMS models, metrics, and status from the EPC, suppliers, and third-tier suppliers (as reported through the primary suppliers).
- **OWNER**: The owners are often the stakeholder who hold the economic responsibility for the plant. Often the Owner is also the plant Operator (O&O) with responsibility for the day-to-day operations of the plant and ensuring that the O&M is being performed in a timely and financially viable fashion.
- **O&M**: The O&M organization, if O&M is not being done by the asset owner or the O&O, must be included as early as possible in the specification and design phase. There should be direct communication between them and the EPC, suppliers, and third-tier suppliers of major equipment. The metrics that they should be tracking include reliability and maintainability predictions, failure reporting software capabilities of the plant electrical and electronic elements, and the capability to fault detect and isolate to the repairable item level. A specific minimum of metrics includes MTBF, MTBCF, MTBSM, MTBR, MTBSCF, MTTR, and Mct. O&M also needs to provide the team with the planned spares approaches, specifically, what may be stored on-site versus a centralized storage and what the MLDT is expected to be. They also need to provide the expected response times to deferred, regular, immediate, and critical (plant or safety) maintenance.
- **Customer (utilities, grid, industry, commercial, and all other end users)**: Customers are looking at generally two parameters. Availability (A_i and A_0) (Chapter 7) of the power to be supplied and the critical failure (MTBCF), reliability metrics, i.e. plant shutdown or inability to meet demand not related to force majeure.

6.7 Reliability Predictions, Analysis, and Assessments

Systems RE provides a foundation for one of the most critical parts of building better, more reliable, robust, and cost-effective PV plants. It requires understanding and use of standard reliability language, techniques, applications, results, and standard terminology (IEC TS *1836 ED 3). Examples abound in everyday life. An individual buys a car and with normal maintenance, the owner expects to go out each morning, start the car, and get to work – preferably on time, until they sell the car. This is the end-user view of the system (car). From the car manufacturer's perspective, in any given year, there is a statistical portion of the population of cars that have failures including tires, water pumps, computer engine controls, and more, starting from day one delivery. This is the "warranty" view of the car's reliability.

And, as time goes on, for a given model and model year a greater percentage of the population will have failed. The auto industry protects itself using warranties such as 100,000-mi power train, or six-year battery life, or five-year limited warranties. All of this does, however, have some specific restrictions that the customer needs to understand before relying on them to get their car repaired by the manufacturer. The owner must have car fluids changed on a regular schedule, tires must be rotated or replaced when they reach a certain level of wear, and brakes must be periodically inspected and linings changed as needed. This also holds true for every product from television sets to laptops, to cell phones, and washing machines.

Performance to specification is based on the stakeholders:

- Defining the functions needed,
- Defining success and failure, and
- Properly specifying the performance measurable metrics under a stated set of conditions against which success (failure) can be assessed.

The functions of the stakeholders are those at the plant level, all of which are based on the necessary elements for the plant to produce power/energy at an optimal cost according to the contract. Specified conditions are that set of operational or induced conditions in which the component, assembly, card, or end item is exposed from manufacturing, transportation, storage, installation, and operation until failure or the end of useful life. A life profile is used to assess the stresses that the item may see over its useful service life. LIFE PROFILE: A time-phased description of the events and environments experienced by an item throughout its life. Life begins with manufacture, continues during operational use (during which the item has one or more mission profiles), and ends with final expenditure or removal from the operational inventory. Figure 6.3 provides a mapping of a set of metrics needed to support all stakeholder needs for a PV plant system.

As shown in Figure 6.3, there are a substantial number of metrics needed to describe the attributes of a system for all of the various stakeholders.

So, depending on the stakeholder there are multiple points of data that can be used to assess the status of the plant and how well it is performing against its metrics and the stakeholder's need for information and status. These metrics and costs defined the health, condtion, and status of the plant at any specific point in the plant lifecycle.

Recently, in 2019, NERC has generated event data recording requirements under EOP-004. EOP-004 makes reliability event data available for study and analysis. Data trends and additional detailed information for the reporting power plants are to identify weaknesses in the suppliers to the grid and, where possible, to avoid similar problems in the future. However, this is currently applicable only to plants over 20 MW. This is a crucial step forward, nevertheless, our view is that it is an initial incremental step that needs substantive expansion in the future to cover all grid-connected power generation.

Note: The left side of the above diagram contains the necessary metrics to provide analysis and assessment that drives the right side of the diagram cost elements ($s). The blue functions are the primary tools that use or generate the metrics. The greater the attention paid to the metrics on the left throughout the plant/component life, the lower the measurable element cost over the same period, resulting in an accurate reduced LCOE.

The life profile for example starts with the simple definition for PV systems, i.e. generate electrical power (energy) for a customer for a minimum of 50 years. For the project's location such as Phoenix, AZ, USA, there will be an average of 3000 hours per year of sunshine

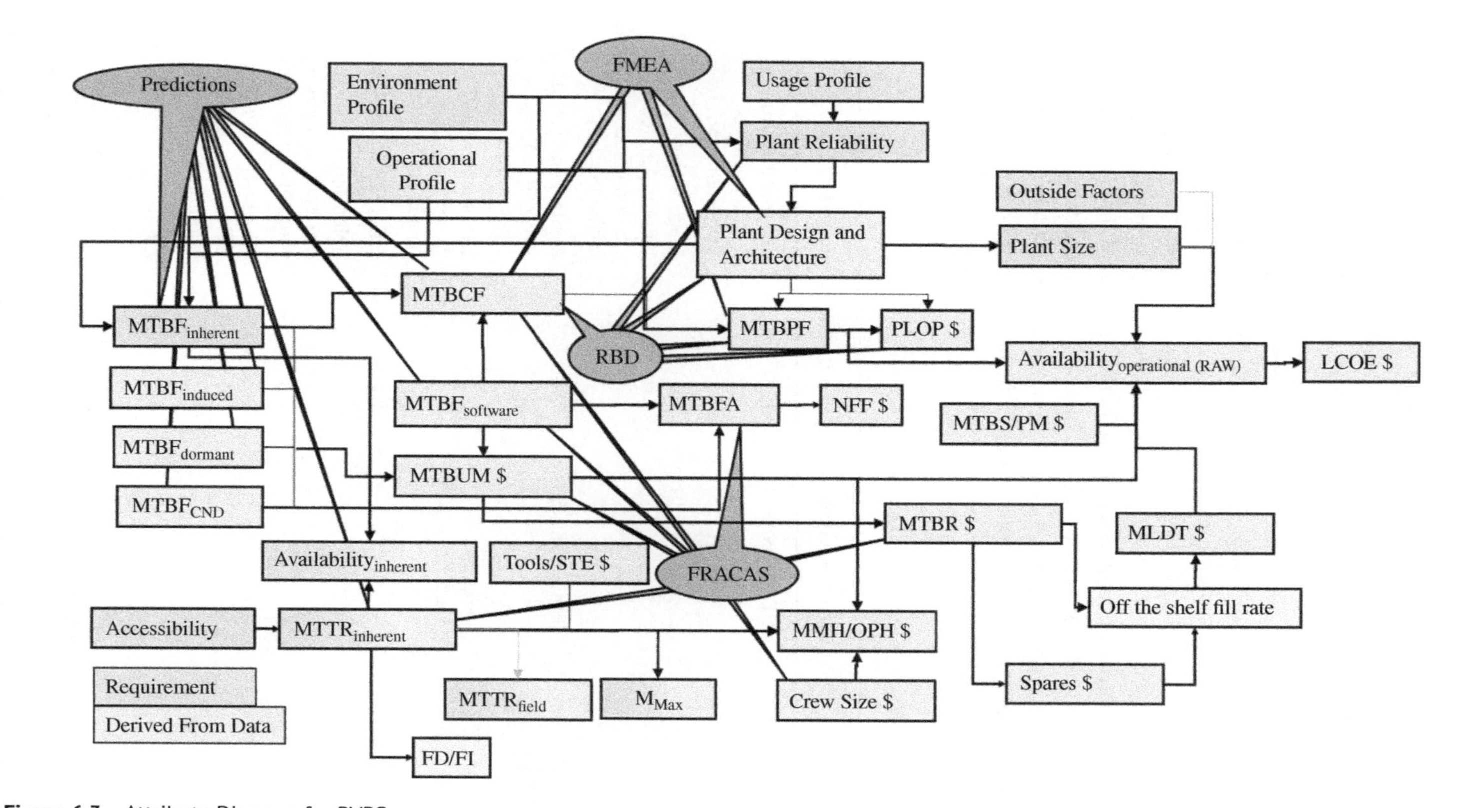

Figure 6.3 Attribute Diagram for PVPS.

to meet contractual energy needs. Additional attributes of the life profile are based on the environmental profile and user/usage profiles. These include estimates of the number of lost power production days due to weather, or failure, or reduced demand by the customer. The life profile provides the basis for performing the statistical analysis and simulations using the various distributions for modeling reliability parameters of the components that make up the PV power plant.

The metrics of Figure 6.3 have found almost universal application in industries including medicine, aerospace, ships (marine), cars, and the PV industry (NREL/TP-7A40-73822, 2018) – (IEC DTS 63265). An additional source of data and definitions useful for PV plant reliability information can be found in IEEE 493.[6] This recommended practice has been in general use for over 30 years. IEEE 493 has reliability data and utility-oriented definitions that are sufficiently broad enough to be applied to much of the PV industry with modification to meet the specific failure modes and mechanisms of semiconductors and other items unique to PV power plants. For predictions, it is better to have data from PV plants of the same size as the equipment will be comparable to the new plant. Other sources of MTBF or FR data used for predictions are shown in Table 6.5.

Table 6.5 Reliability Failure Rate Data Sources.

Source	General Application	Last Update	Comments
Mil-Hdbk-217F and Notice 1 and 2	DOD Military	1995	Conservative, military stress environments Parts count and stress analysis
Bellcore/Telcordia SR332	Telecom	2011	Electronic equipment failures. Limited environments, part count, laboratory, field
Naval Surface Warfare Center (NSWC)	Mechanical, USN	2011	Military environments, material properties, environments
FIDES, TC 56	Commercial, French Military	2009	Development and manufacturing errors and overstress
IEEE 493 Gold Book		2007	Utility grade equipment
IEC 62380 TR Edition 1			Telecommunication electronic equipment, thermal cycling, solder
IEC 61709 Ed 3.0	Electric components	2017–02	REDLINE VERSION Reliability – Reference conditions for failure rates and stress models for conversion
IEC TR 62380 Ed 1	Electronics components	2004–2008	Reliability data handbook – Universal model for reliability prediction of PCBs and equipment

The normal practice is to assess the equipment and sum up the FRs weighted by factors associated with the stresses. These stresses vary from application to application and are also affected by the operating environment. To determine the predicted reliability then, it is necessary to identify the stressing environments and adjust the FR accordingly.

Inverters have mechanical, electromechanical, electrical, electronic, and microelectronic devices. The model data for inverters is found in the sources listed in Table 6.5 and from data each company or supplier has accumulated for their product.

Best Practice: Caution should be exercised regarding handbook data. The source of the information should be questioned to ensure that it is in the same or close to the same components or equipment, and planned environments, and has sufficient data to be statistically significant.

This is important since failure for the inverter is defined as the unit failing to generate AC or generating unusable AC from the DC input thereby reducing the power/energy output of the plant for the duration of the failure. Alternatively, failure for the inverter can be that it produces the power but with excessive harmonics, reducing the actual power quality delivered, or power that is out of phase with the plant resulting in significant distortion in the plant output.

Clearly understanding the value of reliability and its usage can have substantial financial impacts on the PV system delivery process and its results. It shifts the discussion from least-cost systems to an approach of evaluating results that reflect long-term LCC, LCOE, functional operability, output, and revenues over the system's life.

In other words, effective use of reliability in system delivery has an incredibly positive financial element that has not been effectively considered and/or achieved in most projects historically. With PAM, we emphasize the use and understanding of system engineering and RAMS metrics, and their application to make projects and/or organizations healthier and more profitable.

Example: If the inverter MTBF predicted reliability is 20,000 hours, then the probability of reaching the end of year 5 successfully is:

$$R(t:\mu) = e^{-\frac{t}{\text{MTBF}}} = \exp^{\frac{-5*3000}{20{,}000}} = 0.42 \tag{6.6}$$

When this number is combined with the number of inverters in the plant, there may be many failures each year based on random failure modes. If the 100 MW plant is using 1 MW inverters, there are 100 inverters at around 6+ years and 63.7% of the items would be expected to fail. This assumes that the failures are random, and the exponential distribution applies.

Consequently, there is an issue that unlike historical utility power sources, a fraction of the solar cells to be installed will have already failed or are damaged or degraded. Initially, and depending on the type of solar cell, the number of defective units may be between 2% and 20%. Without doing an electromagnetic (EM) spectral scan during the installation with a replacement for defective units, it may not be possible to identify the failed units on installation, resulting in a reduced baseline energy production capability at installation and commissioning.

Best Practice: For the greatest reliability in the PV array, we recommend independent inspection of modules at the factory and upon installation. This can preempt poor-performing modules and provide for greater operational array of health and condition prior to COD.

6.7.1 Reliability Specifications

To accomplish a cost-effective system that meets its performance parameters requires that the specification meet the qualitative and quantitative elements of SMART Engineering specifications. A comparative example of one specification is given below. Note that the difference lies in understanding under what conditions the reliability specification of MTBF is to be measured. Examples of poor/weak and better requirements in specifications are described below:

3.1. Reliability
   a. MTBF; MTBF of the system shall be greater than 10,000 hours.

A better requirement would be written as:

   b. MTBF; MTBF of system X shall be greater than 10,000 hours at 60% Confidence Level when operated within the environmental limits of Section 3.2.1, the life profile of Section 3.2.2, and the operational profile of Section 3.2.3

The limits of an environmental Section 3.2.1 would be the plant or site ambient environments and the induced ones (temperature rise, for example) that can affect the component being specified. The operational profile is the daily and annual power cycling for which the unit is used. SCADA, for example, operates 24/365 while the inverter operates for only the period from sunrise to sunset unless maintenance is performed after normal plant operation hours.

6.7.2 Plant Reliability Drivers

Applying what is learned from a historical review and understanding the attributes of the latest semiconductor and solar photovoltaic technologies results in better component selection and improved reliability (MTBF), reduced maintainability (MTTR), and higher availability (A_0), resulting in higher, more stable economics. Examples of the types of failures common to PV systems are illustrated by the failure Praeto (EPRI 2019) of major failures in several plants as shown in Figure 6.4.

As can be seen in the Praeto diagram, there is a commonality between reports and to a lesser extent the types of specific items that failed and their relative percentage of failures.

For Figure 6.4, the inference of percentage of failures aids in some decisions but does not address the frequency or rate of occurrence at specific sites and environments. In everyday language, reliability needs to be specific and measurable and appropriately applied to the wide variety and number of components in a PV system under the conditions of operation.

Marketing terms of an item being "reliable" and "more reliable" are not measurable and are of no value to the consumer or stakeholder or AHAs and regulatory agencies. One does

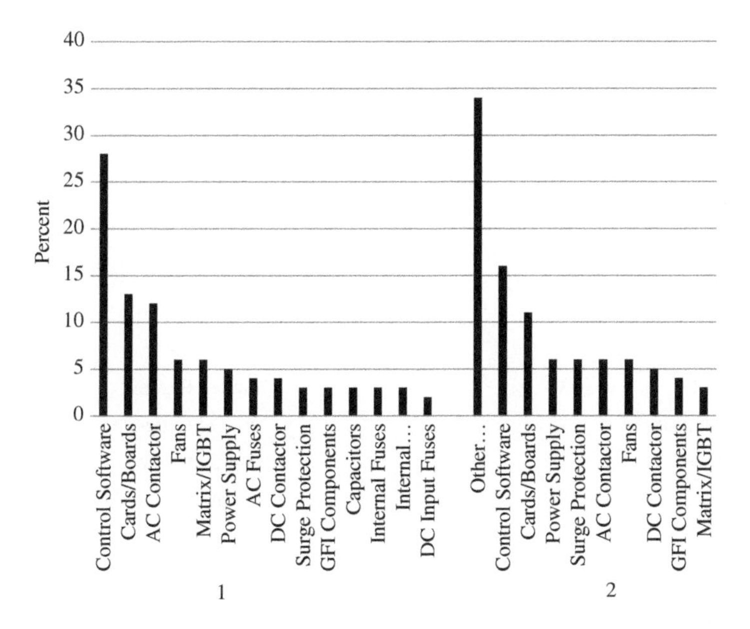

Figure 6.4 Praeto of Two Records by Plant Major Subsystem (Source: EPRI 2019).

not have to be a reliability expert or professional to understand the basic value of applying reliability techniques and practices in the specification, design, and operation of all types of PV installations. Nor does one have to be well-steeped in the reliability universe to understand and see the benefits once properly presented. Much of working knowledge for reliable product development can be found by implementing a reliability program plan as a part of an SE Plan or as a standalone plan.

6.8 Reliability Program Plan

To identify the work to be done by each organization or stakeholder, the project team should establish a reliability program plan. The plan's purpose is to list the tasks to be performed by suppliers (as part of their contracts) and data and information acquired by the project team. RAMS engineer to perform modeling and assessment of the EPC proposed designs. There are many tasks needed to understand the reliability of the system. Although it is recommended to do all of the tasks, this may not be fiscally possible. Table 6.6 is a summary of the tasks and the phase of the project to which they are most applicable. A few tasks are performed in several phases with preliminary work done as early as possible and the task completion with the final closeout of a phase.

Effective application of critical reliability practices is essential for the individual contributor and team stakeholders to understand the elements and tasks of reliability and the associated risk(s).

The reliability tasks above allow a collective understanding and enable the communication, value, and consequences of specific decisions. As the industry evolves it becomes more important that the project team has a basic grasp and understanding of reliability-based

Table 6.6 Task List of Required and Recommended Reliability Tasks.

Task	Required	Recommended	Phase
Reliability Program Plan	X		Concept
Critical Item Identification	X		System design
Component Derating Limits	X		Detailed design
Design Reviews	X		All
Fault Tolerance	X		System, detailed
Part Selection and Application		X	Detailed design
Thermal Design Limits	X		Detailed design
Reliability Modeling, Allocation	X		Detailed design
Reliability Block Diagrams	X		System, detailed
Failure Modes, Effects, and Criticality Analysis (FMECA)	X		System detailed design
Failure Reporting and Corrective Action System (FRACAS)	X		Install, benchmarking, commissioning, O&M
Fault-Tree Analysis (FTA)		X	Safety review
Reliability Predictions	X		Detailed design
Supplier Control	X		Design, install, O&M
Worst Case Circuit Analysis (WCCA)		X	Detailed design
Design for Storage, Handling, Packaging, Transportation, and Maintenance	X		Detailed design
Software Reliability and Security	X		Detailed design SCADA inverter
Accelerated Life Test		X	Manufacturing
Reliability Trade Studies	X		EPC proposal evaluation
Human Reliability		X	Organization, maintainability, O&M, and SCADA design
Develop ESS Criteria	X		Detailed design

choices and their specific impact on the LCC, O&M, and LCOE. Doing so also requires that specific members of the team from concept through O&M must also understand, accurately define, and analyze the data that create the full picture of plant reliability issues. This is especially critical during the specification process prior to EPC bidding.

As a part of the EPC bidding process, the O&O need to ensure that the EPCs provide reliability program plans outlining the tasks and expected results as well as the cost and schedule assessments that include RAMS during the lifecycle. The plan includes identification of all references and standard/regulatory documentation that require or recommend reliability as well as reference documents, e.g. Military Handbooks Mil-Hdbk-338B (1998),

DOD GUIDE (2005), Handbook of Reliability Prediction (2011), Mil-Hdbk 217 (1998), IEC Standards IEC TS 63265 (2022), IEC TS 63019 (2019), and SAE Standards JA 1000, 2012, JA1000/1_201205, JA 1002 200101, JA1003_200401. These provide the information needed to develop a reliability program plan, which includes the task to be completed over the project's life. The RPP also provides the definitions unique to the specific PV power plant design and the national, regional, or local area.

6.9 Reliability Mathematics

To be effective, SE and reliability require accurate data as described in Chapter 10 "Failure Reporting and Corrective Action System (FRACAS)." As previously noted, there are two forms of reliability. The ***predictive*** form is used during concept, design, and development, based on available historical FR data or reliability test data, models, and development of the FMEA, FTA, and trade studies. The second form or ***descriptive (current state)*** form of reliability is based on current field information where the data is acquired and processed to provide for *an assessment* of the current plant health and improvement. The data is analyzed to identify failure trends and using RCA develop corrective actions to remove those failures from the population. Once data has been correlated, it can be used to develop the various reliability factors (Abernethy 2010; Kailash 2014; System Reliability Toolkit 2005; Kececioglu 2002, Elsayed 1996). Application of the use of distributions for PVPS inverters, for example, can be found in Morris and Fife (2009), Fife (2010), Fife et al. (2010).

6.9.1 Weibull

The general form of the Weibull distribution Probability Density Function (PDF), is given by Eq. (6.7):

$$\text{PDF} = \frac{\beta}{\eta}\left(\frac{t-\gamma}{\eta}\right)^{(\beta-1)} e^{(t-\gamma)^{\frac{\beta}{\eta}}} \tag{6.7}$$

where

- o β = slope of the data
- o η = characteristic life
- o t = the operating time of the point of measurement
- o γ = the failure-free operating period

There are two principal data plots for the Weibull distribution. The PDF and hazard functions provide visualization of the three main phases of the equipment's life. There are early failures due to weak components and manufacturing errors and/or design errors. Mid-life is when failures "appear" to occur randomly, and late-life occurs with product aging due to degradation and wear and tear. Figures 6.5–6.7 provide examples of the Weibull PDF for a two parameter (slope [Beta] and characteristic life [Eta]) Weibull. Figure 6.5 illustrates where a $\beta < 1$ represents a decreasing FR with time. This is indicative of early failures due to weak components or manufacturing quality errors.

The underlying ***assumption*** is that the failures are repaired, regardless of source (manufacturing, weak component, design overstress, improper application), and future failures

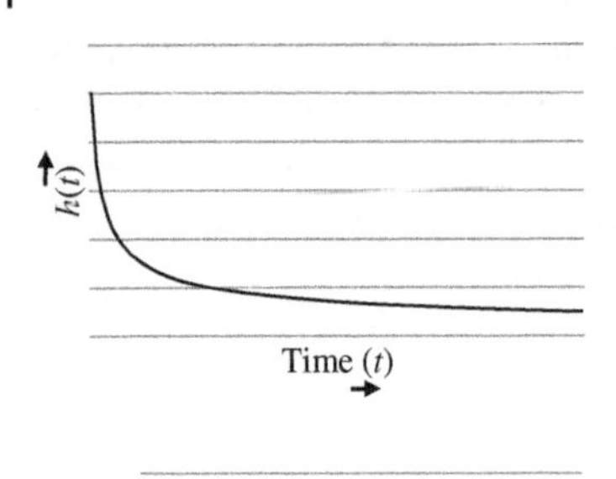

Figure 6.5 Early Population Failures $\beta < 1$ Decreases with Time (Source: Morris).

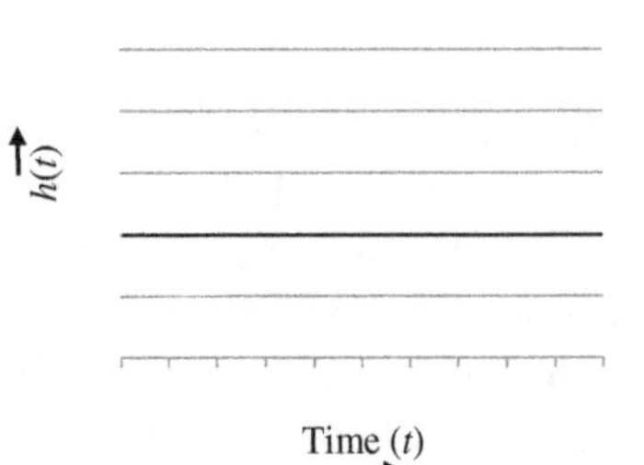

Figure 6.6 Mid-Life Failures $\beta = 1$ Is Constant Over Time (Source: Morris).

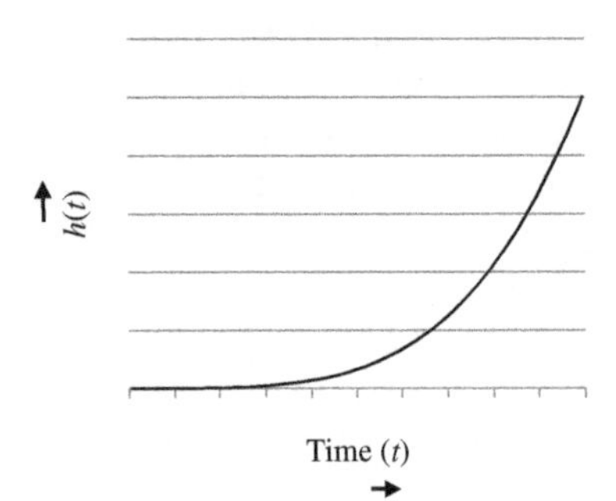

Figure 6.7 End of Life Failures $\beta > 1$ Increases with Time (Source: Morris).

are removed from the population by virtue of corrective action. However, this also assumes that management has budgeted to perform the RCA.

Figure 6.6 represents a random failure. These cannot be predicted to occur with any accuracy. Although there are potentially random failures all through the life of the system, these represent the items for which there is no design or quality processes that can remove them from future failures, e.g. there does not appear to be any failure trends.

The increasing failure population, Figure 6.7, provides insight into the failures that occur as the result of degradation and cumulative damage. This population can be thought of as a wearout mechanism such as that found in ball bearings, mechanical gears, chemical degradation due to excessive heat, and other items where time, temperature, vibration, and other stresses eventually lead to failure.

Each component must be assessed over time to determine if the Weibull distribution, or any distribution for that matter, is applicable and what the appropriate parameters are.

6.9.2 Bathtub Curve

For many, the basic understanding of reliability is based on the bathtub curve (Figure 6.8) and its mathematical modeling based on plotting the hazard rate ($h(t)$). To illustrate this, in

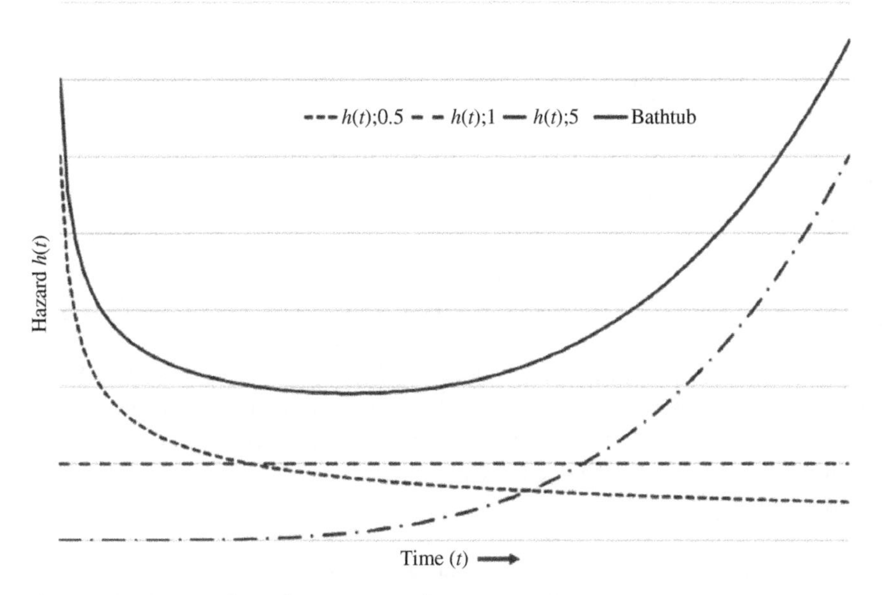

Figure 6.8 Bathtub Curve Development (Source: Morris).

this book, we use the hazard function of the Weibull distribution which is given by:

$$h(t) = \left(\frac{\beta}{\eta}\right)\left(\frac{t-\gamma}{\eta}\right)^{\beta-1} \tag{6.8}$$

where

- $h(t)$ = hazard function
- η, β, and γ are previously described.

Special Note: The γ is generally set at a value of "0" as the population size grows there is a finite probability of failure at *t* slightly greater than 0.

Fundamentally there is no one continuous function that results in the bathtub curve. The curve emulates the early failures associated with weak parts and/or manufacturing-induced failures, the general assumption of a constant FR during its mid-life and an increasing failure rate associated with the effects of aging on components.

β is the slope of data plotted on graph paper extrapolated from like and similar field data, and η is the characteristic life of the component. Figure 6.8 results when the principal periods of failures (early, mid-life, and aging) are determined and plotted as a function of time (t). For $\beta < 1$ the data can be modeled as decreasing failure rate. For $\beta = 1$ the PDF reduces to the exponential form, or a constant hazard rate ($h(t)$) is constant. This is often classified as the completely random failure region of the lifecycle. For $\beta > 1$, the Weibull PDF can emulate the Rayleigh or normal distributions and increase failure rates with time.

$$\eta = t \tag{6.9}$$

The mean time to or between failure is given by:

$$\text{MTBF} = \eta * \Gamma\left(1 + \frac{1}{\beta}\right) \tag{6.10}$$

where:

- o Γ is the Gamma Function (Elsayed 1996)
- o β and η as previously defined

Again, for the exponential reliability form of an inverter when $\eta = t$ the probability of failure is 0.632. If the specified inverter MTBF is 30,000 hours and the average operating time is 3000 hours a year, then it would be approximately 10 years to reach the characteristic life, by which time approximately 63% of the items could be expected to have failed. For the characteristic life (η), the slope of the graph for an exponential population ($\beta = 1$), and the Gamma function of $\Gamma(2) = 1$, the MTBF $= \eta$. If, however, the data shows the number of failures to be increasing and the β is determined to be 5, then the MTBF = 30,000 * $\Gamma(1.2)$ = 27,415 hours or almost a 10% reduction in the predicted/expected MTBF.

This illustrates the need to understand the data and its statistics. With a good FRACAS (Chapter 11) system in place, it is possible to not only identify weak components but also to "see" the actual reliability attributes of the components (equipment) and potential failure trends.

6.9.3 Normal Distribution

As indicated above, the Weibull distribution can be used to model many predictions. A "mean" for a relay, for example, would be a best fit using a $\beta \gg 1$ since its specific failure modes tend to be associated with fatigue. Most semiconduction parts can be modeled with a constant hazard rate ($h(t)$ = c or Weibull ($\beta = 1$).

With this said when one is modeling a large system, the normal distribution is a useful approximation. This is based on the application of the central limit theorem (CLT). The CLT is a statistical theory that states that given a large enough sample with a finite variance, the mean of the sample will approximate a normal distribution with a mean $= x$, a variance σ^2, and a standard deviation of $\sigma/\sqrt{N}$ where σ is the standard deviation and N is the number of samples.

> The distribution of an average tends to be Normal, even when the distribution from which the average is computed is decidedly non-Normal. Furthermore, the limiting normal distribution has the same mean as the parent distribution AND variance equal to the variance of the parent divided by the sample size.
>
> Fife et al. (2010).

A simpler way of stating the above is that for a large population, the average and variance are related to the average and variance of a sample of that population. The sample average being the same, the variance is equal to $1/n$ times smaller. The value of this method is that during the prediction phase of the O&M planning, one can use the normal distribution to approximate the failures in each year regardless of the base distribution. The principal limitation is that the population must have a closed form, and a distribution exists for the main population and its elements.

A simple approximation of the mean is given by:

$$\mu = \frac{1}{n}\left(\sum_{i=1}^{N} t_i\right) \tag{6.11}$$

where

- o μ = mean
- o n = number of failures
- o N = the total number of identical items
- o t_i = operating time of ith item

Additional attributes are that the larger a population gets, the "normal distribution" provides a better fit to their failure data, and this rule applies to any population of failures regardless of the distribution, provided it has a mean or average.

Performing a statistical analysis of how close the data are to a normal distribution is by calculating the median or the value at which one-half of the failures are above and below that value. With notable exception of the central value theorem, if the median and mean are close in value, one can use the normal distribution. The other point to define the normal distribution is if the population of failures when plotted against time (or other measures) is when the mean, the median, and the mode values are equal.

As noted above, when the mean, mode, and median are all the same or close to the same value, the data is distributed normally (Gaussian) and the PDF, Figure 6.9, is given by:

$$f(t|\mu,\sigma^2) = \frac{1}{\sqrt{2\pi\sigma^2}} e^{-\frac{(t-\mu)^2}{2\sigma^2}} \tag{6.12}$$

where

- o μ = mean
- o σ = standard deviation
- o t = time

The standard deviation (σ) is given by

$$\sigma = \sqrt{\frac{\sum_{i=1}^{N}(t_i - \mu)^2}{N}} \tag{6.13}$$

Figure 6.9 illustrates that with a normally distributed function, the mean represents an expectation that 50% of the population may or will have failed by the time the population has reached the predicted or expected MTBF. This is where the reliability modeling for the expected failures in each year can provide a much better estimate of spares and repairs that should occur within each year.

By using the field reliability values and its distribution the maintainer can get a better estimate of the spares needed. Although the plot for these values appears to indicate that the probability of occurrence is 0.000 up till about 20,000 hours, the *reality is there is still*

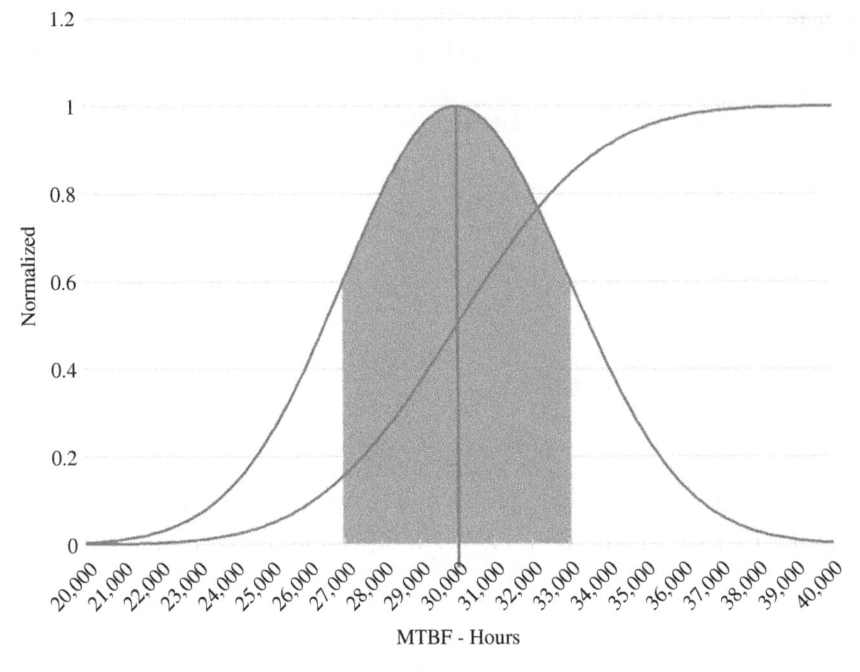

Figure 6.9 Generation Unit Plot of PDF and CDF for an MTBF of 30,000 and $\sigma = 3000$ (Source: Morris).

a finite probability that a failure will occur especially in a large population of components or end items. It is also possible for failures to occur in the first 3000 hours (approximately one year of operation) due to weak components and/or poor manufacturing practices.[7]

This equation can also be applied to the operation within an interval, which requires the distribution to be evaluated from $t1$ to $t2$. The interval of one year's operating time is used for spares. The difference is that the **probability of the unit surviving to the beginning of year 8 is only 0.02275** and the **cumulative probability of survival to the end of year 8 is 0.15865.** This increase from 2% to almost 16% is what the exponential does not cover. For the eighth year, the exponential only provides a value of 5%.

One nuance often forgotten is that it is possible to have an item fail in an end item, but it does not grossly affect the reliability of the end item. If the failed item has an MTBF $\gg$ 50 times greater than the end item, the repair of the end item has a negligible effect on the reliability of the end item's operation after repair.

To assess how the plant is performing, the reliability of each inverter is assessed, and the model will provide an estimate that the O&M can use to determine the potential maintenance and spares to be needed in the next year. If an exponential model is used and applied uniformly to the plant-generating units, a binomial distribution can provide the estimate. In addition, the amount of overdesign or redundancy must be considered. For this example, the plant is assumed to have 10% over design. For the 100 MW example, the probability that at least 100 generating units will be operable over the year without failing at a 60%

confidence can be found using the binomial distribution. Equations (6.14) and (6.15) illustrate the plant's capability based on a number of technical assumptions.

$$R(n, x, r(t)) = \binom{n}{x} r(t)^x (1 - r(t))^{n-x} \tag{6.14}$$

where

- n = the number of independent trials
- x = the number of successful trials
- $r(t)$ = the probability of success of a single unit

$$R(t) = \sum_{x=0}^{n} \binom{n}{x} r(t)^x (1 - r(t))^{n-x} \tag{6.15}$$

The result of taking into account the cumulative wear and tear on the equipment is that during the fifth year of operation (3000 operating hours) that having 100 of 110 inverters operating is 0.866.

From the O&M perspective, the probability of having 105 or more units operating over a one-month time period would be 0.9997. This implies that for a generating unit with an annual reliability of 0.94, the plant can meet its contractual energy demands. This also implies that there is a high probability that the field will have several days where there are four or more inverters down for failures and in need of repair. This assumes there have not been failures in years 1–4. This also assumes that all the inverters are of the same type with the same basic reliability function (MTBF) while operating under the same conditions.

During the fifth year, the MTBF has not changed (for the design), but the probability of a unit failure over the interval of the fifth can be approximated as:

$$F(t_5, t_4 | \mu, \sigma^2) = \frac{1}{\sqrt{2\pi\sigma^2}} \int_{t_4}^{t_5} e^{-\frac{(t-\mu)^2}{2\sigma^2}} \mathrm{d}t = 0.939 \tag{6.16}$$

This does assume that the criteria are based on data that indicates either the generating unit is working or it is not (partial failures or degradation is not counted). From this, the probability of the plant operating over any period of time "t" can be found via Eq. (6.17). Thus, the inverter has a reliability of 0.94 during year 5. For 100 inverters, the probability of all of them working is:

$$R(t\,(\text{plant})) = R(t)^{100} = 0.0025 \tag{6.17}$$

A further example of the application of the normal distribution is in the average degradation rate for solar cells (Figure 6.10). This is also a significant point for commercial plants and utilities and plant O&Os. Degradation is stated as a *mean* value, assuming a normal distribution, which implies that half the units do better than the average of approximately 0.95%/year or less, while one-half of the units have worse degradation rates of 0.95%/year or more. Overall, it implies that the plant output may be lower than expected unless each module has been tested and a degradation rate established with a true mean and associated distribution established. This is an expensive and time-consuming effort. The alternative is to assume at least one standard deviation ($+1\sigma$) higher degradation rate.

This degradation rate will contribute to the reduction of DC power over the course of the module specified life (Figure 6.11).

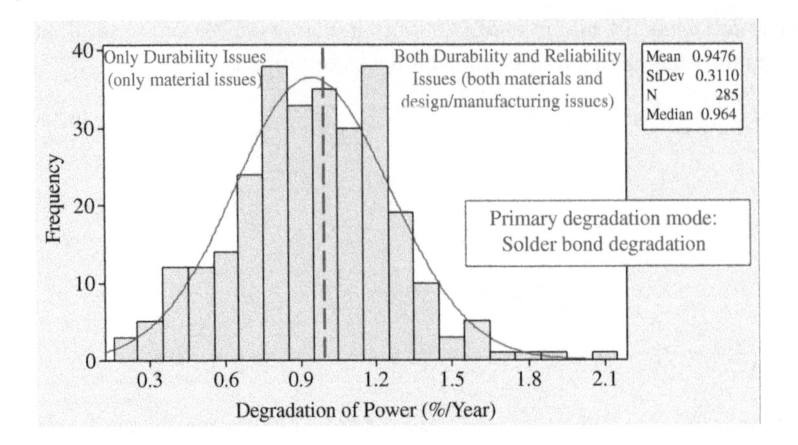

Figure 6.10 Photocell Degradation Rates of G-Modules (Source: TamizhMani (Mani) G.; manit@asu.edu 2014; Jordan and Kurtz 2013).

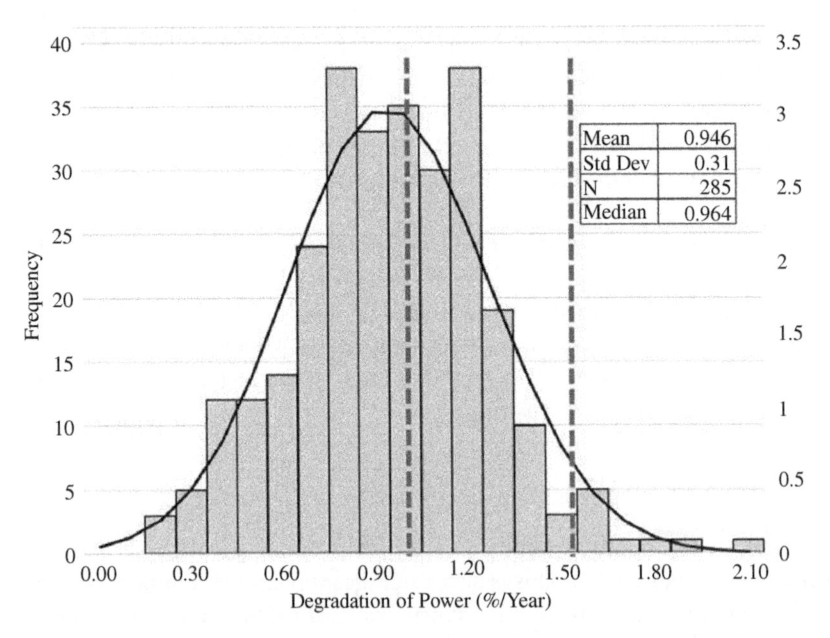

Figure 6.11 Illustration of MTBF PDF with 95% Confidence Level for Cell Degradation. (Source: Adapted from Jordan and Kurtz 2013).

Note that the low end of the integral is $t = 0$ since negative time is meaningless. Even though the components themselves may have had power applied prior to installation and/or manufacturing testing, the greatest accumulated time on an inverter will be after installation in the plant. *All failure time is measured from when the part is built, tested, and verified to be working*. The manufacturing total power on time generally does not exceed 2–300 hours (if there is a burn-in on the equipment during final acceptance testing). If the original part build time is unknown $t = 0$ can be assumed to be the first time that the unit is powered up after plant installation. Annual solar Irradiance varies greatly but can not exceed 4380 hours per year, assuming continuous sunshine and accounting for installation latitude.

6.9.4 Failure Rate

Failure rate is a second common form of expressing failure, which can be developed from data acquired from the field or predictions from historical database. The FR is as failures per unit time or failures per cycle. Its one basic caveat is that FR is a measurement or calculation for a specific point in time or failures per unit of time (cycles).

$$\mathrm{FR} = \frac{\text{Number of failure of identical parts}}{\text{Total operating time/cycles all like parts}} = \lambda_p = \frac{n}{\sum_{i=1}^{N} N_i t_i} \tag{6.18}$$

where

- t_i is operating of the ith item time
- N_i is the total number of identical items in the system
- n is the number of failures from t_0 to current also

$$\mu = \frac{\sum_{i=1}^{N} N_i t_i}{n} = \frac{1}{\mathrm{FR}} \tag{6.19}$$

Equation (6.19) t_i values are taken at a point in time but are a duration for the measurement interval. For example, for a 100-inverter plant, let us say that there are 10 failures of a replaceable item in year one and that the operating time ($t1$) is 3500 hours. This results in a total plant inverter operating time of 350,000 hours per year and an estimated MTBF of 35,000 hours ($\mathrm{MTBF} = t/n$). In the second year, there are 15 failures with the same operating hours ($t2$). The MTBF of the inverter is now 700,000/25 or 28,000 hours. For year 3, again at 3500 hours per year, there are 21 failures and the MTBF drops to 22,800. These values translate to spares requirements of 10, 15, and 21. If one used a constant FR for each plant, there would be a shortage of 5 and 11 spares in the second and third years, respectively. If the failure data are tracked and trend analysis is applied, it reduces the surprises and unexpected cost. Also using a normal distribution in the original model would allow for better projections.

The exception to this rule is when prior failures have an RCA, and a fix is developed and implemented for all the identical plant equipment. If five of the failures were of the same component and manifested the same failure mode, then a FRACAS-based RCA can be implemented and a preemptive fix developed. Provided the repair eliminates the failure and its associated mode, then the failures can be censored resulting in a 35,000-hour MTBF at the end of the second year. The inverter failures need to be continually monitored to document that the failure does not occur again.

The most often used form of FR for electronics reliability predictions is provided by the semiconductor manufacturer (if they provide any at all, is to state it in FITs). Many semiconductor devices have FRs of 5–20 FITS or 5–20/10^9 hours while circuit cards and assemblies are in the 1–1000 failures per million hours. For performing initial predictions where no

PVPS field data exists or the field data is suspect then a rough estimate can be made from Mil-Hdbk 217 (1998):

$$\lambda_P = \lambda_B \prod_{1=1}^{m} \pi_i \tag{6.20}$$

where

- o λ_B = Base failure rate
- o π_i = pi factors that affect the reliability of the part (see example below).

For example, relays are mechanical devices with a connection/no-connection operating cycle. These have MCTF of 1000 to 100,000 or more based on the actual operating conditions. To assess the time equivalent to failure, it is necessary to convert cycles to operating time. For prediction purposes, the MCTF are dependent on the number of cycles per unit of time that the item experiences.

$$\text{MTTF (Relay)} = \frac{1}{\lambda_P(\text{relay})} \tag{6.21}$$

where

$$\lambda_P = \lambda_b \pi_L \pi_C \pi_{CYC} \pi_Q \pi_E \pi_F \tag{6.22}$$

where

- o b = Base failure rate from a database
- o L = Load stress (current, voltage)
- o CYC = Cycle factor (number)
- o C = Contact form factor (C is also used for construction factor)
- o Q = Quality (commercial, military, space, automotive, etc.)
- o E = Environmental factor
- o F = Construction factor

Additional factors exist for a variety of components. One of the more important parameters is the environmental factor, π_E. This has proven to be a true factor in FR adjustments. Items used in controlled 20 °C environments will be more reliable than ones used in more extreme temperature conditions, high and low humidity, and other natural and induced environments. The expected number of operating cycles (mean) is provided by the manufacturer in total cycles, with potential factors affecting the reliability such as voltage, current, and temperature, whereas the operating time per cycle is defined by the usage profile. Note that this uses the operating time as opposed to calendar time. Most of the stress on an item is related to when it is in use and more often specific to the initial application of the stress, e.g. light bulbs switching from off to on tend to fail when power is applied, batteries tend to fail when the demand for power exceeds capacity (e.g. turning the ignition in your car). Cars for example have warranties, but a calculation of the battery life (around 4000 hours) and the engine TTF (about 3500 hours) is less than the 52,020 calendar hours in six years that make up many warranties.

A standard practice has been to assume that reliability parameters are invariant, i.e. the MTBF or FR does not vary with time. But as can be seen in the previous example, some people drive their cars 4–6 hours a day while others drive them 8–10 hours a week. This leads to quite different lifetimes (ranging from 6 to 20+ years) for the individual user. For the car manufacturers, for example, with 10–12 models, a population of more than 500,000 cars for each model built in a year, it is possible to document the failures that a dealership repairs and thus begin to build a reasonable set of statistics. In other cases, light bulbs for example, the reliability or life of the items is based on a single value provided by the manufacturer based on a limited set of testing for which there is an extrapolation. Light bulbs when they fail, they are thrown away and a new one installed to replace it. If data were taken from the populations of light bulbs, the analysis would show some early failures right out of the box, while the bulk of the units would work for several years before failing. This would go on until the last one fails potentially 5–7 (tungsten filament) or 20–25 (LED) years after shipment.

For the relay example, the actual usage profile is affected by operating more cycles per unit time, operating under heavier current and voltage loads or being exposed to higher or more extreme sustained operating environments than planned. Higher temperatures can reduce the strength of the contact arms or below freezing temperatures can cause ice to form in non-sealed units. Upon melting, the water can cause a short circuit. For a relay with operating cycles specified as 100,000 cycles at STP and 365 cycles per year may be good for 3000 years which is obviously unrealistic. However, if the voltage, current, temperature, or cyclic rate is higher, such as 3600 cycles per year and at higher voltages and currents, then the expected life of the relay may be closer to 30 years or less.

For the PVPS, the inverter is a major driver (Jordan and Kurtz 2013; Morris and Fife 2009; Fife et al. 2010) of plant reliability. An inverter[8] may have a listed MTBF of 15,000 hours based on manufacturing testing and/or predictions and many would assume that to be the period of failure-free operation. This is only a partial truth and is misleading. *The application of failure-free operation only applies to an expectation that a specific item will be failure-free. It does not apply to a population of components.* For a large population of inverters, e.g. greater than 30, at $t = 11{,}000$ hours and assuming a constant hazard rate ($h(t)$) or exponential distribution, approximately 63.2% of the population has failed, which is called the characteristic life (η).

6.9.4.1 Confidence Level/Confidence Interval

> A confidence interval is a range of values, bounded above and below the statistic's mean, that likely would contain an unknown population parameter. Confidence level refers to the probability, or certainty, that the confidence interval would contain the true population parameter when you draw a random sample many times. https://www.investopedia.com/terms/c/confidenceinterval.asp.

The question of confidence in the MTBF (or probabilities) is related to the quantity and quality of data (see Chapter 10). One can, of course, make some assumptions when making predictions as to what range of reliability metrics to apply to models. It is important though to understand when modeling the plant that Monte Carlo simulations should be used to assess the potential results of having a highly variable range of values for the equipment.

"Z" is the number of standard deviations away from the mean data point is located. When data are available, the two-sided confidence in the data, if it is normally distributed, can be found using Eqs. (6.23)–(6.25):

$$\alpha = \frac{1\text{-CL}}{2} \tag{6.23}$$

$$\mu - z\frac{\sigma}{\sqrt{n}} < \theta < \mu + z\frac{\sigma}{\sqrt{n}} \tag{6.24}$$

$$z = \frac{x - \mu}{\frac{\sigma}{\sqrt{n}}} \tag{6.25}$$

where

- μ = the mean of the component population
- x = the time to failure of a component
- σ = the standard deviation of the population
- N = the number of failures

Using the above, if the desire is to have an MTBF at 90% confidence the design MTBF would need to be:

$$\text{MTBF (Design)} \geq \text{MTBF(req'd)} + z\sigma = 10{,}000 + 1.2815 * 1000 = 12{,}815 \tag{6.26}$$

Does the data set include enough samples and statistical analysis of weather and climate extremes of those environmental elements? Equation (6.27) illustrates the determination of the confidence interval (CI) associated with a certain confidence level (CL).

$$\text{CI} = \hat{p} \mp z\left(\sqrt{\frac{\hat{p}\,(1-\hat{p})}{\acute{n}}\,\frac{N-\acute{n}}{N-1}}\right) \tag{6.27}$$

where:

- $\hat{p}$ is population proportion
- z is the z-score assumes a normal distribution (Table 6.7)
- $\acute{n}$ is the sample size
- N is the population size

For a 50-year plant, 25 years of weather data and the 95% CL (z = 1.96) then the CI becomes:

$$\text{CI} = \frac{25}{50} \mp 1.96\left(\sqrt{\left[\frac{\frac{25}{50}\left(1-\frac{25}{50}\right)}{25}\right]\left[\frac{(50-25)}{(50-1)}\right]}\right) = 50\% \pm 14\%$$

Taking the same system and using 30 years of data results in 60% ± 11%. In the first instance the CL is 36–64%, while in the second instance the CL is 49–71%. Reducing the CL to 80% (z = 1.28) yields 60% ± 7.3% or 52.7–67.3%. The CI decreases and hence the potential for risk goes up.

Table 6.7 *Z*-Scores Assuming a Normal Distribution.

Confidence Level	*z*-Score (±)
0.70	1.04
0.75	1.15
0.80	1.28
0.85	1.44
0.92	1.75
0.95	1.96
0.96	2.05
0.98	2.33
0.99	2.58

The more data the better. That data must, however, be curated, maintained, and utilized to gain any benefit for the stakeholders or industry.

This confidence assessment is different from that used for predictions. This assessment is based on the comparison of a sample size taken from an actual population. For various stakeholders, this provides a data point on risk. The risk can be better understood with additional high-quality data and a higher CL. A reasonable CL would range from 80% to 90%. Beyond 95% the costs begin to exceed the benefit.

6.9.4.2 *P* Values

Many stakeholders use or are familiar with a *P* value. These *P* values are based on an MTBF for that item and assume a closed form distribution that allows calculating confidence limits. Figure 6.9 illustrates the effects of setting CIs around a data set. In this example, the CI is 60%. This also illustrates the potential cost to a project for high confidence, i.e. if it is desired to have confidence. *P* values, listed as P50, P75, P90, and P99, are a common statement of confidence in many analyses (Dobos et al. 2012). These values are derived as a set of data that assumes the item data is normally distributed.

6.10 Reliability Block Diagrams (RBD)

Reliability block diagrams (RBDs) are used to model the system and allow for an assessment of the reliability and availability. It aids in the identification of weaknesses in the system from the perspective of failures modes and their effects. The initial systems diagram is a functional one that gives an overview of the elements of the generating unit and the system and their interrelationship.

Figure 6.12 is a simplified functional diagram that can be used to develop RBDs. As shown, it starts with the lowest level major items such as modules with structural support

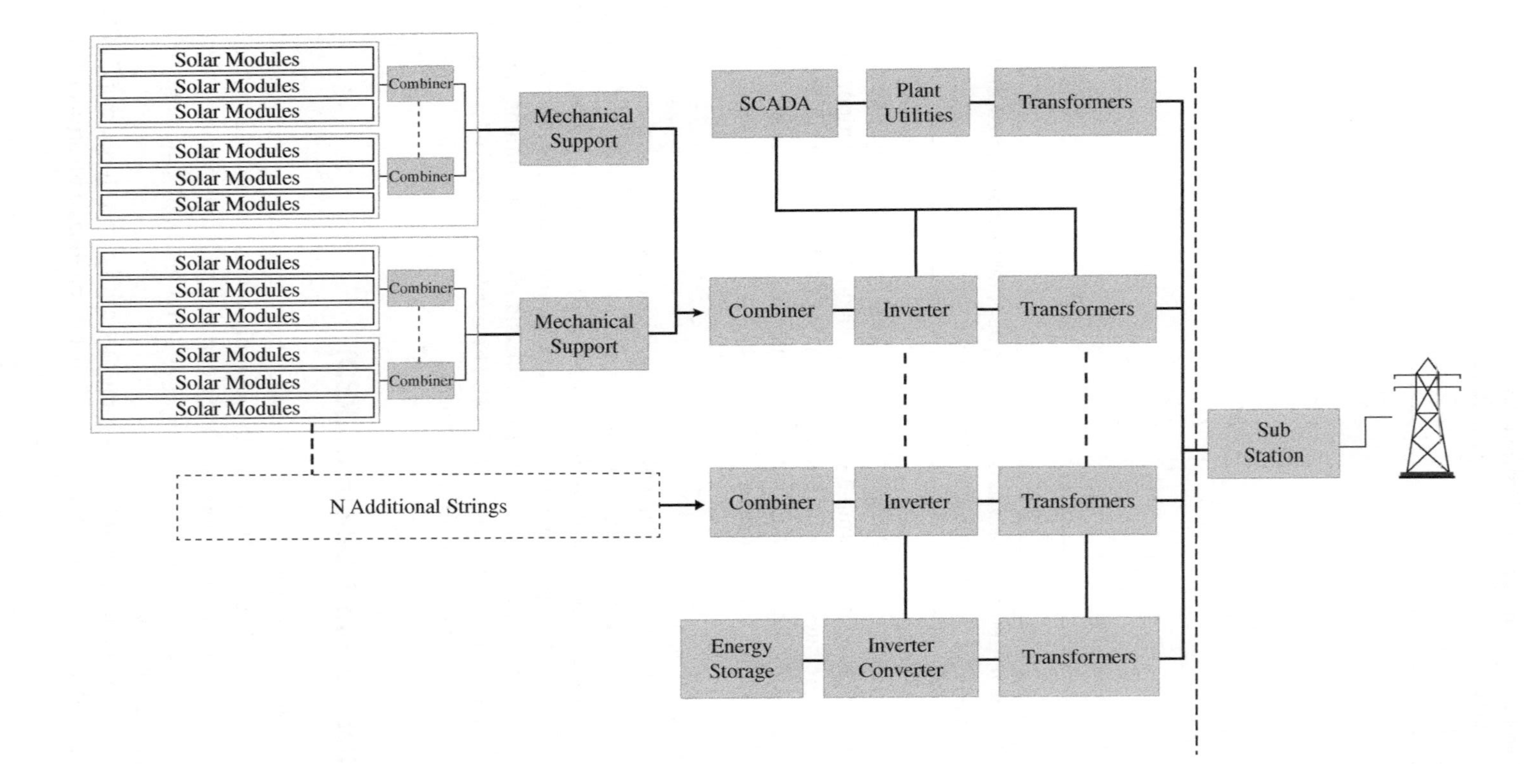

Figure 6.12 Functional Diagram.

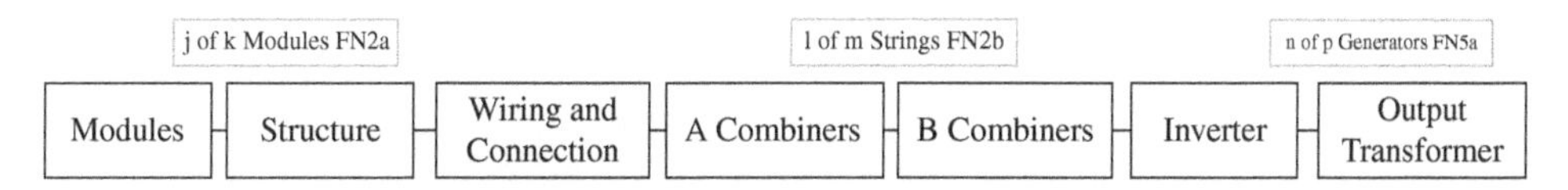

Figure 6.13 Unit Generator Level Diagram.

that are aggregated into combiners and connected to the inverter and transformer connection to the plant output. The solar modules contain all the wiring and connectors needed to generate the input power to the inverter. Included is the SCADA, which controls the inverters and aggregates any sensors, and plant utilities, such as lighting, security, etc. The mechanical support can be a ground-mounted support structure, or a fixture mounted to a single axis pole that is driven by a motor controller, and sun sensor that keeps the sun at orthogonal to the plane of the panel. The functional diagram, combined with the reliability diagrams, is also the source of information for fault-tree diagrams in that it shows the connectivity among the various components.

Figure 6.13 is a functional diagram of a generator unit. Note that it indicates the major subsystem elements for the generator, not how they are interconnected and what the potential effects are for failures.

Depending on the power capability of the inverter, there can be from 1 to 10 or strings or poles resulting in a "*k* of *n*" operating capability at the combiner. For the inverter to function properly, it takes "*r* of *u*" combiners to supply enough voltage and current to operate properly.

Figure 6.14 is the system-level diagram that adds in the SCADA, its sensors, the utility/customer delivery transformers, and the associated safety equipment are added in to cover the rest of the plant. A sensor failure of the SCADA monitoring system could potentially shut the plant down. This is where the FMEA is critical to aid in building the RBDs. It identifies the various components.

The underlying assumption of this partitioning of the plant into identical units allows for the "redundancy" *n* of *m* approach to the analysis of the plant. This oversimplifies the potential problems of failures in the field. For example, the assumption that there is "*o* of *q*" modules required for a single path to work, there is another path of failure. That is, the fact that the failure of multiple strings is required to have a plant failure. The reality is that if enough modules fail across the plant to *represent* multiple string failures the plant would then fail. Thus, another element of the RBDs is the addition of a "g of *h*" element that covers the potential loss of multiple independent modules across the plant.

Mathematical representation of the "*k* of *n*" reliability is:

$$R(t) = \sum_{k=x}^{n} \left(\frac{n}{k}\right) R(t)^k (1 - R(t))^{n-k} \tag{6.28}$$

where $R(t)$ is the reliability or probability of success, and $1 - R(t)$ or $F(t)$ is the probability of failure of the redundant function. X is the minimum number of "*n*" units that must be operable for the function to work to specification.

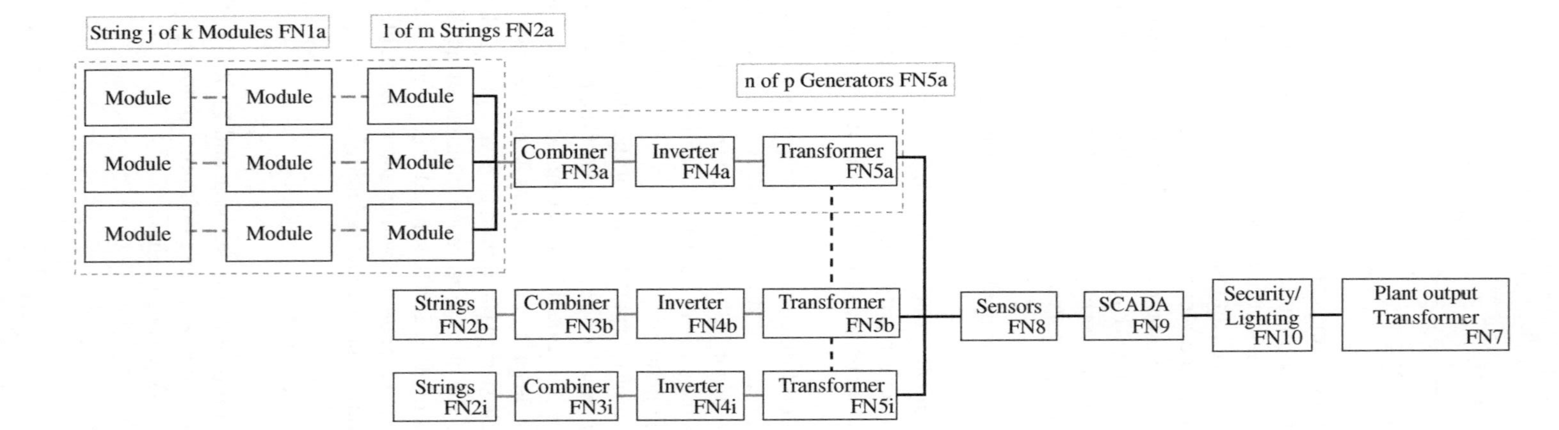

Figure 6.14 System-Level RBD.

6.11 Fault Trees

Fault trees are the domain of the systems and safety engineers and are presented here only to show the concurrency between RAM work and the safety work for a plant. The safety function determines the probability of safety failures that can result in injury or death to plant or even the public with the operation of the plant. Safety issues include floating grounds, open high voltage (HV >50 VDC) DC or AC potential that presents a shock hazard to the maintainers and failures or faults that can lead to serious risk of injury or death due to a combination of events.

The SE function utilizes the fault trees to assess the probability of occurrence of an end item e.g. plant, inverter, and strings) going down. The diagram, Figure 6.15, illustrates the use of logic gates, "and" & "or," to perform the additive or multiplicative properties of the design. The "or" function is additive since any one of the events cause that specific effect, while the "and" function requires all the events to occur before an event can occur.

The combination of FMEAs, RBDs, and FTA provides the project team and suppliers with the required failure rates as well as identifying the weak areas of the design and/or installation.

6.12 Failure Modes and Effects Analysis (FMEA)

An FMEA is first performed at the concept and amended as the design progresses until the product is delivered to the field. For one of the highly recommended tasks IEC 60812, Failure Modes and Effects Analysis (FMEA and FMECA) provides a significant understanding of the system and detailed design of the PV plant.

Figure 6.16[9] illustrates the lifecycle application of the FMEA. With rapidly changing technology, improved manufacturing processes, and changes in climate, the FMEA must be updated to reflect validation of previously identified failure modes and mechanism. In addition, it needs to reflect new, previously unknown failure modes, mechanisms, and effects.

With each step in the lifecycle, the specifications used in the RFP, design, system integration and use, each step is compared with the previous stage and assessments made of the failures and their effects. Differences are documented and fed back with adjustments made in the FMEA. This helps in real time to update FD and isolation and development of corrective actions. Some of the corrective actions may be implemented on the current system while others can be incorporated into future designs. These help in reducing costs through improved reliability and maintainability.

NREL produced a report (NREL/TP-7A40-73822, 2018) on Best Practices for O&M of Photovoltaic and Energy Storage Systems that discusses in detail the various needs for and types of maintenance. Additional resource information can be found in IEC 60300-3-5 ED. 1.0 B:2001 and IEC 60605-2 ED. 1.0 B:1994.

Although no longer a supported handbook, Mil-Hdbk-1629A has continued relevance for outlining and defining the elements of an FMEA and examples of several types of FMEAs. It is intended to identify and eliminate weaknesses in the design, specifically those that

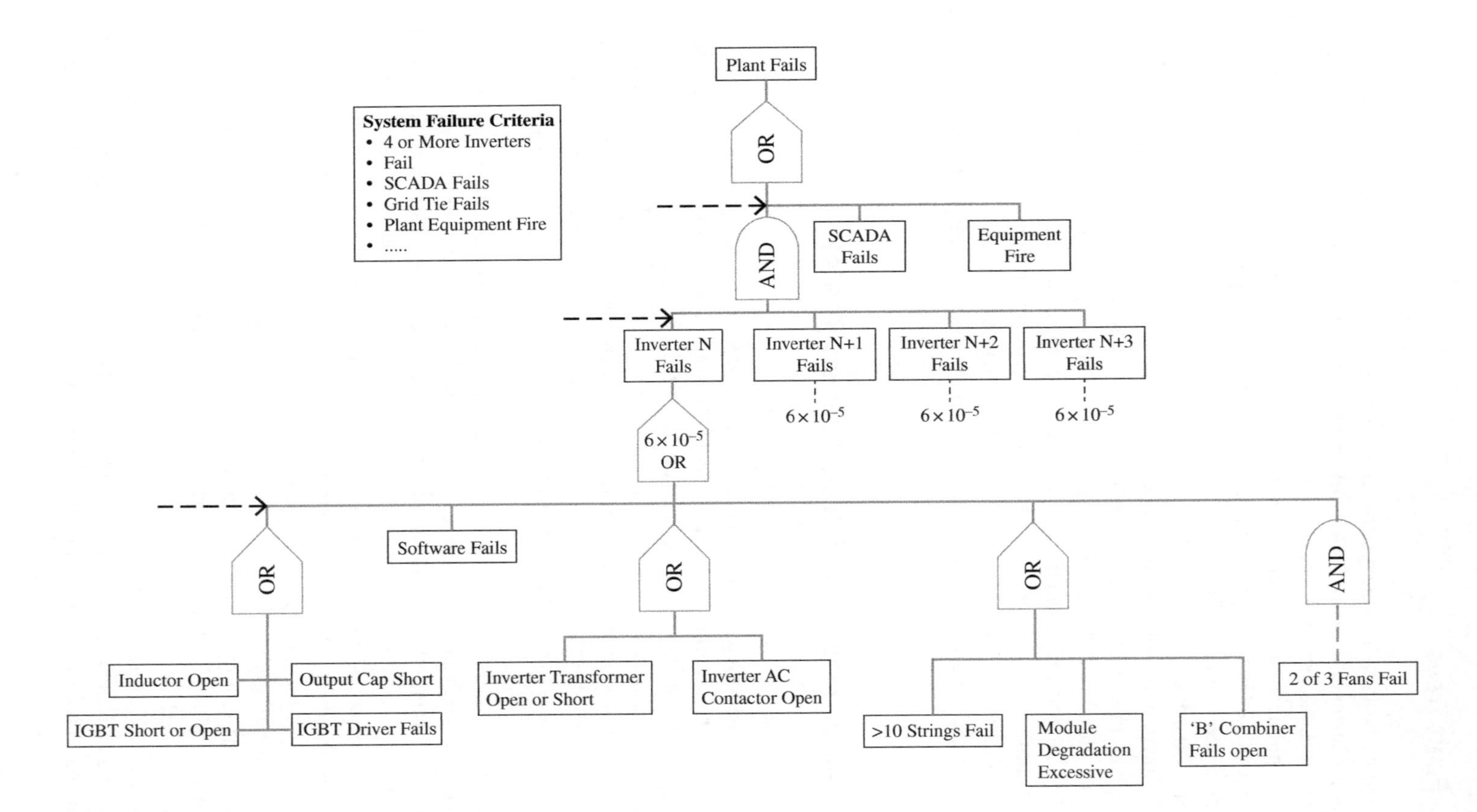

Figure 6.15 Fault-Tree Diagram.

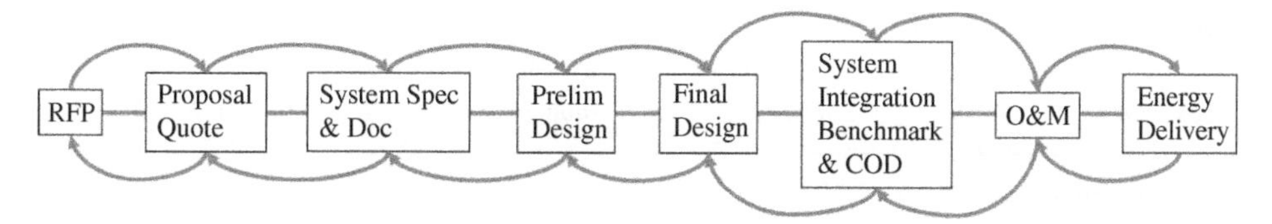

Figure 6.16 Lifecycle Use of FMEA.

would lead to premature failure or whose failure would result in a safety incident. General examples of types of FMEAs include:

- System FMEA. The system FMEA is focused on the top-level elements of the functional subsystems. So, one can postulate the loss of a function such as multiple strings that result in significantly reduced power from the attached inverter. A system FMEA may be considered finished when all the hardware has been defined and the design is declared frozen and is being submitted to the stakeholders for final approval.
- Design FMEA may be considered finished when a release date for production has been set.
- Process FMEAs may be considered finished when all operations have been identified and evaluated and all critical and significant characteristics have been addressed in the control plan.
- Service FMEAs may be considered finished when the design of the system and individual tasks have been defined and evaluated, and all critical and significant characteristics have been addressed in the control plans.
- Software FMEA focuses on software functions that control the inverters and plant.

The FMEA should be developed during the system and functional design phase and added to during detailed design. In addition, it should be available over the life of the product to document failure modes and effects that had not previously been identified. The FMEA then is a tool that can:

- Identify, document, and provide the basis for planning to prevent system, component, and process problems before they manifest themselves in the installed field.
- Reduce LCC by identifying risk early in the concept, system design and detailed design phases.
- Support process improvements through component level Statistical Process Control.
- Utilizing FR and criticality, identify a list of prioritize corrective actions.

The specific elements of an FMEA are given in Table 6.8.

FMEAs fall into three main categories with a couple of subcategories depending on the industry and the level of analysis. These are the detailed, functional, and process control FMEAs. These focus on system design, hardware/software design, concept, and manufacturing, respectively. Below these are the reliability, maintainability, and safety subcategories. The information for the detailed and functional is much the same with the differences being in the level of functional description of the faults, the failure effects, and the potential corrective action.

Table 6.8 Data Field List for FMEAs.

Data Field	Subfield 1	Subfield 2	Subfield 3
Plant/Location			
Failure No.			
Failure Date			
Revision (Letter or Number)			
Revision Date			
Item Description			
Subsystem/Assembly.			
Model Number			
Serial Number			
Installation Date			
Prepared By			
Failure Rate* or MTBF			
Function	Function 1	Function 2	Function 3
Failure Mode			
Failure Cause (Mechanism)			
Operational Phase	Normal operation	Scheduled maint.	Preventive maint.
Failure Effects	Local	Next level	End (system)
Operating Time	Total	Since last maint.	
Severity Classification	Manufacturing	Mission	Safety
Failure Detection Method	Sensor/software	Failure at next higher assy.	Not detectable
Compensating Provisions	Redundancy	Over supply	
Severity Classification	Critical/safety	Plant operation	
Remarks			
Process Improvement			
Responsible Engineer			
Responsible Manager			
Root Cause Analysis			
Corrective Action			

Applying the FMEA process to the system concept, design and detailed design efforts will result in:

- Better product quality and improved resilience through component process improvements and reduced cost. As a "living" document the FMEA also provides for continuous improvement to the design or product reliability.

- Higher reliability and lower maintainability by identifying weak links in the design and compensation for potential SPFs and improving LCC, reducing inventory and reduced maintenance costs. Also, it can significantly reduce the warranty costs.
- Reduce or eliminate the need for redesign within the first 10 or more years.

All the above lead to improved customer satisfaction, reduced O&M costs, improved availability with improved profitability, and decreased production waste, further reducing costs.

The underlying basis for the FMEA is the attributes of a part's failure modes or how something can fail and failure mechanisms or the physics or causes that drive that failure. In the case of software, the PoF is based on the hardware that fails as the result of a software failure since software by itself has failure mechanisms driven by the design and SE process.

6.12.1 FMEA Procedure

The process for conducting an FMEA is a straightforward process but does require that the personnel performing the task have a good grasp on the PoF. The starting point is to describe the component, assembly, or card function. The description of the function provides the basis of understanding the potential failure mechanisms and the modes of failure specific to that part.

To support the analysis of the functions of the plant, a RBD is needed for the FMEA. This diagram shows the lowest level of repair components or assemblies as blocks connected by lines that indicate how the components provide for the success of the design. The diagram shows the functional relationships of the components, parts, or assemblies. This provides the basis around which the FMEA is developed. To simplify the analysis the failure modes and mechanisms should be based on a set of standard codes. This avoids confusion when the FMEA is reviewed by various stakeholders. The RBD is always included as part of the FMEA.

Use the diagram prepared above to begin listing items or functions. If items are components, list them in a logical manner under their subsystem/assembly based on the block diagram.

1. **Identify failure modes**: A failure mode is defined as how a component, subsystem, system, process, and so on could fail to meet the design intent. Examples of failure modes are shown in Table 6.9.

   Not to generate confusion, a failure mode in one component can serve as the cause (a failure mechanism) of a failure mode in another component. So, water intrusion can for example lead to corrosion. Corrosion can lead to structural failure. Each failure should be listed in technical terms, that is: what is the physical effect on the component, assembly, or plant of an identified failure. This includes software that controls hardware. Failure modes should be listed for the function of each component or process step. At this point the failure mode should be identified, whether the failure is likely to occur. Looking at comparable products or processes and the failures that have been documented for them is an excellent starting point.

Table 6.9 Examples of Failure Modes.

Corrosion	Moisture Seal Embrittlement
Electrical Short or Open of a Module	Mechanical or Thermal Cycling Fatigue
Deformation Due to Snow or Ice Loading	PV Cell Cracking
Module Backing Delamination	Bypass Diode Short
Blocking Diode Short or Open	Browning
Loss of Inverter AC Frequency Accuracy	Loss of Inverter AC Phase Lock with Other Generation Units
Loss of Cooling	Overheating

2. **Identify the causes for each failure mode i.e., failure mechanism:** A failure cause (mechanism) is defined as a design, component, use, user, or environmental weakness that may result in a failure. The potential causes for each failure mode should be identified and documented. The causes should be listed in technical terms and not in terms of symptoms. Examples of potential causes include:
   - Improper torque applied
   - Operation outside of min/max specified conditions
   - Contamination
   - Erosion
   - Software failure
   - Improper solar alignment
   - Excessive structural loading
   - Excessive voltage
     - ♦ Utility line transient
     - ♦ Lightning
   - Vibration due to bearing failure
   - Over temperature or excessively high thermal transient

   A short list of failure modes/mechanisms for modules includes those listed in Table 6.10. Delamination, for example, is the result of temperature, humidity, and degradation of adhesives mechanisms. The next-level analysis will list delamination as the mechanism for module failure, overheating, and loss of structural integrity.
3. **Identify the failure effects:** Describe the effects of the failure modes (functional failure). For each failure mode identified, the engineer, working with the SE team, establishes what the failure effects are at each of three levels. These are the local, next

Table 6.10 Failure Mechanisms for Modules.

Delamination	Backsheet Adhesion Loss
Junction Box Failure	Frame Breakage
Cell Cracks	Potential Induced Degradation
High Ambient and Cell Temperature	Cell Hot Spots

level, and end item or assembly. A failure effect is defined as the result of a failure mode specific to the function of the product/process as determined by the customer/user. They should be described in terms of what the customer might see or experience should the identified failure mode occur. Keep in mind the internal and external customer. Examples of failure effects include:

- Injury or death to the user or technician
- Inoperability of the product function or process
- Plant unavailability
- Loss of redundancy
- Degraded performance electrical
- Electrical noise (e.g. third-order harmonics)

Establish a numerical ranking for the severity of the effect. A common standard scale uses 1 to represent no effect and 10 to indicate very severe with failure affecting system operation and safety without warning. The ranking helps the analyst determine if a failure would be a minor nuisance or a catastrophic occurrence to the customer. This enables the engineer to prioritize the failures and address the big problems first.

4. **Determine the probability of occurrence:** A numerical weight should be assigned to each cause that indicates how likely that cause is (probability of the cause occurring). A common industry standard scale uses 1 to represent not likely and 10 to indicate inevitable.
5. **Identify corrective actions:** Corrective actions are the **(design or process)** mechanisms that mitigate or eliminate the cause of the failure mode from occurring. Identify troubleshooting tests, fault record analysis, SCADA fault monitoring processes to detect failures. Each of these methods or processes must be evaluated to determine how well it detects and isolates failures. The FMEA should be periodically updated over the plant's life to address previously unknown failure modes.
6. **Determine the probability of detection:** FD is the system hardware and/or software that detects and flags the system, assembly, or component fault and sends that information to the maintenance organization. The probability of detection can range from 1 to 10 based on the sensors and the ambiguity size. Ambiguity is the number of items in a group that provide the same fault identification number and can require the technicians or maintainers to carry several different spares to affect the repair.

6.12.2 Risk Priority Numbers (RPN)

FMEA RPN (risk priority number) is a numerical assessment of the risk priority level of a failure mode in an FMEA analysis. FMEA RPN can aid the responsible team/individual to prioritize risks and make the decision on the corrective actions. The RPN[10, 11] is a mathematical product of the numerical severity, probability, and detection ratings. The RPN is used to prioritize items that require additional corrective action.

Caution. The RPN numerically provides a number from 125 (weight = 5) to 1000 (weight = 10) as an indication of risk. However, without the attendant driver to the RPN, an item with a high number can have a high severity but low probability of detection and conversely an item can have a high probability of occurrence but a low severity or probability of detection. These have different responses required or needed to correct the design.

FMEA RPN is calculated by multiplying severity (S), occurrence (O), and detection (D) indexes. Severity, occurrence, and detection indexes are derived from the FMEA analysis: *Risk priority number = Severity × Occurrence × Detection*

- *Severity*: The severity of the failure mode is ranked on a scale from 1 to 10. A high severity (10) ranking indicates severe risk.
- *Occurrence*: The potential of failure occurrence is rated on a scale from 1 to 10. A high occurrence (10) rank reflects high failure occurrence potential.
- *Detection*: The capability of failure detection is ranked on a scale from 1 to 10. A high detection rank (10) reflects low detection capability.

$$\text{RPN} = (\text{Severity}) \times (\text{Probability}) \times (\text{Detection}) \tag{6.29}$$

What corrective action(s) can be used to address high to medium risk failures? These actions can include:

- Increased robustness
- Increased scheduled or preventive inspections
- Improved incoming testing or quality procedures
- Material changes
- Derating (lower stress or increase strength)
- Reducing or limiting environmental stresses
- Redesign to eliminate the failure mode
- Better or more monitoring sensors
- Redesign for redundancy.

To ensure that all significant defects are addressed, assign responsibility for the corrective action to an engineer and manager with cost and schedule control. This requires reporting to the team on progress and completion with expected improvements in performance, cost, and customer satisfaction.

Document and curation actions taken: This is a document trail for defining the problem, the recommended corrective action, and the results of testing/deployment to the field to upgrade/update the equipment or software.

Once the corrective actions have been implemented and verified in the factory and/or field, the FMEA should be updated to reflect the elimination or mitigation of the failure or failure mode. The FMEA should be updated as the design or process changes are implemented. Field data/information acquired while the plant is in operation is used to document new, unknown, or unidentified failures.

FMEAs provide a capability to understand not only how the system works but what happens when it fails. The one major limitation is that the analysis only traces the **effects of a single failure** up to the system level (Table 6.11). However, it also allows for a secondary analysis to aggregate all the similar end effects to identify weak areas of the design that result in high criticality to plant operation.

Consequences of failure are the direct, often immediate effects of components, parts, modules, cards, assemblies, or end items failing to perform to specification. The general

Table 6.11 Example Failure Category Definitions for FMEAs.

Severity		Process	Functional	Detailed
V	Catastrophic	Plant production shutdown with loss of revenue and potential injury/death to plant personnel and potential lawsuits	Loss of function with potential loss of one or more lives or major damage with cost exceeding $1,000,000, and/or customer damage claims	Loss of specific equipment with loss of one or more lives or major damage with cost exceeding $1,000,000
IV	Significant	Plant production slowdown with significantly reduced or loss of revenue potential injury to plant personnel	Loss of capability or capacity with potential major injuries or loss of limb or damage exceeding $500,000, and/or customer damage claims	Loss of specific capability or capacity with potential major injuries or loss of limb or damage exceeding $500,000. Backup or alternative may exist
III	Major	Loss of several production days with delays in shipment, possible minor injury, and cost incurred to replace lost items	Reduction in capability with potential injury to one or more persons or damage not exceeding $500,000, and/or customer damage claims	Reduction in capability with potential injury to one or more persons or damage not exceeding $500,000. Redundancy exists with minimal loss of function
II	Marginal	Loss of part of shipment with resulting cost incurred to reproduce lost items	Reduction in capability with potential minor injury or damage exceeding $100,000	Reduction in capability with potential minor injury or damage exceeding $100,000
I	Negligible	Minor cost impacts and/or delays in production – no lost revenue	Reduction in capability with damage less than $50,000	Reduction in capability with damage less than $50,000

categorization shown in Table 6.11 has different meanings depending on the level at which the FMEA is performed and the specific type (process, functional, and detailed) of FMEA. The categorization of occurrence is shown in Table 6.12.

FMEAs are a working document, with the intent that as new failures occur, e.g. a failure that was not accounted for in the previous analysis, the FMEA is updated. The updated document provides information for the first or subsequent **SE/R**epowering™ phases for the current plant and as a foundation for analysis of the next new system. Data is stored in a database (DB), such as Access™ that allows for the automatic updating of the data as well as updating reports generated through a DB tool.

6.12.3 Failure Modes and Mechanisms

Failure modes are the way in which things break or fail while failure mechanisms are the things (physics) that cause the failure. A general categorization[12] of solar cell failure modes

Table 6.12 Probability of Occurrence.

Probability of Occurrence		
Certain	$0.999–10^{-1}$	Requires immediate design or other corrective action
Likely	$10^{-3}–10^{-1}$	Requires design or other corrective action
Possible	$10^{-5}–10^{-3}$	Requires design or process changes, potential for alternative part
Unlikely	$10^{-7}–10^{-5}$	To be corrected if the failure manifests itself or a planned upgrade is to occur
Not Probable	$10^{-7}–10^{-9}$	No action needed

Table 6.13 Example Failure Mechanisms – Precipitate Failures.

	Failure Mechanism	
Natural	**Man-Made (Induced)**	**Effect of Application**
Lightning	Transportation vibration	Over temperature
Earthquake	Drop	Under temperature
Tornado	High/Low storage temperature	Transient voltage
Hailstorm	Handling	Degradation/Aging
Flash Flood	Design	Torquing
Wind	Installation	
Ice		
Snow		
Sand and Dust		

and mechanisms can be found in Jordan and Kurtz (2013), Kurtz et al. (2009), Jordan et al. (2016), Jordan et al. (2016) is shown in Table 6.13.

One note when doing an analysis is that a failure mechanism at one level of analysis can be a failure mode at another. For example: saltwater causes corrosion of a metal plate and a weakening of the metal plate strength. At the next level, a large or growing area of corrosion can result in structural failure. So, the failure mechanism in the former is saltwater or soil chemistry corrosion, while the mechanism for the latter is the reduced strength combined with a normal load.

Thermal considerations consist of several sources: solar irradiance, self-generated heating (often caused by the IR drop across the component), and heating due to ambient temperatures. As shown in Eq. (6.30), most components are simultaneously affected by all three.

Figure 6.17 illustrates that there is a decrease of approximately 30% in the open circuit voltage (V_{oc}) of the PV cell and approximately 5% increase in the short circuit current (I_{sc}). Thus, high temperatures tend to significantly reduce the power available from a cell. Elevated temperatures reduce the MTBF of the parts by accelerating several of the dominant

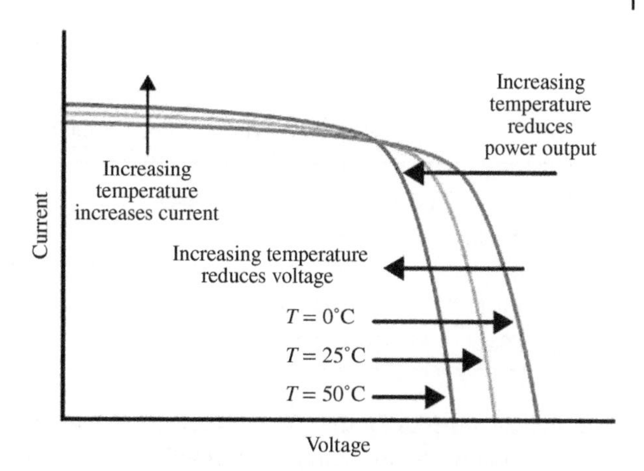

Figure 6.17 Plot of Panel Voltage and Current Dependent on Temperature.

failure modes. This effect on the parts is due to most component reliability values being specified at $T = 25\,°C$ or 298°K (see Section 6.18.1). There are also differences in the module temperatures depending on the type of plant. Fixed tilt and dual axis can have significantly different operating temperature. One example (Eke and Şentürk 2012) had fixed tilt module temperature of 58.5 °C in August and 35.6 °C in January. A two-axis tracker was 24 °C above fixed tilt. The actual cell temperature (Jordan and Kurtz 2013; Kurtz et al. 2009) can be estimated by:

$$T_{\text{cell}} = T_{\text{AMB}} + (\theta_{JA} * E * A * (1 - R)) \tag{6.30}$$

where:

- T_{AMB} = the ambient air temperature in °C
- θ_{JA} = thermal resistance in °C/W
- E = solar irradiance in W/m^2
- A = area of the cell in m^2
- R = reflectance of material

Reflectance (R) of material is the percent of incident irradiance reflected off a surface. Table 6.14 provides a summary of general reflectance numbers while Table 6.15 is a listing of the thermal conductivity and resistance of metals.

Tables 6.14 and 6.15 provide a basis of understanding the increase in temperature of objects exposed to the sun, such as PV modules, combiner, and inverter boxes. Table 6.15

Table 6.14 Values of Emissivity of Metals.

Material	Emissivity (%)	Reflectance (%)
Metallized and optical coated Plastic	75–97	3–25
Processed Anodized and Optical Coated Aluminum	75–95	3–25
Polished Aluminum	60–70	30–40
Chromium	60–65	35–40
Stainless Steel	55–65	35–45

Table 6.15 Material and Thermal Conductivity and Resistance.

Material	Thickness In/mm	Thermal Conductivity W/(mK)	Thermal Resistance θ (°C/W)
Air		0.024	41.66667
Aluminum	0.05/1.2	205	0.004878
Copper	0.05/1.2	401	0.002494
Steel, Carbon	0.05/1.2	43	0.023256
Stainless Steel	0.05/1.2	16	0.0625

specifically provides for the effect of transmission of solar energy in the form of heat to the inside of an inverter for example. Total solar reflectance is a percentage (0–100%) of a material's ability to reflect solar irradiance back into the atmosphere. ASTM C-1549[13] or ASTM E-903.[14]

Infrared Emittance (also called Thermal Emittance) is rated between 0 and 1 and rates the product's ability to cool itself by releasing thermal radiation back into the atmosphere. Painted finishes often maintain a lower emittance value than bare metal products. Many inverters, for example, use gray paint schemes.

Water, rain, or flooding can cause several failures such as corrosion of metals and potential terrain slumping or settling that puts strain on the structure or causes it to "warp" and lose sun facing accuracy. If not sufficiently protected, water can also intrude into buried cables and cause short circuits in connectors.

Installation failure mechanisms include plastic wire ties and grommets, which can break or pinch wires; exposure to sunlight can cause degradation of the insulation; exposure to wind and the weight of ice (center) can lead to flexure and failure with cycling; and finally cabling can fail due to rodents chewing access by chewing rodents (right).

Soiling can readily result in a power reduction of 5% or more which is well beyond any annual degradation rate. This necessitates periodic cleaning to maintain the designed power generation. In addition, the soiling components can contain acids or bases that will eat away sealants and adhesives (Figure 6.18).

A major concern for PV plants in coastal areas and in some suburban settings near seacoasts, lakes, landfills, and other areas is birds where, they may feed or nest the acquisition of bird feces. Birds tend to congregate in these areas resulting in large areas of bird excrement, Figure 6.18, on the front surface of the modules/panels. Like soiling this can reduce the power output of modules or panels by 5–90% or more. To complicate matters, those dropping can be acidic, thereby etching the glass surface. Cleaning is the appropriate corrective action based on preventive maintenance inspections; however, this soiling can be very resistant to cleaning and require multiple initial and more iterative cleaning maintenance.

Figure 6.19 illustrates the effect of excessive winds, microburst, tornado, hurricane, and so on. Wind damage is complicated by improper design, application, understrength mounting hardware, too few securing nuts and bolts, and incomplete installation. It should be noted that to verify that the equipment loss was due to winds in exceedance of the specification, the specification itself must be reviewed to ensure that it correctly specified wind speeds and site conditions based on a statistical analysis with a high enough CL.

Figure 6.18 Bird Excrement Can Cover Large Areas and Block Irradiance (Courtesy Andy Walker, NREL)[15].

Figure 6.19 Excessive Winds above Specification (Courtesy of Andy Walker, NREL).

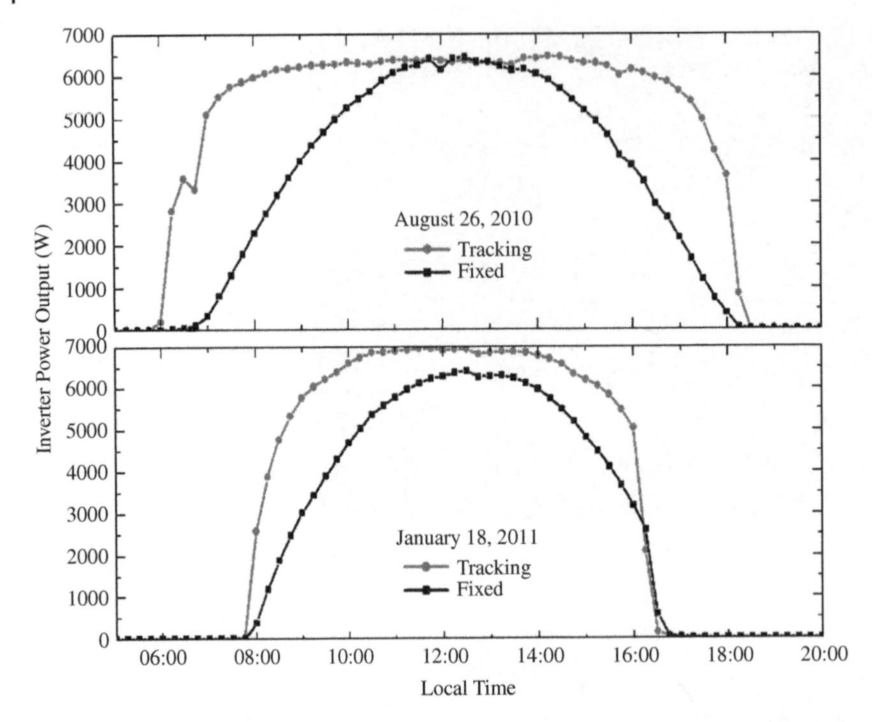

Figure 6.20 Comparison of Fixed and Two-axis Tracking Systems Power/Energy Production of a Standard Day (Source: Eke 2012/with permission of Elsevier).

As shown in Figure 6.20 (Eke and Şentürk 2012), there is more energy to be harvested from the sun using trackers. The significant addition of high-strength poles, tracker motor and motor controllers, sun trackers, and the associated power and control wiring needed far out weights many of the benefits of adding sun-tracking capability. These components have a significant impact on system reliability and maintenance, increasing the cost of O&M, spares, and maintenance overhead. Wind and wind gusts can cause significant failures.

6.13 Failure Reporting and Corrective System (FRACAS) and the PV SCADA

FRACAS is a tool using a database either manually or electronically populated when failures are detected. To be blunt, failure to gather failure and operating data will result in unknown causes/effects of system failure at some point. This is the main mechanism of identifying failure trends and sources of failures that cost the plant excessive O&M events and can grossly affect the LCOE for the plant.

The easiest aspect of the FRACAS process is that it can be included in the SW for the SCADA and the inverters. In addition, with minimum loss of power, it is possible to measure the power from each string to identify early string failures due to degradation, or module/panel failures or loss of tracking for one-axis arrays.

The FRACAS process consists of:

a. Acquiring failure data associated with the operation of the equipment. This includes identifying additional failed parts that may have been the cause/result as fault propagation.
b. Performing a root cause analysis to identify the problems, which are candidates for corrective action.
c. Convening a failure review board with the authority and responsibility to track, identify responsible organization, and responsible engineers to affect the analysis of the failed item and propose resolution.
d. Investigate each failure report and working with the responsible organization and engineer determine the root cause failure mode and mechanism
e. Determining corrective action and proposed implementation (cost and schedule) to mitigate or ameliorate the fault from the population of items with the same failed items.
f. Continued tracking of the failure to ensure that the failure has indeed been fixed and no new failures have been introduced as a result of implementing a corrective action.

Chapter 10 addresses the form and format for data collection.

6.14 Root Cause Analysis

RCA is used to determine what failed (component) and how (mechanism). When combined with additional plant information, the failure mechanism can also be determined.

RCA is the process of finding the cause or causes of equipment failure, including the failure mechanism where possible. The objectives of an RCA are to:

1. Determine the cause of failure
2. Propose a solution or solutions to correct the failure and
3. Monitor the solution to ensure that equipment repair works and that it does not introduce any further problems.

RCA is a crucial tool to work problems in real time and to avoid potentially difficult trends in failures leading to contractual default or liquidated damages due to failure to meet demand. Analysis of field data to assess whether the failure is within the expected range of failures. Using the information from the field determine potential source of the failure(s). The source or mechanism of failure should lead to testing that can verify or validate the failure. The process of an RCA is outlined below:

1. Gather any extra data about the failure not documented automatically by the FD and Isolation system in the SCADA or inverter.
2. Perform nondestructive and destructive analysis of the failed item to determine failure mode.
3. Verify that the failure mode (using the mechanism identified by the field data) occurs for a portion[16] of the items as currently manufactured.

RCA is also a lifecycle issue and should be considered as a tool for the life of the plant.

6.15 Data Analysis

Field reliability analyses are initialized using design predictions starting on the first day of a commissioned plant, and then monthly, the field failure data are added to the database and the true MTBF begins to emerge. It can take as much as 12 months of operation for the trend data to begin to settle out. The first 12 months allow time for any warranty or poor manufacturing practices to be corrected.

Data statistical analysis is the process of taking information from the FRACAS system and breaking it down into information that can be used to find weak components, designs, or manufacturing problems and apply these lessons learned to optimize the operation of the plant and reduce its maintenance costs. Several tools to perform the analysis include:

1. Excel™ provides a tool to perform some early mathematical analysis of data and the data trends.
2. Access is one of several commercial databases for data storage and analysis.
3. Weibull++™ is a tool that provides for the analysis of data using Weibull techniques.

Access can function as the database for the SCADA system to dump sensor results into a structured database and allow for easy report generation of individual events as well as providing reports with the data needed to perform the ongoing reliability data analysis. It also has the capability to export in an EXCEL™ format.

The most common error made by RAMS analysist is to assume that the equipment has a constant hazard rate over the equipment's life. This grossly overestimates early failures and grossly underestimates mid-life failures and does not predict at all the knee in the curve for aging failures.

Data analyses take two basic forms:

1. Raw data without parsing into equipment categories. This provides the top-level system RAM assessment. Take the cumulative failures from the field and generate the field MTBF and the MTTR. Although these are gross estimates, they do provide insight into where the maintenance money is going and what needs work.
2. Top-level analysis of the data at the subsystem level, e.g. inverters, transformers, combiners, strings or tracker system structure. This is parsing the data into end item categories such as transformers, inverters, strings, and so on.

6.15.1 Example Weibull++™ Analysis

The authors make use of EXCEL™ and Weibull++™ to analyze field data. Here raw data is generated by a capability within Weibull++™ tool. For this book's purposes, no failure attribute such as the equipment type or usage profile is inferred or implied. A generated plant data example allows some initial assessment of the plant health, condition, reliability, maintainability, and availability. At the plant level, one can take the number of failures in a month and the number of plant operating hours, which will result in the point estimate of the MTBF for the plant.

Table 6.16 provides a randomly generated data set of 100 failures, sorted from lowest to highest time. Without looking at any other data, the failures can be plotted in Weibull++™ or in EXCEL™.

Table 6.16 Example Table of Field Data Values after Ranking by Time to Failure (TTF).

Failure Number and Time to Failed (hours)									
1	193.4	21	2849.9	41	5436.2	61	10,788.8	81	27,486.1
2	202	22	2977.3	42	5538.2	62	11,048	82	27,586.2
3	255.3	23	3056.8	43	5556	63	11,170.9	83	27,939.2
4	257.8	24	3103.3	44	5648.3	64	14,579.9	84	28,570.1
5	518.9	25	3164.6	45	5660.5	65	14,885.3	85	29,702
6	593.9	26	3177.2	46	5758.2	66	17,274.1	86	30,497.4
7	919.6	27	3275.2	47	5801.3	67	17,872.9	87	30,643.5
8	933.1	28	3390.6	48	6069	68	18,445.9	88	31,357.2
9	1053.7	29	3504.9	49	6340.1	69	19,262.7	89	31,379.5
10	1124.5	30	3753.2	50	6771.2	70	19,601.9	90	31,935.4
11	1162.4	31	3781.7	51	6944.4	71	20,330.8	91	32,077.4
12	1239.6	32	3878	52	7118.7	72	21,414.3	92	32,312.8
13	1432.9	33	4250.6	53	7373.5	73	24,070.1	93	32,517.1
14	1475.6	34	4291.4	54	7453.6	74	25,142.5	94	33,072
15	1756.3	35	4364.1	55	8417.8	75	26,050.1	95	33,397.2
16	1959.5	36	4484.2	56	8984.2	76	26,087.5	96	33,820
17	2305.9	37	4611.9	57	9151.1	77	26,178.6	97	34,106.3
18	2408.3	38	4799.3	58	9215.1	78	26,632.2	98	34,429.7
19	2508.1	39	4856	59	10,165.2	79	26,667	99	36,280.5
20	2752.4	40	5417.7	60	10,687.9	80	27,349.5	100	36,931

This data is TTF data, with no reference to what failed. This example's intent is to show that even without specific equipment identified, it is possible to analyze the data and identify failure trends. This leads to opportunities to perform maintenance prior to failure and/or ensure that spares are available to repair the items and maintain the plant at optimum energy production.

The data as plotted in Figure 6.21 indicates that the entire dataset is not a Weibull distribution but a mixed distribution of failures. This is an attribute that you would expect to see from a plant where there are multiple items that can fail, and they have different failure attributes including modes, mechanisms, and TTF. Reviewing the data from Table 6.16, the first four or five failures appear rather early in the life of the plant. These may be the result of transportation, handling, manufacturing error, weak component, or induced during installation. Once these are repaired, they can be removed from the population as they are probably not related to the design or application of the parts operating in the site environment.

The generalized steps using the Weibull++ tool is to look at the plots and then parse the data to "pull-out" that data associated with a single failure mode. Figure 6.21 is a Weibull Plot of the uncensored data and represents an example of raw assessment. Although a line

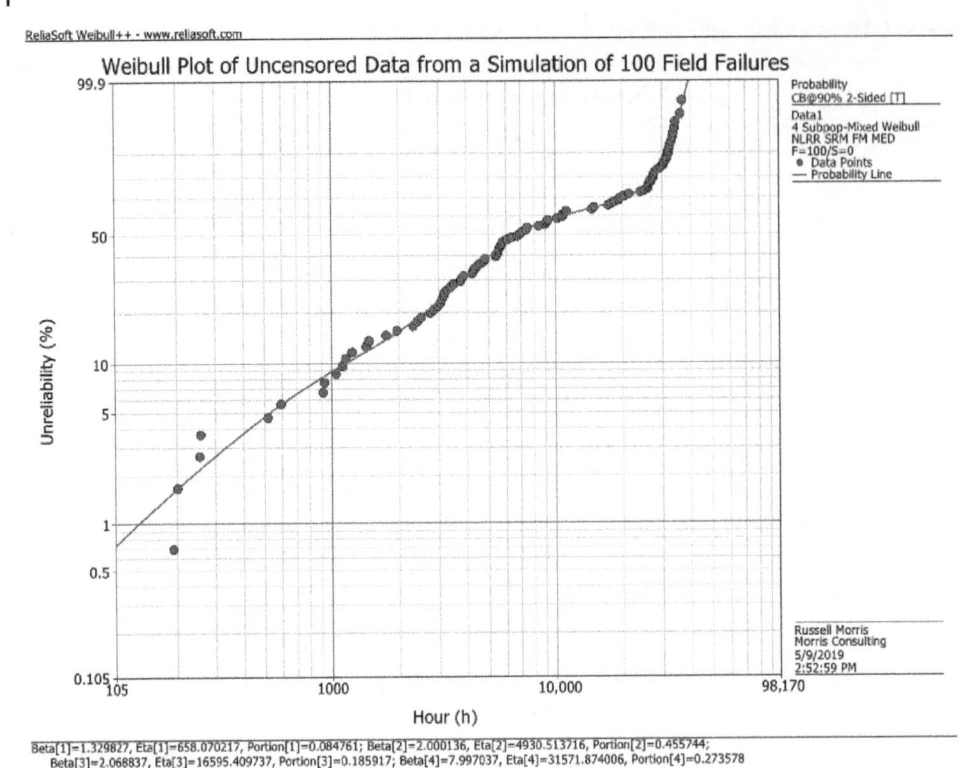

Figure 6.21 Plot of Uncensored Field Failure Data (Source: Morris).

can be fit to it, the analysis of the data would show that a CI (upper and/or lower) cannot be developed, and even a visual review of the data indicates that there are several failure modes represented.

As an example, the first four data points are analyzed separately since there appears to be a straight-line fit through this data. Figure 6.22 is indicative of failures that might be attributed to Manufacturing Defect or Weak Part as they arrive within the first week or so of operation.

Figure 6.22 is the second data set that occurs during the first months of operation and may be weak parts or poor handling during installation. These failures can include poor wire crimping, cold solder joints, incorrect component insertion, and poor part quality.

Figure 6.23 shows early failures that are more characteristic of weak parts or ones that are being overstressed and consequently failing early. These may be capacitors, resistors, and some of the power components if not properly derated. Figure 6.24 represents the general class of failures that occur during the first year of operation. These can be weak parts or improperly selected parts that fail during the warranty period and are the responsibility of the manufacturer. The FRACAS effort supports identifying the root cause and assigning it to the proper responsible party.

The first-year failures occur within the 90% CL line (upper) but may have a meandering rate of occurrence due to a second failure mode such as combined heat and humidity.

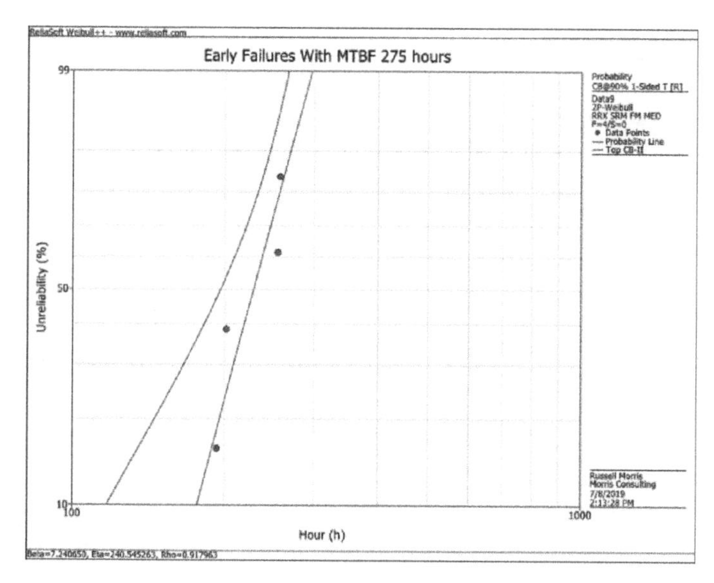

Figure 6.22 Early Failure Illustrating Manufacturing Defect or Installation Damage (Source: Morris).

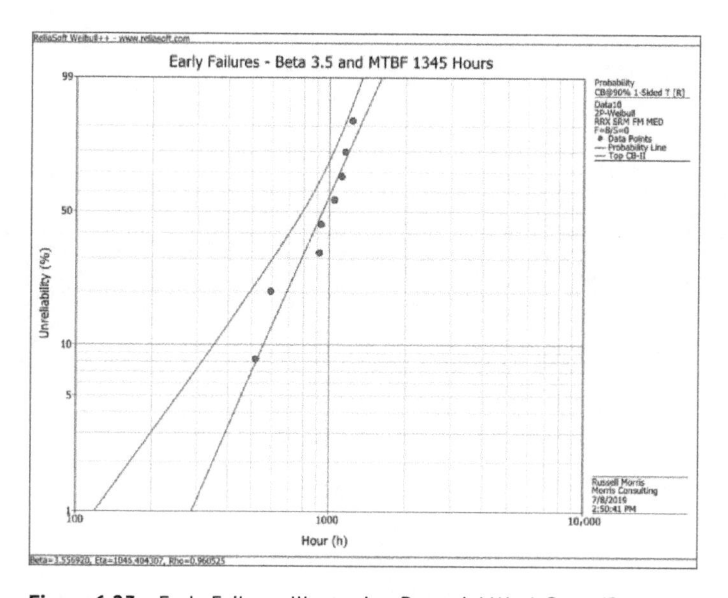

Figure 6.23 Early Failures Illustrating Potential Weak Parts (Source: Morris).

It is also possible that there are multiple parts or components failing but the data allows for planning as well as investigation into what is driving this large population.

Figures 6.25–6.27 illustrate wearout type conditions as they have extremely high beta values but significantly different MTBFs. These are the types of failures that would be related to cooling fan failures or mechanical fatigue associated with thermal cycling or relay switching where 8–10,000 cycles of power relays may result in pitting and increased contact resistance leading to failure or the contact arm wears out due to power cycling.

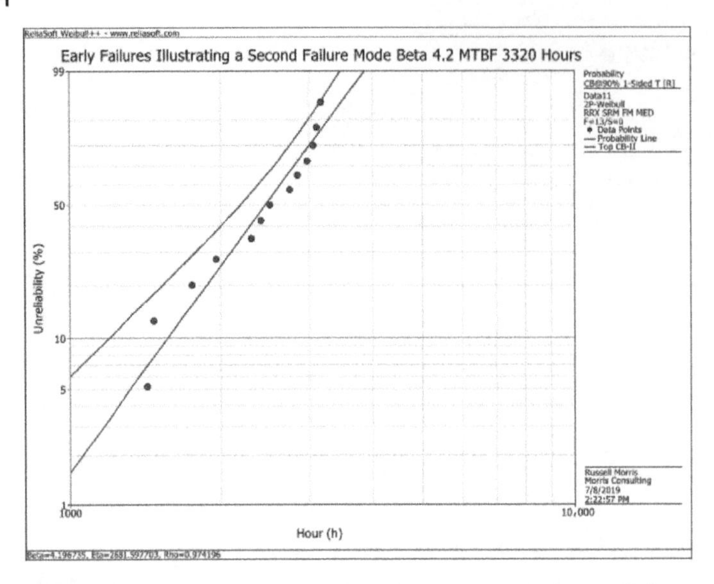

Figure 6.24 Early Failures (Less Than One Year) with a Second Failure Mode (Source: Morris).

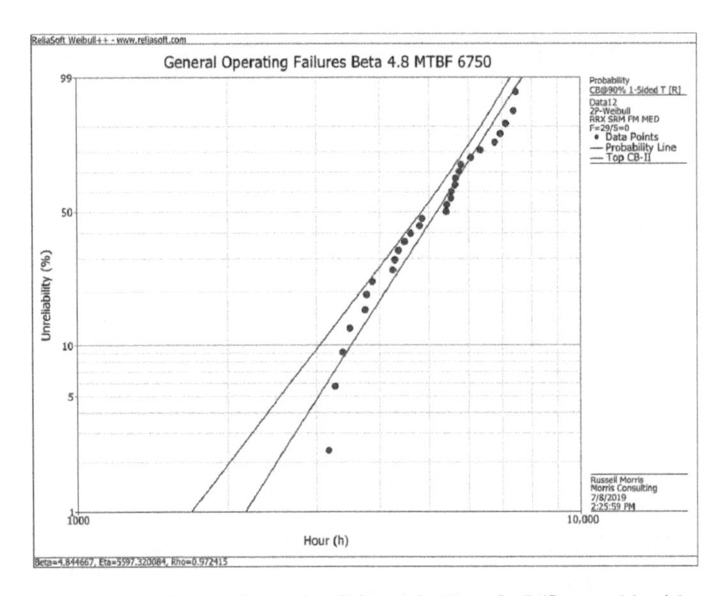

Figure 6.25 General Operating Failures in Years 2–3 (Source: Morris).

Figure 6.28 illustrates what would be random failures occurring over the mid-life of the plant from year 4 to 10+ years. These have an increasing FR but one for which failures are distributed over an extended period. This can also be data from more than one or two components.

Figure 6.29 illustrates the rapid wearout of an item. This failure mode has a very high Beta and the close boundary of the lower confidence line. These failures occur within one

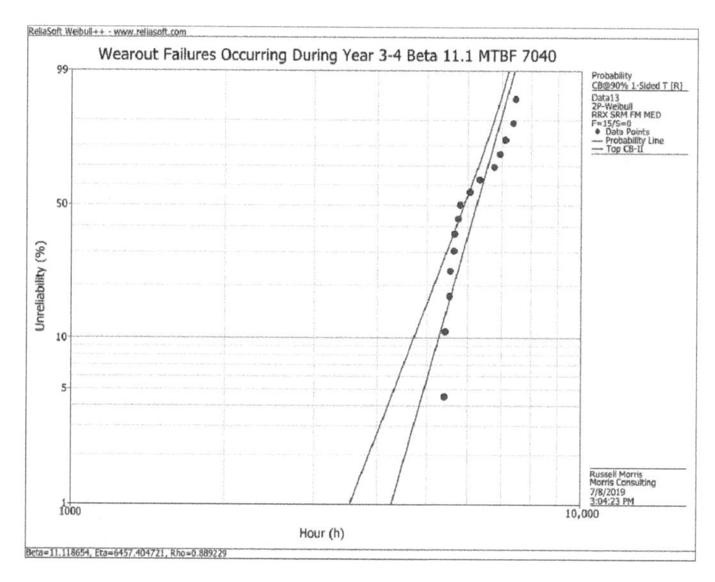

Figure 6.26 Wearout Failures Occurring in Years 3–4 (Source: Morris).

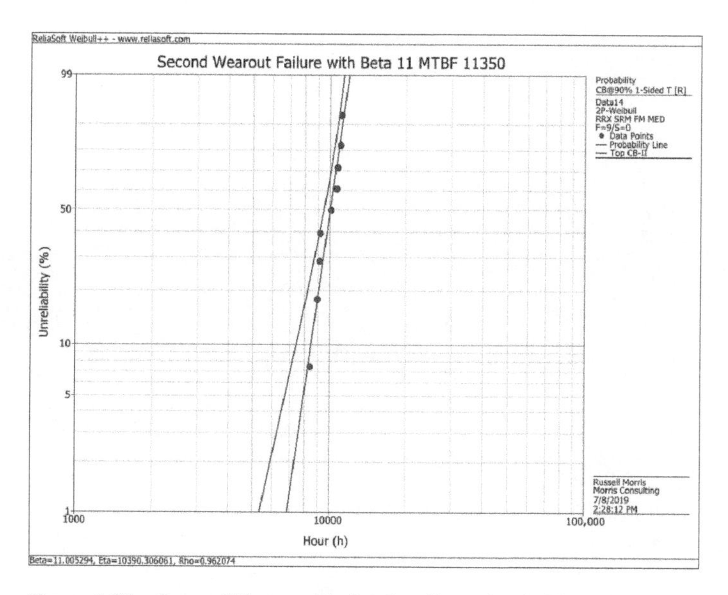

Figure 6.27 Second Wearout Mechanism Occurring in Years 4–5 (Source: Morris).

year with an MTBF of 35,000 hours. The part(s) are probably mechanical and all within the same environmental exposure and usage profile.

Top-level analyses provide the data that supports the general system metrics for the plant. The data comes from a variety of sources including failure data, energy generation, energy supplied, energy days lost, total energy delivered annually, and so on. Most of the top-level analysis are done to meet all of the various reporting requirements to NERC, Financial Institution, Utility (customers), and the O&O.

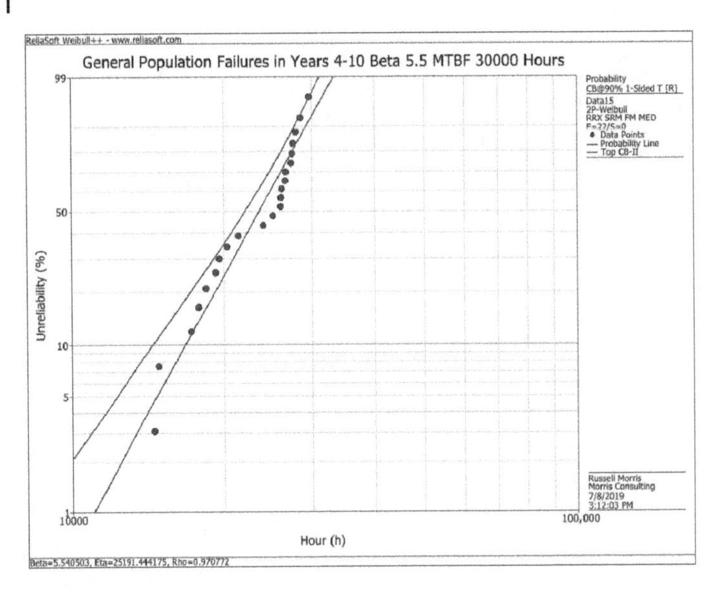

Figure 6.28 General Population Failures Occurring in Years 4–10 (Source: Morris).

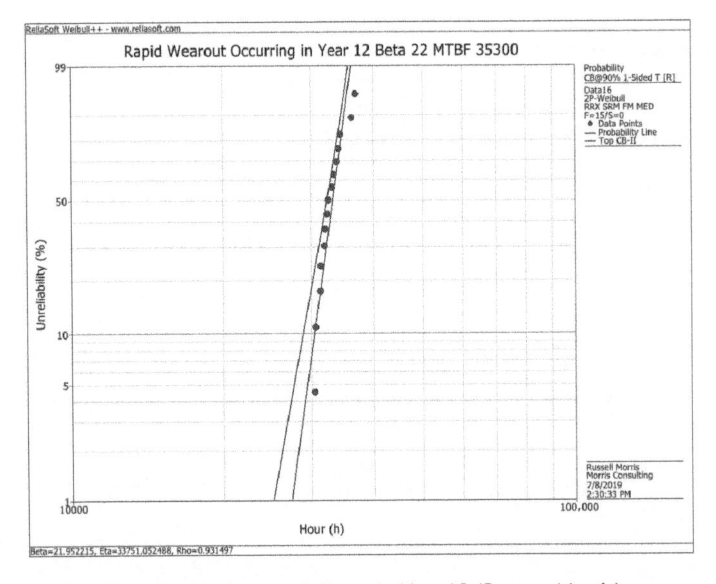

Figure 6.29 Rapid Wearout Failures in Year 12 (Source: Morris).

Detailed analyses are used to find, document, track, and recommend or enforce corrective actions to correct the root cause(s) of problems and include all tests and test results (even those later to be deemed nonrelevant). The data sources include the SCADA system, results of manual or automated testing, supplier responses to failure reports, and their testing results. The data are then transferred to the RAM tracking models to demonstrate the current state (health and condition) of the system.

The current state of the system is then compared to the predicted or modeled state and determinations are made on the trends that may be appearing in equipment or plant failures. When the detailed failure data are combined with the cost to effect repair, the result is a global view of the O&M costs, potential trends in the MTBF of the equipment, and how well the maintenance teams are meeting expectations for repair times and spares availability.

6.15.2 Praeto Analysis of Data

A useful analysis that can be done in a spreadsheet such as EXCEL™ is to bin the part/component data within blocks of time. Figure 6.30 is using one-year (3000 operating hours) bins to illustrate the variable nature of failures with a selected part.

The value of the Praeto analysis lies in the ability to quickly identify the "heavy hitters" in the maintenance and operation of the plant. Figure 6.30 also illustrates the double mode for failures at the plant level. Taking data from individual failure reports and classifying these according to the failed item or maintenance event for plant inverters is shown in Figure 6.31.

A major driver for maintenance based on apparent failures is the category of No Fault Found (NFF). Figures 6.31 and 6.32 illustrate that there is a large disproportion of NFF. This is as many maintenance calls as the top three failed items combined. *Given this high percentage it implies that the FD system is not operating efficiently or has excessively tight error bands or even that the sensors used to detect faults are too sensitive.* This is one area where the O&M should work with the equipment OEM and determine if there is something that can be done to improve the FD/FI and reduce NFFs.

A second useful Praeto is to assess the fault documentation against the kWh lost due to inverter down time. Down time in these cases is an automatic shutdown due to a detected, but not confirmed, fault.

As shown in Figure 6.32, the NFF contributes significantly to the lost energy, but the largest energy losses are due to the card/board, AC contactor, fuses, and power components (IGBTs, capacitors, and power supply). Again, this data should be used to work with the OEM to determine if there are form fit and function upgrades or improvements that can be made to reduce the cost of the maintenance of the plant.

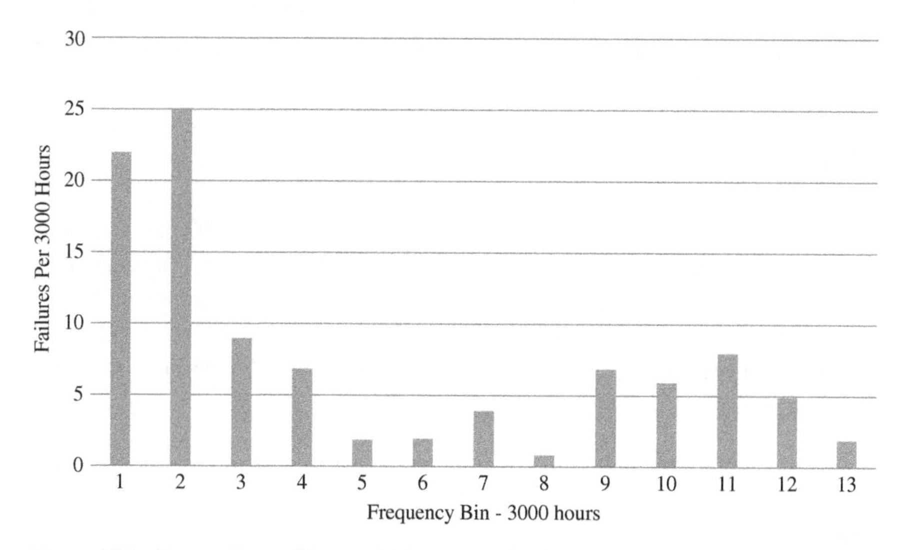

Figure 6.30 Praeto Chart of Annual Maintenance Based on Simulation Data.

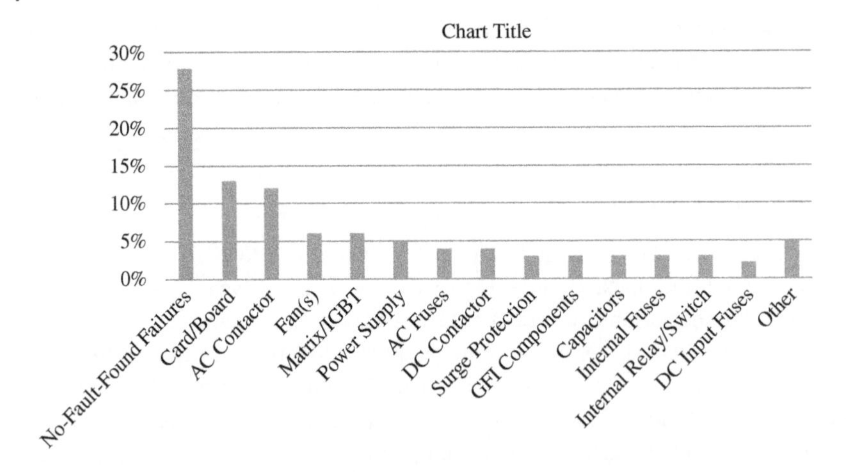

Figure 6.31 Praeto Failure for Inverters (Source: Formica et al. 2017).

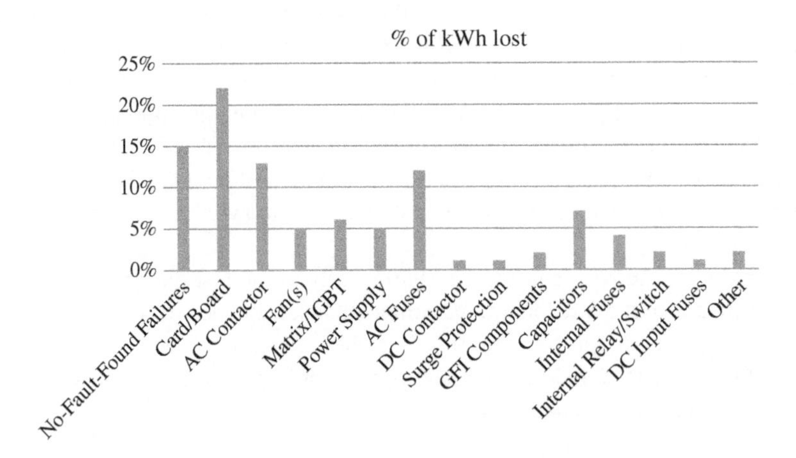

Figure 6.32 Praeto of Energy Lost Due to Inverter Component Failure (Source: Formica et al. 2017).

6.16 Reliability Predictions

There are two basic types of predictions, design, based on historical data gathered from field failure reports for which an RCA has been done (Mil-Hdbk-217F, IEC TR 62380, First edition, 2004–2008), and field, based on failures occurring in current equipment. Design reliability predictions are used to provide the first look at how the system can be expected to work over its life. The problem with past usage of this information can be narrowed down to four major points:

- Supplier MTBFs or FRs based on analysis alone using handbooks or other data sources that may be out of date, inaccurate, or not applicable to the current technology and its application(s).

- Supplier models that are based on incorrect assumptions of the actual environmental operating conditions and environments as well as little or no understanding of the usage or user profiles or the system that the equipment is being built into.
- Supplier failure to perform stress or life testing to verify and validate their prediction models. This is more expensive on the front end of a program, but for each problem found and corrected the usual cost benefit ratio is about 10 : 1 or more. Put another way, every dollar spent on fixing the problems before production delivery avoids 10 times that cost to perform redesign, changes in manufacturing processes, field repairs, potential retrofits, loss of customer goodwill, and loss of future sales due to a low product reliability.
- O&O failure to perform integration testing and verifying that the equipment, when integrated into the system, meets *all* specification performance metrics and that there are no weak or poor-quality parts or, poor installation, or there does not exist interface problems between equipment.

An established but controversial approach to predicting the reliability of a product is to use FRs based on existing equipment. The controversy is that the "existing" equipment can be 5 or more years old, and its technology can be out of date as the data may not cover the latest technologies. This handbook approach should always be taken as a method of comparing two or more designs against each other and not necessarily the actual field MTBF or FR that the unit will achieve in the field. This handbook approach can also, to a limited extent, provide a good comparison to the impact of operating a single design in multiple ambient environments.

The general form that the handbook calculation takes is to estimate the FR on an item based on Q-factors that adjust a base FR. Table 6.17 provides a summary of one set of Q-factors that are common, but the reader is encouraged to acquire the various data sources and apply them judiciously. An example case for calculating the FR of an EEPROM is shown in Eq. (6.31).

$$\lambda_{\mathrm{p}} = (C_1\pi_{\mathrm{T}} + C_2\pi_{\mathrm{E}} + \lambda_{\mathrm{cyc}})\pi_{\mathrm{Q}}\pi_{\mathrm{L}} \tag{6.31}$$

- π_{T} = Temperature factor
- π_{E} = Environment factor
- π_{cyc} = EEPROM Read/Write cycling-induced failure rate
- π_{Q} = Quality factor
- π_{L} = Learning factor

Table 6.17 Summary of General Pi Factors for ICs Used to Modify Base Failure Rates.

Factor	Pi	Factor	Pi	Factor	Pi	Factor	Pi
Environment	π_{E}	Temperature	π_{T}	No. of Pins	π_{Pin}	Complexity	π_{C}
Quality	π_{Q}	Package	π_{P}	Cycles	π_{Cyc}	Stress (Mech)	π_{S}
Learning	π_{L}	Cycles (Op)	π_{Op}	Stress (Electrical)	π_{E}		π_{E}
Die Complexity Failure Rate	C_1	Package Failure Rate	C2				

- C_1 and C_2 are factors associated with the EEPROM. Other complex microelectronics require different "*C*" factors.

Equation (6.32) is the generalized form of the parts count method of calculation for the FR of a part based on it attributes such as quality, complexity, and packaging. Of bigger interest with today's modern electronics is the number of solder connections, as these are proving to be the first failure mode for many ICs (Integrated Circuits).

$$\lambda_{\text{part}} = \sum_{i=1}^{n} \left[N_i \lambda_{Bi} \prod_{j=1}^{m} \pi_j \right]_i \tag{6.32}$$

where

- n = Number of part categories
- m = Number of quality factors
- N_i = Quantity of ith part
- λ_{Bi} = Base failure rate of ith part
- π_j = jth quality factor of ith part

Based on the BOM and assigning the appropriate distribution to the repairable item level, it is possible to build an initial model of the maintenance events and the plant availability for each year. This model can serve as the foundation of a Monte Carlo analysis that will converge to general values that can be used for planning (Figure 6.33).

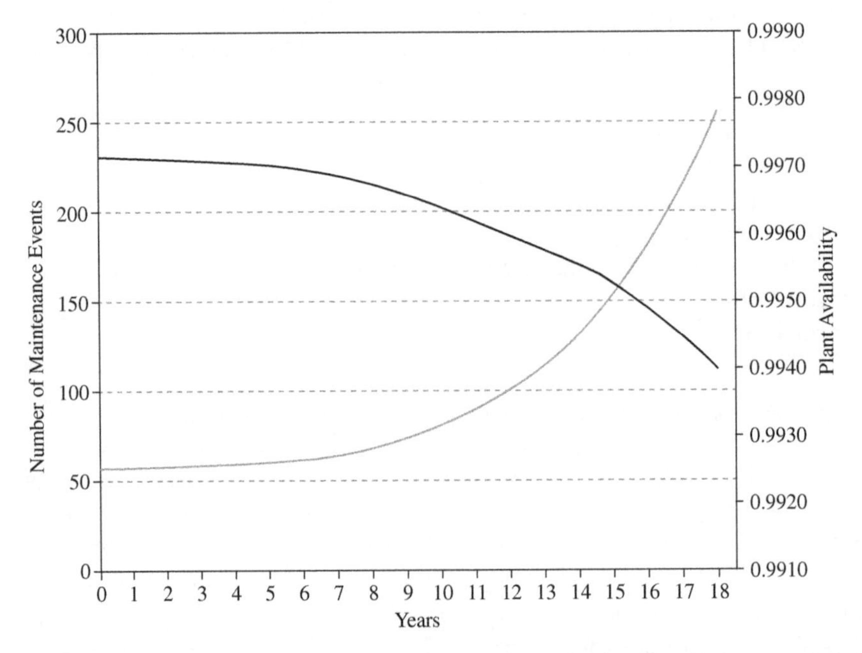

Figure 6.33 Simplified 20-Year Monte Carlo Simulation of Availability and Maintenance Events without Repair (Source: Morris).

Software predictions tend to be fairly difficult unless there is focused data acquisition that supports not only the defects and faults found but the development attributes of the software developers. A general guide to predictions can be found in IEEE 1633 2016.

6.17 Derating

Derating is a design and application tool for reducing the stress on components, improving their life/reliability. Derating is defined as the practice of limiting electrical, thermal, and mechanical stresses on parts to levels below their maximum specified ratings to provide additional safety margins and improve RAM (DOD GUIDE 2005; GEIA STD 0008 2011).

GEIA STD 8 provides design-related derating guidelines for electrical and electronic components and a sample is shown in Table 6.18.

A major point regarding derating is that ALL components including mechanical or metal ones are sensitive to temperature, vibration, humidity, and other stressful environments. Some materials can be weakened by elevated temperature resulting in deformation, while extremely low temperatures can result in the material falling below its "glass temperature" and becoming brittle and breaking on impact under stress.

Metals can also perform like plastics and glass at very high and very low temperatures. At high temperatures, metals will lose strength and can, in some instances, exhibit inelastic deformation, while at low temperatures they can become brittle and subject to breakage under stress.

The purpose of derating is to reduce stress on the parts and materials, thus improving their reliability, which reduces the number of maintenance events per year that must be performed to maintain the plant's availability.

Table 6.18 Example of Derating for Electrical and Electronic Parts.

Component	Parameter	Derating Factor (%)
Resistor	Power	80
Resistor Variable	Power	75
Transistor	Power	75
Diode	Voltage	50
Diode Signal	Voltage	85
IC Linear	Current	85
IC Digital	Fan-out	80
Thermistor	Power	50
Capacitor	Voltage	75
Transformer	Power	80
Relays	Contact current	50
Switches	Contact current	50

6.18 Reliability Testing

Reliability testing IEC 61215-1 through -4 :2016, IEC 61701:2011, IEC 62670-1:2013 should be an ongoing effort by putting a large sample, greater than 30 items, on test at the start of production and at least 5 replaced on every production run. These 30 units provide data for reliability analysis of the modules/panels over time and assure that the quality manufacturing standards (ISO 9000 series) continue to provide reliable units. The IEC has many standards for testing PV cells and modules/panels. Most of these are for the manufacturer of the cells/modules/panels and are intended to provide a high assurance of a reliable product design, but do not address the reliability of the product under field conditions.

Caution – A manufacturer who states that they meet the intent of these standards is not the same as them saying that they meet or test to these standards. Standards also change with time as newer technology arrives and/or new manufacturing processes are developed.

Many of the IEC Standards for PV relate to specific testing that should be done to ensure the integrity of the product design. This does not, however, deal with manufacturing. Sample testing, which is used in many industries, provides some level of confidence that the products meet a set of quality attributes, for manufacturing. Reliability testing, when performed prior to manufacturing or with the initial manufacturing run, provides an assessment at a point in time.

Although IEC 61215-1:2016 does not require UV testing regarding the life of a module/panel,[17] it is highly recommended as UV accelerates the degradation rates of polymers and adhesives before it is necessarily visible. The degradation of polymers leads to seals leaking, which can lead to water intrusion and short circuit of a cell(s). These tests provide higher assurance of a reliable product since they precipitate failures by degradation of materials with an acceleration factor of 4x–10x or more depending on the specific test.

The types of reliability testing address different failure mechanisms and aid in finding the actual failure modes. Basic PV module/panel testing should include the following while exposed to 1400 W/m^2 of simulated solar radiation:

- Thermal cycling (200 cycles) with illumination cycling
- Humidity freeze (40 cycles) with illumination cycling
- Damp heat (1000 hours) with illumination cycling
- Salt mist
- UV with illumination cycling
- Erosion (Wind with sand particulate >500 micron)
- Hail

As noted above the basic testing should include the additional effects of thermal cycling effect of generating electricity. Remember, reliability is a physics problem that is seldom if ever a single failure mechanism at work.

6.18.1 Electronic Components

The effect on semiconductor reliability of elevated temperature is a well-known and demonstrated fact (Elsayed 1996). The specific impact on reliable operation can be determined by applying the Arrhenius equation. All materials have an activation energy expressed in electron volts (eV), and this is particularly useful for semiconductor materials used in solar cells, transistors, microprocessors, memories, IGBTs, and other power electronics.

$$K = \mathrm{A}\exp^{\frac{-E_a}{kT}} \tag{6.33}$$

where

- E_a = Activation energy in eV
- k = Boltzmann's constant or 8.6e−5 eV/K
- T = Operating temperature in °K (278°K+T_{op}°C)
- A = Constant dependent on semiconductor or component type

The main element of the Arrhenius model (Bernstein and White 2008; Pecht et al. 2006; Mil-Hdbk 217 1998; Kailash and Pecht 2014) – is the activation energy (E_a), which is unique for all materials and can have a wide variance even for various types of semiconductors. For Eq. (6.33), "A" is a constant unique to each type of semiconductor or part. Table 6.19 provides a sample summary of the activation energy for a number of common failure mechanisms used in electronics.

Most manufacturers of semiconductor's reliability data are taken at or are based on a 25 °C ambient. However, operational ambient of photovoltaics and power devices (HV FETs, IGBTs) tends to be between 35 and 50°C, with PV cell temperatures ranging from 50 to 100 °C. The effect of this is that the stated failure rate of these devices is higher than expected. An approximation of the temperature's impact can be found by using the Arrhenius equation by taking the ration of the value of k at the ambient temperature and the operating temperature. For the case where one is determining the impact of temperature on the component, the calculation of the K_{T1} factor at ambient is given by

Table 6.19 Semiconductor Failure Activation Energy.

Failure Mechanism	Accelerating Factors	Activation Energy (E_a)
Dielectric Breakdown	Electric field, temperature	0.2–1.0 eV
Corrosion	Humidity, temperature, voltage	0.3–1.1 eV
Electromigration	Temperature, current density	0.5–1.2 eV
Au–Al Intermetallic Growth	Temperature	1.0–1.05 eV
Hot Carrier Injection	Electric field, temperature	−1 eV
Slow Charge Trapping	Electric field, temperature	1.3 eV
Mobile Ionic Contamination	Temperature	1.0–1.05 eV

Eq. (6.34). The ratio of k_a (ambient) and k_{op} (operating) leads to an acceleration factor, "AF," that can be applied to supplier reliability data:

$$AF = \frac{k_{op}}{k_a} = \frac{A\exp^{\frac{-E_a}{kT_{op}}}}{A\exp^{\frac{-E_a}{kT_A}}} = \exp^{\frac{-E_a}{k}\left(\frac{1}{T_{op}} - \frac{1}{T_A}\right)} \tag{6.34}$$

Also, the actual operating ambient conditions can range from −40 to +55 °C. The basic reliability of semiconductor parts is often listed by the manufacturer as FITs or failures per billion hours. A good first model to use is 1.2 eV for activation energy (E_a) which is for electromigration – a dominate failure mechanism for most IC's and power semiconductors. For complex semiconductors, the activation energy is closer to 0.9–1.1 eV. These values of activation energy indicate and has been demonstrated to increase the failure rate by a factor of 2 for every 10 °C rise in temperature above the qualification temperature of 25 °C.

6.18.2 Mechanical Components

The acceleration factor for some mechanical components can be found by increasing the number of cycles per unit time for the testing under actual load conditions. For relays, the number of cycles per year of expected operation may be 3000. For accelerated testing under load conditions, maybe run it 3000 cycles per day. This would achieve "30 years of testing" in one month. Caution should be taken as just increasing the rate of relay operation does not account for other failure mechanisms such as corrosion, solder creep (if soldered), and increased potential for fatigue failures due to high operating temperatures.

6.18.3 Test Time – MTTF

One factor helping to reduce the cost of reliability testing is that it is possible to accelerate or shorten the test time by increasing the test parameters. Table 6.20 provides a basic approach to assessing testing attributes using the number of units in test, test time, test acceleration factors (AF), CL desired, and the acceptable number of failures during testing. Equation (6.35) is used to determine the test time for a part, component, assembly, or system.

$$MTTF = \frac{2tAF}{\chi^2{}_{CL,2(n-1)}} \tag{6.35}$$

where t = test time; n = degrees of freedom; CL = confidence level; AF = acceleration factor.

The "AF" factor based on the equation is derived from Eq. (6.34). Kurtz et al. (2009), Habte et al. (2014), Wilcox and Marion (2008) cover some aspects of the impact of temperature on operation based on the TMY 3 data set. The limitation of this analysis is that it is based on a mean value, while the effect of temperature in such places as Arizona, California, and New Mexico (USA), Saudi Arabia, Egypt, and Kuwait may have extended days (weeks) of extremely high temperatures that permanently reduce the MTTF for the components.

Table 6.20 Estimating MTTF/MTBF Based on n-Item Testing with 0.90 Confidence.

Goal MTBF	10,000	Hours										
Test Time	2000	Hours	CL	0.9								
No. Failures		0	1	2	3	4	5	6	7	8	9	10
UUT	*n*	Test MTBF										
No. of Test Units	5	4343	2571	1879	1497	1251	1078	949	850	770	704	649
	10	8686	5142	3758	2994	2502	2156	1899	1699	1539	1408	1298
	15	13,029	7713	5637	4490	3753	3235	2848	2549	2309	2112	1947
	20	17,372	10,284	7516	5987	5004	4313	3798	3398	3078	2816	2596
	25	21,715	12,854	9394	7484	6255	5391	4747	4248	3848	3520	3245
	30	26,058	15,425	11,273	8981	7506	6469	5697	5097	4617	4224	3894
	35	30,401	17,996	13,152	10,478	8757	7547	6646	5947	5387	4927	4543
	40	34,744	20,567	15,031	11,975	10,008	8626	7596	6796	6156	5631	5193
	45	39,087	23,138	16,910	13,471	11,259	9704	8545	7646	6926	6335	5842
	50	43,429	25709	18,789	14,968	12,510	10,782	9495	8496	7695	7039	6491
	55	47,772	28,280	20,668	16,465	13,761	11,860	10,444	9345	8465	7743	7140
	60	52,115	30,851	22,547	17,962	15,012	12,938	11,394	10,195	9235	8447	7789

6.19 Summary

Improving the reliability of the plant through accounting for component variation, degradation, stress reduction, and applying the effects of the stressing environments will always raise the initial investment but provides for a much lower LCC. There is always the management decision to reduce front end costs, but as shown in Chapters 5 and 6, there is a significant improvement in the duration and quality of the power generation.

To achieve this improvement requires traceability of the system-level requirements, proper allocation to the EPC and their suppliers, oversight by the SE team, and data collection and analysis of field data. Reliability testing should be performed as early as possible to identify the weak design elements and fix the design deficiencies as they are precipitated. One essential element is to use a systems reliability program plan as the project controlling document for the above improvement. Examples of reliability program plans are available from a variety of books.

One thing of note that the reader should have noticed is that one can predict the reliability of a population of components or end items (systems) but predicting the reliability of a single or even a few components is not easy or accurate. Accurate predictions can only be achieved if the population is large enough and the population is subjected to the same environmental, user, usage, and mission stresses. The result is still subject to the rules of statistics; however, the standard deviation of the population will be smaller.

Bibliography

Patton, A.D. (1968). Determination and analysis of data for reliability studies. *IEEE Transactions on Power Apparatus and Systems* PAS-87: 84–100.

IEEE Committee Report (1974). Report on Reliability Survey of Industrial Plants, Parts I–VI. IEEE Transactions on Industry Applications, March/April 1974, pp. 213–252; July/August 1974, pp. 456–476; September/October 1974, p. 681 (See Annex A and Annex B.). IEEE.

IEEE Committee Report (1979). Reliability of Electric Utility Supplies to Industrial Plants. IEEE-ICPS Technical Conference Record, 75-CH0947-1-1A, 5–8 May 1979, pp. 70–75 (See Annex D). Toronto, Canada.

McWilliams, D.W., Patton, A.D., and Heising, C.R. (1974). Reliability of Electrical Equipment in Industrial Plants—Comparison of Results from 1959 Survey and 1971 Survey. IEEE-ICPS Technical Conference Record, 74CH0855-71A, 2–6 June 1974, pp. 105–112. Denver, CO: IEEE.

Dobos, A., Gilman, P., and Kasberg, M. (2012). P50/P90 Analysis for Solar Energy Systems Using the System Advisor Model. NREL Conference Paper Preprint No. CP-6A20-54488 (pdf 432 KB).

Habte, A., Lopez, A., Sengupta, M., and Wilcox, S. (2014). Temporal and Spatial Comparison of Gridded TMY, TDY, and TGY Data Sets. NREL Report No. TP-5D00-60866 (pdf 17.4 MB).

National Solar Radiation Database (1992). National Solar Radiation Data Base User's Manual (1961–1990) [TMY2 format description].

Sengupta, M., Habte, A., Kurtz, S., et al. (2015). Best Practices Handbook for the Collection and Use of Solar Resource Data for Solar Energy Applications. NREL Report No. TP-5D00-63112 (pdf 8.9 MB).

Wilcox, S., Marion, W. (2008). User's Manual for TMY3 Data Sets (Revised). 58 pp. NREL Report No. TP-581-43156 (pdf 1.7 MB).

Pickerel, K. (2019). Latest PV Module Reliability Scorecard finds that solar panels with PERC are unexpectedly degrading quickly, 4 June 2019, Solar Power World. https://www.solarpowerworldonline.com/2019/06/latest-pv-module-reliability-scorecard-finds-that-solar-panels-with-perc-are-unexpectedly-degrading-quickly/.

Raheja, D. (2019). Reliability is More Art than Science, The Journal of Reliability, Maintainability, and Supportability in Systems Engineering Winter 2018–19, © RMS Partnership, Inc., 9461 Shevlin Court, Nokesville, VA 20181.

Gevorkian, P. (2011). *Large Scale Solar Power System Design*. McGraw Hill.

Jordan, D. and Kurtz, S. (2013). PV degradation rates – an analytical review. *Progress in Photovoltaics: Research and Application* 21 (1): 12–29.

Kurtz, S., Miller, D., Kempe, M., and Bosco, N. (2009). Evaluation of High-Temperature Exposure of Photovoltaic Modules. National Renewable Energy Laboratory, K. Whitefield Miasole, J. Wohlgemuth, BP Solar, N. Dhere, Florida Solar Energy Center, T. Zgonena, Underwriters Laboratories Inc., 34th IEEE Photovoltaic Specialists Conference. Philadelphia, Pennsylvania.

Jordan, D.C., Deceglie, M.G., and Kurtz, S.R. (2016). PV degradation methodology comparison — a basis for a standard. 43rd IEEE Photovoltaic Specialists Conference. Portland, OR, USA. https://doi.org/10.1109/PVSC.2016.7749593.

Jordan, D.C., Kurtz, S.R., VanSant, K.T., and Newmiller, J. (2016). Compendium of photovoltaic degradation rates. *Progress in Photovoltaics: Research and Application* 24 (7): 978–989.

Hasselbrink, E., Anderson, M., Defreitas, Z. et al. (2013). Validation of the PV Life model using 3 million module-years of live site data, 39th IEEE Photovoltaic Specialists Conference, pp. 7–13. Tampa, FL, USA. https://doi.org/10.1109/PVSC.2013.6744087.

DOD GUIDE (2005). For Achieving Reliability, Availability, and Maintainability, Active, August 3.

GEIASTD0008 (2011). Derating of Electronic Components 2011-08-01.

IEEE Std 493™ (2007). IEEE Recommended Practice for the Design of Reliable Industrial and Commercial Power Systems.

Kececioglu, D. (2002). *Reliability Engineering Handbook*, vol. 1. Lancaster: DEStech Publications, Inc.

Kececioglu, D. (2002). *Reliability Engineering Handbook*, vol. 2. Lancaster: DEStech Publications, Inc.

Elsayed, E.A. (1996). *Reliability Engineering*. Reading: Addison Wesley Longman, Inc.

Formica, T.J., Hassan Abbas Khan, H.A., and Pecht, M.G. (2017). *The Effect of Inverter Failures on the Return on Investment of Solar Photovoltaic Systems*, October 25, 2017. Doi. 10.1109/ ACCESS. IEEE Access. 10.1109/ACCESS.2017.2753246.

Klise, G.T., Lavrova, O., and Gooding, R. (2018). PV System Component Fault and Failure, Compilation and Analysis, SANDIA REPORT, SAND2018-1743, Unlimited Release.

Klise, G and Balfour, J.R. (2015). A Best Practice for Developing Availability Guarantee Language in Photovoltaic (PV) O&M Agreements, SAND2015-10223. Sandia National Laboratories.

Dr. Deline C., Dr. Jordan D., Dr. Stein J. Photovoltaic Lifetime Project. https://www.nrel.gov/pv/lifetime.html.

Mani, G.T. (2014). Reliability Evaluation of PV Power Plants: Input Data for Warranty, Bankability and Energy Estimation Model, PV Module Reliability Workshop Golden, CO.

Kapur, K.C. and Pecht, M. (2014). Reliability Engineering, Wiley ISBN: 978-1-118-14067-3 April 2014.

System Reliability Toolkit (2005). Reliability Information Analysis Center & Data Analysis Center for Software, SRKIT, Contract No. HC1047-05-D-4005, Latest Version dtd 2015. https://www.quanterion.com/product/publications/system-reliability-toolkit-v/.

Naval Surface Warfare Center Carderock Division (2011). Handbook of Reliability Prediction Procedures for Mechanical Equipment, NWSC-11.

Abernethy, R. (2010). *The New Weibull Handbook*, 5e. Robert Abernethy Publisher.

Mil-Hdbk-338B (1998). Military Handbook Electronic Reliability Design Handbook, active, October 1998, DOD.

JA 1000, 2012-05-07 (2012). SAE Reliability Program Standard.

JA1000/1_201205 (2012). SAE Reliability Program Standard Implementation Guide.

JA 1002 200101 (2001). SAE Software Reliability Program Standard.

JA1003_200401 (2004). SAE Software Reliability Program Implementation Guide.

IEC TC 88/WG 26, Availability and reliability for wind turbines and wind turbine plants.

Morris, R.W. and Fife, J.M. (2009). Using probabilistic methods to define reliability requirements for high power inverters. Proceedings of the SPIE 7412. https://doi.org/10.1117/12.826528.

Fife, J.M., Scharf, M., Hummel, S.G., and Morris, R.W. (2010). Field reliability analysis methods for photovoltaic inverters. 35th IEEE Photovoltaic Specialists Conference (20–25 June 2010), pp. 2767–2772. Honolulu, HI, USA: IEEE.

Fife, J.M. (2010). *Solar Power Reliability and Balance-of-System Designs*. Renewable Energy World.

Amin, S.M. (2011). *U.S. Electrical Grid Gets Less Reliable*. IEEE Spectrum.

North American Electric Reliability Corporation (NERC) (2016). Reliability Considerations for Clean Power Plan Development.

IEC TR 62380 (2004). First edition, 2004-08, Reliability data handbook – Universal model for reliability prediction of electronics components, PCBs and equipment.

IEC 61709 (2017). Electric components – Reliability – Reference conditions for failure rates and stress models for conversion, Edition 3.0 2017-02, REDLINE VERSION.

PV O&M Best Practices Working Group (2018). National Renewable Energy Laboratory, Sandia National Laboratory, SunSpec Alliance, and the SunShot National Laboratory Multiyear Partnership (SuNLaMP).

NREL/TP-7A40-73822 (2018). *Best Practices for Operation and Maintenance of Photovoltaic and Energy Storage Systems*, 3e. Golden, CO: National Renewable Energy Laboratory https://www.nrel.gov/docs/fy19osti/73822.pdf.

IEC 60904-3 (2019). RLV, Photovoltaic devices – Part 3: Measurement principles for terrestrial photovoltaic (PV) solar devices with reference spectral irradiance data.

IEC 60904-7 (2019). RLV. Photovoltaic devices – Part 7: Computation of the spectral mismatch correction for measurements of photovoltaic devices.

IEC 60904-8 (2014). Photovoltaic devices – Part 8: Measurement of spectral responsivity of a photovoltaic (PV) device.

IEC 60904-10 (2009). Photovoltaic devices – Part 10: Methods of linearity measurement.

IEC 61215-1 (2016). Terrestrial photovoltaic (PV) modules – Design qualification and type approval - Part 1: Test requirements.

IEC 61215-1-1 (2016). Terrestrial photovoltaic (PV) modules – Design qualification and type approval - Part 1-1: Special requirements for testing of crystalline silicon photovoltaic (PV) modules.

IEC 61215-1-2 (2016). Terrestrial photovoltaic (PV) modules – Design qualification and type approval - Part 1-2: Special requirements for testing of thin-film Cadmium Telluride (CdTe) based photovoltaic (PV) modules.

IEC 61215-1-3 (2016). Terrestrial photovoltaic (PV) modules – Design qualification and type approval - Part 1-3: Special requirements for testing of thin-film amorphous silicon based photovoltaic (PV) modules.

IEC 61215-1-4 (2016). Terrestrial photovoltaic (PV) modules – Design qualification and type approval - Part 1-4: Special requirements for testing of thin-film Cu(In, GA)(S, Se)2 based photovoltaic (PV) modules.

IEC 61701 (2011). Salt mist corrosion testing of photovoltaic (PV) modules.

IEC 61730-1 (2016). Photovoltaic (PV) module safety qualification – Part 1: Requirements for construction.

IEC 61730-2 (2016). RLV, Photovoltaic (PV) module safety qualification – Part 2: Requirements for testing.

IEC 62670-1 (2013). Photovoltaic concentrators (CPV) – Performance testing – Part 1: Standard conditions.

IEC 62670-2 (2015). Photovoltaic concentrators (CPV) – Performance testing – Part 2: Energy measurement.

IEC 62670-3 (2017). Photovoltaic concentrators (CPV) – Performance testing – Part 3: Performance measurements and power rating.

IEC TS 63019 (2019). Photovoltaic power systems (PVPS) – Information model for availability.

IEC TS 63265 (2022). Photovoltaic power systems – Reliability practices for operation.

IEC 62446-1 ED 2 (2016). Photovoltaic (PV) systems – Requirements for testing, documentation and maintenance - Part 1: Grid connected systems – Documentation, commissioning tests and inspection.

Hobbs, G.K. (2005). *Accelerated Reliability Engineering*. Westminster, CO: Hobbs Engineering Corporation.

Mil-Hdbk 217 (1998). Military Handbook (Discontinued), used for reference method only.

Kurtz, S., Whitfield, K., Miller, D. et al. (2009). Evaluation of high-temperature exposure of rack-mounted photovoltaic modules. 34th IEEE Photovoltaic Specialists Conference (PVSC), 2009, pp. 002399–002404. https://doi.org/10.1109/PVSC.2009.5411307.

Bernstein, J. and White, M. (2008). Physics Based Reliability Qualification, NASA NEPP JPL Publication 08-5 2/08.

(2004). *Reliability Qualification of Semi-Conductors Devices Based on Physic of Failure Risk and Opportunity, Assessment JEP148*. JEDEC Solid State Technology Association.

Pecht, M., Lall, P., and Hakim, H. (2006). *Influence of Temperature on Microelectronics and System Reliability: A Physics of Failure Approach*. Lavoisier.

Ramasamy, S. and Govindasamy, G. (2008). A software reliability growth model addressing learning. *Journal of Applied Statistics* 35 (10): 1151–1168.

IEC TS *1836 ED 3 (2016). Solar Photovoltaic Energy Systems – Terms, Definitions and Symbols.

SAE J2816 (2009). Edition, December 2009.

Eke, R. and Şentürk, A. (2012). Performance comparison of a double-axis sun tracking versus fixed PV system. *Solar Energy.* 86: 2665–2672. https://doi.org/10.1016/j.solener.2012.06.006.

EPRI (2019). Application of machine learning to large-scale PV plant faults and failures. EPRI, Palo Alto, CA: 2019. 3002013671.

Gunda, T. (2020). Inverter Faults & Failures: Common modes & patterns, Photovoltaics Reliability Workshop, February 27, 2020.

IEEE 1633 (2016). IEEE Recommended Practice on Software Reliability.

TamizhMani (Mani) G.; manit@asu.edu (2014). Reliability Evaluation of PV Power Plants: Input Data for Warranty, Bankability and Energy Estimation Models, PV Module Reliability Workshop 2014, Golden, CO (25 February 2014).

Pecht M., Dasgupta, A., (1995), *Physics-of-failure: an approach to reliable product development* IEEE 1995 International Integrated Reliability Workshop. Final Report, published by IEEE 1995.

Notes

1 NERC January 2016.

2 NERC, GADS Solar Generation Data Reporting Instructions, Effective Date: January 1, 2024.

3 https://blog.greensolver.net/en/the-five-most-common-problems-with-solar-panels/.

4 Thomas Bayes (c. 1701–1761) English statistician, philosopher and Presbyterian minister who is known for formulating a specific case of the theorem that bears his name: Bayes' theorem.

5 It is highly recommended to implement SCADA systems or functions that provide for monitoring of the physical plant security as well as failure reporting and secure communications.

6 See IEEE Gold Book for latest update.

7 This assumes that the design has all the reliability attributes incorporated.

8 https://www.renewableenergyworld.com/articles/print/rewna/volume-2/issue-5/solar-energy/solar-power-reliability-and-balance-of-system-designs.html.

9 "System Reliability" Presentation, R. W. Morris, Dependable Systems Applications (DSA) Conference, Beijing China, October 2017.

10 Risk Priority Number (RPN) for Warranty Claims, Quantifying Risk for Every Defect, G. TamizhMani (Mani); manit@asu.edu.

11 Reliability Evaluation of PV Power Plants: Input Data for Warranty, Bankability and Energy Estimation Models, G. TamizhMani (Mani); manit@asu.edu, http://www.nrel.gov/pv/pdfs/2014_pvmrw_04_tamizhmani.pdf.

12 https://www.solarpowerworldonline.com/2017/06/causes-solar-panel-degradation/.

13 ASTM C1549 – 16 Standard Test Method for Determination of Solar Reflectance Near Ambient Temperature Using a Portable Solar Reflectometer Active Standard ASTM C1549 | Developed by Subcommittee: C16.30.

14 ASTM E903 – 12 Standard Test Method for Solar Absorptance, Reflectance, and Transmittance of Materials Using Integrating Spheres Active Standard ASTM E903 | Developed by Subcommittee: E44.20.

15 Photo Courtesy of NREL and Andy Walker.

16 The project must set a limit on the probability of occurrence to minimize the impact to the plant. This can also be listed as a risk and managed based on cost of redesign versus the cost of field maintenance and spares.

17 Michael D. Kempe, "Ultraviolet light test and evaluation methods for encapsulants of photovoltaic modules, Solar Energy Materials and Solar Cells," Elsevier, February 2010 https://www.sciencedirect.com/science/article/pii/S0927024809003274#!

7

Maintainability

7.1 Introduction

Among the many attributes that affect the cost of a PV system and the levelized cost of energy (LCOE) is maintenance:

- How much, how often, and what resources are needed?
- What personnel and support equipment are needed?
- What and how many spares will be required?
- When should the various maintenance tasks be performed?
- What design attributes are necessary to provide a reasonable fault detection and fault isolation (FD/FI) capability?

These are just a few of the attributes that must be decided over which the project has control, through performing systems engineering (SE) and reliability, availability, maintainability, and safety (RAMS).

To be cost effective, these decisions must be determined and executed at the earliest possible time in the lifecycle. This chapter addresses how to determine the actual maintenance task times from field data, which of four types of maintenance may be required, and some of the general considerations for improving maintainability. This includes the application of reliability data to determine maintenance rates and the identification of the items that improve the mean time between maintenance (MTBM) through scheduled maintenance. All of this is to reduce the loss of revenue. (There are a number of references that can be used for establishing maintainability programs. Housel et al. 2013, Reliability Analysis Center 2000, SAE JA1010 2011, SAE JA1010-1 2011, DOD GUIDE 2005, Blanchard et al. 1995, Smith 2011, Tortorella 2015.) These references were written as the result of learning hard lessons in reliability and maintainability over the course of more than 60 years.

Of note is that the metrics for maintainability as with reliability (Chapter 6) are based on when failure occurs and how events occur, the criticality of that event, and what is needed, such as time, tools, and spares, to return the system back to specification.

Why maintainability?: It is easy to say the operations and maintenance (O&M) contract addresses the maintainability attributes of the plant. Actually no! As stated above, maintainability, like reliability, is an attribute of the plant all the way down the indentured parts list to the lowest level of repair. This attribute must be developed during the system

Photovoltaic (PV) System Delivery as Reliable Energy Infrastructure, First Edition.
John R. Balfour and Russell W. Morris.

specification and detailed design of all the components and then managed after installation through O&M that is properly funded, contracted, and applied (Chapter 11). It is the project team's responsibility to develop a set of maintainability specifications and assure that they have been accounted for in any proposals and bidding quotes from Engineering, Procurement, and Constructions (EPCs). The Owner or the Owner Operator (O&O) must budget and contract for a solid O&M effort and track the results as a continuous monitoring of the condition, status and health of the system. There is a direct relationship between maintainability, the total LCC (https://energyeducation.ca/encyclopedia/Life-cycle_cost), Levelized Cost of Energy DOE OFFICE OF INDIAN ENERGY, 2015, Xueliang Yuan et al. 2021, the LCOE (https://corporatefinanceinstitute.com/resources/valuation/levelized-cost-of-energy-lcoe/), total energy production capability and the revenues, i.e. profitability of the plant. These critical fiscal and operational issues are primarily determined at system concept and specification, yet tend to be poorly addressed, if at all, unless an effective **SE/R**epowering™ RAMS approach is defined and implemented. This is a critical foundational element of the system delivery process.

For stakeholders, maintainability is a critical attribute for system and stakeholder success that is determined at concept and specification.

For maintainability, the mean corrective maintenance time (M_{ct}) which begins with notification of an event requiring maintenance until the repair action is completed and all "paperwork" is done, or data entry is complete. As stated in Chapter 6, mean time between failure (MTBF) is measured between failures which includes downtime awaiting repairs, i.e. its measurement is based on operating time. That is, it is the point in time from when the item is installed or its last failure to when the item fails.

Maintenance attributes are measured during the downtime or from when it fails to when it is restored. Inclusive in this definition is the need to have all the resources needed to perform the above tasks, including training, toolkits, support equipment, transportation, special test equipment, safety equipment, manpower, and *special* support equipment. All these maintenance actions need to be documented with time and material accounted for, to track the total cost of maintenance (Chapter 11). Figure 7.1 provides an example of the simplified impact of maintenance events (reliability) and availability over a 30-year life at a 95% confidence level.

The plot in Figure 7.1 was generated by using EXCEL™ Macro that simulated a plant availability based on maintenance models using an exponential failure distribution and a mixed population of repairable/replaceable parts using normal and Weibull distributions. This figure assumes that the O&M is done on a demand basis and repairs affected as soon as possible. Included is the average mean logistic down time (MLDT) and crew sizes adjusted for by the size and type of equipment and repairs being done. Logistics can have a major impact on the mean time to repair, thus affecting lost power generation (Reliability Analysis Center 2000, Tortorella, M. (2015), *MIL-HDBK-338B,* 1998).

The major decrease in availability in years 8–22 is the result of a larger number of failures associated with the increasing failure rate of components and assemblies with time. Mechanisms causing increasing failure rates are related to wear, degradation, effects of power cycling, etc. Even with design margin and redundancy, it is possible to run into maintenance issues that result in reduced availability and increased maintenance. As with reliability, the

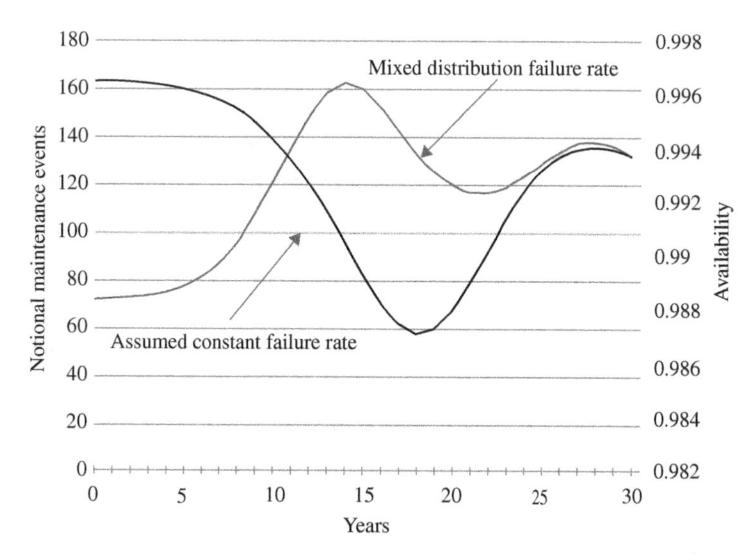

Figure 7.1 Simulated Effect of Failure Distribution on Availability and Repair on a Large PV Plant Over 30 Years (Simplified) (Source: Morris).

owner and other stakeholders need to understand the various attributes associated with maintenance to understand where the major cost elements lie within an O&M contract, other than spares.

The assumption of a constant failure rate, as shown in Figure 7.1, results in projected excessive spares costs in the first years of operation and grossly underestimates the spares starting around year seven or eight. The use of appropriate failure distributions (mixed) results in the more likely scenario of the new plant or newly installed plant equipment operating at lower failure rates. As time and stress accumulate, the failure rates increase, resulting in many more unplanned maintenance events, and required increases in the LCOE and LCC. At the same time, the plant availability to generate energy decreases, resulting in lower or decreasing income. These hits to the viability of the system affect an earlier than expected need for major unbudgeted maintenance. Additionally, there are higher spares costs, higher personnel costs, potential for liquidated damages, higher risk of demand for insurance payments, and requirement for additional funding. These early costs are significantly greater than those that could be incurred with a detailed **SE/R**epowering™ plan and implementation.

With an assumed constant failure rate, the result would be that the number of failure events per year would be a constant and the resulting availability of the plant would be a constant. The truth and/or reality is that there are some components that can claim a constant failure rate but the majority of components have other characteristics. Mixed distributions contain constant, Weibull, normal, and other general distributions. Figure 7.1 illustrates a very simple plot of a constant failure rate and a mixed distribution. One thing we learn from studies of what field data is available is that the mixed distribution curve is far more accurate than the constant failure rate one.

7.2 Responsibility for Maintainability

Often, the project team leaves maintenance up to an O&M organization and may not even attempt to address maintainability requirements. The project team must determine what a reasonable set of maintenance attributes is as well as who is responsible for maintenance actions. One specific job that is allocated to all personnel with access to the plant is one of identifying or seeing a potential problem which requires that they say something and document that observation as part of lessons learned.

All stakeholders and O&M organizations that are bidding on maintenance related services should be provided the information contained in this chapter. How will they address maintenance delay times, what kinds of spares do they propose (including the percent coverage), and how do they propose to document all maintenance actions, responsibilities, and root cause analyses?

7.3 Types of Maintenance

Maintainability consists of four basic types with unique metrics that must be specified at concept/system specification. These apply to all types of PV systems and must be defined and agreed to prior to EPC bidding, and/or when a major upgrade such as a planned **SE/R**epowering™ event is implemented during the life of the plant.

7.3.1 Unscheduled Maintenance

Unscheduled maintenance, also commonly referred to as corrective maintenance (CM), is maintenance performed to restore the system to its previous capability or use to meet specification due to one or more failures in the system. These repairs are documented as the mean time to repair (MTTR) or M_{ct} of the equipment. The various elements of MTTR and M_{ct} are discussed later in this chapter.

7.3.2 Scheduled Maintenance

Scheduled maintenance consists of two forms, scheduled maintenance (SM) and preventive maintenance (PM). SM includes tasks to be accomplished specifically based on a time to failure history for which simple maintenance can prevent failure. Examples of SM are timed replacement of fuses, transformer insulating oils, filters, and batteries that are used as power backup for the SCADA system. PM is a task of visual or other specified inspections with procedures to find potential problems and perform maintenance *before* a failure actually occurs. Examples include checking solar panels for soiling, dirt, or other accumulation, painted surfaces for erosion due to sand and dust, and erosion due to water saturating the ground around the supporting structure that may occur as the result of prolonged or significant rainfall. PMs are performed on scheduled intervals and are one of the ways to reduce the impact of failures.

The principal difference between the two is that the PM is looking for known and unknown failure modes for which there may be widely varying time to failure based on site environments or usage profiles. SM is based on *known failures* that will occur at regular intervals or under a defined set of conditions and for which a preemptive action (removal and replacement) will prevent loss of power generation or remove a potential safety problem. The SM metric is mean time between scheduled maintenance (MTBSM).

7.3.3 Software (SW) Updates Maintenance

Software has both unscheduled and SM attributes. In the former, the unit has been operating but suddenly fails under a set of specific operating conditions or profiles. Root cause analysis (RCA) finds that the software code failed because of:

- Lack of specification,
- Incorrect specification, or
- A set of conditions not previously identified for which SW is responsible.

The latter is code corrections, improvements, or changes, not deemed to affect the operation of the plant's ability to produce power (see Chapter 11). Software has a several draw backs among which is that different companies use different languages, applications specific programs, compliers, and input/out applications (Morris 2008).

7.3.4 Ancillary Maintenance (AM)

AM covers all the maintenance that may be specifically excluded in the O&M contract and yet must be accounted for in the budget (Chapter 11). This illustrates that the cost of maintenance and operation is generally what is cited in the O&M contract. The real costs, however, must include not only the contracted O&M but any and all uncontracted and unexpected maintenance and overhead to generate power to meet customer needs and to determine the "all in cost" of energy produced. This requires the Owner/Operator (O&O) to specifically budget for this "separate (ancillary) maintenance" (see below). To understand the true cost of the system over its life, one needs to account for all costs. Failing to do so misrepresents the total cost of maintenance of the plant and negatively affects the cost per kilowatt hour and LCOE. O&O should be budgeting for this maintenance and should be included in the O&M contract, whether the O&M organization performs this maintenance or it is done through subcontracting to a third-party vendor. It is up to stakeholder team to decide who "owns" AM (Chapter 11).

Examples of AM from field experience include:

- Pest control,
- PV module washing if not part of SM,
- Vegetation removal,
- Snow or ice removal,
- Other less common but often overlooked maintenance items such as road maintenance that affect budgets.

At PV plant specification and prior to EPC bidding, knowledgeable and experienced system Owners and O&M organizations need to determine specifically what technical expertise is required to perform each maintenance task.

To facilitate maintenance, there is a fundamental need for automated data acquisition (SCADA and/or DAS) as well as manual data entry performed by maintenance/site personnel. For automated data, the system needs to incorporate the ability to detect a fault (FD), identify the type of fault, notify and/or warn operators and maintenance personnel of the fault in the system. This must also provide sufficient information to the maintainers to allow for rapid, accurate, and timely fault isolation (FI), trouble shooting, and repair/return to service.

Best Practice: It is strongly recommended that maintenance/site personnel have access to automated failure data information.

Servicing an item requires consideration of the item's physical attributes as well as maintainer safety concerns since PV modules generate electricity at voltages and currents sufficient to cause injury or death if improperly handled. Table 7.1 contains a summary of the various maintenance attributes that are used to assess the support requirements of the equipment. The importance of these metrics is that they separately and together form the mathematics and data needed for modeling and analyzing a system for operational attributes as well as the financial (cost) impact of failure.

For new plants, along with the reliability and availability metrics, the maintenance attributes need to be defined clearly in the specification and the O&M contract. These metrics form the basis of performing an analysis to act as a part of the concept development via a cost–benefit trade and as a basis for establishing expected operational costs and/or O&M contract details. They also establish the level of maintenance to be accomplished (lowest repairable item level) and the maintenance priority, as well as supporting the spares analysis. Table 7.2 is an example of maintenance priority for a plant. The maintenance priority description for repairable/replaceable equipment is established through the failure modes effects analysis (FMEA) and fault tree analysis (FTA) (Chapter 6).

Table 7.2 is usually more typical of large commercial or utility systems, although the same principles apply to off-grid, commercial, and residential PVPSs. For off-grid conditions, the criticality may be somewhat more loosely defined unless there is a safety hazard, e.g. major system damage or downing event, risk of injury or death, which dictate immediate response.

Depending on the contract, the O&M may aggregate a number of priority 4 and 5 events and dispatch a team to go out and perform preventive or corrective maintenance. Chapter 11 contains a more detailed discussion of the O&M function from the administrative and operational side as well as the basis for making decisions on the front end of a project to establish how system maintenance will be handled over its lifecycle.

Issuing work orders to perform repairs is criticality dependent on the a priori classification of criticality and priority. Safety issues for example are always a Priority 1 concern. In addition, there is a power generation impact that if it exceeds an established cost–benefit threshold, it will result in potential significant repair costs and/or liquidated damages for failing to generate the power contracted.

Specific definitions of priority, such as emergency (medical versus power), should be clearly defined and included in the O&M contract, as well as the plant written processes and procedures. The procedures should identify the responsible party for approving or applying the appropriate action.

Table 7.1 Maintenance Attributes.

Metric (See Glossary)	Measurement	Attributes
Accessibility	Descriptive	Ease of access; tools, special tools, safety
FD/FI	Percent Faults Detected/ Isolate to one fault, 2 faults, or 3 or more faults	Automatic versus manual, safety, maintenance aids. Ambiguity groups (section 7.5.6 and Chapter 6)
FD/FI Time	Time	Automatic versus manual, safety, maintenance aids. Ambiguity groups
MTTR	Time	Supplier-provided MTTR, FRACAS, data analysis, assumes all equipment, spares, personnel, and Special Test Equipment (STE) are available
MTBM	Time	Supplier data, site/spares location, and planned maintenance generally applied to end items such as inverters and combiners (Chapter 4)
MTBSM	Time	Supplier data
MTBPM	Time	Supplier data during warranty, site experience after warranty
MTTIWO	Time	Delay time from notification of a needed maintenance event to when the responsible organization issues the authorizing work order
MTTO	Time	O&M/Contractor delay to get personnel on site to perform maintenance
MLDT	Time	The total delay time to begin maintenance to fully repair an item
MTBUSM	Time	FRACAS data (Chapter 4)
MTBF/MTTF	Time	Supplier MTBF/MTTF or FRACAS data (Chapter 4)
Mmax	Time	FRACAS, data analysis
M_{ct}	Time	FRACAS, data analysis, includes all of the logistical elements associated with the MTTR
Crew Size	Numeric	Number of and speciality of personnel required to perform specific maintenance
Weight	Pounds/Kilograms	Requires additional personnel or special support equipment to perform maintenance

Note: Maintenance aids are tools and software that facilitate performing maintenance.

Table 7.2 Maintenance Priority Classifications.

Priority Description	Priority Level	Example Criteria Based on Failure Consequences	Work Order	Response	Repair	Overtime
Emergency	1	Serious and/or immediate impact on DC/AC power generation or serious risk to personnel safety or damage to plant property	Immediate (same day)	ASAP (<24 h)	ASAP (<24 h)	Y
Urgent	2	Serious and/or impending impact on continued power generation of facility operation is threatened or potential for serious injury to personnel	Immediate (same day)	1 day	1 day	Y
Priority	3	Significant and adverse effect on the plant is imminent. Degradation in quality of power generation or injury to personnel	Immediate (same day)	1 business day	2 business days	N
Routine	4	Insignificant impact on power generation due to redundancy	Current or next business day	2 business days	3 business days	N
Discretionary	5	Resources are available. Impact on operation of the plant is negligible	Current or next business day	6 business days	The next mobilization for other work	N

Note: The first three priorities require that supervisory personnel must be available 24/7 to make the required timely decisions and approve the necessary work authorization or emergency actions.

For many of the initial stakeholders in a project, e.g. developer, financial, customer, insurance, owner, and safety/regulatory, the maintainability information may appear as interesting, but not useful, or believe it does not impact their area of responsibility. As we will show below, there are several elements that both the financial and insurance institutions should be aware of and track as predictions, specification, and design and development continue.

Failing to address maintainability at concept, SE specification, and design phases can result in equipment that is difficult to maintain, repair, and inspect within reasonable timelines. This adds to the downtime of the system resulting in excessive loss of power/energy production, increased maintenance costs, with an increased LCOE and O&O risk for liquidated damages.

An additional factor to be addressed early in the project concept is to identify the O&M plan and maintenance costs. These are the principal costs associated with the operation and maintenance of the plant after installation. This provides for the establishment of the "sanity" checks that should be performed to track actual cost and potential areas of cost improvements. There are several alternatives for O&M, which include the use of in-house support (self-maintained) or opt for a third-party O&M organization or to even split the O&M function into operations and have a separate company provide the maintenance (Chapter 11).

7.4 Maintenance Cost

Cost metrics of interest for all stakeholders are based on but are not limited to:

a. M_{ct},
b. MTTR,
c. MLDT,
d. Maximum maintenance time (CL),
e. Spares,
    i. Spares forecast – types and numbers,
    ii. Spares cost,
    iii. Ambiguity MIL-STD-1388-2B (1998) Group Size.
f. Average repair cost,
g. Maintainer costs including overhead (burdened labor rates),
h. PM interval and time to perform,
i. SM interval and time to perform,
j. AM interval and time to perform,
k. Maintenance material storage costs,
l. Overhead not accounted for in burdened labor rates,
m. Reserve for unplanned costs (as a percentage of the maintenance budget).

Of the above, the ambiguity group size can have a significant effect on the type and number of spares as well as the cost of a repair action. If the testability attribute of the plant does not provide for FI to the lowest level of repair, then to minimize downtime, it is necessary to take a number of possible spares and perform multiple unit replacements to get the plant back to full operation. The alternative is that not having the requisite spares necessitates multiple repair trips and/or delays awaiting a spare. Design for maintainability (DFM) (Gullo and Dixon 2021, Smith 2011, Blanchard et al. 1995) requires that the plant, as a whole, be able to minimize the number of spares on hand needed to affect repairs. As a general rule, by specifying that the plant ambiguity groups are capable of FI to the following lowest level of repair, it is possible to lower overall repair costs. Examples of FI requirements are:

i. Isolate to One (1) 60% of the time,
ii. Isolate to Two (2) 30% of the time,
iii. Isolate to Three (3) or more 10% of the time.

The FI requirement then provides a way to minimize the number of spares needed to keep the highest availability of the plant. All the above have either direct or indirect impact to the cost of maintenance, the LCOE, and LCC. The inclusion of the M_{ct}, MTTR, and MLDT for the nontechnical reader is that these are the foundation of many currently missing cost metrics.

7.4.1 Run to Failure

There is one approach to maintenance that has a fatal flaw in a fiscally sound system, and that is letting the plant accumulate failures until the plant or a portion thereof fails. At this point, the O&M organization would go out to the plant to repair *all* the *known* failures and return the plant to operation. This approach is flawed from two underlying assumptions:

1. That by aggregating all of the work into a single maintenance event (which it is not) that money can be saved, and downtime reduced through a concerted effort potentially using overtime to reduce downtime, and
2. That all the needed spares and equipment resources are available to perform the maintenance in an expedient manner.

This approach results in the notion that all the maintenance can be classed as a single event while the reality is that it cannot. An extreme example of this approach is the commonly referred to repowering, revamping, or refurbishment, where substantial elements of the plant or the whole plant are allowed to deteriorate or fail before action is taken for correction. After the plant has deteriorated or run to failure, pouring massive resources into bringing the plant to full operation is required to make it fully operational.

Each repairable/replaceable item on which maintenance is performed must be documented as a separate job with the time, resources, and equipment documented along with the fault or actual fault and its consequences. The need for individual documentation for each failure is required to assess failure trends, maintenance trends, spares forecasting, and maintenance planning for the planned maintenance period, most often a year.

The significant impact to the costs is that:

1. It is possible and common to run out of spares. This will result in delays in repair until the needed components can be purchased and delivered. This necessitates at least one additional maintenance event thus no real savings.
2. Working crews on standard shift and working on overtime tend to lead to maintenance-induced errors that result in more broken equipment needing repairs. Overtime premiums are typically 50% or more, especially if that work exceeds either maximum hours per week or occurs over a weekend.[1]
3. The cost in lost revenue while the plant is down for extended time, can lead to potential penalties, including liquidated damages. The penalties can be imposed by the entire affected energy chain including regulatory, depending on the local, regional, state, or even national entities that have an AHJ.

The cost of and the specific maintenance terminology used in the maintenance domain revolves around the actual tasks that are mandated by contract or specification to be performed due to failure or the need to perform maintenance to prevent failures. IEC 63019 (Availability) has a basic timeline that should be used as a minimum to establish the time to repair. Figure 7.2 is an expansion of the IEC timeline to include more detailed tasks making up a maintenance action.

7.5 Typical Maintenance Flow

Although Figure 7.2 provides a typical maintenance flow, it is not the only flow that the maintenance can take. Most references and books provide versions of this flow. Repair sequences are set by the type of organization that is performing the maintenance as well as what the Owner has contracted for in terms of maintaining plant performance.

7.5.1 Fault Detection and Acknowledgment

Fault detection (FD) – the clock time when the fault is detected either visually, aurally, or by remote detection such as temperature, current and/or voltage detection, or sensors connected to the SCADA/DAS, visual, IR, or other spectral sensing, and the failure is reported by maintenance personnel. Maintenance in this case can be a centrally located management center, or directly to a shop or the home office of the technicians/engineers that

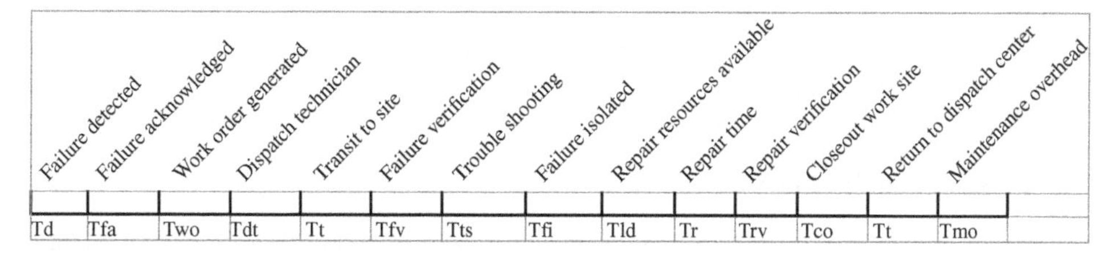

Figure 7.2 Typical Maintenance Flow for Field Repair (Source: Morris).

perform the plant maintenance. This is the start of the repair clock. For systems with automated detection and reporting, this time can be measured in seconds. Where the fault is detected through other maintenance activities, it can vary from minutes to hours or even days.

7.5.2 Work Authorization Delay

Work authorization delay or mean time to issue work order (MTTIWO) (7.1) is the time it takes for the work center to be notified of an event and generate the work authorization for the O&M organization to perform the maintenance. In its simplest form, the O&M minimizes this by having all the requisite tools, spares, and personnel working for them.

However, it is not uncommon for the O&M to control the administration of the contract and to subcontract the work to third-party maintainers. These third-party personnel can be electricians, computer technicians, mechanical technicians, etc.

$$\text{MTTIWO} = T_{\text{ND}} + T_{\text{AW}} \tag{7.1}$$

where

T_{ND} = notification delay time
T_{AW} = time to generate a work authorization

The notification delay is generally but not exclusively related to the time it takes the O&M or maintenance organization to be notified that there is/are item(s) that requires maintenance. The work authorization time is the management overhead time that covers the need for a maintenance supervisor or even the O&O to authorize the work to be performed. Failing to have a maintenance supervisor available 24/7 can result in delays of 8–16 hours or more for weekend unavailability.

7.5.3 Mean Time Till Onsite (MTTO)

Mean time till onsite (MTTO) (7.2) is the measure of the maintainer's response time from when they receive a work order until they are on site and includes the time needed to pick up any support equipment, order and receive spares (if not readily available), and travel time.

For cost accounting, the repair team cannot perform maintenance until on site. Final closeout of the event is when the team is back at the shop or point of dispatch.

$$\text{MTTO} = T_{\text{R}} + T_{\text{T}} \tag{7.2}$$

where

T_{R} = response time to notify and have maintenance personnel enroute to site
T_{T} = Transit or travel time to the site

7.5.4 Equipment Delay Time (EDT)

Equipment Delay Time (EDT) – the time it takes for the replacement part for the failed item and any additional time needed to get necessary test equipment to the site to start the fault verification and FI.

7.5.5 Fault Verification (FV)

Fault verification is the time required to gain access to the equipment and verify that the equipment itself is reporting/displaying a failure. This may include the use of test equipment to take measurements such as voltage, current, and temperature.

7.5.6 Fault Isolation (FI)

Fault isolation time T_{FI} – the time needed to isolate the failure to the failed component(s) or group of components[2] of which the failed item will be included. It includes performing a fault verification. It is possible – especially on computer-based systems – that it is difficult to determine precisely which one of two or more units are defective or failed. A group of components in this context is called an ambiguity group. (MIL-STD-1309D, 1992 – A definition of an ambiguity group as (i) a group of replaceable items which may have faults resulting in the same fault signature, and (ii) the group of items to which a given fault is isolated, any one of which may be the actual faulty item.)

Therefore, the repair action (needed to minimize downtime) is to swap out parts of the ambiguity group to determine repair. Upon repair they take all the units back to an intermediate repair facility to determine which unit has failed. This requires the technician to have test/support equipment that allows them to disconnect portions of the path and then perform tests to verify that it is working or not.

7.5.7 Mean Logistic Delay Time

MLDT (7.3) is a major element to be considered for maintenance in the power plant. This is the average amount of time it takes to get an item needed to perform the repair. A contributor to this is the number of spares that are kept on hand to perform maintenance efficiently. The MLDT considers all the *non-maintenance time* such as lead time from a vendor, delay time awaiting spares, and overhead needed to support all the maintenance (scheduled and unscheduled) performed on the plant.

$$\text{MLDT} = \frac{\sum_{i=1}^{N} (T_c + T_T + T_W + T_{WS})_i}{N} \tag{7.3}$$

where

T_c = time to contact supplier or warehouse
T_T = time to transport
T_W = incoming receiving and additional storage time if needed
T_{WS} = time to transport unit to work site
N = the number of maintenance events in a specified period that required logistic support

7.5.8 Repair Time

Repair time T_R – the time needed to remove the failed part and repair it with the replacement part. The repair time is specifically related to the average repair time for an item at the lowest level of repair. Repair time can be the supplier's MTTR provided it is supported by documentation. Otherwise, the repair time can be acquired from data entered by the technician at the time of the repair from the field.

7.5.9 Repair Verification Time

Repair verification time T_{RV} – the amount of time required to power on the unit and verify that the repair was successful. With completion of the repair verification, the unit is then connected back into the DC or AC circuit path to restore power generation.

7.5.10 Overhead Time

Overhead time T_{OH} – the total maintainer time needed to document and accept that the repair is complete and secure the facility and return the repair personnel to their station(s). This can be further extended by special events such as safety, medical, or other special situations. Stations in this context refer to the point of dispatch, such as maintainer's residence, a hotel room, or a central shop located in an urban area.

7.5.11 Minimum Maintenance Time

As previously mentioned, there is no such thing as zero maintenance time. The minimum maintenance time (MMT) (7.4) includes all initiated maintenance actions starting with the notification of a failure whether automatically through the SCADA or by field personnel reporting.

For the Immediate Response Category, there is potentially the need to shut off some portion of the power generation or supply capability as well as performing other overhead maintenance functions needed to clear the maintenance event. If there are no onsite personnel or storage of parts needed for repair, then every maintenance event has an element of delay built in to travel to the location where the spares reside[3] and pick up the anticipated spares based on the level of detail in the failure report.

$$\text{MMT} = t_{\text{RPT}} + t_{\text{Verify}} + t_{\text{Doc}} \tag{7.4}$$

where

t_{RPT} = time to review failure report
t_{Verify} = time to perform verification
t_{DOC} = time to document

7.5.12 Time to Repair

TTR_i (7-5) is the time to repair for each maintenance event (i). This equation represents the maintenance repair time at the lowest level of repair. It should also be noted that like reliability calculations, they can change over time because of plant degradations that affect various aspects of the repair time.

MTTR is generally information that the supplier, OEM, or manufacturer supplies that identifies the time needed to perform the repair with all parts, support equipment, and personnel available to perform the task. This manufacturing analysis can be based strictly on prior historical data for a prediction or by using data collected during prototype design and development. To ensure consistency, the data from the manufacturers or suppliers should include the elements in Eq. (7.5).

$$\text{TTRi (I)} = T_{\text{FV}} + T_{\text{FD}} + T_{\text{FI}} + T_{\text{Access}} + T_{\text{Remove}} + T_{\text{Replace}} + T_{\text{Verify}} + T_{\text{Close}} \tag{7.5}$$

where

T_{FV} = time to perform fault verification
T_{FD} = time to perform FD
T_{FI} = time to perform FI
T_{Access} = to gain access to the defective item
T_{Remove} = time to remove the defective item
T_{Replace} = time to replace the item with a spare
T_{Verify} = time to verify the repair
T_{Close} = time to close up the system/subsystem

Note that TTRi (I) is not weighted by the failure rate or MTTF/MTBF of the item. This is strictly a measure of how much time it takes to return to service, an item needing repair assuming that all required personnel, support equipment, spares, and test equipment are immediately available for a single event. MTTR (7.6) is the mean repair time for single items such as specific components or end items. It does not account for spares, personnel, and other logistic factors.

$$\mathrm{MTTR} = \frac{\sum_{i=1}^{m} \lambda_i \mathrm{TTRi\ (I)}}{\sum_{j=1}^{m} \lambda_j} \tag{7.6}$$

7.5.13 Mean Corrective Maintenance Time

The mean corrective maintenance time (M_{ct}) (7.7) and (7.8) for "m" items accounts for the response and transit times and can be calculated by determining the time to repair (7.7) for "I" items:

$$\mathrm{TTRi\ (T)} = T_{\mathrm{response}} + T_{\mathrm{transit}} + T_{\mathrm{FV}} + T_{\mathrm{FI}} + T_{\mathrm{EDT}} + T_{\mathrm{LDT}} + T_{\mathrm{R}} + T_{\mathrm{RV}} + T_{\mathrm{OH}} \tag{7.7}$$

where

TTRi (T) = total time to perform corrective maintenance for the ith item
T_{Response} = the time for a repair team to acknowledge the maintenance call and obtain the relevant data
T_{Transit} = the time for the maintenance team to reach the plant/to perform the repair
T_{EDT} = Equipment delay time or the time it may take for special test equipment (STE) or support equipment to arrive at the plant
T_{FI} = FI time
T_{LDT} = Logistics delay time
T_{R} = Repair time
T_{RV} = Repair verification
T_{OH} = Overhead time per documented maintenance event

And this results in (7.8):

$$M_{\mathrm{ct}} = \frac{\sum_{i=1}^{m} \lambda_i \mathrm{TTR}_i\mathrm{(T)}}{\sum_{j=1}^{m} \lambda_j} \tag{7.8}$$

In this book, the authors use the terms MTTR and M_{ct} to indicate different values of maintenance. M_{ct} covers *all* aspects of a maintenance action including the response times, delay times, etc., while MTTR covers only the hands-on time to actually repair the item and return it to operation. Some texts and books use the terms interchangeably. For the stakeholders, the difference is important as the MTTR can be predicted and documented by suppliers, while the suppliers cannot predict or even guess what the various time elements of M_{ct} is, determined by field data for a specific site.

An example of calculation of the data in Table 7.3 is a simulation of total maintenance times with a minimum maintenance time of 20 minutes. The M_{ct} is 2.7 hours with an M_{max} (90% confidence) of 11.7 hours. Data from a supplier may indicate that the repair time (MTTR) is on the order of 45 minutes to an hour, while the field data models would indicate 2.7 hours. This is illustrating that maintenance costs based on a supplier-provided MTTR can be off by a factor of 3 or more.

O&M organizations can work off their history, provided they have data to support their maintenance time predictions. The data for a specific plant will of course provide a more accurate maintenance cost assessment. The simulation data in Table 7.3 illustrate the attributes of MTTR and M_{ct}, i.e. the data are log-normally distributed and skewed to the left to account for maintenance times that far exceed the mean. Figure 7.3 is a plot of the data in Table 7.3.

As shown, the major area of interest is for maintenance at less than 10 hours. This truncation only provides a better understanding of the general maintenance time attribute. As plotted, the mode is approximately 1.3 hours while the MTTR is over twice the mode and as expected the value of M_{ct} is greater than the MTTR. M_{max} (95) or the longest expected maintenance time at 95% confidence is approximately 14.5 hours.

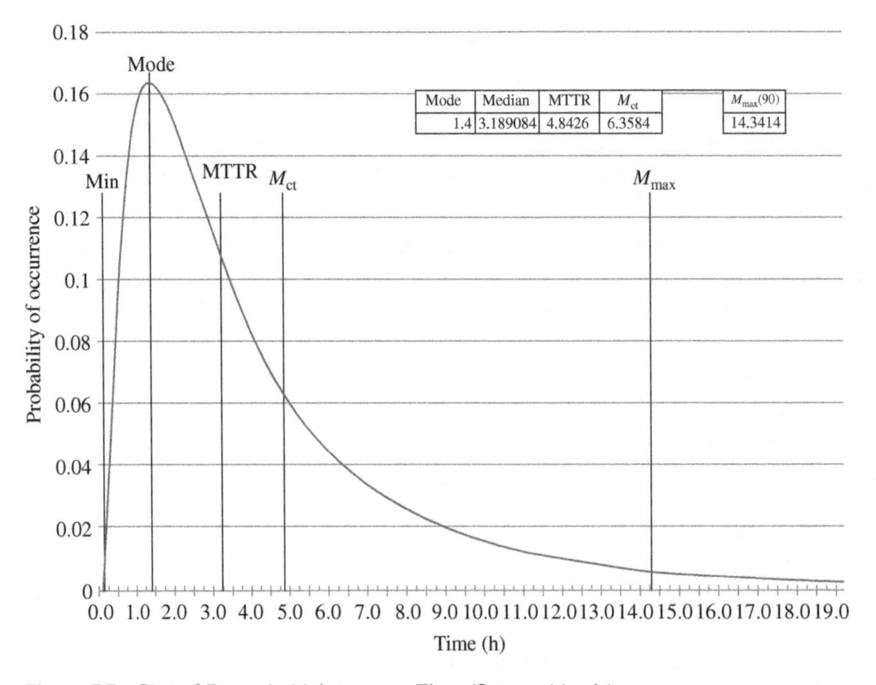

Figure 7.3 Plot of Example Maintenance Time (Source: Morris).

Table 7.3 Example Data for Maintenance Analysis.

Item	TTR-Hour			Frequency		
j	t_j	$\ln(t_j) = t'$	$(t_j')^2$	n_j	$n_j t_j'$	$n_j(t_j')^2$
1	4.6	1.5261	2.3288	4	6.1042	9.3154
2	1.2	0.1823	0.0332	3	0.5470	0.0997
3	9	2.1972	4.8278	4	8.7889	19.3112
4	4	1.3863	1.9218	2	2.7726	3.8436
5	1.8	0.5878	0.3455	1	0.5878	0.3455
6	6.2	1.8245	3.3290	2	3.6491	6.6580
7	6.2	1.8245	3.3290	5	9.1227	16.6449
8	0.7	−0.3567	0.1272	4	−1.4267	0.5089
9	8	2.0794	4.3241	5	10.3972	21.6204
10	2.1	0.7419	0.5505	1	0.7419	0.5505
11	3.1	1.1314	1.2801	5	5.6570	6.4004
12	6.7	1.9021	3.6180	5	9.5105	18.0901
13	6.1	1.8083	3.2699	4	7.2332	13.0796
14	2.7	0.9933	0.9865	2	1.9865	1.9731
15	2.6	0.9555	0.9130	2	1.9110	1.8260
16	3.2	1.1632	1.3529	2	2.3263	2.7058
17	7.2	1.9741	3.8970	5	9.8704	19.4850
18	2.6	0.9555	0.9130	5	4.7776	4.5650
19	0.1	−2.3026	5.3019	3	−6.9078	15.9057
20	0.4	−0.9163	0.8396	2	−1.8326	1.6792
21	2.3	0.8329	0.6937	2	1.6658	1.3875
22	1.8	0.5878	0.3455	3	1.7634	1.0365
23	2.6	0.9555	0.9130	2	1.9110	1.8260
24	3.7	1.3083	1.7117	2	2.6167	3.4235
25	4.4	1.4816	2.1952	3	4.4448	6.5855
26	7.4	2.0015	4.0059	2	4.0030	8.0118
27	8.1	2.0919	4.3759	5	10.4593	21.8795
28	2.7	0.9933	0.9865	5	4.9663	4.9327
29	0.3	−1.2040	1.4496	3	−3.6119	4.3487
30	2.8	1.0296	1.0601	1	1.0296	1.0601
		29.7363	Sum	94	105.0648	219.0996
				47		
		M_{ct}	Median	M_{max} (90)	st	$/t'$
		2.3701	3.0578	11.6774	1.0456	1.1177

7.6 Additional Maintenance Metrics

A major attribute of maintenance for any system is that there is no such thing as zero maintenance time and when plotted, maintenance time appears to be log-normally distributed (7.9).

Thus, to assess the median (50th percentile) time to repair it can be found as

$$\text{MRT}(50\%) = \exp\left(\frac{\sum_{i=1}^{n} \lambda_i \ln M_{CT_i}}{\sum_{i=1}^{n} \lambda_j}\right) \tag{7.9}$$

The 50th percentile implies that half of the maintenance can be less than the median while the other half can be greater than the median. This is the actual values of the maintenance times needed to calculate costs, but it does provide an idea of what the actual values may be.

In addition, a good maintenance plan will include the worst-case maintenance time, based on an acceptable risk confidence level, such as 90% or 95%, that can be expected, excluding the potential for external factors, but including some elements of the MLDT (7.10). This can generally be defined as

$$M_{\max}(90\%) = \exp^{(M\text{ct}+z\sigma_{\text{CM}})} \tag{7.10}$$

where

T_{CM} = mean of the log of the maintenance times
σ_{CM} = the standard deviation of the mean of the log of the maintenance time
z = a numerical measurement used in statistics (Haldar, A., Mahadevan, S., 2000) of a value's relationship to the mean (average) of a group of values, measured in terms of standard deviations from the mean based on the confidence level desired. See Chapter 6.

The specific point is that the failure distribution is assumed to be log-normal with the median (50%), mean (average), and mode approximately equal. These values and the minimum and maximum (assuming a certain confidence level) maintenance times are based on data provided in the FRACAS reports (Chapters 5, 10, and 11) to generate the statistics.

For the case shown in Table 7.3, the mean corrective maintenance time is assumed to be log-normally distributed and the maximum maintenance time assumes that 90% of the maintenance can be done in less time. The remaining 10% of maintenance times can be anywhere from 10s of hours to weeks or months. A special note for Figure 7.3 is that it illustrates that there really is no such thing as zero maintenance time. Just answering the phone for a complaint or making an actual maintenance call consumes some finite amount of time that needs to be accounted for in a plant's analysis.

7.7 Available Maintenance Time

Although maintenance can be done any time during a 24-hour day, the nighttime hours are generally the safest time for working on modules, electrical wiring, electrical panels, and combiners. Provisions need to be made for sufficient lighting in work areas. Available maintenance time is the time scheduled for general maintenance and is set by those performing the maintenance contract. Emergency maintenance will of course be performed when required. It varies over the year which is developed from the plant's latitude, i.e. how far north or south one gets from the equator in degrees. Performing maintenance at night does not imply that there are no maintainer safety risks. Night by general definition is the time from dusk to dawn. It occurs when the sun reaches 18° below the horizon and no longer illuminates the sky. An aspect of performing maintenance at night is that if there is solar illumination of some form (twilight hours), there is the potential of power being generated by the solar cells. There are three distinct named sunsets (twilights) and a specific definition of dusk:

- **Civil** twilight, which begins once the sun has disappeared below the horizon and continues until it descends to 6° below the horizon.
- **Nautical** twilight, between 6° and 12° below the horizon.
- **Astronomical** twilight when the sun is between 12° and 18° below the horizon.
- **Dusk** occurs at the darkest stage of twilight, or at the very end of astronomical twilight after sunset and just before nightfall.

Of these, nautical and astronomical twilight should be used for maintenance on module/panels unless modules can be covered and easily disconnected from the string. Noncritical maintenance can be done during plant off hours, which can vary significantly depending on latitude.

The various twilights result in differing amounts of solar energy, all of which can generate the potential energy (voltage) of the modules or strings and actually only really differ in the amount of current generated as a result of that exposure. Once the sun is 18° below the horizon (not to be confused by being below or behind a mountain range), then power can be generated but full power from the modules/panels is not possible. When the sun is behind a mountain range (or building that casts a shadow), there is a potential for strong sky glow or reflection that will still allow for a high level of power generation.

Best Practice: Regardless of the maintenance to be performed, the modules/panels and combiners should always be considered electrically HOT and daylight safety measures applied.

This includes the use of PPE as well as insulated tools and proper equipment grounding as needed. This is true even when maintenance is done at night as the lighting needed to perform such maintenance may provide enough luminance to generate minimal power (1 Watt/M^2). Inverters may not receive enough DC power to operate; however, it is possible to create enough power to electrically shock someone.

7.8 Maintenance-Driven Availability

Availability of the plant, that is the operational availability, is optimized based on its design, manufacturing, and installation reliability and maintainability attributes. The optimized design uses the design for reliability (DfR) and design for maintainability (DfM). The plant maintenance attributes have several features that include those previously mentioned of good FD/FI, ease of access, and common tools, and having sufficient human resources, and spares readily available.

Plant over-design, often assumes incorrectly, that plant degradation is the only or primary failure mode related to PV modules. The reality is degradation affects all components from inverters (power components), fuses, circuit breakers, and connections. A common error is to presume that the modules are degrading at published cut sheet values. Degradation alone is not the only source of reduced module power output over time. There are other quality/reliability issues, such as component selection and mismatch that affect system output. For example, assuming a 1% degradation rate, a 25-year module event replacement would require a minimum 25%+ over-design to compensate just for the module degradation. This however does not address degradation of the inverter and other components nor the early failure of these components from overstress, under or incorrect specification and of course damage from installation and handling. Spare availability that covers the future, greater than 5 years, makes sparing exceedingly difficult as degradation of one form or another also occurs for items sitting unpowered on a shelf awaiting installation.

From an operations standpoint it is important to have an effective **SE/R**epowering™ plan that identifies and describes the time elements, triggers or conditions to detect the occurrence and hooks to enable a **R**epowering™ event. Maintenance planning must also account for site conditions, and major events such as force majeure, as well as accelerated wear and tear, and design (inherent) induced failure modes. This requires collecting and analyzing component field data that must be used to update the FMEA and adjust the **SE/R**epowering™ plan as needed.

7.8.1 Crew Size

Crew size (CS) is one area where many Owners and O&M organizations attempt to reduce costs. Instead of sending out a two- or three-person team, which is necessary for maintenance and safety, they only send a single person or maybe two. Even where more than one person is required (e.g. NFPA 70E, Utility requirements, OSHA, O&M contract, etc.), these requirements are often ignored, resulting in safety and legal risk. The crew size should be dependent on tasks to be accomplished safely and identified in the O&M specification and contract, including personnel, property, safety, and protection, etc. We have been part of many discussions with numerous O&M companies and self-providers, where they have stated that they will only send out a single person to perform some tasks such as changing out cooling filters or to perform framework exterior rust or corrosion inspections. This can be short-sighted as the problem may be more complicated and the organization will have to send them back to complete the maintenance – or worse, have an employee injured on the job. Or, in some cases, the maintenance organization will have to send a second person to the plant to help the first if there are problems found requiring two people to perform the task. Failing to properly staff qualified maintenance personnel presents a certain level of risk for maintenance personnel and risk of damage to property as well as civil and/or criminal liability for the system and property Owners, including the O&M organization. In the process of attempting to reduce costs, unforeseen and unwanted expenses (e.g. additional mobilizations, legal, safety, etc.) may be triggered, which can complicate and increase costs instead of reducing them.

If these issues are not completely addressed during specification in the concept and SE phases, there are a whole series of maintenance (labor and component) costs that will accrue with time. These are not generally accounted for in the planned budget or even covered by a contingency budget. The result is a loss of revenue due to reduced energy production, or increased unplanned maintenance costs or both.

Crew size is also dictated by the maintenance being performed. In areas where winds are generally expected, then the wind loading on a module may make it impossible, or at least very difficult, for a single maintainer to remove and replace a module. When working on uneven or sloped surfaces, the difficulty of repair is exacerbated.

7.8.2 Plant Maintenance Time

Plant maintenance time requires considerations such as the priority/criticality of the maintenance, the available human resources, spares availability, and consideration for adding teams of people to handle any aggregated maintenance. Combined with a need to perform maintenance at the lowest cost, the O&M or maintenance organization is faced with a tough decision. Do they perform maintenance during daylight hours when there is a higher probability of electrical safety issues; or perform the maintenance in the off hours (nighttime) when visibility issues can increase the risk of improper or incorrect repair or missing a problem completely due to reduced visibility? Haight (1967).

7.8.3 Accessibility

Accessibility is an attribute of the specification and design that directly impacts the maintainability of an item (Reliability Analysis Center, 2000, Blanchard et al. 1995). Accessibility is the ability to access the equipment for repair and upkeep (cleaning[4] for example) using standard tools and without having to remove other equipment or parts to affect a repair. An example of improper cleaning that can induce failures is shown in Figure 7.4. In addition, accessibility, cost, and safety are interrelated, as the more difficult it is to reach a failed part for repair, the greater the chance for equipment damage and/or maintenance personnel injury or death. Figures 7.5 and 7.6 illustrate the principal differences between maximizing the plant for power/m^2 and a more balanced plant design allowing for ease of maintenance.

The value of including accessibility in the plant specification and design affects one-site attribute, namely plant size in area (acreage). For a given contracted power and systems design, the space needed for the module solar collection area (kW/m^2), inverters, combiners, operations shed, and maintenance access drives the overall plant size which needs to be accessible to sunlight (no or limited shading) and space for the maintenance crews to work in safely. For the maintainers to work inside the inverter unit, the output transformer cabinet for example, inverter shell, and cabinet designs will need to get progressively larger to provide adequate working space to perform maintenance actions safely. This is especially true for the large (>250 kW) units.

At the same time, to provide access to the PV module requires that they be separated sufficiently to allow for large equipment access. Although single modules of less than 100 cells can be handled by a single maintainer, larger modules will require two or more field service technicians to perform removal and replacement or repair to the structure. There can be a real cost–benefit in the proper balance of technicians, especially if single technician performing maintenance, resulting in additional damage to the modules handled (glass or

Figure 7.4 Improper Cleaning Resulting in Induced Failures (Water) (Courtesy of Andy Walker, NREL).

Note from the picture that this design has poor maintainability attributes (changing modules and accessing power wiring for example), there appears to be morning/evening shading, the structure appears weak (susceptible to wind damage) and during rain storms the lower modules will receive dirt splatter from close proximity to ground.

Figure 7.5 Low-Latitude Row Separation (Courtesy of NREL – Andy Walker).

cell cracking, etc., or injury) or damage to other equipment. The US National Institute for Occupational Safety and Health (NIOSH) has recommendation[5] for lifting limits to prevent injury to people who must lift heavy objects as a part of their jobs. This is further complicated by bifacial modules that have glass on the front and back, resulting in much heavier modules that exceed the one person lifting limits. These also have the potential for breakage on both surfaces of a module. Consideration must also account for the future potential of larger inverters, which may require greater special equipment access and could become a shading issue, while the solar collection area efficiency or fractional average of irradiance may decrease.

Fixed tilt module installations generally have retention clips holding the modules or panels in place on the framework. Designing the framing/support for standard modules measuring a meter or so in length and with good human engineering should provide ease

Note: This structure can handle snow load, it is high enough to not have rain splatter, and it has ease of maintenance access to modules, inverters, and combiners.

Figure 7.6 High-Latitude Row Separation. (Courtesy of NREL – Andy Walker).

of access. However, when 3–100 modules are placed side by side and two or more rows deep, access will be difficult, affecting safety and raising the potential for damage to the equipment. This also requires space for support equipment such as hard stands or cranes capable of attaching to and lifting modules to/from the frame.

Accessibility for maintenance is impacted by the module or panel angle of inclination[6] from a ground reference. Ground reference considerations are required as many plants are built on flat ground while others are not. This compounds the accessibility for removal and replacement, but also affects the ability to perform preventive and SM. The higher latitudes require greater separation between rows to reduce or eliminate as much of the module shading effects as possible. Reducing shading on PV modules also extends their life while providing additional energy/cashflow. The case worsens for maintainability access in lower latitudes where shading from higher sun angles allows for greatly reduced separation and designers are tempted to create higher densities of modules to improve the ground coverage ratio (GCR).[7]

As noted in Chapter 5, bifacials have two energy collecting surfaces on the front and back of the module. This requires additional row separation and clearance for maintenance handling and shading from not only surrounding man-made and terrestrial obstructions but from undergrowth that can affect the albedo. Care during maintenance must be taken to avoid putting excess pressure on either surface or damaging the back side of the module during PM and cleaning. For maximum performance over time, bifacials require that cleaning be done on both the front and back surfaces with attendant accessibility and capability to verify cleanliness.

Based on this information, we can make an estimate to determine the minimum acceptable and functional spacing for maintenance. It should be noted that principle spacing and accessibility requirements are derived to meet shading requirements (also see Chapter 2) and the need to account for the variability in the size of maintainers and the tools and special equipment needed to perform maintenance. This latter item is based on using the 95% tile male/female population ranging in sizes from 5′ to 6′2″ and 100 to 220 pounds. A good source of information on for these attributes is found in Mil-Hdbk-1472. This topic is often not adequately addressed during the concept or even system specification stages and can create situations where only limited stature field service technicians can perform certain jobs.

GCRs, or the total area of photovoltaic array to the total ground area, cause problems for maintainer accessibility in general and also impact the need for special support equipment to facilitate maintenance. High GCRs imply restricted access to module/panel frame mounting hardware and the need for extreme care and caution so that the maintainers do not induce failures in adjacent equipment during routine maintenance. This has the potential to increase annual maintenance costs by tens to hundreds of thousands of dollars just for the special support equipment let alone the added maintenance time to access and repair the equipment.

Potential for clarifying determinant for minimum shading obstruction, for all PVPS (fixed tilt, one, two and three axis tracking), is to perform a site shading study using the winter solstice for the northern hemisphere, the summer solstice for the southern hemisphere at "solar noon" to define the site exposure angles at sunrise and sunset and throughout the day. This is one area where any calculations need to have an onsite visual inspection/review and survey to ensure that all sources of potential shading have been identified. This includes consideration of things such as trees that continue to grow and the construction of new buildings that can produce shading years into operation. The winter solstice angles provide for the module/panel worst-case angles needed for minimizing the effects of shading around the prime power generation hours based on solar noon.[8] During the summer, failing to shade the inverter at many plant sites results in excessive heating, resulting high work surface touch temperatures as well as reducing the reliability of the unit. Prime power generation hours need to be clearly defined early on in the concept and system engineering phases. It is up to the owners to determine which hours are actually going to meet customer needs and to not use convenient AM/PM terminology. This has an impact on the SE effort as well as the financial energy production values. Solar noon is seldom coinciding with 12 PM noon local time.

Other considerations may include extended "shading" due to early morning or afternoon fogs or marine layer clouds. For fixed tilt or even tracking PV installations, the spacing requirement is to minimize shading to maintain power generation. The common mistake is for plant designers to optimize for winter solstice using standard or daylight saving time zone noon instead of solar noon. Fixed tilt installations properly sited for solar noon at the winter solstice.[9]

Note: Some aspects of maintenance accessibility are driven by the design requirements themselves. Recommendations for spacing based on maintenance:

1. All support and STE have been identified and sizes accounted for in the row spacing and ground clearance, including potential future equipment.
2. Support equipment may all fit into a Ford F250, for example, with a clearance of at least 10 feet needed for movement in and around the pickup.
3. Sufficient space has been allocated for 95% tile personnel to gain access to plant components.
4. For one- and two-axis trackers with central and string inverters for example, a crane is necessary; therefore, space needs to be allocated for the crane and any stabilizing legs

used when lifting large, heavy loads. Some prefabricated panels mounted as a subarray used on two-axis tracking systems measure 5 × 10 m or more and can weigh several thousand pounds.[10] Modules alone can exceed 60 pounds or 27 kg.
5. High-voltage AC for example requires safety barriers and special shutdown and work area isolation procedures. These take more space and special tools. High-voltage DC connections are often more accessible but still require special handling and tools to ensure safe handling of the equipment. Insufficient spacing can result in creating a "kill zone."
6. When troubleshooting and performing emergency repairs to speed up the return to service, there is a tendency to take the most probable spares that would address the fault or failure data. This can require greater working space or multiple special tools for the different components. This is defined as an ambiguity group repair.

An ambiguity group is the set of components that a fault can be isolated to and, unfortunately, generally requires swapping out all the members of the group to affect repair with a minimum of downtime. This generally stems from a lack of FD or in designs with functions distributed over several cards or assemblies. Using multiple spares from an ambiguity group also allows for "manual" FI such as a split-half method.

This is where siting is critical. If the plant can be situated on the south (north in the southern hemisphere) side of a hill shading, separation distance can be minimized while allowing space for maintenance. It should be noted, however, that maintenance (preventive, scheduled and/or unscheduled) becomes difficult for personnel and equipment with much more than a 3% grade. The principal concern for heavy equipment such as lifts or cranes is the potential for a tip-over hazard and/or damage to other system elements.

7.9 Preventive Maintenance (PM)

Preventive maintenance is the leading method to identify *developing problems* in the plant when sensor or automated FD is not possible or practical. These may be identified and reported by the owner operators or other site personnel as well as during the performance of PMs by O&M technicians. With few exceptions, during the performance of PM's, there is no repair/restore maintenance actions unless they can be fixed quickly and safely completed with the personnel available in the time allocated.

Table 7.4 contains an example set of PM routines and their intervals that can be used in a plant. Scheduling PM routines are based on the plant size, geographic distribution, type of plant (fixed tilt, one or two-axis), and plant features such as energy storage.

7.10 Customer-Generated Maintenance

Customer-generated maintenance is the result of the customer (plant supervisor [on site], service personnel or visitor, owner, or O&O) seeing something that does not "look right" or raises questions. We encourage taking a photo and sending it to the O&M contractor or site contact. Often, without additional context, this will generate a maintenance event that may or may not be real. Rewarding this behavior may result in identifying something hiding in

Table 7.4 Inspections and Intervals for Preventive Maintenance.

Material	Inspections	Interval
Bolts	Indications of being loose, on the ground, loss of paint, bare metal showing, or corrosion	Annual or major storm event
Welds	Indications of pitting (with/without paint fill), bubbles, streaks/stains, cracking	Annual or major storm event
Interfaces: Metal-Concrete	Streaks or stains, bubbles, appearance of micro-cracking around base, loss of structural integrity with bending at base	Annual or major storm event
Interfaces: Al-Fe	Look for advanced corrosion that is at dissimilar metal interfaces	Annual or major storm event
Interface: Seals – Rubber	Inspect for micro-cracking, or stiffening of the rubber (loss of rebound) – Inspect for cold flow (bulging or thinning that was not previously noted) – Also note any material that has been pinched	Annual or major storm event
Interface: Seals – Silicon	Inspect for bubbles, cold flow (per above), change of color	Annual or major storm event
Paint	Inspect for bubbles, cracking, micro-cracking, pitting, significant changes in color, erosion, or scratches	Annual or major storm event
Filters	Check any flow/pressure monitors to determine if there has been a decrease in mass flow rate. Inspect for clogging or loss of filter integrity	Annual or major storm event
Temperature sensors	Verify and perform calibration, connectivity, and readings	Annual or major storm event
Pressure Sensors	Verify readings, appearance of any unexplained transient data	Annual or major storm event
Hinges/Doors and Frames	Inspect for mechanical damage and corrosion	Annual or major storm event
Dehumidifiers/ Coolers	Inspect for water pooling, corrosion, or mold growth in dark or sun-shaded areas	Monthly
Module or Panel front glass	IR Scan panels and note by panel serial number and location all "dark" cells or "light" or hot spot cells	Annual or major storm event
Inverter Parameters	Check all unreported meter measurements	Monthly

Table 7.4 (Continued)

Material	Inspections	Interval
Connectors, Power	Inspect all bolts, attachments, lugs, and crimps for corrosion, potential fatigue, worn or broken wires, and nicks, scrapes, or scratches in insulation	Annual
Connectors, Communications	Check for worn or loose connections or indications of damage	Annual
Fuse/Holders	Inspect for corrosion, damaged/discolored pitting, or looseness	Annual
Transformer Fans	Properly operating, shut down to clean grease/dirt from blades	Semi-annual or annual
Transformer Cooling Fins	Inspect for corrosion, salt, oil leaks accumulation, weld cracking, and loss of paint integrity	Annual
Skids	Inspect for corrosion, salt accumulation, weld cracking, and loss of paint integrity	Annual
Panels/Modules	Inspect for dust, dirt buildup, bird droppings, cracked or compromised surface elements	Annual and after major storm events
Panels/Modules Connections	Visually inspect each connection for looseness or degradation	Annual

plain sight that needs to be addressed. This necessitates formalizing communications channels and data formats to minimize false alarms, incorrectly ranking of maintenance actions, and confusion over which part(s) of the plant are needing maintenance. There should be a standard numbering system for the plant, even for inverters that may be maintained by the manufacturer.

7.11 Energy Storage

Energy storage systems for PV include a broad range of technologies (see Chapter 9) all of which require scheduled and PM. All forms of energy storage degrade with time and occasionally fail catastrophically. The discussion here is primarily limited to batteries as it is currently the dominant form of energy storage. Chapter 9 discusses the various types of energy storage. For maintenance, the considerations are sufficient automated monitoring and reporting of cell voltage and currents (charge and discharge), temperature, maintenance intervals and PMs required, and the changeout date and/or conditions for aging effects (Figure 7.7). Temperature monitoring may require multiple sensors depending on the size of the battery, the specific chemistry, and the number of cells in each unit.

Figure 7.7 Example of a Battery Energy Storage Design (Module & String). Source Andy Walker NREL.

Battery PM includes visual inspections for corrosion, loose connections, bulging, and IR scanning for potential over heating in cells and connections. As shown in Figure 7.7, space is needed for maintenance access as well as accessibility for special support equipment such as hand trucks or hydraulic lift devices to remove and replace weak or dead units.

Batteries also require special training to understand the chemistry involved and the effects of any leaks due to cracks or failure of the case material around the terminals and how to identify and document them. Terminals need to be torqued to specification to ensure the minimum resistance across them. This is where, again, the FMEA can provide a solid foundation for the development of the maintenance instructions, trouble shooting, operating manuals and planning for training. To allow for reduced direct oversight, all batteries should have testability elements built into the system. Also, there are facility elements that include:

- Monitoring and management of the ambient temperature in the battery facility in potential extremes of ambient environments with at least a 95% confidence, and
- Keeping the humidity under control especially in areas where salt air may be present.

It is essential to have a detailed and complete energy storage, maintenance, and hazard response plan for all storage systems and devices. Maintainers also need special handling training and instruction to reduce and/or respond to potential or actual fire conditions that occur when they are present as well as the design accommodating the special chemicals needed to suppress battery fires. Lithium batteries, for example cannot use water for fire suppression. Since most PV plants are well away from firefighting resources, the maintainers may well be the first and/or only people at a fire scene.

7.12 Spares

Sparing parts for a large power plant can require a large floor space warehouse or a distributed warehousing schema to lower the overhead for plants in or near urban and suburban areas. The cornerstone for the estimated number of spares required is tied directly to the reliability, cost, and lead-time attributes of an item. The higher the reliability (MTBF, $R(t)$, MTBM, etc.), the fewer maintenance actions and a consequent lower number of spares are required. Table 7.5 provides an example of cost quantity and MTBF for parts used in a PVPS.

Sparing, or the purchase and storage of components, assemblies, or electronic circuit cards for a plant (or region), is based on the need to keep the highest availability of the

Table 7.5 Example Spares Calculation for Three MTBFs.

Average Down Time =	36 hours	Average Cost per Spare =		$5,000	Average Cost of Maintenance call =		$1,000
	Hours 1/MTBF	Units	E(x)	Number of Spares	Average Annual Cost	Operating Down time	Lost Power Generation (Whr)
MTBF 1	150000.000067	100	20	23	$120,000	720	20,000,000
MTBF 2	200000.000050	100	15	18	$90,000	540	15,000,000
MTBF 3	300000.000033	100	10	12	$60,000	360	10,000,000
Op Hours	3000						
Confidence	0.8 %						

Spares	MTBF 1		MTBF 2		MTBF 3		Determine Minimum number of spares for confidence level		
	PDF	CDF	PDF	CDF	PDF	CDF	MTBF1	MTBF2	MTBF3
0	0.000000	0.000000	0.000000	0.000000	0.000045	0.000045	0	0	0
1	0.000000	0.000000	0.000005	0.000005	0.000454	0.000499	0	0	0
2	0.000000	0.000000	0.000034	0.000039	0.002270	0.002769	0	0	0
3	0.000003	0.000003	0.000172	0.000211	0.007567	0.010336	0	0	0
4	0.000014	0.000017	0.000645	0.000857	0.018917	0.029253	0	0	0
5	0.000055	0.000072	0.001936	0.002792	0.037833	0.067086	0	0	0
6	0.000183	0.000255	0.004839	0.007632	0.063055	0.130141	0	0	0
7	0.000523	0.000779	0.010370	0.018002	0.090079	0.220221	0	0	0
8	0.001309	0.002087	0.019444	0.037446	0.112599	0.332820	0	0	0
9	0.002908	0.004995	0.032407	0.069854	0.125110	0.457930	0	0	0
10	0.005816	0.010812	0.048611	0.118464	0.125110	0.583040	0	0	0
11	0.010575	0.021387	0.066287	0.184752	0.113736	0.696776	0	0	0
12	0.017625	0.039012	0.082859	0.267611	0.094780	0.791556	0	0	12
13	0.027116	0.066128	0.095607	0.363218	0.072908	0.864464	0	0	0
14	0.038737	0.104864	0.102436	0.465654	0.052077	0.916542	0	0	0
15	0.051649	0.156513	0.102436	0.568090	0.034718	0.951260	0	0	0
16	0.064561	0.221074	0.096034	0.664123	0.021699	0.972958	0	0	0
17	0.075954	0.297028	0.084736	0.748859	0.012764	0.985722	0	0	0
18	0.084394	0.381422	0.070613	0.819472	0.007091	0.992813	0	18	0
19	0.088835	0.470257	0.055747	0.875219	0.003732	0.996546	0	0	0
20	0.088835	0.559093	0.041810	0.917029	0.001866	0.998412	0	0	0
21	0.084605	0.643698	0.029865	0.946894	0.000889	0.999300	0	0	0
22	0.076914	0.720611	0.020362	0.967256	0.000404	0.999704	0	0	0
23	0.066881	0.787493	0.013280	0.980535	0.000176	0.999880	23	0	0
24	0.055735	0.843227	0.008300	0.988835	0.000073	0.999953	0	0	0
	0.035605		0.037637		0.045545				

system to provide power when demanded during operating hours. This also includes keeping the downtime to a minimum by lowering the logistic delay time and having a clear understanding of what is in stock (see data collection in Chapter 10). Elements, such as the cost (present value versus future value), quantity, reliability, and lead time, factor into the decision. To properly establish the spares needed for the plant, it is necessary to identify the criticality of failures through an FMEA (Chapter 6) and from prior history, a failure trend analysis[11] and average cost per maintenance call, as well as the cost per unit. To determine the criticality, the FMEA provides a risk assessment by establishing the consequences to the plant. The criticality, major effect, then provides the initial source information to establish the spares list. Since 100% spares is impractical, the Owner and/or the O&M must establish what items must be available for immediate repairs and also establish the confidence level to calculate the number of spares.

The tools for determining spares are generally based on the exponential distribution, i.e. constant failure rate (λ). For PVPS, it is our recommendation that the spares forecasting be based on a one-year period using a failure rate or MTBF/MTTF based on the appropriate distribution. For most items, this represents about a 3–10% portion of the MTBF which allows for this assumption, even for components with other distributions. However, this can only really work if there is sufficient data to provide for a trend analysis that permits the understanding of which direction the component MTBF is going, e.g. higher, lower, or the same (Note: PDF is Probability Density Function, CDF is Cumulative Density Function.)

7.12.1 Spares Calculation

For the purposes of calculating spares, the general approach is to assume that the failure rate is constant over the maintenance interval "t" used for planning and budgetary purposes, typically a year. Provided that the MTBF is large enough and t is much less than a standard deviation from the mean, then the constant failure assumption will hold.

The calculation is based on Poisson,[12] Haight (1967), process, i.e. all failures are independent of each other and have an exponential distribution (λt). Equation (7.11) is the estimator of having *n or fewer failures* over the time interval specified, typically one year.

$$\mathrm{CL}(t) = \sum_{x=0}^{n} \frac{(\lambda t)^x (e^{-\lambda t})}{x!}, \tag{7.11}$$

where

$\mathrm{CL}(t)$ = the probability (confidence level) of being able to repair a unit within the M_{ct} (time): $t = M_{\mathrm{ct}}$
n = number of units available to ensure filling a demand for a spare from the plant,
t = the time interval covering the maintenance period to be assessed,
λ = the failure rate (failures per hour), and
$x!$ = is the factorial of x.

Thus, for a specified failure distribution probability (such as exponential or exp(-λt)), then the confidence of having sufficient spares to handle a years' worth of spares to maintain operation of the plant with the lowest practical downtime is calculated in Eq. (7.11).

There are several factors that play into the time t element. For this example, we chose 3000 hours for roughly a year's operation. This is based on:

a. Lead time to obtain parts (components, assemblies, PC card, etc.).
b. Cost of storage (facility, heating and air conditioning, personnel, taxes, and overhead. These are LCOE elements often not covered by the developers and/or owners).
c. O&M maintenance bundling (accumulating a number of maintenance actions to be performed to reduce maintenance "trips" to the site).
d. Criticality of the failure and need to get the equipment back online as soon as possible.

Of course, shorter time periods can be chosen, but the above factors should be considered when doing so.

The example in Table 7.5 refers to a single component item with three different MTBFs. It illustrates that there can be a need for many spares to keep the plant availability (energy and physical) as high as possible. From the table with a 15,000-hour MTBF and 100 items requires 26 or 27 spares (based on an 80% CL) using Eq. (7.12).

For a demonstrated MTBF of 30,000 hours, the logistics requires only 14 spares. This represents a potentially significant cost difference just based on the MTBF alone. There is an attendant reduction in labor cost, the number of unscheduled maintenance events, as well as the savings in energy generation gained by avoiding excessive downtime and the reduction in costs for spares and their storage.

The probability and cumulative density functions (PDF/CDF) are used to illustrate that with each incremental assessment period the value of the density functions changes. Again, assuming that the equipment sees relatively consistent operational and environmental profiles and that failures are random, the cumulative effects can be estimated with a specified CL.

$$P_{\text{Spare}}(n) = \sum_{k=0}^{n} \frac{N\lambda T^{k}}{k!} e^{-N\lambda T} \tag{7.12}$$

$P_{\text{Spares}}(n)$ = probability of using n spares
n = largest number of spares allowed
N = number of assemblies deployed
k = an index variable for the number of spares used
λ = failure rate of the assembly

The true cost of maintenance in Table 7.6 needs to be based on actual equipment bought and O&M/contract labor costs with overhead.

It is also important to understand that there is a need to make sparing decisions based on the cost today, but also for planning a budget set aside for spares bought in the future on then dollars. If one is assuming that present costs will be constant over the planned period, the actual costs may be anywhere from 5% to 40% off over the span of just a few years. In financial terms, the money spent now for spares has a value over spending money in the future to purchase a spare. A simple future value of money calculation can be found using Eq. (7.13).

$$\text{FV} = \text{PV}(1 + r)^{n}, \tag{7.13}$$

where

FV = Future value
PV = Present value
r = the interest rate in decimal form
n = the number of periods on which the value of "r" is based

Table 7.6 Example Data Needed for Calculating Spares.

	Repair or Replace	Notional Cost/Unit	Cost of Repair	QTY per Plant	MTBF	Lead Time Months
Solar Module/Panel	Replace	$300		450,000	5,000,000	1
Inverter	Repair	$100,000	$5,000 15,000	110	20,000	6
Inverter Components						
• Controller Card	Replace	$10,000	$500	1	250,000	2
• Frequency/Phase/ Sync	Replace	$10,000	$500	1	750	2
• Driver Card	Replace	$5,000	$500	1	250,000	2
• Input Card	Replace	$5,000	$500	1	250,000	2
• Wiring and Harness	Replace	$5,000	$500	1	1,000,000	4
• Power Driver Card	Replace	$5,000	$500	1	250,000	2
• Power Card	Replace	$10,000	$500	1	250,000	2
• OV/SC Sensors	Replace	$1,000	$750	10	5,000,000	0.25
• Temperature Sensors	Replace	$1,000	$750	10	5,000,000	0.25
• Fans	Replace	$1,000	$750	3	50,000	0.25
• Filters	Replace	$1,000	$250	3	PM Item	Off The Shelf
• Fuses	Replace	$1,000	$100	6	100,000	Off The Shelf
• Circuit Breakers	Replace	$1,000	$250	6	100,000	Off The Shelf
• Wiring	Repair/Replace		>$3000			
• Safety Interlocks	Replace	$500	$250	5	100,000	Off The Shelf
• V/I DC and AC Sensors	Replace	$5,000	$500	15	5,000,000	0.2
• Power Contactor	Replace	$3,000	$1000	1	50,000 cycles	0.2
• Emergency AC/DC Disconnect	Replace	$3,000	$1000	2	50,000 cycles	0.2
• Delta-Wye Transformer	Replace	$10,000	$1000	1	50,000	2–6
• AC Inductor	Replace	$2,000	$1000	1	50,000	2–6
• Power Capacitors	Replace	$10,000	$1000	4	50,000	3
Combiner	Repair	$2,500	$1,000	500	250,000	0.5
SCADA Computer	Replace	$2,000	$500	2	30,000	0.25
Sun sensors	replace	$5,000	$1000	15	50,000	0.25

Note: Assumes that the Inverter supplier is the prime source for replacement parts for an inverter.

If it is assumed that the annual inflation rate is 3% and the O&M organization is stocking a five-year supply, the savings value on a $100.00 cost at the end of five years is:

$$\text{FV} = \$100(1 + .03)^5 = \$115.93 \text{ or } \$15.93 \text{ per } \$100 \text{ unit cost}$$

From the above example, to wait and purchase spares on a replacement basis only, i.e. buying only when needed, has, at five years, approximately a 16% penalty in cost. There are other factors that affect the actual value, and these include potential cost increases/reductions due to improved or updated manufacturing processes and/or cost increases in materials or the increased interest rate on a bank loan. More dramatically, the supplier goes out of business or no longer produces the item. This results in a potentially whole new purchasing cycle, delay in obtaining replacements, and higher costs due to requirement to be form, fit, and function capable. The difficult question to answer at the beginning of a project is what the impact of technology, manufacturing, and raw materials will have on the cost of replacement parts. It should also be noted that the cost of the spare may include additional modification and infrastructure costs if the future replacement parts are not a form, fit, and function replacement for the failed item.

7.12.2 Spares Storage and Availability

Component and equipment needed as spares for the plant are required to be readily available to minimize plant downtime and stored in controlled environments. Availability of spares, based on calculating the number required, is also driven by where and how the spares are stored. If the plant is more than 15–30 minutes away from maintainers' departure point, then consideration should be given to have a warehouse on site to store a major portion of the spares list. This storage should be environmentally controlled to reduce the chances of the spares developing failures while in storage.

The warehouse should be covered by a good security system as well as controlled access. Unfortunately, the cost of base materials such as copper has increased the incidents of theft and now necessitates that the warehouse and plant security be capable of notifying the proper authorities of any breaches.

7.13 Testability

Testability is a function embedded in the equipment when designed or added as part of the system design as monitoring and is generally incorporated as part of the maintenance requirements and considered as part of the design for reliability. These may include test points, temperature sensors, voltage or current monitoring, and air flow monitoring. Being able to detect, verify fault, and isolate the failed part significantly reduces the amount of direct maintenance time needed to repair.

From a cost standpoint, a general rule is that components added to facilitate testability should not exceed 5–10% of the failure rate of the equipment. The rationale is that adding the parts needed for testability aids in M_{ct} reduction, but adds to the base cost and increases the potential for false positive and false negative failure events. A false positive is when a failure is reported but actually did not occur, while a false negative is that an actual failure was not reported.

7.13.1 Function of Testability

To reduce the repair time, wasting of spares and increasing the uptime of the system require that to the greatest extent possible, there must be an ability to report the repair level status to the SCADA and the O&M organization. This design requirement is developed during the system concept, specification, and initial design phase of the plant. As plant design progresses, all items must be assessed to determine if the items qualify as the lowest level of repair component. Components at that level need to have the critical parameter(s) defined to be effectively monitored and reported on.

However, for PVPSs, there are practical limitations on how far down the indentured parts list the FD/FI design needs to extend. Table 7.7 is a simplified example of an indentured parts list. The value of this listing is that if the maintenance is done by someone other than the manufacturer, there is an understanding of what is repairable or replaceable and the potential number of spares required. An additional attribute for each item is the product nomenclature or number to identify the failed or faulty item and what the critical parameters to be monitored are. This listing of the critical parameters provides guidance to manufacturers and suppliers as to what items are expected to have some level of fault detection/isolation.

The first questions that a project team should ask are:

1. How to flow fault detection and fault isolation (FD/FI) requirements down to equipment that already exists?
2. Should the project or the manufacturer pay for changes?
3. What percentage of the circuitry should be allocated to the capability to fault detect and fault isolate?

Table 7.7 Example Indentured Inverter Parts List Example.

Level									
1	2	3	4	5	Nomenclature	Mfg.	PN	QTY	SN
X					Inverter	Solar Magic	SM100-05	1	3001
	X				Cabinet	Metal Inc	SS513002-1	1	
		X			Doors	Metal Inc	SS513002-2A	3	
			X		Handles	Handle Inc	HR513672-2L	3	
			X		Hinges	Handle Inc			
				X	Screws, Mounting	IRC	PHMS-6-1	30	
	X				Transformer	Electrical Inc	TR1MW-321-5	1	100501
	X				Inductors	Electrical Inc	IN5KA-10KV-506-5	1	1564
	X				Controller Assm		CA1MW-12	1	
		X			Screws, Mounting	IRC	PHMS-6-1	12	
		X			Cable Harness Assm	Electrical Inc	SMCHA-1MW	1	
			X		Power Harness			1	
			X		Control Harness			1	

4. How much of the FD/FI should be reported automatically versus locally on the equipment?

If the specifications do not cover or only marginally cover testability, the project design engineer needs to identify the areas that are lacking specificity and request additional information. Since many projects have a lead time of up to 12–15 months or more, there is an opportunity to review the currently available product and to open discussions with the manufacturer on making testability improvements. This may well increase the purchase cost of an item. This cost increase must be viewed with a LCC benefit analysis to assess the value of paying for improved FD/FI.

This is where a trade study needs to be done on any improvements in the design to improve FD/FI. The higher the percentage of FD/FI, the lower the maintenance costs over the expected life of the product. Automated FD/FI allows the maintainers to find the defective item and replace it more quickly. In any case, the ability to fault detect/isolate reduces the repair time of the equipment. This in turn reduces the loss of some fraction of the power generation capability of the plant improving availability and reducing annual maintenance costs.

7.13.2 Design

Referring to number 3 above, there is a practical limit on the cost of parts, design, and manufacturing of circuitry just to support testability. Testable components imply some form of measurable attribute, which in turn requires some form of capability to add "sensors." This also implies the need for being able to take the measurement and perform some conversion to a warning, caution, or alarms. Present day capability of both hardware and software allows for some sophisticated sensing and reporting.

Some general measurements that can be made for electronics/electromechanical systems include:

- Voltage
- Current
- Temperature – ambient and equipment
- Vibration (often needed on one- and two-axis trackers due to wind induced vibration)
- Solar irradiance (e.g. global and direct normal)
- Wind
- Humidity
- Rainfall/Snow/Ice
- IR scans
- UV scans
- Other scans

Using specialized semiconductor hardware, most of the above measurements can be converted from an analog sensor to a digital value, via an analog to digital converter, and then compared to stored thresholds that provide the point of detection.

7.13.3 Percent Coverage

One consideration for testability is the percent coverage of the system that the built-in testability element can cover. This has impact on nonsolar power generation cost as well as potential impact to the cost of O&M. As a rule, the percent of plant/function design that is dedicated to collection and dissemination of the fault/status data should not exceed 10%. Aside from the cost of designing and testing the FD circuitry, there is the possibility of the circuitry providing false positive or false negative.

A false positive is the indication of the system having a fault where no fault actually exists. A false negative is the system not indicating a fault when one actually exists.

This latter case becomes particularly important where there are high-voltage/high-current and high-temperature components. Chapter 11 addresses the issues of fires, whether internal or external to the plant.

False positives, found after the warranty period, generate additional maintenance actions with the associated costs of a no fault found (NFF). If there is ambiguity in the fault reporting and equipment is removed and returned to a manufacturer for repair, the false positive can result in a cannot duplicate (CND). The CND may result in costs between $5,000 and $20,000 in labor and transportation costs alone.

7.13.4 Requirements

To understand and apply testability requirements, the project team needs to have a detailed FMEA report in hand describing the mode, mechanism, failure rate, method of detection, and effect of each failure. Describing how a fault is detected provides the basis for the maintenance manuals and training for the field service technicians. Also, there is a need to tabulate what percent of faults are detectable and how many require some form of automated response to the equipment beyond flagging a fault such as performing a safety shutdown of an inverter. Also, this analysis will indicate what the total coverage for the equipment is as a basis for determining the value of increased components and complexity to reduce maintenance costs.

The specification of testability requirements is addressed in Chapter 7. These are the design specifications, "the specific requirements," that are needed to ensure a minimum maintenance time. What is seldom covered is:

1. What specialized training is needed for the maintainers?
2. What type and form of any Special Test Equipment (STE) are needed?
3. What type of Special Support Equipment (SSE) is needed?
4. What is the standard and specialized crew size?

The maintainers, once the equipment is out of warranty, must be trained on all the plant equipment, from utility electrical interface equipment and inverters to solar modules. This training also must include the operation, use, and limitations of any STE. However, an evaluation of what actual training is required to meet the system needs is often inadequately addressed. *This lack of training is often cited as a cost consideration when the reality is the value of effective training reduces time and money and reduces plant personnel hazardous risks.*

SSE needed to perform testing may include anything from special covers for module power connectors to heavy cranes needed for installing/removing inverters or dual axis solar trackers that may have 60 or more modules in a special framing unit. The need for the SSE also leads to identifying the different number of personnel needed to crew for certain jobs.

7.13.5 Special Test Equipment

Today, a few new technologies are available to PV plant owners, operators, and/or maintainers to aid in identifying existing or emerging failures within the plant that were not available in the past. Specifically, they are remote sensing data collection technologies in the electromagnetic spectrum within a variety of wavelengths. They include visible, IR and associated thermal, UV, electroluminescence, LiDAR, and standard measurement using remote RF technology for electrical (AC/DC volts, amps and resistance), acoustic, and other capabilities (see Chapter 10). These allow PV professionals to scan a plant and other specific electrical infrastructure including PV cells, modules, inverters, energy storage, conductors, racking and tracking, and other issue faults and failures. They can be handheld, delivered by air or mobile automation, and include technologies that are or can be imbedded in the system components.

In Figure 7.8, we see basic IR scanning delivered in an aerial mode. This can identify hotspots, damaged cells, strings, and J-boxes. As effective as this use of basic technology is, it only scratches the surface. New transferred technologies can take these forms of imagery, deconstruct then reconstruct that data to include the identification of additional critical information such as voltages, current, cracked cells, and damaged EVA. This has a substantive impact on safety, time, costs savings, and accuracy both in collecting the data and analyzing it. By incorporating the capability to sense remotely without physically touching the equipment, health and condition, monitoring is dramatically improved. It allows operators to see issues as they develop versus failed conditions only.

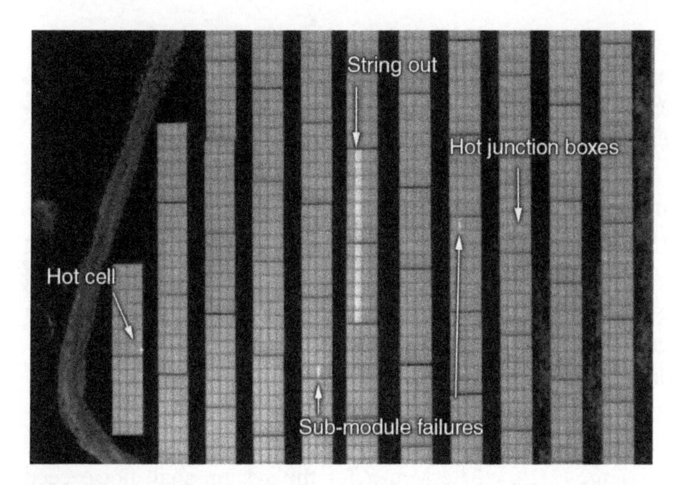

Figure 7.8 Aerial IR Photo of a PV Plant. (Image by Rob Andrews, Heliolytics).

7.14 Maintenance and Testability Specifications

Maintainability for many plants is often not specified prior to EPC bidding or during engineering, or if it is, it is applied to only a few pieces of equipment such as the inverters. This is due in part to the tendency to avoid addressing lifecycle issues early if at all, perform a lifecycle cost analysis, and then outsource the O&M to a third-party and assume that they are covering the maintenance attributes in a cost-effective manner. As a result, many RFPs are based on a less detailed and accurate specification often resulting in system bids that are inconsistent with the specific needs of a PVPS and how it will operate.

To properly maintain and operate the plant for the actual lowest cost (a real and accurate LCOE), all the repairable/replaceable items must have an MTTR assigned to the design, verified then validated through SE techniques. For the Owner/Operator, modeling additional assumptions should be listed for M_{ct} (MTTR is an attribute of the equipment only). These can be in the specification as a statement of condition for measurement or in the contract as a deliverable logistic model. In addition, there should be specified metrics on the maximum expected lead time for items not included in the spares list or stored locally (regionally). These time and cost numbers could be improved and updated through equipment sharing agreements, discussed in Chapter 10. Some items can, for example, take three to six months to receive on dock after placing the order.

7.14.1 Recommended Specification

The recommended MTTR specification, for example, should be written as follows:

3.4 Mean Time To Repair (MTTR)
3.4.1 MTTR Definition
MTTR contains, at a minimum, all the following maintenance times:

- Fault verification (FV)
- Fault detection (FD),
- Fault isolate (FI),
- Trouble shooting (TS),
- Accessing the defective unit (A),
- Time to remove and replace or repair the defective unit (R&R),
- Performing a test to verify the repair (VR),
- Closing the unit, and (C)
- Restoring power to the unit (RP).

3.4.2 Mean Time To Repair (MTTR): MTTR of system X shall be less than 3.2 hours, using the plant-specific standard tool kit Appendix (xx), required STE/SSE, with the prescribed trained manpower from Appendix (yy) and the calculation methodology specified in Appendix (zz) of the contract.
3.4.3 The minimum repair time for any maintenance event shall be less than *45* minutes.
3.4.4 Confidence Level (CL): Unless otherwise specified the confidence level for maintenance calculations shall be at least *60%*.
3.4.5 Maximum Maintenance Time (M_{max}): The Mmax for the system shall not exceed 4.1 hours at 95% confidence.

3.4.6 Preventive Maintenance: MTBPM, the supplier shall provide a list of all PM items with the time to perform the maintenance action and the required interval. PM items include all general inspections for corrosion, erosion, clogging, cleaning, and the associated maintenance such as filter replacement/cleaning, rust removal, and repainting.
3.4.7 Scheduled Maintenance: MTBSM, the supplier shall identify all items that require repair or replacement to preempt a failure(s) from occurring that can cause a loss of power generation.
3.5 Testability
3.5.1 Fault Detection Time
Automated FD shall occur within *five seconds* of fault occurrence

3.5.2 Fault Isolation

The plant shall be capable of FI to the repairable item level for all items.

3.5.3 Ambiguity Group

The plant shall be capable of isolating to an ambiguity group of:

3.5.3.1 Isolate to One (1) 60% of the time
3.5.3.2 Isolate to Two (2) 30% of the time
3.5.3.3 Isolate to Three (3) or more 10% of the time
3.6 Spares: System X shall spare to an 80% CL.
4 Verification

Verification shall be performed per the matrix in Table 4 (Table 7.8).

7.14.2 Specification Notes

Note that there is no manufacturing specification for O&M or maintenance (M)-controlled items at the plant. MLDT is, for example, solely controlled by the O&M maintenance plan, budget, spares availability, resource availability, and other elements outside of the supplier control. For a system specification, the only method to including an MLDT in the plant specification is to have the maintenance plan available from the maintenance organization.

Table 7.8 Sample Verification Matrix.

Metric	How Verified	Method	When
Voltage	Measure	AC RLC Load	As installed
Current	Measure	AC RLC Load	As installed
Harmonic Noise	Measure	AC RLC Load	As installed
Environments	Measure over One year	Site Data Collection	One year prior
Reliability	FRACAS, Field, Test, Vendor	Analysis	Annually
Maintainability	Demonstration	Vendor	Up to 2 days – Benchmark
Testability	Demonstration	Vendor	Up to 2 days – Benchmark

The additional elements for MLDT and M_{max} include:

- Delay time for STE or SSE to arrive on site.
- Delay time (transportation) in receiving replacement parts from the warehouse. This includes the time it may take to build or assemble a replacement part.
- Overhead time to perform management or clerical functions to order parts.
- Delay time for manufacturer to supply the part or end item. The manufacturer may have the item on the shelf, or it may need a special build.

7.15 Conclusion

Maintainability as an attribute grossly affects the financial aspects of the plant through annual maintenance cost of type and number of spares, their ease of access, having STE and SSE available, and maintainer dispatch time. And, as indicated in section 7.3, it is able to minimize downtime through an effective FD/FI schema. As such, all the stakeholders must, acting in their own interest, pay attention to these attributes and in fact select the attributes that they are interested in tracking. Maintenance costs includes all aspects of the plant maintenance from crew size to spares availability. In addition, it is also in the best interest of the stakeholders, including the O&M or maintenance organizations, to ensure that adequate capability to fault detect and isolate to the repairable item level is a part of the specification.

Maintenance as a function, in turn, is driven by the reliability of the product as well as its ease of access to components or parts and the testability attributes that are designed into the equipment.

Bibliography

Blanchard, B.S., Verma, D.C., and Peterson, E.L. (1995). *Maintainability: A Key to Effective Serviceability and Maintenance Management*, 1e, ISBN-13:9780471591320. Wiley.

DOD guide (2005). DOD guide for achieving reliability, availability, and maintainability, Active.

Gullo, L.J. and Dixon, J. (2021). *Design for Maintainability*, ISBN: 978-1-119-57851-2.

Haight, F.A. (1967). *Handbook of the Poisson Distribution*. New York, NY, USA: Wiley, ISBN 978-0-471-33932-8.

Housel, T.J., Hom, S., and Guertin, N.H. (2013). *The Impact of Maintenance Free Operating Period Approach to Acquisition Approaches, System Sustainment, and Costs*, 93943. Monterey, CA: Naval Postgraduate School, Acquisition Research Program.

Levelized Cost of Energy DOE OFFICE OF INDIAN ENERGY (2015), https://www.energy.gov/sites/prod/files/2015/08/f25/LCOE.pdf.

MIL-STD-1309D (1992). Definitions of Terms For Testing, Measurement, and Diagnostics.

MIL-STD-1388-2B (1998). DoD Requirements.

Haldar, A. and Mahadevan, S. (2000). *Probability, Reliability and Statistical Methods In Engineering Design*, John Wiley & Sons, isbn 0=471-331199-8, 2000.

Morris, R. (2008). Software Reliability – 40 Years of Avoiding the Question, ISSRE, November 11–13, 2008.

Reliability Analysis Center (2000). *Maintainability Toolkit*. Rome, New York: Reliability Analysis Center.

SAE JA1010 (2011). Maintainability Program Standard.

SAE JA1010-1 (2011). Maintainability Program Standard Implementation Guide

Smith, D.J. (2011). *Reliability, Maintainability and Risk: Practical Methods for Engineers including Reliability Centered Maintenance and Safety-Related Systems*, 8e, ISBN 978-0-08-096902-2. Elsevier Ltd. All rights reserved.

Tortorella, M. (2015). *Reliability, Maintainability, and Supportability: Best Practices for Systems Engineers*, ISBN:9781118858882. Wiley.

Yuan, X. et al. (2021). Life cycle cost of electricity production: a comparative study of coal-fired, biomass, and wind power in China. *Energies* 14 (12): 3463. https://doi.org/10.3390/en14123463.

Notes

1 Although highly variable, overtime is generally anything past 8 hours in one day or 40 hours in a week. In some instances, exceeding 48–50 hours per week results in double time pay. Excessive work hours also increase the probability of an induced failure. This often holds for work on weekends or holidays resulting in higher per maintenance event repair costs.

2 Ambiguity Group is the population of items whose fault signature is the same

3 Spares and support equipment may be located at a shop, warehouse, operations center, or the Field service technicians' home or residence.

4 Utility-Scale Solar Photovoltaic Power Plants. In partnership with a project Developer's Guide, © International Finance Corporation 2015.

5 https://www.cdc.gov/niosh/topics/ergonomics/default.html#lift

6 Inclination is an aggregation of azimuth and altitude of the modules with respect to the ground reference.

7 GCR or the ratio of vertical solar collection area to ground area.

8 Solar noon refers to the specific point in time when the sun is at its zenith for that day. Depending on the time zone, it is possible to have as much as an hour's difference between clock time and solar time.

9 Check on solar time.

10 Utility-Scale Solar Photovoltaic Power Plants. In partnership with a project Developer's Guide, © International Finance Corporation 2015.

11 A failure trend analysis provides information on the potential failure distribution(s).

12 Poisson distribution developed by Siméon Denis Poisson is a discrete probability distribution that expresses the probability of a given number of events occurring in a fixed interval of time or space if these events occur with a known constant mean rate and independently of the time since the last event.

8

Availability

8.1 Introduction

Availability is the most common form of plant performance used for contracts today, generally expressed in the form of a posterior assessment of the energy output of the plant over a defined measurement period. This measurement, however, fails to provide an assessment of the current or real-time capacity of the plant. In the 2015–2019 timeframe, an IEC committee worked to define not only the availability of the plant in the contractual sense but also in the physical sense. IEC TS 63019 provides a detailed discussion and description of "availability," which is summarized in this chapter.

For PV systems, contracts are often based on performance and availability guarantees. The Operations and Maintenance (O&M) or Maintenance team are the principal stakeholders responsible for acquiring, analyzing, assessing, and curating the necessary data (Hill et al. 2015) to verify compliance with the contract. This includes gathering the data from the point of interconnection (POI) meter readings of energy generated at agreed-upon time intervals. These are typically separate line items, gathering data and meter readings, in the contract and may have nothing to do with each other. The POI provides information about the plants power/energy generation. PV energy performance is based on solar irradiance, which has a high degree of variability over time spans of minutes, hours, days, and years. Solar irradiance data is needed to understand the efficiency and effectiveness of the modules/strings and the rest of the plant. Plant generation performance is based on equipment availability, which uses reliability (IEC TS 63265:2022) and maintainability attributes.

Assessment of contractual compliance requires the acquisition of data (Chapters 10 and 11) related to the equipment performance and energy output over the period of performance. The lack of a data collection system to track failures down to the lowest level of repair, can result in challenges in understanding the actual health, condition and status of the plant, the cost associated with unreliability (failures), and the impact of plant equipment availability to meet power/energy demands of the customers.

Contractually, in the current market, failure to meet the availability guarantees are conditions under which liquidated damages (LDs) can be imposed. Since most plants have third-party O&M contracts, the company providing the O&M service will be required to pay LDs for a breach of contract. The LD can be a specific amount, or in some cases, the LD is based on a tiered percentage of how much the "availability" varies from the contract

Photovoltaic (PV) System Delivery as Reliable Energy Infrastructure, First Edition.
John R. Balfour and Russell W. Morris.

value. However, it should be noted that the LD depends on the contract exclusionary clauses or conditions for which the O&M/Owner is not held responsible.

An additional project stakeholder requirement is to understand whether the equipment performs to specification. Failure to maintain equipment to specifications (while accounting for degradation) and to perform automated or manual periodic checks on component performance against specification, can result in unknown performance degradation. This only becomes visible when the energy measurements do not match expectations given the irradiance measured at the plant. A 1000 VDC string, rated at 10 A, can only generate 7.5 A or less due to low irradiance caused by sun angle, clouds, or even overly dirty modules. If this attribute (string level) is not measured and tracked, the plant is partially derated for an unknown cause. Without bypass diodes on each cell, it is also possible that the 1000 V string may be running closer to 970 VDC, for example. A 1000 VDC string may require some 42 modules in series at STC, but the design would be for some 50 modules to allow for V_{oc} reduction at elevated temperatures. Assuming that these are 350 WDC modules, then the power in a 1000 VDC string is approximately 10 kW. A shorted cell decreases V_{oc} by 0.6–0.7 VDC. An open cell reduces the voltage by n*cell V_{oc}, where n is the number of cells bypassed by a shunt diode to protect the cells from open circuits and shading. (Bypass diodes allow a series (called a string) of connected cells or modules or panels to continue supplying power at a reduced voltage rather than no power at all.) Some manufacturers use $n = 30$, while others may use $n = 2$. Adding discreet diodes to a cell or module takes up real estate and can add failure modes to the modules or panels. Cells are limited in their current capacity to around 10–12 A (dependent on the cell size). A 1000 VDC string produces approximately 10 A at maximum exposure or 10 kW for a 100 MW plant that is 12,500 fifty module strings.

When a 1 MW inverter is employed, it requires 100, 1000 V, and 10 A strings. It is guaranteed that the 100 strings will not have the same voltage measured under ideal conditions, and the variance can be as much as 2–5% ***at installation***. After several years the variance can be as much as 10% or more, and at a 1%/year degradation rate, the strings are now generating at a reduced power, as shown in the simplified power Eq. (8.1):

$$\text{Plant Capability} = \left(1 - \left(\text{DR}_{\text{initial}} + (\text{n} - 1) * \text{DR}_{\text{avg}}\right)\right) * \text{STC Capacity} - \text{m} * (\text{kW}_{\text{failed}}) \tag{8.1}$$

where $\text{DR}_{\text{initial}}$ is the degradation rate during the first year; DR_{avg} is the average degradation rate for the modules as measured at the string or substring level; Capacity is the rated capacity of the plant; n = number of years since installation; m = the number of failed modules (modules rated in kW).

As shown in (8.1), the true production capability of the plant is decreased by both failures and degradation. Thus the plant's capability to produce power to meet contractual the energy availability is grossly affected by time (degradation and failures).

Understanding the true capability is necessary at the lowest level of repair to require some form of measurement that provides string performance data over time for analysis and is reflected in a plant equipment reliability, maintainability, and availability metrics and guarantee(s).

For most plants, the lowest level of repair is needed to establish the spares. Chapter 7, Table 7.1 illustrates a basic indentured list to a possible lowest level of repair.

The determination of the confidence level that the spares will be readily available when needed typically ranges from 70% to 90%. Higher confidence, i.e. 95%, would generally only be used on remote sites with excessive travel time (delay) to obtain the parts from central storage and move them to the site. Also, spares cost, inventory control, storage, and possible failures while in storage can affect the type and quantity of a particular spare. Eq. (8.2) provides a very simplistic calculation of *energy* availability

$$\text{Energy Availability} = \frac{\text{Energy Delivered}}{\text{Energy Required}} \tag{8.2}$$

Energy delivered is as measured at the POI energy (W-hr) meter. Energy required is that as specified in the contract with the accepted reductions due to force majeure. Note that this does not provide any insight into the actual state of, nor capability of, the plant. How much energy is delivered is in fact transmitted is generally only measured at the output of the plant at the POI. Eq. (8.2) is expanded on in Eq. (8.3) section 8.4.1 (IEC 63019:2019).

8.2 Why Measure Component Availability

Availability hinges on the requirement of ***available when needed***. In aerospace, for example, the availability of aircraft to fly a planned mission has an aircraft that (i) can fly, (ii) can carry designated weapons, passengers, or cargo, and (iii) do so regardless of when called upon to perform. Another application of everyday availability is having one's car start when leaving for work in the morning and again in the evening when going home. When parked, and assuming that there were no problems with the car on the drive home, the ***expectation*** is that it will be available to start and operate the next morning. PVPS available when needed begins when the sun rises. Is the plant, from structures to output transformers, ready to provide power to the customer(s)?

NOTE: Availability, like Reliability (Probability of Success), can never exceed 1.0. The fact that there is additional generation capacity in excess of demand in the plant is not reflected in the Availability calculations but is reflected in the calculation of the reliability of the plant.

Availability has in the past been predicated on meeting a contracted amount of energy over a specified time period, generally annually. Failure to meet contractual energy production due to a plant-level issue can result in increased LCOE, increased maintenance costs, and the potential for LDs. An interesting aspect of contractual language is that it is possible that some exclusions have been appended and will alleviate the O&M, for example, from paying damages for failure to meet power generation (energy delivered). The contract can become contradictory with difficulty in enforcement if the availability and reliability (Collins et al. 2009) are not tied to performance.

A more specific systems and engineering description of Availability is found by assessing the stakeholders needs, as there are several different forms of Availability depending on the parameters of interest. These include Inherent, Operational or Mission, Achieved, and

Energy Availability. All of these are generally based on some period of performance such as quarterly, semi-annual and or annual, which also applies to modeling of the expected plant performance when assessed under a stated set of conditions. Also, some owners and O&Ms currently track a plant's daily and weekly availability of any or all of the four above major metrics.

Availability, as a metric from an equipment perspective, includes Inherent and Operational Availability, translated to the number of hours of lost power generation capability when demanded. For example, assuming that the local plant site receives about 3,000 hours a year of usable sunshine, then a 0.98 inverter availability (regardless of the daytime solar irradiance) translates to a loss of 60 energy hours a year per inverter. For a one-megawatt inverter that amounts to approximately 60 MW-hours lost for each inverter. For a 100 MW plant (100 1-MW Inverters with no operational spares) that is 6,000 MW-hours of lost energy per year. This assumes that the actual downtime for the field averages only 30 operating hours/inverter/year. This assumes that the actual down time for the plant averages only 60 operating hours/inverter/year.

Here there is a variance to the usual measurement of availability as the importance of the factor is based on the generation of electricity, and that requires that the assessment period is essentially from sun up to sun down.[1]

Establishing availability specifications requires identifying the equipment that directly and indirectly affects the reliability and maintainability of the system as well as the tools needed to determine the level of power generation or energy delivered. IEEE-STD 762-2006, the standard used for non-PV systems, e.g. hydroelectric, coal or oil-fired, and nuclear power plants, defines a "generating unit." The generating unit is the equipment, components, and energy source needed to generate power/energy at its maximum capacity. A PV system generating unit, in this book, is defined as the modules/panels, wiring, combiners, and circuit protection all tied to an inverter with its associated grid connecting transformer. As of 2021, there is no comparable database of information on PV components, partly driven by the speed of technological change and the lack of making critical data available in a double-blind public database. In addition, PV plants can have a variety of module, string, and central inverter configurations that further affect the capability of the plant in operation or modeling.

As noted in the Hill et al. (2015), the language is often unclear, and the definition often does not account for the effects of solar irradiance (energy source) variations. This latter item may be covered at the plant level by under-specifying the annual energy production.

As with success and failure, the PPA and O&M contracts must include the proper set of definitions and either include the mathematics to be used or reference the source material for the mathematics. According to Hill and Balfour, "The mathematical representation of availability for the reviewed contracts and equations focuses primarily on how the equipment uptime and downtime is to be tracked. Four of the eight equations include performance in the availability calculation as a measure of available irradiance or performance ratio." This begs the question as to what the rest of the industry is doing.

Since many contracts utilize energy generation measured over the specified period (typically a year), it relies on the POI energy meter. For the plant component availability, it is measured using uptime and downtime. It is necessary to define the threshold as to when power production or uptime starts and ends. This can be tied to the irradiance; the values vary from 50 to 100 W/m^2 or 5–10% of the STC AM1.5 value. Most energy contracts reviewed for O&Ms set the irradiance at 50 W/m^2.

This requires solar sensors that can measure irradiance to establish when the threshold has been crossed at sunrise and sunset and power measurements to verify meeting the threshold.

Contractual Availability in its current form, Energy, provides a form of measurement, but it is limited in the amount of information that it conveys about the PV plant(s) and has a penchant for significant error bands around the value (typically ±3%). As contracts focus on the delivery of energy over a specified time, generally a year, they also include exclusionary clauses for meeting the requirements. At issue is that the contractual availability is related to a single measurement of energy delivered over a time period, typically a year, and assessing that against the contractual requirement. For many stakeholders, this single value does not provide any insight into the plant health, status, and condition and failure trends (see Chapters 6, 7, and 10).

So, Energy Availability is a posterior view of the capability of the plant. As pointed out in the first five chapters, there is substantially more to a plant than the total energy delivered per year as a function of solar irradiance. True, the amount of solar radiation received is an important factor, but it is not the only factor affecting the power/energy output of the plant. The reliability, maintainability, and supply chain (logistics) greatly affect the ability of a plant to produce energy. In addition, there are factors that can decrease the capability of the plant to produce energy due to external events (Force Majeure) as well as failure events that take a part of the plant down for maintenance (Figure 8.1).

8.2.1 Capability and Capacity

As previously discussed in Chapters 2–5, in the PVPS context, *Capability* refers to the fact that the plant can produce power, while *Capacity* is the sum of the module's nameplate

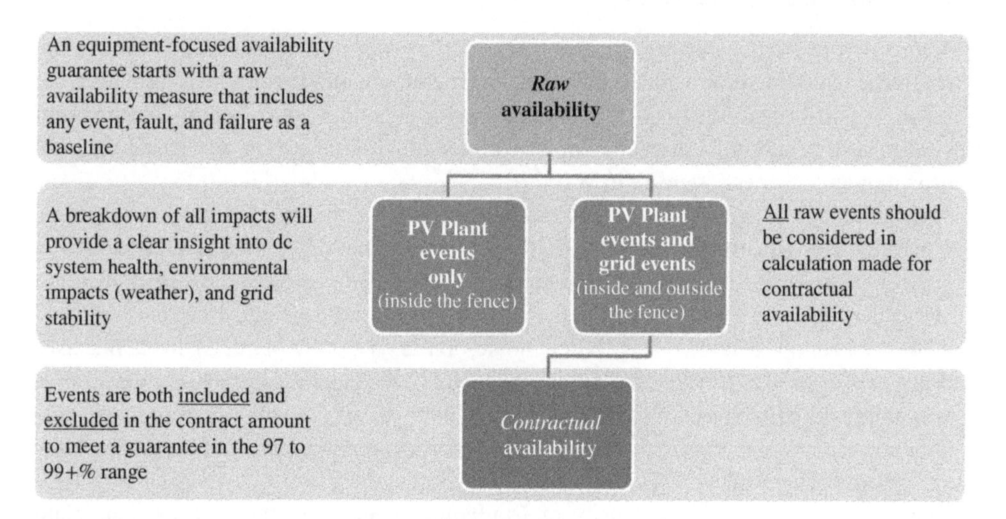

Figure 8.1 Basic Practice Flowchart for Defining Equipment-Focused Availability (Hill et al. 2015).

power rating. The capacity rating will not change over the life of the plant, while the capability rating determines how much power is available to the customer. Availability of the components (strings, combiners, inverters, etc.) significantly affects the capability of the plant.

8.2.2 Force Majeure

A significant factor impacting power generation is the external effects of a Force Majeure. The elements of Force Majeure (IEC TS 63019:2019 and Chapter 6) are those events that can cause a reduction or loss of power or energy generation capability that is outside the control of the owners and O&M. These include:

- Hail,
- Lightning,
- Snow,
- Ice,
- High winds with loss of solar alignment,
- High winds, infrastructure damage,
- Earthquake (ground movement and damage to infrastructure),
- Excessive rain resulting in erosion (ground movement and damage to infrastructure), and
- Fire started outside the plant but traveled across the plant resulting in fault propagation and subsequent damage.

One additional factor for power generation is smoke in the air. In the summer of 2020, for example, California wildfires had such large amounts of smoke in the air that some plants near the fires were reduced to one-half capability.[2] Others further away from the fires were affected by 10% to 30% between 1200 and 1600 local time.

8.2.3 Annual Solar Irradiance

Fluence is the "radiant energy received by a surface per unit area, or equivalently the irradiance of a surface, integrated over time of irradiation, and spectral exposure or is the radiant exposure per unit frequency or wavelength, depending on whether the spectrum is taken as a function of frequency or of wavelength."[3] This is typically based on local measurements of energy density in W/m^2.

Measurement of solar fluence or irradiance can be done by two types of sensors. A pyranometer (measuring global radiation) includes the effects of sky glow and/or a pyrheliometer (measuring direct radiation), which only measures the direct incident energy density.

The total energy generated daily is dependent on the type of PV plant, i.e. fixed, one axis, or two axes, as well as the type and amount of cloud cover. Having said this, the question of Availability becomes one of: If the sun is shining, how much of the plant is available to generate power/energy to the customers? This is a physical plant assessment of the operating equipment in the plant.

Variance in the solar flux density or irradiance depends on the type of solar plant (Chapter 3). Fixed installation power/energy varies with sun angle in two dimensions (azimuth and elevation), while a single axis is sensitive in one dimension (azimuth) and the two-axis tracker compensates for both azimuth and elevation.

As a percentage, the weather and cloud cover will further vary the irradiance hourly. In addition, accumulations of snow, ice, dust, and dirt reduce the energy solar cells see.

8.2.4 Customer Demand

Customers (anyone that is an off-taker[4] of plant energy) can demand all or only part of a plant's output capability as well as potentially require the plant not to be online at all. Energy demands vary from time of day as well as time of year. Extremely hot and cold seasons require more energy for cooling and heating. Homes and places of business and commerce have overlapping demands during morning and evening hours. Many homes and apartments are empty during the day, while schools and businesses are the most active.

These variations in demand and supply are compensated for by the grid distribution system. The United States and European reliability and resilience of the grid (Marion et al. 2005) is an ongoing topic of discussion and engineering improvement under the various governing agencies.

This leads to the need for the plant to be physically available when demanded and the consequent need for different measurements.

8.3 Information Categories for Plant Availability (Unavailability)

IEC TS 63019:2019 was developed as a consequence of the need to better understand what is happening with the PVPS health, condition, and status. Table 8.1 is an excerpt from that standard to illustrate that operating capability and, thus, availability is affected by the components.

8.4 Types of Availability

There are generally four types of availability metrics that can be used by different stakeholders for validation and verification (V&V) of the plant performance. These include Energy, Inherent, Operational, and Achieved. Raw Availability in the PAM method gathers all factors, including equipment, maintenance, and force majeure. The project stakeholders use it to see not only what the is doing when demanded, but the impact of plant generation events resulting in loss of energy production. These are categorized as information, e.g. data are being taken, and the system health and status can be determined.

When the system information is unavailable, it is actually unacceptable, as essentially none of the stakeholders can determine any risks, hazards, or the basic capability of the plant.

Table 8.1 breaks the events affecting the Availability into two main groupings. The plant is Available when it is in service. The plant is then Unavailable when it is out of service or nonoperative.

Table 8.1 Information Categories of PV Plant Operating Capability – IEC 63019.

<table>
<tr><th colspan="5">Information Categories</th></tr>
<tr><th>Mandatory level 1</th><th>Mandatory level 2</th><th>Mandatory level 3</th><th>Mandatory level 4</th><th>Optional level 4</th></tr>
<tr><td rowspan="11">Information available</td><td rowspan="6">Operative</td><td rowspan="3">In Service</td><td>Full capability</td><td></td></tr>
<tr><td>Partial capability</td><td>Derated
Degraded</td></tr>
<tr><td>Services set points</td><td></td></tr>
<tr><td rowspan="3">Out of Service</td><td>Out of environmental specification</td><td>Irradiance received below Threshold
Other</td></tr>
<tr><td>Requested shutdown</td><td></td></tr>
<tr><td>Out of electrical specification</td><td></td></tr>
<tr><td rowspan="4">Non-Operative</td><td colspan="2">Scheduled maintenance</td><td>Specific services, scope</td></tr>
<tr><td colspan="2">Planned corrective action</td><td>Retrofit, upgrade</td></tr>
<tr><td colspan="2">Forced outage</td><td>Response, diagnostics, Logistics repair</td></tr>
<tr><td colspan="2">Suspended</td><td></td></tr>
<tr><td colspan="3">Force majeure</td><td></td></tr>
<tr><td colspan="4">Information Unavailable</td><td></td></tr>
</table>

As shown, there are factors from scheduled and unscheduled maintenance as well as environmental and customer impacts to availability.
Source: Adapted from IEC 63019.

8.4.1 Energy

For the customers, there is the need to understand whether or not the plant is meeting the contractual energy production, within the limitations of solar irradiance and force majeure

$$\text{PVPS}_{\text{Energy}}\ \text{Availability} = 1 - \frac{1}{H_{\text{ttp}} X\, kW_{\text{np}}} \left[\sum_{\substack{i=1 \\ \text{Incident}}}^{n} H_{i(\text{un})} X\, kW_{i\,(\text{dr})} \right] \tag{8.3}$$

where:

- H_{ttp} = Hours of theoretical total production time: The hours in the period when the solar irradiance meets the minimum specifications for the inverters to operate. Depending on the type of array, this is around 200 W/m^2 Direct Normal Irradiance (DNI). This is derived from data from the sun sensor that measures the level of power density at various points in the plant and passes that information to the SCADA system for logging.
- kW_{np} = Array power. The expected DC power of the array for the entire solar generating facility is determined by the sum of each module DC nameplate (np).

- H_{un} Component unavailability time: The hours in the period when solar irradiance is sufficient to power (produce exportable energy) the inverters, yet a component within the plant is not available to generate power due to equipment damage, fault, or failure or other issues such as shading.
- N = the number of failures Incident: Every outage event during the measurement period.
- Note: KW_{dr} describes the capacity reduction because of an outage.

The component unavailability can be measured at different PVPS levels, depending on the granularity of measurement desired or available. But for this equation to work, it focuses on the whole plant. Since availability can never exceed 1.0, it is possible for the plant to operate for several years with no apparent loss of energy due to failures depending on the level of inverter masking by overdesign and redundancy.

The voltage produced by each cell is approximately 0.6 VDC for Silicon PN junctions. The current is approximately 0.035 A/cm^2. As of 2020 the average cell size is around 150 cm^2 and the module has a collection area of 0.87 m^2. Panel sizes vary, but this combination fits a form factor of 1.25 m × 0.69 m (~4′ × 2.25′).[5]

Many currently manufactured 60-cell, 72-cell, and 96-cell modules are rated from 270 to 300 Watts DC (~37 VDC) for 60 cells and 500 Watts DC (~44 VDC) for 144 half-cut cells. (Note that the 144-cell half-cut is 72 cells cut in half.) To reach an appropriate power level to support a 250 KW inverter for example, requires approximately 625 ninety-six-cell modules. Also, approximately 49 VDC is generally insufficient to compensate for the IR drop in the String wiring. Thus, strings consist of series/parallel combinations of modules or panels (two or more modules mounted structurally into a common frame). The panel or array forming the string will have, for example, 30 or so modules in series and to increase the voltage sufficient to provide up to 1000 VDC and three to five 30-module strings to reach up to 30 A, or 10 KW at STP conditions. Note that there is a general limitation to the maximum voltage a string can have. Mostly, the limitation is 1000 to 1500 VDC, depending on the technology. All of this will be determined by the wide variety of information entered into the sizing software for a project.

The factor affecting the availability of a module is dependent on the failure mode of each module. A cell short circuit will result in a module voltage decrease of approximately 0.6 VDC, while an open circuit will result in a module decrease of n*0.6 VDC, where n is the number of diode bypassed cells. A module open circuit will result in a 30-module string, for example, not producing any power.[6]

As noted above, the model only provides for the specific loss of power due to some event causing an outage. The equation has no definition of the source of that reduction in power and only really documents that there was a loss of power sufficient to affect the customer.

8.4.2 Power

Power Availability is limited by the following three conditions:

1. The equipment is available but is on standby due to a customer demanding power reduction.
2. The equipment is operating but in a degraded state due to degradation of the equipment, through defects, soiling, damage, or other condition or some units have failed and not yet been repaired.

3. The equipment is operating at a less than actual capability due to external customer or owner decisions.

In addition, there are the conditions at the time of measurement:

$$A_P = \frac{P_{\text{actual}}}{P_{\text{STC}}\left(\frac{\text{GHI}}{I_{\text{ref}}}\right)(\eta_{\text{BOS}} * \varepsilon * (1 - \delta(T_{\text{ambient}} - 20\,^{\circ}\text{C})))} \tag{8.4}$$

where STC = Standard Test Condition of a modules power measurement (see Chapters 1 and 5) at 20 °C.

The power actual (P_{actual}) is adjusted by:

- δ = ambient (at time of measurement) temperature coefficient of power (1/ °C), which is usually on the order of 0.004/ °C for silicon PV modules. The range of this coefficient is approximately 0.3% to 0.5%, depending on the number of die parameters.
- η_{BOS} = efficiency of the balance-of-system; typically, this can vary from 80% to 90% based on inverter efficiency variance over time and other power losses such as wiring and connector losses and failures as well as any plant side (prior to the POI meter) power losses (Figure 8.2) or reductions such as passive component losses, such as power inductors and transformers.
- ε = the cumulative amount of degradation of the module power capability from installation to point in time of calculation.
- Global Horizontal Irradiance (GHI) = PV Irradiance, the sum of direct normal irradiance (DNI), diffuse horizontal irradiance, and ground-reflected irradiance measured at the surface plane of the modules/panels (W/m^2), GHI = DNI*cosθ + DHI.
- I_{ref} = the insolation or irradiance as measured at the time of manufacturing at STC conditions of AM 1.5 at orthogonal angle of incidence.

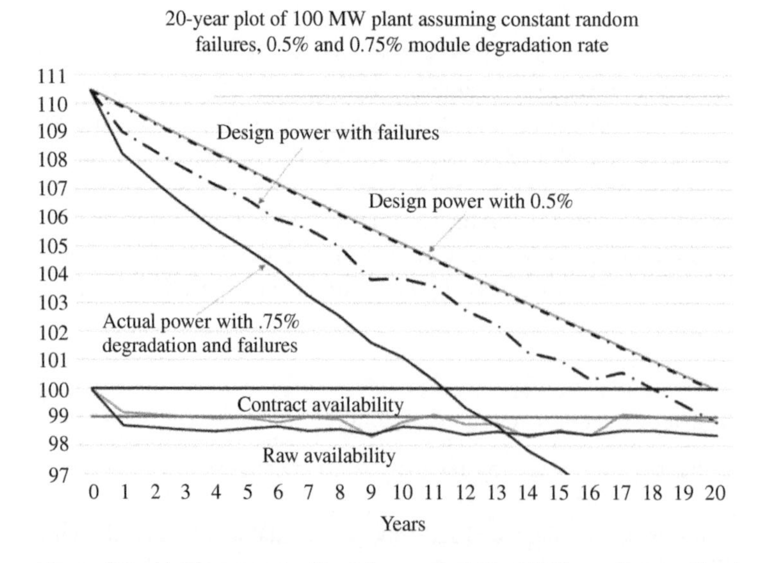

Figure 8.2 Multi-parameter Plant Assessment Over 20 Years. Source: Morris.

Table 8.2 Summary of Potential Losses in a PVPS.

Cause of Energy Loss	Percentage Loss (%)	Design or Maintenance
Shading	7	Both
Dust and dirt	2	Maintenance
Reflection	2.5	Design
Spectral losses	1	Design
Irradiation	1.5	Design
Thermal losses	4.6	Design
Array mismatch	0.7	Design
DC Cable losses	1	Design
Inverter losses	3	Design
AC Cable losses	0.5	Design
Total loss	23.8	

https://www.solarempower.com/blog/10-solar-pv-system-losses-their-impact-on-solar-panel-output/.
This an example of losses, but in the field, they are highly variable in many instances and often higher.

- Tambient = ambient temperature (°C) as measured near the PV modules/panels/arrays. Table 8.2 provides a summary of potential losses in a PVPS plant. The values should not be taken as absolutes as these values will vary from site to site.

8.4.3 Inherent

Inherent Availability[7] is the attribute of availability based solely on the physical attributes of the equipment MTBF and MTTR as designed. This is often just a calculation based on MTBF and MTTR predictions from the manufacturer. It assumes that all of the test equipment, spares, technicians, and all other resources are immediately available at the work site to perform the maintenance tasking.

Inherent Availability (A_I) is related directly to the design and as-built attributes of the plant equipment at the level of repair being assessed. $A_I(x)$ is based on the specific component (e.g. fan), assembly (e.g. capacitor and inductor filters), cards (e.g. controller cards, driver card), and/or end items such as an inverter or combiner. Inherent availability provides insight into the theoretical amount of time it takes to repair/replace (R/R) an item and return it to service. The usual source for the inherent attribute of the item is from the supplier as a calculation based on their predictions or their actual measurements under laboratory conditions, often at STC.

It *does not, however, cover all* maintenance attributes. So why bother to calculate and verify the A_I of an item? A_I of an item is the absolute minimum amount of time it would take to repair an item and return it to service, thus forming the basis for performing comparative assessment of competing designs.

A_I is a single value, and if being compared to field data would be done against the mean value of the MTTR and MTBF. This illustrates the application of the log-normal distribution characteristics that need to be accounted for in the maintenance of a plant (see Chapter 7). The ground rules for the calculation of the Inherent Availability are that:

- All tools, personnel, test, and special test/support equipment needed for the job are available at the job site and at hand to perform the maintenance.
- The components, parts, cards, or assemblies needed for the repair or replacement are available at the site when maintenance begins.
- The time starts when the maintainer verifies that a fault has occurred and first opens the item or gains access to it, and the time stops when the maintainer closes up the item, verifies a successful corrective action, and puts the equipment back in service:

$$A_i = \frac{\text{MTTF(or MTBF)}}{\text{MTTF(or MTBF)} + \text{MTTR}} \tag{8.5}$$

A short example of A_I is when a repair team is on-site, and a call is received that Inverter 14 is reported as being inoperative. The team also received information that the reported defective unit code is for the controller card, which they have as part of their standard kit. The team goes out to Inverter 14 and verifies that the unit is not producing power by checking the output meter of the inverter and also checking the inverter diagnostics for any additional fault information.

When the team verifies the fault and shuts off the DC power input to the inverter the repair clock is started on the repair. They access the electronics section of the inverter and remove and replace the controller card. They close up the unit, restore DC power and check the inverter output meter for proper operation. Confirming this, they then reconnect the inverter to the plant grid tie.

Note: None of the time that the unit was inoperable before the maintainer's arrival and the lack of any delay in maintainer or part availability is counted as downtime. For this example of a simple card remove and replace, the repair time was 40 minutes in total. This can be considered the minimum repair time possible for the inverter, assuming that other items require the same or more time to perform the maintenance. This minimum repair time does not cover the maintenance efforts associated with a false positive (failure reported, but none actually exists) such that the "repair time" for this event is restricted to the time it takes to verify that the equipment is actually working.

8.4.4 Operational

Operational Availability[8] (A_O) provides the basis for understanding the true cost (time) for a repair. In the above A_I example, this only accounts for direct hands-on maintenance. Operational Availability accounts for the total time that the equipment is nonoperative, i.e. from the time it fails (whether reported or not) to when it is returned to operation. For PV systems A_O is determined as shown in (8.6)

$$\text{Availability (Operational)} = \frac{\text{Uptime}}{\text{Uptime} + \text{Downtime}} \tag{8.6}$$

Operational availability is defined as "the probability that a system is operating satisfactorily at any random point in time t, when subject to a sequence of 'up' and 'down' cycles which constitute an alternating renewal process."[9] For PV plants, this would generally be based on an annual basis to account for all of the operational and environmental conditions experienced.

Uptime is the cumulative time over the measurement period, typically a year, that the equipment is fully operational and meeting its specification. This does not include any of the induced external commands such as curtailment or shouldering that the items may be subject to. This is a measurement of the ability of the item to meet the specified demand when requested.

Downtime is the amount of power generation hours lost if the item cannot meet specifications on demand. This can be minutes to hours, depending on the failure and the time it takes to repair the item.

Downtime (Chapter 7) consists of the following items as applicable:

1. Logistic delay time for the item to arrive at the site installation.
2. Maintenance delay time or the time it takes for the maintainers to depart the shop area and arrive at the site and stand in front of the failed items.
3. STE or SSE delay (if not covered under the two above delays) for the support equipment to reach the site to perform maintenance.
4. Administrative delay in a dispatch to the job site.
5. The actual time to repair the equipment.

The above items are added to the Mct or MTTR to reflect the actual downtime of the item more accurately as well as being able to track the manhours expended to affect the repair and return the item to service. For each event, the above items must be added. At the end of the year, the total downtime is:

For Operational Availability of A_O, where the unit is measured against what is specified, to achieve the specified energy output, all units must be work, then

$$A_O = \prod_{i=1}^{N} A_i = 1.0 * 1.0 * 1.0 * 0.88 * 1.0 * 0.98 * 1.0 * 1.0 * 1.0 = 0.862$$

This reflects the randomness of the Poisson process. Even for the case where $E_{produced} \geq E_{demand,}$ the ***plant*** $A_E = 1.00$. However, since there is equipment down and the $E_{Produced} < E_{Capacity}$ the ***equipment*** has an operational availability, $A_O = 0.862$.

If the Availability of each unit is the same (identical MTBF and MTTR), then system availability is given by (8.7). This is, however, not a realistic number, as the failures have to be random to meet the IID criteria

$$A_o = \frac{\text{MTBM}}{(\text{MTBM} + \text{MMT} + \text{MLDT})} \tag{8.7}$$

8.4.5 Achieved

Achieved Availability[10] (A_A) and A_O are different in the items measured for the metric. Operational availability refers to the total of the plant equipment, needed to produce energy, being available to meet demand when required, while achieved availability is based on ALL maintenance events including unscheduled, scheduled, and preventive maintenance.

The general form of A_A is:

$$A_A = \frac{\text{MTBM}}{\text{M}} \tag{8.8}$$

MTBM is the meantime between all maintenance and $\overline{M}$ (8.9) is the total average downtime for all maintenance:

$$M = \frac{\text{Unscheduled Maintenance Down Time} + \text{Scheduled Maintenance Downtime}}{\text{Number of system Failures} + \text{Number of System Downing Events}} \tag{8.9}$$

The number of system failures is all failures (including lighting, fencing, etc.) or anything that requires maintenance. System downing events are all unscheduled and preventive and scheduled maintenance events that require all or some part of the plant to be down in order to perform that maintenance.

8.4.6 Raw Availability

Raw System Availability (A_R): The sum of the components tracked in the availability guarantee is rolled up into a raw system-wide availability and can be used for comparing against the raw availability at the PV plant boundary to separate out any grid availability issues and contractual availability calculations. This requires having a sufficiently accurate accounting (data collection) of the performance down to the lowest level of repair. This differs from A_E as A_E only accounts for the items specifically covered in the contract, while A_R covers all maintenance performed at/on the plant. Figure 8.2 presents one view of the impact on availability accounting for module degradation only and degradation with failures (cumulative).

This term can be used to provide a gross assessment of the plant and should not be used for lower-level indentured items.

There are additional items depending on the size of the plant, its tie to a grid, and the end customer. There may also be some equipment associated with safety such as Personal Protect Equipment (PPE), interlocks, degradation of the plant solar cells, and fire/smoke detection equipment on the edge of the plant or within the plant at the string level. The net effect is that with degradation and other cell/module/panel failures, the actual capability of the plant is decreasing with time and the A_R with it.

The calculation of A_R provides for trend analysis year over year that aids in defining when the data support the potential liability for LD. From Figure 8.2 it is in years 4–6 that there are indicators that the plant availability is decreasing. This should be a flag to the management team that they need to find out the specifics of what is causing the decrease and address it.

Using the combination of Energy, Inherent, Operational, Achieved, and Raw Availabilities, the data are now consistent and can be distributed to all interested stakeholders on the state of the plant.

This approach is further expanded in IEC TS 63019:2019 Annex A to address operational and technical availability perspectives for varying stakeholders and areas of PVPS operations. IEC 63019 Annex B also extends this energy assessment of unavailability and energy assessment methodology. The problem with Eq. (8.1) is that it does not address the real impact of the failure in terms of specific failure(s), the actual downtime associated with those failures, and the reduction in capability of the plant.

This approach is not a measure of the power capacity of the PVPS, and users are referred to IEC 61724-2:2016 if the methods of IEC TS 63019:2019 Annex A indicate capacity varies from that assumed in the performance model and expected energy calculations.

8.5 Confusion With Availability Metrics

The four main availability metrics have different attributes. This is illustrated by looking at the example in Figure 8.3.

The raw availability is shown as 0.88 for the DC Combiner. What this number is not telling you is; what is the source of the below-expectation energy delivery? Is it multiple strings that have gone out? Or is one of the combiners (A or B) failed? Or is there an inverter out? At what point in time was the measurement made, i.e. is this a spot number, a daily, weekly or monthly number or even a yearly number? If it is given that the plant is planned to deliver 400 MW, in 100 MW blocks, and it uses 1 MW inverters, a more realistic model would have

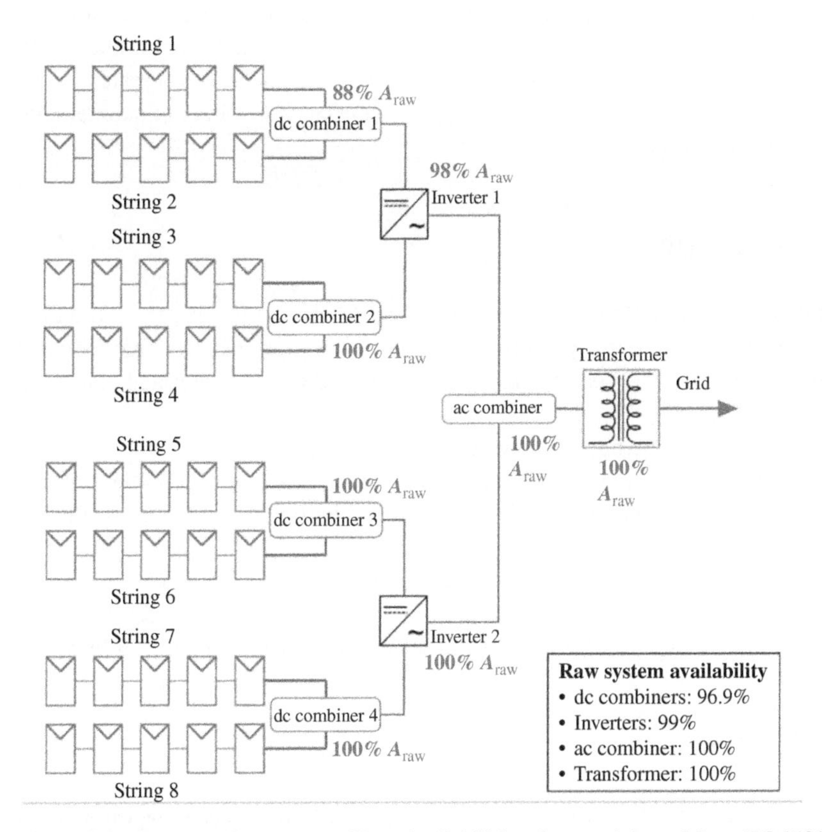

Figure 8.3 Example Two Inverter Plant Availabilities. Source: Adapted from IEC 63019.

the failures distributed over the plant wherein there would be "x_i" inverter failures, "y_j" combiner failures, and "z_k" string failures.

If the demand is given by energy required of each unit and the demand is for maximum production, then energy availability is 0.97.

If, however, the plant is overdesigned, i.e. with generation unit redundancy to compensate for potential major outage due to inverter or other failure, then from a hardware perspective, the plant has a 1.0 availability for meeting contractual demand.

Note that if the Energy production capability exceeds demand, the value of availability cannot exceed 1.0. The underlying assumption is that the plant is meeting the power demand from the customer even with failures.

For example, in Figure 8.3, if the generating units have Availabilities of 0.88, 0.997, 0.991, and 0.98, and the plant is expected to operate at capacity, then the $A_O = 0.852$.

One way to look at Availability of multiple generating units is by asking the question: at any given point in time, what is the inherent availability of each generating unit?

Contract Availability requirements are generally specified as the energy availability achieved over a specified period, such as one year. Note: Contracts are established with a variety of conditions outside the scope of this book.

However, since there is known degradation in the plant's generation capability with each year and allowing for some projected user and environmental profiles, these requirements are often low. For environmental conditions, for example, shown in Chapter 6, Table 6.4, there is a potential for a fair degree of variability from year to year. This also affects the demand from customers. During especially hot years, the demand will be greater for air conditioning and during cold years for heating. The contracts, however, fundamentally cover forecasted energy production.

8.6 Grid Availability

Grid Availability, Hunt et al. (2015), relates to the tie between the PV plant and the final customer. Although it does not happen often, the plant substation may be down for some reason, or the utility may have decided that it does not need the energy production from the plant.

One impact of the grid tie being shut down is that the PV plant may not be able to get the parasitic power needed to run the plant lighting and security. This necessitates the installation of a backup generator sized to handle all the security, lighting, motor controls (for one or two-axis plants), and the SCADA/FRACAS system.

This also points to the need for an Uninterruptible Power Supply on the SCADA and the remote terminal connections for monitoring plant operation. Although this fault may not occur often, failure to maintain plant control and security may have a significant impact.

8.7 Specifications

Specifying availability for the various organizations and EPC(s) is determined in part by contractual availability.

8.7.1 Allocation

Clearly specified Availability (A_O, A_E, A_R, and A_A) cannot be applied to suppliers or even the EPC as it is a derived attribute dependent on the reliability and maintainability attributes of the equipment and the system design. The EPC is, however, on the hook for developing a reliable plant design, and the suppliers are responsible for designing reliable equipment. The O&M has the responsibility for the maintenance of the plant including maintaining a sufficient quantity of spares to keep the plant at its best performance.

8.7.2 Requirements

8.7.2.1 Suppliers and Third-Tier Vendors

Inherent Availability requirements should be developed during the concept phase prior to EPC bidding, refined through the systems engineering, proposal, and design phases, and applied to specifications. This requirement should, however, only be levied on components for which repair is needed to maintain plant operation. So, putting such requirements on the inverter helps to ensure that the equipment is not only reliable but also can be easily maintained. Putting such a requirement on PV modules, however, is not enforceable as modules are not repairable end items, but similar to a computer card resistor.

In general, specifications for availability of a third-party vendor are given by the physical attributes of items MTBF and MTTR. Availability for a third party is tough to do for anything else, in that they are not responsible for having spares available or the manpower and/or tools available to do the job. If the MTBF and MTTR have been properly specified, then determining the downtime becomes easier to model. However, A_O as a specification, is not recommended to measure downtime. For example, A_O depends on not only direct maintenance but also spares, tools, human resource availability, travel time, logistic delay time, and the potential for variability with any of the prime attributes of maintenance due to contractual O&M activity.

8.7.2.2 O&M

Plant O&M Availability takes the form of the operational availability of the system at each of the lowest level of repair items. There is a physical plant operational availability, that is, the plant can produce any power demanded up to its current physical capacity regardless of the solar irradiance.

Operational Availability specifications are applicable to the plant designer, the EPC or contractor, and the O&M provider. The EPC's responsibility is to ensure the equipment, as designed and installed, is easily maintainable. The O&M's responsibilities are to ensure that the active maintenance has all the tools, spares, and human resources quickly and easily available to minimize the total downtime.

Achieved Availability is the sole responsibility of the O&M. The MTBM is affected by scheduled and ancillary metrics, and the M_{ct} is driven by the logistics that the O&M has put in place.

8.7.2.3 Owner

The owner's responsibility for availability is properly funding and contracting for the O&M of the plant. A full discussion of contracts is outside the scope of this book, but it should be noted that despite the best efforts of systems engineering, EPC design, and proper installation teams, if the owner does not properly contract for the O&M, it is only a matter of time before the plant will fail to meet customer performance demands.

8.7.3 Standards

There are few standards regarding all the Availability metrics covered here. Most derive one value from the basic reliability and maintainability metrics. Two of the relevant PV standards are IEC TS 63019 and IEC 61703. In addition, several IEC and IEEE standards are applicable to PVPS Availability, IEEE-STD 762-2006, IEC TS 63019:2019, and IEC 61703:2016. Hunt et al. (2015), summarize the current state of availability of Utility-Scale photovoltaic power plants as essential for developing and financing these projects. This provides a reference for years 2010–2014 measured availabilities that were in the range of 98.6% to 99.5%. More recent measurements are in the same range. Neglecting, force majeure, requested shutdowns, and inverter derating, leave operating downtime (lost production) up to 30 hours[11] a year per end item due to failure.

From Chapter 6, there is the mean time between failures which implies that the items are repaired by an actual repair action such as replacing a driver card in the inverter, while the mean time to failure is the attribute of the repair by replacement of a part, component for which the only return to service is based solely on a nonrepairable part, i.e. replacement.

Inverters, combiners, SCADA, and other end items can be tracked by an MTBF attribute. Transformers, fans, capacitors, PV cells, modules, and other large components will generally fail and are nonrepairable but replaceable and are tracked using the MTTF.

8.8 Conclusion

Energy availability has been a prime factor for PVPS for decades. It does not, however, provide the information needed to assess the health, condition, and status of the plant to generate that power. By considering the component level availability metrics that include the component's reliability and maintainability, it is possible to provide a much better understanding of how the plant is performing over time. The trend analysis of the plant RAM metrics also allows for a better understanding of the costs of the plant O&M impacting Life Cycle Cost, IEC 60300-3-3:2017, and LCOE.

Bibliography

Availability factor. Available: https://en.wikipedia.org/wiki/Availability_factor.

Bufi, C., Elasser, A., and Agamy, M. (2015). A system reliability trade-off study for centralized and distributed PV architectures. In: *2015, 42nd IEEE Photovoltaic Specialist Conference*.

Collins, E., Dvorack, M., Mahn, J. et al. (2009). *Reliability and availability analysis of a fielded photovoltaic system*. In: *34th IEEE Photovoltaic Specialist Conference*.

Hill, R., Klise, G., and Balfour, J. (2015). *A Precursor Report of Data Needs and Recommended Practices for PV Plant Availability, Operations and Maintenance Reporting* SANDIA REPORT, SAND2015-0587X, Unlimited Release. Albuquerque, NM: Sandia National Laboratories.

Hunt, K., Blekicki, A., and Callary, R. (2015). Availability of utility-scale photovoltaic power plants. In: *2015, 42nd IEEE Photovoltaic Specialist Conference.*

Hunt, K., Blekicki, A., and Callery, R. (2015). Availability of utility-scale photovoltaic power plants. In: *IEEE 42nd Photovoltaic Specialist Conference (PVSC)*, 1–3. https://doi.org/10.1109/PVSC.2015.7355976.

IEC 60300-3-3:2017. Dependability management – part 3-3: application guide – life cycle costing.

IEC 61724-1:2021. RLV photovoltaic system performance – part 1: monitoring.

IEC 61724-3:2016. Photovoltaic system performance – part 3: energy evaluation method.

IEC TS 61836:2016 RLV. Solar photovoltaic energy systems – terms, definitions and symbols.

IEC TS 63019:2019. Photovoltaic power systems (PVPS) – Information model for availability.

IEC TS 63265:2022. Photovoltaic power systems – Reliability practices for operation.

IEEE-STD 762-2006 (2006). IEEE Standard Definitions for Use in Reporting Electric Generating Unit Reliability, Availability, and Productivity. Revision of IEE Std. 762-1987.

Klise, G. and Balfour, J. (2015). *A Best Practice for Developing Availability Guarantee Language in Photovoltaic (PV) O&M Agreements*, SAND2015-10223. Albuquerque, NM: Sandia National Laboratories.

Marion, B., Adelsten, J., Boyel, K. et al. (2005). Performance parameters for grid-connected PV system. In: *Proceeding of the 31st IEEE Photovoltaic Specialist Conference*, 2005.

Notes

1 Sun up and sun down applies to tracking PVPS, while for fixed installations it is generally sun up plus an hour to sun down minus an hour. This roughly corresponds to the plant being at some level of generation (such as 60%) of current capacity.

2 https://phys.org/news/2022-12-california-wildfire-dimmed-solar-energy.html#:~:text=The%20smoke%20from%20intense%20California,for%20Atmospheric%20Research%20(NCAR).

3 https://en.wikipedia.org/wiki/Radiant_exposure.

4 What is a Power Purchase Agreement? In a PPA, a solar purchaser or "off-taker" buys power from a project developer at a negotiated rate for a specified term without taking ownership of the system.

5 This is the solar cell collection area for 0.035 A/cm^2 output. Including the frame structure for the module the overall dimensions will be approximately 1.4 m × 0.75 m.

6 Loss of module depends on whether or not the module has bypass diodes that allow for effectively shunting around the short.

7 https://www.weibull.com/hotwire/issue147/index.htm.

8 https://www.weibull.com/hotwire/issue147/index.htm.

9 https://www.eventhelix.com/RealtimeMantra/FaultHandling/system_reliability_availability.htm#Availability%20in%20Series.

10 https://www.eventhelix.com/RealtimeMantra/FaultHandling/system_reliability_availability.htm#Availability%20in%20Series.

11 This assumes that a plant can have 3000 production hour a year.

9

Energy Storage System (ESS)

Energy storage is fast becoming a major consideration for new and existing PVPS. The general categories of storage are introduced with the focus on battery technologies. High-energy density, long life, cost, ease of maintenance, and ability to handle transient loads are a few of the requirements needed to pick the right technology. The need for the system to include the design and installation of additional facilities is also discussed as well as the safety features to be considered.

Key Chapter Points

- Energy storage development is complex – from cell chemistry to system architecture to dispatch, there are nuances to energy storage that are not fully understood by many if not most experienced energy storage developers.
- Care should be taken to effectively understand and address the complexity of paired PV, grid, electric vehicle (EV), and other energy storage systems ESSs. These are not plug and play; treating them as such will result in underperforming assets or, worse, unsafe conditions.
- The use case for energy storage is arguably the most important and complex aspect of a system deployment. Without explicit understanding of how the battery/ESS is to be used in the field, the proper system design, cell selection, effective operations, and revenue projections cannot be made.

Key Chapter Impacts

- Addressing critical issues at specification to save time, missteps, and unrecoverable costs.
- Differentiating lifecycle results from marketing claims.
- Increased awareness of product lifecycle based in charging, discharging, and the broader range of environmental conditions, depth of discharge (DOD) including number, and depth of usage cycles and charge.

As with the rest of this book, we believe that for the best results in addressing energy storage and all other PV plant or system elements, it is critical to remember that PV and ESSs are not simple. Stakeholders must not be drawn into thoughts that this is all plug and play. *It is not!*

Photovoltaic (PV) System Delivery as Reliable Energy Infrastructure, First Edition.
John R. Balfour and Russell W. Morris.

The general electrical segments, in the United States today, are classified as approximately utility (52%), residential (28%), commercial and industrial (19%), and transportation (<1%). EVs in the transportation category are an extremely small segment, but growing rapidly as of 2024. For our purposes, we further modify this demand by grid-connected and off-grid systems. In North and South America, Asia, and Europe, the largest majority of customers are grid-connected, with residential, California for example, and some commercial (also military) being off grid. For the most part, there are three dominant technologies used today: lead-acid, lithium, and flow batteries.

It is by addressing all of the systems details where the greatest benefits and value are revealed, by reducing the broadest range of common and uncommon errors, mistakes, and negative results. The Office of Energy Efficiency and Renewable Energy (2019) has a good general description of energy storage for the newcomer to the industry. As we have illustrated in Chapters 4–8 in this book, it is up to the stakeholders to develop the technical wherewithal to support the design, installation, operation, and retirement of a PV plant.

Table 9.1 summarizes some of the more well-known capabilities for storing energy. Of interest is the attributes listed for each. The list below focuses on technologies that can currently provide large storage capacities (of at least 20 MW). Technologies such as super capacitors and super conducting magnetic systems are excluded here as they are still too experimental to discuss.

Table 9.1 Energy storage technologies for greater than 20 MW.

	Max power rating (MW)	Discharge time	Max cycles or lifetime (Wh/l)	Energy density (watt/kg [lb.])	Efficiency (%)
Pumped hydro	3000	4–16 h	30–60 years	0.2–2	70–85
Compressed air	1000	2–30 h	20–40 years	2–6	40–70
Molten salt (thermal)	150	Hours	30 years	70–210	80–90
Li-ion battery	100	1 min–8 h	1000–10,000	200–400	85–95
Lead-acid battery	100	1 min–8 h	6–40 years	50–80	80–90
Flow battery	100	Hours	12,000–14,000	20–70	60–85
Hydrogen	100	Mins–week	5–30 years	600 (at 200 bar)	25–45
Flywheel	20	Secs–mins	20,000–100,000	20–80	70–95

Characteristics of selected energy storage systems.
Source: The World Energy Council. https://www.eesi.org/papers/view/energy-storage-2019/Public Domain/CC BY 4.0.

9.1 Introduction Energy Storage Systems (ESSs)

A common refrain in print, among panelists at conferences, and in the media by industry associations is that "energy storage is like a Swiss Army knife, it can do just about anything." That phrase is a reference to the multitude of capabilities that ESSs have and the various ways they can be deployed and dispatched. They can deliver value to all stakeholders including grid operators, homeowners, commercial and industrial facilities, and even developers. Yet, not everyone understands what is meant by this statement nor how to squeeze the most "value" out of a battery in the field. Batteries have been around since 1800 (Alessandro Volta), although batteries used as energy storage for electricity systems is relatively new. Energy storage technology is rapidly developing, and many do not know how to develop these systems to a level of engineering diligence RAM, and safety that is necessary for any grid-connected or off-grid system. It is not enough to consider that ESS is complex, the technology and products continue to evolve at an incredible pace.

In addition, the actual capabilities of batteries are not always readily apparent, especially when one considers that ESSs can only operate in three states – charge, discharge, and idle. Yet, even with just these three states, energy storage can be deployed to solve some of the most difficult technical challenges of transients and peak demand on the grid which we address. This, in turn, can generate significant financial returns for the developers, owners, and operators of these systems. This is a burgeoning lucrative new market, with favorable incentives in certain locations, as well as the ability to support residential needs. Energy storage is ever increasing in popularity and growth in the "job" and career arena for energy storage engineers and technicians. This will continue to grow as technology matures, costs fall, markets expand, and operating strategies are refined.

While the popularity of energy storage is clear, the technology remains shrouded in unknowns, risks, and marketing buzz to a large number of specifiers and other stakeholders. Many in the field will be under pressure to maximize the value generated by the use of batteries. Often, this is without in-depth consideration for the safety, operations, maintenance, accessibility, and other costs that are associated with operating batteries within the bounds of the operational limitations and performance guarantees. Stakeholders must take time upfront to better understand their needs and requirements for the application and operational use of ESS. This includes battery chemistry, products, and proper installation and maintenance techniques. This ensures that developers, engineering, procurement, and construction (EPCs) companies, and owners/operators will maximize the value from the ESS without compromising safety or overall product reliability or capability.

The purpose of this chapter is to shed some light on the areas of energy storage that are not often discussed at conferences and in marketing materials. The intent of this chapter is to provide the reader with the understanding and hopefully lead them to develop lists of proper questions to ask, details to seek, and answers to find before designing and deploying an ESS in conjunction with their PV systems.

ESSs are defined as: "Simply put, energy storage is the ability to capture energy at one time for use at a later time. Storage devices can save energy in many forms (e.g. chemical, kinetic, or thermal) and convert them back to useful forms of energy like electricity."[1]

The method of storing electrical energy can take many different forms including kinetic, potential, electrochemical, and even thermal. A few examples of each form of energy storage are:

- Kinetic energy storage – spinning flywheels,
- Potential energy storage – pumped hydro, stacked concrete blocks, train cars filled with gravel that move up and down a hillside,
- Thermal energy storage – molten salt, ice storage, and

- Electrochemical energy storage (batteries) – lead-acid, sodium sulfur, sodium nickel chloride, lithium ion, aqueous hybrid ion, vanadium redox, and zinc air are some of the more common ones.

Each of these energy storage forms has its advantages and disadvantages in practical applications. Fly wheels are kinetic energy storage requiring significant motor-generators, gears, bearings, inverters/converters, and potentially clutches. The use of energy to raise and then recover energy using gravity has additional electrical and mechanical conversion inefficiencies as well as the motor/generator and conversion losses. The two-way efficiency of these types of systems is not conducive to actual implementation for small plants and the sheer size of the components for larger systems would be limiting.

Electrochemical storage systems (that is, batteries) have many specific advantages that the PV technical markets need. And within the realm of electrochemical energy storage, batteries based on lithium ion are presently the most widely deployed form of storage (Office of Energy Efficiency and Renewable Energy 2019).

9.2 Applications of Energy Storage

The primary motivation for pairing energy storage with a PV system is the ability to take PV energy generated during one time of the day and moving it to another time of day where the value of that energy is higher or in greater demand. From Chapter 4, the duck curve from CAISO indicates that there are times in the day when there is increased demand, but periods of time earlier in the day from which stored energy can be utilized. This can be done at the request of a utility company, which needs capacity during peak hours of the summer or for a behind-the-meter project where the battery can be discharged to manage time of use pricing or reduce demand charges for "peaky" load profiles. The battery can also be charged from the grid, but in this scenario some of the benefits of a PV-coupled system may be lost including significant tax benefits from the investment tax credit (ITC) or production tax credit (PTC), or other incentives in different countries.

Combining energy storage technologies with PV is gaining significant market share; therefore, it is critical for the stakeholders to understand their strengths and weaknesses when integrating them.

Understanding the details of battery charging, discharging, and site-driven operational lifecycle limitations must be comprehensive to be realistic. This is needed to complete an accurate financial analysis and achieve relatively accurate predictable results over the service life. This means that the use case for the battery (the application and its financial consequences in a cost-benefit analysis) should be well defined from the start of a project, i.e. at specification. For ESSs in the United States, there are several core applications to be considered, such as bulk energy shifting and capturing clipped PV plant energy (see Chapter 4). This can also limit the ability of a DC-coupled system due to the complexity of the wiring and distribution. There may also need to be a combined PV+ESS O&M and/or power purchase agreement (PPA), for providing capacity, providing ancillary services, and demand charge management to the grid in a behind-the-meter application or as virtual energy storage.

As the name implies, bulk energy shifting, for geographies such as California and many other locations, is used to combat the growing "duck curve" problem. An ESS is used to move PV plant over generation from the mid-day hours to the peak time in the late afternoon or even the late evening hours to compensate for the higher demand in the afternoons and evenings. Bulk energy shifting requires the battery be suited for long-duration charging and long-duration discharging (more than four or six hours). These batteries will be expected to deliver this energy shifting day after day, year after year, meaning that cycle and service life is crucially important as well as a solid understanding of the degradation, defect, or use attributes of the battery.

The use of batteries to capture unused (clipped/masked) PV energy is not dissimilar from bulk energy shifting. There are, however, a few considerations that are technology sensitive depending on what is being clipped (Balfour et al. 2021).

Part of the clipping impact takes place from energy the inverter cannot process from the array. That voltage and current may be too low to turn on the inverter for operations and/or meet the MPPT input requirements. This may also be due to overbuild of the array where additional energy could be produced but not processed by the inverter. In these scenarios, potential energy production is outside the inverter MPPT window. (Too much which is limited by the inverter or too little because there is insufficient irradiance to raise voltages sufficiently.) The inverter may respond by shifting the MPPT voltage and/or current, which can have a negative impact on the inverter and modules in high or low temperatures. An ESS that can take some of this DC energy to be stored reduces the negative impacts of the MPPT window losses, and monetizes that potential energy. Analyzing the opportunity to reap a potential loss requires a substantive understanding of how the entire system operates, especially when marginal losses can be turned into gains. The mathematics can be complex; however, there are some instances where the value over time can be defined in a profitable manner.

In essence, the PV+ESS plant is engineered in such a manner that these forms of energy from the facility are gleaned as usable energy and stored in the ESS and moved to later times in the day when the PV power is below the MPPT window or during transients due to partial shading due to clouds. This use of energy storage will still require the same level of lifecycle storage capability from the battery system but may have different sizing as compared simply to the bulk energy shifting use case. In other words, gleaning another 1–5% in additional energy on an annual basis may impact the size of the ESS while requiring electronics which may or may not be available in products at this time. In addition, this duty cycle is likely to be more variable than the bulk shifting of energy as the daily/annual variance in site irradiance will have a direct impact on the amount of energy which is clipped and must be stored.

One note on the application of storing clipped and/or excess energy, all PV systems degrade over time and at varying rates depending on the types of components and overall design and reliability of the system. Nevertheless, when delivering PV as energy infrastructure, addressing these issues and how they may be addressed with augmentation can provide additional economic benefits. Great care and diligence should be taken when designing all aspects of a PV plus storage system in order to ensure that the financial projections made can be more accurately realized by the technology being deployed.

Gaining more value and popularity in the US market, combined PV and storage PPAs allow the offtaker the ability to call upon the stored energy when needed, up to a certain number of times and/or hours a year. The owner of the PV+ESS plant must attempt to ensure that the battery remains charged, but the actual use of the energy in the system is only when called upon by the offtaker. (Note – like other parts of the PV industry, energy storage depends on having enough sunshine to provide the additional charge.)

In a very similar manner, PV plus storage can be used to provide capacity to the grid when called upon by the utility company themselves. While this Virtual Storage has a different commercial structure than the more common PV plus storage PPA described above, the application remains the same – store energy in the battery for use during peak periods of demand (typically during the summer months in hot climates and winter when additional heating loads are required) when extra capacity is needed to maintain system stability.

9.3 Batteries

A battery is a device that converts the chemical energy contained in active materials directly into electric energy by means of a chemical oxidation/reduction reaction. Batteries are first categorized as either a primary or secondary battery (Buchmann 2016). The main trade-off in battery development is between power and energy: batteries can be either high-power or high-energy, but not both. Often, manufacturers will classify batteries using these categories. Other common classifications are high durability, meaning that the chemistry has been modified to provide higher battery life at the expense of power and energy.

9.3.1 Primary Versus Secondary

A primary battery is a device that can only provide one discharge and is unable to be recharged, i.e. single use. That is, a primary battery is unable to absorb electrical energy in order to convert its active materials back to the original state of "charge."

Secondary batteries, otherwise known as rechargeable batteries, are able to convert active materials back into stored chemical energy, which can then be discharged once again. Like all forms of energy generation, there are operating service life and material life concerns to be addressed early. At the current time, all ESSs that are connected to the grid, either with or without connection to a PV facility, are rechargeable in nature.

9.3.2 Selection Criteria

To properly select the type of device for energy storage, it is important to consider that there is great variability in batteries (Table 9.2). Battery metrics allow us to talk about what the battery can do and how it can be applied in a specific use case. With battery products, their capabilities include:

1. The potential size and environmental and usage profiles of the ESS including time (battery life);
2. The amount of energy to be stored (power and duration);
3. The charging capacity and rate (at varying voltages and currents);
4. The discharge rates (at varying voltages and currents) and DOD;
5. The variance of characteristics of and between each different type and manufacturer.

Table 9.2 Example metrics for batteries.

Efficiency	Power-to-Weight ratio (kW/kg)
Weight	Volume
Temperature limits	Charge/discharge rates
Cycle life	Chemistry life
Self-discharge rate	Environmental sensitivities
Voltage/voltage range	

Although batteries may seem to be very simple, their application and capabilities are complex. Selecting the best specific product for a range of applications requires more than just the basic product information. It also requires an integrated operational engineering and management knowledge of how that particular product will operate.

Batteries are not an "Install and forget technology." They must be carefully operated, maintained, monitored, and evaluated constantly.

To be successful in this application requires working with professionals who fully understand the capabilities, risks, dangers, and benefits of the technology. They must display these capabilities as a team to be able to envision, communicate, specify, design, install, and provide all the operational information and support necessary for the lifetime of that particular system.

The internal ohmic and ionic resistances have a direct impact on the performance of the battery in any application. All of these different fundamental electrical, physical, and electrochemical attributes must be carefully considered when selecting the right battery for an application.

Today's technologies are advancing extremely fast, which will affect the data below. The reader is strongly encouraged to perform continual due diligence in researching the current state of the art for their planned applications. This also requires clearly identifying and gauging the level of technical support the team and eventual owner may receive from the battery manufacturer engineering and other OEM support services. This ripples through the lifecycle and includes all stakeholders who deliver, install, manage, or own that ESS system whether they be plant owners or others.

There are a variety of chemistries that are used in today's modern batteries. Lead, nickel, and lithium form the core of bulk energy storage for the different sectors of the industry (Table 9.3). The various lithium types have expanded to almost every niche application and play a major role in PV with cost being one of the primary selection criteria. As always consideration for the cost (initial and life cycle) potentially negative side effects such as toxicity, handling, recycling misuse, lifecycle, functionality, and disposal must be included.

All batteries have some level of toxicity and thus recycling and disposal require special handling!

Table 9.3 Basic types of batteries.

Type	General	Safety
Lead acid (Pb acid)	Relatively inexpensive, low self-discharge, fast discharge, long charge time, degrades with deep discharge, sensitive to overcharge	Toxic, H_2
Nickel cadmium (NiCd)	Rugged, long shelf life, cheap, low specific energy, memory effects, high self-discharge, low cell voltage	Toxic
Nickel-based, nickel metal hydride (NiMH), nickel hydrogen (NiH), nickel iron (NiFe)	Less prone to memory, cost-effective recycling, deep discharge reduces service life, sensitive to overcharge, high self-discharge – NiH and NiFe have some benefits more expensive	Mild toxins
Lithium-ion batteries	High specific energy, high capacity, simple charge, low self-discharge, requires protection from thermal runaway, degrades at high temperatures (storage), cold temperature reduces performance, transportation restrictions, service and cycle life temperature sensitive	Mild toxic, overheat, and fire
• Lithium iron phosphate ($LiFePO_4$)	Safest, specialty markets, low capacity	
• Lithium cobalt oxide ($LiCoO_2$)	Expensive, small applications	
• Lithium manganese oxide ($LiMnO_4$)	High power	
• Lithium nickel manganese cobalt oxide ($LiNiMnCoO_2$) (NMC)	High capacity, high power	
Flow –	Long service life, low cell voltage, electrolyte recyclable	Temperature
• Sodium sulfur (NaS)	Requires high temperature for electrolyte (14% of battery energy), load leveling, moderate service life	Toxic
• Sulfur (S) acid, vanadium (Vd) salt	Low leveling capability, expensive, volume of electrolyte governs capacity	Toxic

9.3.3 Types

9.3.3.1 Battery Metrics

Batteries have a number of metrics or specifications that affect the design of an ESS in PV system applications (Buchmann 2016; EPEC Engineering 2023; Sila 2022). These include specific energy, DOD, cycle life, service (calendar) life, voltage (V), current (I), charge/discharge rates (C-rates), and chemistry life. The more generally recognized metrics and terminology include:

a. Energy capacity (Watt-hours)
   Energy capacity is the measure of how much energy in watt-hours a battery will deliver in an hour, and it is the standard of measurement for a battery.

When dealing with large amounts of energy, like with batteries in PV ESS applications, capacity is typically measured in kilowatt hours (kWh) which is 1000 watt-hours, or megawatt-hours (MWh) which is one million watt-hours.

b. Energy density and specific energy
Specific energy density is the amount of energy a battery contains relative to its size. The more common specific energy density definition is stated in watt-hours per kilogram (Wh/kg). Energy density is the amount of energy a battery contains relative to its volume and is typically measured in watt-hours per liter (Wh/l).

c. Charge rate (The C-rating)
Battery charge rate (C-rate) refers to the rate at which a battery can be discharged without potentially damaging a battery.
For example, a 1C discharge rate describes the current at which the battery will discharge in one hour, while a battery discharged at a 5C discharge rate, being five times faster, will discharge in 12 minutes (1 hour divided by 5).
Not all PV applications have the same power needs. Some require quick bursts of energy, such as transient protection, while others need a steady flow of energy over an extended period of time such as supplying power to a relatively steady load, such as a computer UPS.

d. Run time
Run time refers to how long a battery or battery pack will run on a single use. This is directly related to the discharge rate.

e. Shelf life
Table 9.4 as an example of shelf life, you may buy a pack of 9 V batteries, but you do not use them all. You put the unused ones on the "shelf" for later use. How long they will hold a charge while on the shelf refers to "shelf life." If the battery pack is rechargeable, for instance, a two-way radio battery or a sealed lead-acid battery, shelf life takes on a little bit of a different meaning. In the case of rechargeable batteries, shelf life refers to how long the battery pack can sit on the shelf without going bad, before you charge it.

f. Cycle life
Cycle life refers to how many complete charges and discharges a rechargeable battery can be used before it no longer holds a charge, Table 9.4. Currently, there is no single

Table 9.4 Comparison of basic characteristic life by chemistry.

Chemistry	Shelf life	Cycle life
Alkaline	5–10 yrs	None
Carbon zinc	3–5 yrs	None
Lithium nonrechargeable	10–12 yrs	None
Nickel cadmium	1.5–3 yrs	1000+
Nickel metal hydride	3–5 yrs	700–1000
Lithium rechargeable	2–4 yrs	600–1000
Lead acid	6 mo	200–300

standard test for cycle life as each manufacturer generally specifies their cycle life characteristic based on chemistry, application, warranty margin, etc. Generally, the lifecycle of a battery is the number of usable charge/discharge cycles before you cannot recharge the battery to greater than 80% of the original capacity. The general standard for lithium battery manufacturers is to not discharge below 80% of its capacity as its DOD, although there are exceptions.

As an example, for a 100 Ampere hour (Ah) battery with a specified 500 cycle life, it should be able to recharge to greater than 80 Ah for at least 500 cycles of an 80% DOD. For most design consideration then, an ESS would be designed with at least a 20% over capacity to ensure meeting the full long life. Thus, a 10 MWh (ability to deliver 1000 VDC at 1000 A for 10-hour requirement) would be designed with an installed capacity of 12 MWh.

g. Calendar life

Calendar life is the degradation amount that occurs over years (time not cycles) while the battery is either inactive or stored, while still maintaining its energy capacity. The degradation caused by calendar life is considered to be independent of the degradation from cycle life; however, the degradation is additive (e.g. 10% drop in capacity due to calendar life + 10% due to cycle life = 80% remaining capacity).

Calendar life is one of the metrics used to quantify battery lifetime and it is a marker for the efficacy of the chemistry, cell design, and materials within the battery. Today's calendar life does not typically allow for the 30-year use a PV on- or off-grid energy application might require. Most chemistries necessitate commercial use replacement approximately every 10 years regardless of usage and replacement as necessary after the specified DOD limit has been hit.

h. Cell swell rate

The swell rate of a lithium-ion battery is the amount that the anode material within the battery expands when charged. Both traditional graphite anodes and next-generation silicon anodes swell when charged, and contract when discharged. While all batteries swell, controlling the swell rate of silicon anodes has been one of the biggest challenges in advancing lithium-ion battery technology. Being able to control or accommodate silicon anode swell impacts the cycle life, function, and safety of the battery.

i. Battery impedance

Impedance is the amount of resistance within a cell when stimulated by an electrical current. Elevated levels of impedance mean that there is a weakness within the battery, which can lead to stored energy being converted to heat rather than a useful current when the battery is used.

Tables 9.3, 9.4 and 9.5 compare the various general categories for batteries and the generally more important attributes needed for selection and system design. For high-power applications, there is a need for high specific energy, low internal resistance and self-discharge rate, high C-rates, and of course low cost. As shown, it is not possible to get all of these needed attributes in any particular battery type (Tables 9.5 and 9.6).

Today, the most prevalent class of large energy storage batteries is based on lithium-ion chemistry. While there are differences among the family of lithium-ion chemistries, they share the same basic core attributes, which make it an extremely attractive chemistry.

Table 9.5 Comparison of various battery types (A).

Type	Op temp range	Internal resistance	$/€ kW-hr	Self-discharge %/mn (25 °C)	Voltage (VDC) typical	Current (I)
Lead acid	−20–50 °C −4–122 °F	<100	$270 €253	5	2	5C
Zn-air					1.65	
NiCd	−20–50 °C −4–122 °F	100–200		20	1.2	20C
NiMH	−20–50 °C −4–122 °F	200–300		30	1.2	5C
$LiCoO_4$	−0–45 °C 32–113 °F	150–300	$ 140 € 132	<10	3.6	>3C
$LiFePO_4$				<10	3.3	>30C
$LiMnO_4$				<10	3.8	>30C
$LiNiMnCoO_2$		25–50		<10		>30C
Flow					1.15–1.55	

Table 9.6 Comparison of various battery types (B).

Type	Energy density (Wh/kg)	C-Rate	Depth of discharge	Cycle life 80% discharge	Service life years
Lead acid	30–50	8–16		200–300	3–6
Zn air	300–400				
NiCd	45–80	1		1000	5
NiMH	60–120	2–4		300–500	
$LiCoO_4$	150–190	2–4	80%	500–1000	5–10
$LiFePO_4$	90–120	≤1	80%	1000–2000	5–10
$LiMnO_4$	100–135	≤1	80%	300–700	5–10
$LiNiMnCoO_2$	40		80%	1000–2000	5–10
Flow				10,000	20

Lithium is situated at the top left corner of the periodic table as atomic number three. It is the lightest known metal as well as the lightest solid element with high reactivity. Lithium occurs naturally and is relatively abundant in both the Earth's crust and oceans and while it is never found in its pure elemental form, the processes needed to extract lithium from natural compounds are well understood and economically scalable. The very low weight of the metal leads to the high specific energy (energy per unit weight) of batteries based on lithium – critical for motive applications where weight is a detriment. The high reactivity of lithium leads to higher cell voltages than batteries based on other metals. This leads to higher energy density (energy per unit volume) that is attractive in both motive and stationary applications. At its most basic level, a lithium-ion battery works through the shuttling (often called "rocking") of lithium ions back and forth between the anode and cathode of the battery. The various families of lithium-ion batteries will have different chemistries for both anode and cathode, but still fundamentally work in the same way. When the battery is fully charged, the lithium ions are contained within the anode and move to the cathode during discharge. When the reaction is reversed to recharge the cell, the lithium ions move out of the cathode and back to the anode. This rocking process of the lithium ion between electrodes is highly reversible and, while certain anodes and cathodes are better than others, the cycle life of the electrode pairs for stationary applications is adequate for daily cycling for thousands of cycles.

As noted, within the family of lithium-ion batteries there are various subcategories of different anode and cathode chemistries. The most common form of lithium-ion batteries contains a cathode made of nickel manganese cobalt (NMC) or lithium iron phosphate (LFP). The anodes in most commercial cells today are comprised of graphite with varying, very low percentages, or silicon doped into the anode. Each of these two different types of cells have advantages over the other in certain respects. For the purposes of this chapter, we will focus on generic NMC and LFP batteries that have been tailored primarily for stationary, long-duration applications. We will also address batteries for EVs and their potential usage as grid-connected energy storage. As compared to other electrochemical batteries, lithium ion offers this highest energy density (energy per unit volume), higher cell voltages, longer shelf life, a wider range of power-to-energy ratio performance, and longer expected cycle life when operated within manufacturers' recommended parameters.

Not only are lithium-ion batteries the reference product for the stationary marketplace, but they also continue to undergo significant research and development efforts, which have the promise to improve the performance, life, safety, and costs of this chemistry. Advances in new cathode technology promise to reduce, or even eliminate, the high-cost metal cobalt from the structure while simultaneously improving the energy density of the batteries. Significant research has been performed on doping anodes, which have been traditionally made from graphite only, with silicon that has the potential to increase the energy density of the battery even further. Advancements in electrolytes hold the potential for moving cell voltage higher and reducing, or eliminating, the potential for thermal runaway.

Improvements in lithium-ion battery technologies are not the only thing driving interest in energy storage today – more important is the previously unforeseen reductions in the cost of these batteries. After seeing cost reductions of 10+% year over year in the past decade, lithium-ion cells have now settled into a more predictable cost down path of ~4–5% annually. Added to vastly reduced prices are reductions in balance of plant costs, increased market experience, learning rates, and market incentives promoting the deployment of large-scale ESSs. An ESS of 50 MW, 200 MWh would have consumed the world supply of batteries not more than a decade ago. Yet today, these size systems are becoming the norm for utility-scale projects showing that this is not an aberration, but a real marketplace not driven only by regulation.

9.3.4 Flow Batteries

Another category of ESS gaining market share in the marketplace is flow batteries (Qi and Koenig 2017) (Figure 9.1). These systems are very different from lithium-based systems with the primary differences come in two areas. The first is that of chemistry and the second is that of charge and discharge efficiencies. On the chemistry side, the energy is held within dissolved materials within two liquids separated by a membrane in an electrochemical cell. Charge is transferred between the two liquids (electrolytes) in the electrochemical cell through a variety of means depending on the chemistry composition of the electrolytes, but the result is that one electrolyte will lose "charge" to the other electrolyte during discharge and the process is reversed during charging. One of the most interesting aspects of a flow battery is the ability to size the energy and the power of the battery independently from each other. The energy of the system is determined by the volume of electrolyte which is contained in tanks, while the power of the system is dictated by the design and surface area of the electrode stack. This attribute of flow batteries allows for greater customization of the battery for the application and can help minimize costs in some instances.

While a great deal of space can be dedicated to introducing various flow battery chemistries and discussing the technical attributes of each, that will be left for another time. Flow batteries are a "younger" class of battery systems than lithium ion and, while they show great promise, in today's marketplace lithium-ion carries 90%+ of the market.

9.3.5 Battery Configuration

At the most basic level:

- Battery cells are connected together in series/parallel into modules.
  - Modules are then connected together in series/parallel into a battery rack.
  - Battery racks are then connected together in parallel to form a DC block.
  - The DC block is placed within a container – oftentimes an ISO type of shipping container or battery building/structure – that has the necessary security, heating, ventilation, and air conditioning (HVAC), safety disconnects, and fire suppression built in.

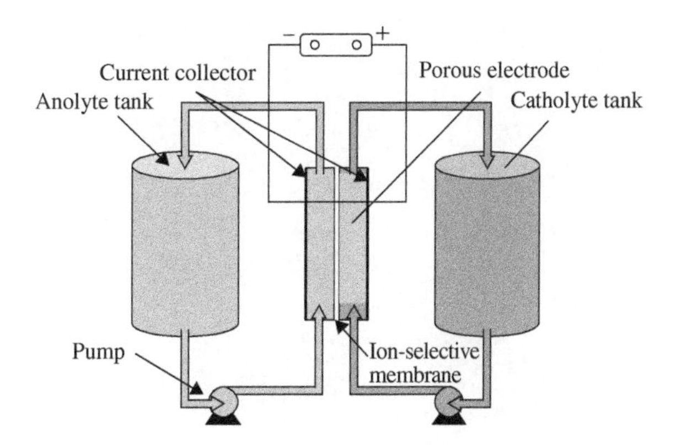

Figure 9.1 Flow Battery (Source: Qi and Koenig 2017 AIP Publishing/CC BY 4.0).

  - DC blocks are connected together in parallel to reach the power/energy necessary to then be connected to a bidirectional battery inverter.
- The battery inverter has communication ties to the battery management system (BMS) within the DC block – which itself communicates with battery management (and/or monitoring) systems at the rack level, which in turn pull data from BMS at the module level.
  - The battery inverter communicates with an outside site controller which manages scheduling of the battery, responding to utility signals, or has connection to the wholesale or retail markets depending on the application for the battery.
  - The inverter is then connected to a common point of AC – coupling (POC AC) for multiple inverters or, most likely, through a transformer to bring the voltage/current up to the final voltage at the point of interconnection (POI).

While the above figure presents a simplified version of a full ESS, in practice these are highly complex integrations of electrochemical, mechanical, thermal, electrical, and data systems. Great care is taken in sizing all of the current carrying components of the system to reduce resistance, manage safety concerns, and mitigate the need for thermal management.

These systems are not simply strings of batteries connected, but highly engineered products where the smallest of details and variances can make significant differences in the final performance in the field. They must be fully understood to achieve the highest value for the storage system's Reliability, Availability, Maintainability, and Safety (RAMS) while being designed for compatibility with the rest of the PV system.

It does not matter if the lithium-ion cells are based on NMC or LFP cathode chemistry or if they have the format of cylindrical, prismatic, or pouch cell. Each cell has a voltage of roughly 3.3–3.8 VDC, and these individual cells are packaged within a module in a specific configuration of series and parallel connection to reach the desired module voltage potential and current capacity (Figure 9.2). In some cases, this could be the final DC bus voltage for the entire ESS (typically around 800–1000 VDC). In other designs, the voltage of the module could be lower and multiple modules will be connected together in series to reach the necessary DC bus voltage. The overall system design and architecture is based on many factors including the performance of the cells, the overall system control, and battery management architecture as well as long-term maintenance and augmentation strategy.

Figure 9.2 Example Battery Powerpack (Courtesy of Andy Walker, NREL).

9.4 Components of an Energy Storage System

While ESSs that are deployed as part of a grid or off-grid connected PV system are often considered "black boxes" by most developers, these are highly engineered systems with a huge number of electrical and communication interfaces contained within. It is not expected that the reader needs to understand every single aspect of an ESS. However, there should be a basic understanding of the components of the system, which leads to an appreciation for the complexity and respect for the reliability of the system.

9.4.1 Battery Interface

The input power conversion unit is dependent on the point in the system that the power is tapped to be stored. If the power is coming directly from the modules or panels, then there is a need for a DC–DC converter under control of the BMS. The input voltage can vary but the output voltage must be stable to prevent damaging the batteries. If the input is tapped from the output of an inverter or the plant (before the POI meter), then the conversion must be from an AC–DC power supply. Again, stability is crucial.

The output conversion is through an inverter (DC–AC) and is placed before the POI. This is generally the same size inverter used by the rest of the plant, but can be a lesser or greater power output. Table 9.4 illustrates the basic configurations for connection of the energy storage to the distribution grid.

9.4.2 Storage Architecture

Figure 9.3 is a simplified diagram for a battery storage unit container/building used for energy storage. In addition to the component selection, it is also necessary to allow room for or access to the modules and racks for maintenance. This would be a requirement established in the maintenance portion of the specification and addresses not only access but also dimensions for access by general maintainers. This same general configuration can be used for an ESS building.

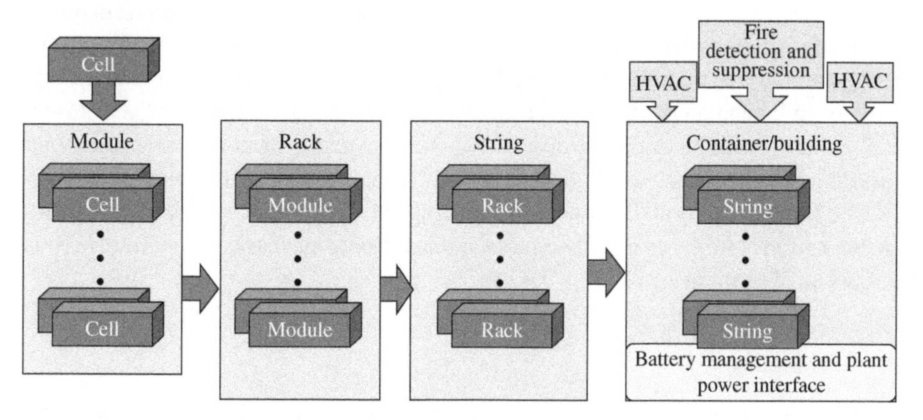

Figure 9.3 General Configuration of a Battery Storage Unit.

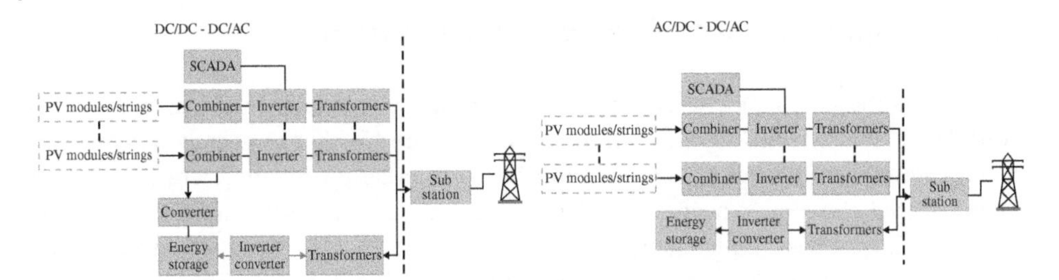

Figure 9.4 Example Storage Architectures, DC/DC–DC/AC and AC/DC–DC/AC.

Figure 9.4 illustrates two of the more common configurations for utility-scale energy storage. The DC/DC–DC/AC configuration derives its energy from a designated section of the plant as well as potentially from the utility if it allows a large amount of parasitic or recharge power to be pulled from the grid. The AC/DC–DC/AC can only draw energy from the plant if it is in operation or it can draw energy from the grid, again as a parasitic load.

The energy storage unit(s) contain their own BMS as well as *redundant* HVAC units on each storage container or building. This is especially critical for lithium ESS as the loss of cooling in high-temperature locations and conditions can result in fire or other system damages and/or disruption.

9.4.3 DC- and AC-Coupled PV plus Storage Systems

While it might not seem obvious that the exact spot where the battery and the PV array are connected, it has an impact on the overall system use. The type of connection or coupling requires several details that should be considered before making these decisions. AC-coupled storage systems are dominant in the market today and have the advantage of being able to be relatively easily implemented for new as well as retrofit systems.

In a DC-coupled system architecture, a bidirectional DC–DC converter sits between the energy storage DC bus and the DC bus and behind the typical inverter. This bidirectional DC–DC converter acts as the charge controller for the batteries, drawing power from the DC bus and charging the batteries appropriately. When the batteries are to be discharged, they discharge into the common DC bus through the DC–DC converter before moving through the inverter to be connected to the grid (or local load). The advantage here is in overall efficiency since this reduces the number of DC–AC conversions, but there are also disadvantages. A main one is having a single point of failure that, in the event of this fault removes both the ESS and the array at the same time resulting in lost energy as well as the inability to charge the batteries from the grid. Therefore, a redundancy mechanism may resolve some of the associated challenges.

9.4.4 DC Coupling

First, the decision to couple on a common DC bus or an AC bus will be dependent on the location of the ESS – behind the meter or in front of the meter. For a behind-the-meter installation, the reasons to tie PV and storage together really come down to three use cases:

1. Increasing the total consumption of (PV) energy,
2. Addressing time of use rates, and
3. Demand charges and future proofing the site against changes in rate structure or utility rules.

In two of these use cases, the PV and the battery are primarily used to offset the power that is being sourced from the local utility. This is addressing the time of use rates and/or potential outages. The battery is able to capture any excess energy generated throughout the day and move that energy to the late afternoon or evening hours when the PV array is not producing as much. The battery, as it is a dispatchable asset, can also be used to specifically discharge during times of high rate prices or to offset demand charges by managing peak loads.

And lastly, the system will be an insurance policy for changes to the utility rate structure, which may make it less economically attractive to sell any excess power back to the utility. In all three of these cases, where the battery is being primarily (if not exclusively) charged by the array, a DC-coupled system may make more sense. These DC-coupled systems have a higher efficiency overall (between 5% and 7%) since you are eliminating multiple inversions to charge the battery. There are fewer components that typically have a beneficial impact on the overall reliability and maintenance of the system. Depending on the size of the system, this can also translate into a lower total installed cost. Considering all of these factors, when using a PV and storage system to primarily reduce the power pulled from the grid, then a DC-coupled system could be a good way to go.

9.4.5 AC Coupled

In this architecture, the ESS is connected with a bidirectional inverter attached to the POI. In this scenario, the ESS can be charged from both the PV plant and/or the grid. This provides flexibility in application as well as the ability for future upgrades. In addition, these systems are typically more reliable if there is a failure on either the energy storage or the PV system, where they can be isolated from one another.

Conversely, in a front-of-the-meter application where the battery can operate as a merchant asset for the grid (providing grid services such as spinning reserve, frequency regulation, even capacity) then a DC-coupled system is not going to be as attractive of a system. The battery needs to be able to operate independent of the array – both charging from the grid and from the array. This is much more difficult in a DC-coupled system.

The second factor in determining whether to move forward with an AC- or a DC-coupled system will come down to the commercial agreements in place and the term of those agreements. An AC-coupled system where the ESS as subject to the ITC has more stringent monitoring and reporting requirements to demonstrate that the battery is being charged with solar power. In a DC-coupled system, this is much easier to demonstrate as the charge for the battery can only come from the PV system. For a commercial application where the battery is expected to participate in grid services to generate additional revenues, an AC-coupled system is a better choice. A system with a limited PPA term may also benefit from an AC-coupled system to allow the owner/operator of the system additional operational flexibility at the end of the PPA term. Behind-the-meter applications where the system is being sold outright to the end customer may benefit from the higher efficiency of a DC-coupled system. Each application has its own considerations; however, the main point is that DC- and AC-coupled systems have very different operational limitations and value propositions to the marketplace. In either case, the best strategy should be to be prepared for options in the future – after the term of the PPA has expired – and design the system appropriately to deliver the performance and reliability necessary for extended project life.

9.5 Battery Management System (BMS)

The BMS must be specified to meet system energy storage requirements, of control and safety and to support data collection, analysis, and storage, from installation to replacement.

Battery storage systems can be exceptionally well designed and used safely, or they can be an engineering, management, project, and long-term cost nightmare. Above all, take energy storage seriously and *focus on safety first*, prior to any assumptions and requirements about how a particular battery will be used in the system or specific application. The right battery system, management, and monitoring will define the overall system life cycle cost.

The BMS must control the charging and discharging of the batteries by monitoring the voltage and current as well as monitoring the ambient temperature and humidity of the storage unit and the temperature of each module. If the design utilizes the lead-acid technology, then there must also be a capability to detect GH_2 (gaseous hydrogen). This requires venting of the battery storage unit to avoid toxic and explosive buildups of hydrogen.

Battery management is required to monitor the input and output voltages, including the frequency and phase of the supplied power, and extensive monitoring of additional points in the storage units such as:

1. Temperature of each module or cell pack;
2. Room/container ambient temperature;
3. Room/container ambient humidity;
4. GH_2 monitoring for facilities using lead acid;
5. Fire and smoke detection at multiple points;
6. CO detection at floor level at multiple points.

In addition, an effective BMS collects and provides data that informs O&M of the state of the equipment (health, condition, and status) and in many applications, the metered value of energy entering or being drawn from the battery.

The BMS should also be capable of triggering the fire suppression system, and reporting that condition directly to O&M and the fire department if not already provided by the fire/smoke detection.

9.6 Battery Thermal Management

For the size and location of ESS, it is necessary for health and safety to monitor specific attributes of any battery being used for deep discharge.

Moving from single cells to a module also entails adding embedded thermal and energy management system (Buchmann 2016; VanZwol 2021) and the ability to monitor system health and safety at the cell level. The module reports its performance, condition, and state of health up to a higher level controller within the rack, which is the next level up of the system architecture. Racks contain anywhere from 6 to 12 modules typically and will either be connected in series or parallel depending on the architecture of the cells and the modules. Each rack will be at the desired DC bus voltage so that racks can be connected together in parallel to reach the overall desired energy for the system. A typical commercial/utility rack will be 100 kWh of energy or greater – which means that a 1 MWh system in this example would need 10 racks connected in parallel. Each rack will also contain a BMS controller – aggregating the data (battery metrics) that it receives from each individual module to present an overall rack status to the master battery controller.

Once racks are populated with modules (connecting both power and communications), the rack is now the minimum useful block of battery. It contains the necessary communications and is at the right bus voltage that could be connected to power electronics.

9.6.1 Housing Batteries

Most often, utility ESSs are built within standard size ISO shipping type containers (2.438 M × 2.591 M × 12.192 M or 8′ × 8′6″ × 40′). These steel containers, when combined with appropriate environmental controls, offer protection from the operational elements, are robust from an installation standpoint, and are relatively inexpensive.

Companies have implemented different design configurations on the containers to differentiate their product offerings from others. While the engineering specifics may vary, there are several key elements in common among all containers. Today, these containers are most often designed to be serviced through external opening doors – as compared to legacy systems which may contain a hallway designed for in-container servicing of the battery racks. Energy density is a premium, so the move to outside opening doors and battery maintenance from the outside of the container allows for a denser packing of the system as well as safety considerations for the O&M personnel.

If the plant is intended to be of infrastructure specification, design, and O&M consideration, it may alternatively be given to housing the ESS in a building versus containers. Buildings can and do last for 100 years or more so it is a onetime construction cost. As many energy storage technologies are continuously shrinking in size and increasing in specific energy density, a building is easily upgraded. A single building can be sized for the total stored energy needed, as well as replacement storage, and maintenance equipment to meet performance and contract. Since a building can contain the equivalent of many containers, there is a simplification of the HVAC, wiring, battery monitoring, and battery maintenance.

With extremes in environmental conditions becoming more common, whether in hot or cold locations, HVAC redundancy considerations are required to provide additional protection against these extremes.

One major consideration when dealing especially with lithium products but not exclusively is to address the risk of a building fire, site access challenges, environmental extremes including hurricanes, tornados, flooding, and terrorism placing the entire storage system at risk. This addresses extremes that hopefully will never happen. It should consider more potential issues carefully and safely spread out containers or other structures.

Properly sited containment may be designed, specified, and positioned to reduce the losses to one or two containers or other structures in a facility, reducing damage and/or not the endangering entire loss of the ESS. A good place to start is to work closely with the local Fire Marshal's office, fire safety organizations, and manufacturers to explore the best alternatives to protect life and property, while providing plans to reduce and preferably eliminate fire and other issues.

9.6.2 Containment of Failures and Safety

Figure 9.5 is a picture of the Boeing 787 battery damaged by fire (Kolly et al. n.d.). The module had multiple cell ruptures due to a thermal event in one of them. It was a difficult fire to put out in part due to the nature of the lithium-water interaction. In addition to the fire itself, there was significant release of toxic gases. One of the design changes implemented was a containment vessel to limit the spread of fire and a vent port for the release of any gases. Due consideration must be given for these effects of cell/module failure. This is one aspect that is covered by a detailed failure mode and effects criticality analysis (FMECA) of the plant design where there are safety considerations.

9.6.2.1 Fire Suppression

The reader is encouraged to familiarize themselves with the latest in regulatory and certification requirements and ensure systems meet both these requirements and the needs of the local jurisdiction. In addition to an environmental control system, each container/building must have an embedded fire suppression system and additional chemical health and safety

Figure 9.5 B787 Battery Failure, Excerpt from NTSB Report (Source: Kolly et al. n.d./National Transportation Safety Board).

equipment. Fire suppression and safety is fundamental to saving expensive equipment and providing for personnel health and safety while controlling insurance rates. Significant variances in local requirements can lead to varying system designs, which in turn may lead to large differences in system costs. A battery system installed in a very dense city will have different requirements than a battery system installed inside or outside of a PV plant in an uninhabited region. The specific authority having jurisdiction (AHJ) will specify what the fire suppression must contain. Specific research into local regulations and preferences should dictate the strategy for fire suppression systems. By addressing these issues early, they will define specification and other resources as necessary.

When considering what the totality of fire prevention and control options to be specified include, it is best to consider the impact on the entire system and the cost of eliminating a potential fire or incident. Avoiding an incident has a value on energy and production losses, insurance rates (before and after), legal disputes regarding products, property, injury and/or death, and the loss of organizational credibility. It also can have a major impact on the plant in meeting its contractual obligations, which is often underestimated.

The costs in time, energy, corporate angst, and disruption should not be taken lightly!

Lithium fires can exceed 2000 °C/3632 °F which is capable of melting the steel in a container. Lead-acid batteries generate GH_2 (gaseous hydrogen) which, if it accumulates and is ignited, can generate a significant explosion. Nickle-based, nickel metal hydride (NiMH) has a similar problem and requires temperature sensing, gas detection, and alarm.

9.6.2.2 Operating Ambient

All battery systems operate optimally at a certain temperature – most often several degrees around 23 °C (21 °C – 25 °C or 70 °F – 77 °F). A properly sized HVAC system must be installed on the container to maintain the batteries within manufacturer-specified temperature ranges. Almost every battery warranty requires the batteries to be maintained within a very tight temperature window. In addition to an external HVAC system, each container must have an embedded fire suppression system as discussed above.

Issues that affect the reliability of the battery and may be a cause of a safety event, i.e. venting noxious or toxic gas, cell failure leading to open circuit, or cell rupture leading leaking electrolyte and/or cell/module fire.[2,3]

9.7 ESS Cost

ESS costs consist of just the categories, installation, O&M, and site restoration. Engineering costs would of course be imposed for plants undergoing **SE/R**epowering™. For new systems with ESS designed it is straightforward. For all applications, there is the battery replacement cost as the batteries reach end of life (EOL). Since the EOL is highly variable, driven by degrading chemistry, number of deep DOD events, and variability environmental profiles, it can be a planned event. This will a have large degree of variability often to the negative side, such as 8 years versus 10 years.

9.7.1 ESS Installed Cost

It is easy to be distracted by the ever falling US$/kWh metric that battery manufacturers use to describe the price of their battery systems. However, this is only one element of the overall price to install and operate an ESS. There are other questions that one should ask when looking to design and then price their ESS. Certainly, price per unit energy is one of the key elements, but one also needs to consider the allowable DOD for the system as well as the cost of warranty for the intended application, flexibility in changing the application, any future augmentation costs, auxiliary loads, round trip efficiency numbers, etc. For example, many companies will state a price of US$250/kWh for their battery system, but that is for the nominal rated energy of the system. When looking at the details of how the battery can be cycled, one often finds that the limit is between 90% and 10% state of charge (SoC). This means that US$250/kWh is actually US$312/kWh (=250/0.8) per unit of usable energy. Some manufacturers may claim 6000 cycles on a spec sheet at an implied 100% DOD, but in practice only will warranty 10 years of daily cycling (3650 cycles) at 80% DOD. This has a material impact on the top line economics of installing a battery system and must be carefully considered.

9.7.2 O&M Costs

O&M costs are generally in two categories for new systems. These are the scheduled and preventive maintenance (SM/PM) and unscheduled maintenance. The greatest cost element is the SM/PM. To ensure continuous operation, the SM/PM needs to be performed on a monthly basis. The PM element is visual inspections, general cleaning, filter changes, and checking the pressure levels in the HVAC system. Except for the disposables, the costs are generally limited to the burdened overhead cost of sending maintainers to the site to perform the tasks. Scheduled maintenance on the other hand generally includes taking parts that are to be replaced before failures can occur.

ESS facility SM include replacing fuses, circuit breakers, periodic leaning and/or replacement of terminal connections (especially in lead-acid applications), and annual recharging of coolant lines in the HVAC. The costs then for SM are burdened overhead maintenance costs and the cost of the parts being replaced.

Both of these cost elements vary greatly from region to region; a general number for solar/photovoltaic maintainer labor rates is $41.20 (New York State).[4] Consideration for overhead can be roughed out as 50% more or a burdened rate of $61.80. Additional cost needs to be added for the truck (fuel, lubrication, and depreciation) and any parts used.

This is much more important when comparing different battery chemistries against each other. Each battery chemistry has a different ideal cycle profile, different usable DOD, and different rate capability. All of these attributes should be weighed against the expected duty cycle and use case to determine what will be the best product at the optimum price and lifecycle, to meet the needs of the job.

Lastly, and perhaps most complex, is the use case where the battery is used as a merchant resource in the grid to provide ancillary service offering. This can take the form of frequency regulation – responding to a signal to charge/discharge every few seconds to maintain the overall frequency stability of the grid; or spinning reserves – ready to be called upon to deliver up to several hours of fixed power. The below lists are the most common grid service applications with brief definitions of each for reference. Energy storage can be configured to perform to all of these applications – some technologies and systems are better than others, but this list shows the flexibility of energy storage as a tool for grid operators.

- **Resource adequacy**: Instead of relying on new thermal peaking plants to meet generation needs during peak hours, utilities are able to pay for other assets – including energy storage – to reduce the need for new generation capacity. The end result of using energy storage to manage these peak demands is that the utility is able to avoid overinvestment in peaking resources which may not be needed in the long term.
- **Transmission and distribution deferral**: Delaying or deferring the need to invest in upgrades to either the transmission or distribution system in order to serve future load growth.
- **Transmission congestion relief**: Some utilities are charged a congestion fee during certain seasons and time of day. Strategically located energy storage can be deployed downstream of congested transmission lines and discharged during peak times to avoid these congestion charges and reduce overall stress on the transmission network.
- **Energy arbitrage**: Buy wholesale energy at low prices (typically overnight) and sell back when the prices are higher.
- **Frequency regulation**: Providing both charge and discharge to the grid in order to manage second-to-second imbalance between supply and demand in order to stabilize the frequency of the grid and ensure stability.
- **Spinning/nonspinning reserves**: The ability to inject power to the grid in the event of an unexpected outage or a contingency event. Spinning reserves are expected to come online immediately while nonspinning reserves typically have between 10 and 30 minutes to come online. ESSs are always online and can provide power immediately.
- **Voltage support**: The voltage of both the transmission and distribution system must be maintained within an acceptable range to ensure that real and reactive power needs are balanced. This is similar to frequency regulation in that it is needed to ensure grid stability.
- **Black start support**: When an outage occurs, black start assets are needed to provide power to traditional power generation assets to bring the grid back online.

This means that a full-scale replacement of a battery system may be necessary in year 10, or less, if not effectively managed or operated, of a 20-year contract. Consideration of backward compatibility for the future energy storage must meet the contracted application needs.

Both of these scenarios must be carefully considered at specification with the various elements of the Reporting Process. Based on what is discovered, realistic cost analyses must be developed and applied to the LCOE to deliver real and more accurate lifecycle system costs.

9.8 Reliability

As noted throughout this chapter, the use of the battery in the field is the single largest factor in determining the overall life of the battery. A good analogy for this is to think of the tires on a passenger vehicle. There are many different types of tires – designed for many different types of vehicles and use cases. If you outfit a two-door sports car with tires designed for an off-road SUV, you would not expect the sports car to be able to perform to the same standard as if it were outfitted with the factory-specified low-profile tires. Taking this even a step further, outfitting a vehicle with the wrong tires will also have cascading impacts on the other elements of the system – causing premature wear and failures to suspension components and even the drivetrain, which would result in an extremely expensive repair bill. Practically speaking, putting those low-profile tires on the SUV is a guarantee to get stuck at the first sign of mud in the path. Even putting the right tires on the right car is not a guarantee of performance and life. The tires on a sports car that is driven on a race track will have a very different life than tires that are used by daily drivers. Hard acceleration from each stop light and screeching to a halt is going to wear down the tread of a tire significantly faster than driving within the typical rules of the road.

The same is true for energy storage batteries. Using a high-energy battery in high-power applications will not result in the consistent and instantaneous power delivery needed for those applications. In fact, pushing energy cells into power applications can often result in excessive heating, unequal usage of the cells, and potentially higher changes of thermal runaway due to overall misuse, all factors that could reduce the life of the equipment and compromise the system. Putting a high-power lithium-ion cell into an application tied to renewables for bulk energy shifting is not going to deliver the needed runtime and would be too costly due to the design for power. When thinking about the life of a battery – full cycling of the chemistry from 0% state of charge (SoC) to 100% SoC each and every day is going to be more deleterious than cycling from 20% to 80% SoC. Attempting to "fast-charge" cells and butting up against upper limits of charge voltage will cause undesirable side reactions to occur much faster within the cell than slow and steady charging. Operating the batteries near the upper end of their temperature limit will only accelerate all of these undesirable side reactions.

One must always remember that each and every cycle of a battery degrades the battery – more or less depending on the actual cycle, but degradation of a battery is like gravity – it is unavoidable. The key is to maximize the value generated by the battery on each and every cycle, or performing cycles that have the lowest possible degradative effect on the battery, or ideally – a combination of both. Knowing that the batteries will always degrade when in use, it is critical for the engineer and team designing the project or the developer and owner of the system to have a strategy on how they plan to maintain the overall energy level of the battery throughout the contracted life of the asset. A battery that is contracted for 20 years of use to a utility must maintain a certain threshold of specified energy available for the application. It cannot simply be allowed to degrade down to a paperweight over the course of the contracted life.

As with any other complex product, ESSs have inherent wear out mechanisms and also suffer from premature failures and other unexpected issues. Unlike other systems, the lifetime of a battery is much more dependent on the application the battery has been exposed to. Certain battery systems are designed for long-duration discharges that are used several hundreds of times a year. Other systems can be designed for tens of thousands of micro-cycles that happen very frequently in a highly stochastic manner. As a point of definition, a "stochastic duty cycle" is one that is more or less random – while one could analyze it statistically after the fact, it is virtually impossible to predict the behavior ahead of time. For example, in a PV-paired application where the battery is expected to smooth out the peaks and valleys of the plant, one might be able to predict that a day will be partially cloudy. Nevertheless, the actual pattern of clouds impacting the PV array (and thus determining the battery duty cycle) will be rather random and not precisely predictable.

Other battery systems are designed to be held in a standby mode for years between discharges for emergency backup. As with everything in the energy storage space, it is critical to understand the details of the application and pair the right technology that has the inherent attributes which will allow it to perform the emergency backup when needed.

When considering reliability and product wear out of a battery system, the primary point of emphasis must be the battery cells themselves. Other elements of the system will certainly experience failures and wear out, but their overall cost is typically much lower than the cost of prematurely replacing the batteries in a large-scale installation. As noted above, selecting the right battery for the application is the first step in ensuring proper reliability of a battery system.

9.9 ESS Maintenance and Operational Considerations

9.9.1 Misconceptions, Myths, and Assumptions About ESS

As stated at the start of this chapter, with the rapid expansion of both PV and ESSs, many people believe that these are standardized building blocks that can be easily and quickly configured to whatever specifications are needed without risk. Unfortunately, this is not the case – significant risk exists if the details of a project and the technology are not carefully considered and tailored to the needs of the project and the application. This means that clearly defined battery, battery specification technical and operational, design and integration must be completed prior to EPC bidding. Many of these risks are simple misconceptions that exist in the market – the result of outdated information, incorrect assumptions that have been perpetuated by word of mouth, and rapidly changing market dynamics and product performance. The list below is not all inclusive, but a high-level overview of some of the biggest misconceptions about PV and ESSs.

9.9.2 Plug and Play – PV+ESS

Not exactly!!!

While it is possible to design a wide range of PV plus ESSs, the design of both the PV system and energy storage, and the product selection and the engineering approach for integrating the two, can vary wildly and cannot be considered plug and play. As described in this chapter, not all lithium-ion batteries are the same – the considerations needed for a LFP are different from a NMC battery. Designing a solar plus ESS assuming NMC batteries and switching at the last moment to an LFP battery will likely cause significant redesign of the entire site. LFP batteries are naturally less energy dense than their NMC cousins and a system that needs 5 containers for NMC might need as many as 10 containers for LFP. This leads to more land being needed, additional high-voltage DC wiring trenching, more copper wiring, additional auxiliary loads (from the additional containers), more communication lines to be run (potentially leading to latency issues), and the list goes on. Batteries from various suppliers even within the same class of lithium-ion chemistry can have different operating limitations (and physical/operational/environmental characteristics) that should be accounted for. A NMC battery system from supplier A may allow for discharges between 10% and 80% SoC, while supplier B limits the discharges between 20% and 90%. This might represent the same spread of discharge window, but could have differences in control strategies and performance results. Similarly, the auxiliary load requirements for a container of batteries from a certain supplier could be different than another, qualified inverters may vary, the method of installation and therefore the labor needed, and on and on. Everything is simple while in the design phase of a project, but when it gets down to the specification details of installation and operation, there can be wide differences between very similar products. These small details can lead to some projects being profitable while others are not. This is a fundamental part of the structure we have discussed and why the specification should be determined prior to EPC bids.

9.9.3 ESS Dispatched Grid Service

Again, not exactly! This may seem obvious on the surface, but there is still a sentiment in the market that one battery can do the same applications of any other battery. This can result if battery component issues are not specified early or left up to the EPC. While technically true that all ESSs can charge, discharge, and sit idle, how they accomplish these three states can vary, which means certain applications simply are not viable. Some battery systems, flow batteries most notably, cannot switch instantaneously between charge and discharge which makes them difficult to dispatch in applications with a high degree of stochastic charge/discharge behavior. A long-duration battery, while technically capable of providing frequency regulation services, has the potential to perform much worse than a battery designed for the application. A power-focused lithium-ion battery may be able to provide a full 100% DOD cycle, but this could have a major impact on the life expectancy of the battery. As noted throughout this system, the details are of the utmost importance when selecting a battery for a specific project application, and care should be taken to select the right battery. Even in a PV plus storage system where the battery is to be charged exclusively on PV power, an installation in a geography like the northeastern United States – with more instances of intermittent cloudy days, as compared to the same system in Nevada, with nearly cloudless skies – one battery may perform better than another.

It is a misconception that batteries can do all of these applications well, in some instances certain batteries should not even be considered for certain installations. The specification and design team must clearly define all of the ESS uses and quantity choices carefully.

Where this misconception is most notable is in PV+ESS installations where the battery is charged predominantly from "solar" for the first five years (the term to used previously to qualify for the US federal ITC) and is then expected to move into a different application after this time period.

When the application for the battery changes, it can have significant impact on the lifetime of the battery and potentially void the original warranty or performance guarantee provided by the ESS provider or OEM.

Volumes have been written on the US federal ITC, so we will not attempt to repeat all of the details that are already publicly available. However, there are a few salient points to make as related to this chapter. First, it is important to understand that the primary requirement of the ITC as related to energy storage is that the battery must be charged from renewable sources. At minimum, 75% of the energy that charges the battery must come from renewables; if that "cliff" is not achieved, then the battery system cannot claim the ITC. In order to claim the full ITC tax benefit, 100% of the charge going into the battery must come from a renewable resource. For charging energy between 75% and 100%, the ITC that can be claimed is prorated. Note that this requirement for charging from renewable resources is only enforceable for a period of five years – starting with year six, there is no limitation on where charge energy can come from and there is no longer any recapture of the ITC benefit that the IRS can claim.

What does this mean in practice? For PV+ESS, where the battery will be claimed under the ITC, the job that the battery can deliver for the first five years is limited by the requirement to charge from the attached PV plant. Practically speaking, this limits the battery from delivering grid services that require the battery to absorb power (frequency regulation) and adds an element of risk to any application where the battery needs to be at a full SoC to deliver (capacity, resource adequacy, and spin reserve). Either the system operator needs to assume that there will be adequate solar energy to charge the battery before it could be called upon or must augment the charge with grid power, which will impact the ITC. This is not to say that no grid services can be delivered by a battery during the five-year ITC period, just that there are risks that must be considered and managed appropriately by the system owner.

Moving out of the five-year ITC period opens up the application space for the battery system, but as discussed throughout this chapter, great care should be taken to ensure that the battery – from a technical standpoint – can deliver the application that it is shifting into and that the warranty/performance guarantee contracts allow this shift in application.

9.9.4 Stack Value – Value Stacking

Value stacking of multiple applications is a very interesting strategy for operating an ESS; however, it is not a panacea. This is particularly true when one considers that it must be grid-connected and available to participate in the market. There are both practical and technical as well as regulatory barriers that can prevent energy storage from fully realizing their ability to provide multiple services to the grid and being compensated fairly for all of these applications. Ignoring the technical, use, and regulatory challenges that exist in varying degrees across the country, the ability of a system to monetize all of the value will come down to the ability of the owners to predict what service/s to participate in (and when) as well as the status of the battery at the start of providing that high value service.

As noted in Fitzgerald et al. (2015), it is important to understand the potential entire range of benefits that grid-tied PV+ESS systems offer to a broad range of stakeholders. Those services provide stakeholders with a broad range of "energy value" in reliability, availability, power quality and consistency, cost control and customer access.

Those shared benefits serve all energy users including Independent System Operator (ISO), Regional Transmission Organization (RTO), utilities, and Customer services. Service examples include:
ISO/RTO Services

- Energy Arbitrage,
- Frequency Regulation,
- Spin/Non-Spin Reserves,
- Voltage Support,
- Black Start.

Utility Services

- Resource Adequacy,
- Distribution Deferral,
- Transmission Congestion,
- Relief Transmission Deferral.

Customer Services

- Time-of-Use Bill Management,
- Increased PV Self-Consumption,
- Demand Charge Reduction,
- Backup Power.

Therefore, specific use cases and stakeholders receive a complex, intertwined set of fiscal and energy quality benefits, not just the ability to use energy at a later time.

For example, an ESS with perfect foresight into future value streams may call for the battery to provide frequency regulation for the morning hours and then be ready to provide resource adequacy (capacity) during the afternoon peak. In practice, a battery typically is around 50% or 60% SoC when providing frequency regulation, so that it is able to both adequately discharge and charge. However, when the battery needs to move to providing resource adequacy, the system should be fully charged. This means that the battery cannot simply shift from frequency regulation to resource adequacy and expect to receive full value for providing capacity (a 10 MW, 40 MWh battery operating at 50% SoC for frequency regulation will have 20 MWh of stored energy in the system. When moving to resource adequacy, the battery must be able to provide four hours of continuous power, which means this system will only be able to bid 5 MW into the market thus only realizing half of the total potential value that the system has.). It could be possible that the battery will be recharged between providing frequency regulation and resource adequacy, but the costs to recharge could be prohibitive.

In another example, an ESS that primarily relies on energy arbitrage to capture value from critical peaks (in a market like the system operator in Texas – Electricity Reliability Council of Texas, ERCOT) cannot afford to miss any of those peak prices or only be partially charged going into those peaks. This is where the battery system can actually only perform as well as the operator has foresight of what is going to happen in the market. Economic stories that are built on the assumption of perfect foresight for value stacking energy storage applications rarely are realized.

9.10 Considerations

As with any new technology, mistakes are inevitable as the market continues to learn realistic expectations for ESS as well as the market rules continue to evolve to allow batteries to participate in the market in new ways. It is impossible to fully retire all risks (technical and commercial), but those below are some of the most common mistakes that we see made in the market by developers, specifiers, engineers, and system owners new to energy storage.

Battery system wears out prematurely because the developer and/or downstream stakeholders forgot that "a cycle is not a cycle is not a cycle" or they do not have control of the battery system.

As described at length in this chapter, the operational profile that the battery is exposed to can be the single biggest driver in the overall lifetime and performance of the system. A "cycle" that takes a battery from 100% SoC to 0% SoC will have a very different impact on the lifetime of the battery as compared to an application where the battery charges and discharges rapidly between 55% and 45%. Moving from one application to another throughout the lifetime of a system can increase the value that the system generates, but can lead to much quicker wear out and replacement. Having a precise definition of the application/usage that the battery will perform is one of the most critical elements of designing a PV plus storage project.

Depending on the commercial structure of the project, the owner of the ESS should also be very cautious not to allow the system operator to use the battery outside of the intended (and warranted) duty cycle. For installations where the owner of the system is responsible for the battery performance and augmentation, setting appropriate limits on cycles per day, total energy discharged per day, upper and lower limits on SoC, and application type is critical to consider at the start of the project. By making these decisions early (at specification), the system operator can avoid "irreversible damage" on the battery system. This requires addressing all stakeholder interests as they are impacted by the ESS without making any assumptions that have been clearly vetted. These application limits should be applied and considered back-to-back with the warranty details provided by the battery system provider. Additionally, it is not sufficient for just the battery professionals on a team to understand these details. They must be clearly presented in a clearly understandable manner to O&M and management to avoid serious negative consequences.

Project economics do not look as attractive because the auxiliary loads and the energy efficiency losses were misunderstood, underestimated, or forgotten altogether.

All ESSs are net consumers of energy – no matter what claims a new battery company makes, they cannot circumvent the second law of thermodynamics. The conversion of energy from AC to DC and back again has an efficiency loss. There are also electrochemical losses that occur within the battery during recharge in addition to conversion and wiring losses. Even the most efficient energy storage chemistry is only 92–93% efficient – meaning that to store 10 MWh of energy, 10.75 MWh will need to be put to the DC terminals of the system – approximately 11 MWh to the AC side of the bidirectional inverter. Other energy storage products can have AC-AC round trip efficiencies as low as 70% or lower.

In addition to these efficiency losses in the system, there are other auxiliary loads that must be served separately and considered in the operating expenses for the project. Inverters, fans, BMS and other control systems, monitoring devices, fire suppression, HVAC, etc. all require separately metered power from the grid to keep the system running or in a standby mode. While these loads may not be considered significant, over the course of a 20-year project lifetime, they must be considered.

That being said, when delivering PV+ESS systems as energy infrastructure, the project lifetime and potential usages can be dramatically expanded. This means that applications, technologies, and product capabilities will change more substantially, which needs to be considered at concept and specification. It can have a serious impact on space allocation for system expansion, changes, and O&M efforts.

Lastly, all ESSs suffer from some degree of self-discharge. This is when a battery sits idle (neither charging nor discharging) and loses stored energy from internal secondary and tertiary side reactions. While self-discharge is very low compared to efficiency losses, for an ESS that is expected to sit idle during the winter months, for example, the energy lost will need to be recharged back into the battery before use.

It is important to remember that every MWh of energy that has to go to the ESS that is not stored (for efficiency losses and auxiliary power needs) cannot be sold back and therefore is a lost opportunity cost to the project.

When the time comes for a common version of repowering (extreme maintenance/rebuilding) the ESS, the costs are significantly higher than modeled because there was not a qualified augmentation, growth, recycling, or replacement strategy. In other words, there was no **SE/R**epowering™ planning or strategy!

This is another area where adopting our PV **SE/R**epowering™ planning processes and procedures can address details that are often overlooked or incompletely addressed. Many concepts and options that were originally considered must be documented along with the decision processes for the initial specifications, product, and serviced choices. That initial documentation and its curation can improve decision-making choices. At times, it may uncover site or other project peculiarities that may not be remembered, yet have a positive or negative energy, dollar cost, or time impact if ignored.

Battery technology costs have dropped significantly over the past decade – for several years achieving double digit percent reductions. However, it is not reasonable to expect this same pace of cost reductions into the future and, at some point, lithium-ion batteries will be reaching the practical limits of cost reductions without major improvements to the energy delivery of the cells themselves.

Modeling battery augmentation (or replacement) costs in the future at artificially low levels can allow projects to meet initial investment hurdle rates, yet may not be reasonable in the future. If project owner/s are looking to sell the plant in the future, this could cause a project to be unattractive to other buyers taking a more realistic view on battery costs.

The warranty of the battery system was voided because the application changed and the owner was locked into a long-term warranty or performance guarantee contract.

In both behind-the-meter and front-of-the meter installations, there is the possibility that the application that the battery needs to deliver will change. Behind the meter, this could be due to changes to the load profile of the building or a change to utility structures that will make it more advantageous to dispatch the battery in a manner different than originally designed. Front of the meter, there is risk that lucrative markets today will not be as financially advantaged in the future. Installing an ESS to provide capacity for shifting solar energy today will not carry with it the right warranty and guarantees if the frequency regulation market of the future becomes very attractive. Battery warranties and performance guarantees are written for the expected life of the battery (~8–10 years), under specific stated conditions. Therefore, it is necessary to be very certain that the application the system is designed for on day one will be applicable by the end of the life of the system.

The market or policy changes in an unanticipated way that no longer favors energy storage.

A risk that is difficult to quantify and hedge against are future regulatory and policy changes that would render ESS paired with PV to be less valuable in the future than they are today. ESSs are a unique asset on the grid due to their ability to absorb excess power as well as provide power on demand. When thinking about the need for energy storage to balance the intermittency of renewables – both at the local level (attached to the PV system) as well as at the macro level (large-scale, stand-alone energy storage) – at some point the value that energy storage brings can begin reaching diminishing returns. Batteries used to provide frequency regulation for the grid diminish the value of each successive MW of batteries connected to the grid for the same application. The great wave of ESSs is coming, but it is not clear what the market of the future will value and/or accept.

9.11 Electric Vehicles as Grid Storage

While the focus of our book is not on EVs, they will have a major impact on the use of electric motive power, the need for massive amounts of new levels of electrical energy production, distribution, and storage and the technologies to meet those needs.

This usage is in good part due to the shift from fossil fuel electrical generation and the replacement of hydrocarbon fuels for cars, trucks, heavy equipment, and a myriad of other common uses today. This transition is one of the many critical reasons for why building PV+ESS+EV infrastructure is critical and the waste from existing practices must be replaced by a new business and test of technical models.

Considerations in preparation for the policies necessary to deal with rapid policy and tsunami of technology change:

- Who regulates the cost and availability of energy?
  - Federal, state, or local?
  - PUC, state legislature, AHJ?
  - Private individual/s or corporations?
  - Other

- Who owns the EV ESS, the available energy, the charging BMS, the charging station (plug in charger)?
  - Utility
  - Private or public business
  - Home, commercial, or other owners
- What is the rate structure for residential, multi-family, commercial (offices, retail, others, etc.), industrial, utility?

It has been postulated that EVs can act as energy storage for utilities and homes by using them to store energy during the day, at the office or while shopping. The assumptions for these scenarios are diverse and, combined with human behavior, tend to complicate those assumptions. …And of course, the results!

Today, those assumptions include:

- Depending on the energy generation resource, personal, commercial, or utility grid, there will be sufficient energy at a reasonable price on demand.
- The EV battery is partially to fully charged during the day at home or potentially another location like at an office or retail.
- The EV is charged at night when rates are supposedly low.
- Determination of which scenario, rate structure, and energy availability will drive energy charging availability versus the available rate structures.

In our professional work, research, and endless study, PV/related standards development, and committee work, we continue to see and hear more questions than answers. How we as an industry and community respond will drive the level of success and cost of electricity. At this nexus in time, the good news is that the entire planetary community does not have the answers on how to deal with these issues.

We know this is a fact, because we, our friends, organizations, elected officials, policy makers, manufacturers, and the plethora of users of today, do not have sufficient information to identify the most much, less least pertinent questions.

Without developing a better grasp of the myriad questions and use cases, addressing them socially, technically, and fiscally, and addressing the path (RoadMap) to the future, many incredible opportunities can be lost.

To minimize significant impacts to service life and life cycle costs, "Assume nothing"?

Simple assumptions like someone drives home and, connecting to their home system, boosts the home system with the "extra" energy for the evening hours, because the rates are lower, may be a bridge too far. This is an interesting idea yet there are some obvious and not so obvious problems:

- Energy stored during the day in the car and then used at home may leave the car without any energy to drive to work the next day. Even if the home has two-way capability (grid to home and home to grid), EV batteries cannot be charged and discharged too quickly. DOD also plays heavily into the life of the battery, which becomes an additional cost burden to the home owner.
- Energy stored in the EV during the day at the office or other location requires a substantial ($Billions/€Billions) investment in infrastructure to boost the capability of the business area to support charging hundreds to thousands of personal vehicles.

- In cities like New York that rely heavily on public transportation (buses, taxis, and trains for commuting), EVs would not work as they are continuous users of electricity with limited time each day for charging. Towns below a certain threshold may never reach that threshold. Small towns represent a substantial portion of the population of the United States. The same holds for many parts of the European Union and the rest of the world, where there are large numbers of small towns, and in some countries, there are more bicycles than cars.
- The issues must be effectively identified and hopefully community, local, regional, and national roadmaps will be developed to aid in a transition that is far greater than the initial introduction of electricity.

We do not have the answers. At this time and like the rest of the planet, we all need to ask the questions that will populate that new future energy roadmap.

Or, we can get the results one might if they treated the questions like this:
Alice: "Would you tell me, please, which way I ought to go from here?"
The Cheshire Cat: "That depends a good deal on where you want to get to."
Alice: "I don't much care where."
The Cheshire Cat: "Then it doesn't much matter which way you go."
Alice: "...So long as I get somewhere."
The Cheshire Cat: "Oh, you're sure to do that, if only you walk long enough."
Alice's Adventures in Wonderland, Lewis Carrol, 1865

9.12 Summary

In the past 100+ years, various chemistries have come and gone in the marketplace and the overall performance of the battery systems that have remained continue to press to higher and higher levels. The reader will note that the authors have not discussed specific performance attributes of these ESSs – no life cycle claims, and no discussion of energy efficiency.

With the rapid pace of development of energy storage chemistries and typologies, as soon as numbers are published with "state of the art" performance, they are outdated. Instead, the purpose of this chapter is threefold:

Introduce the reader to the attributes behind these battery performance metrics, and leave the reader to seek the latest performance numbers and the attributes that best match their application needs.

With ever-increasing interest in battery technology, we should expect that advancements will be discovered at an ever-increasing rate over the next hundred years as well. With these changes and the accelerating perspectives shifting across the industry, ESS education will continue to require continual updating and refreshing.

Bibliography

Balfour, J.R., Hill, R.R., Walker, A. et al. (2021). Masking of photovoltaic system performance problems by inverter clipping and other design and operational practices. United States: N. p. https://doi.org/10.1016/j.rser.2021.111067.

Battery Universe Blog Tips, news and musings about batteries and the battery business https://www.batteryuniverse.com/blog/tags/cycle-life/.

Buchmann, I. (2016). *Batteries in a Portable World*, 4e. Cadex Electronics Corp.

EPEC Engineering website. https://www.epectec.com/batteries/cell-comparison.html (accessed 08 August 2023).

Fitzgerald, G., Mandel, J., Morris, J., and Touati, H. (2015). *The Economics of Battery Energy Storage: How multi-use, customer-sited batteries deliver the most services and value to customers and the grid*. Rocky Mountain Institute, September 2015. http://www.rmi.org/electricity_battery_value.

IEA reports website: https://www.iea.org/reports/solar-pv (accessed 08 August 2023).

Kolly, J.M., Panagiotou, J., and Barbara, A. *The Investigation of a Lithium-Ion Battery Fire Onboard a Boeing 787 by the US National Transportation Safety Board*. Czech of the National Transportation Safety Board https://data.ntsb.gov/Docket/?NTSBNumber=DCA13IA037.

Office of Energy Efficiency and Renewable Energy (2019). Energy.gov. Solar-plus-storage 101. https://www.energy.gov/eere (accessed 08 August 2023).

Qi, Z. and Koenig, G.M. Jr., (2017). Review article: flow battery systems with solid electroactive materials. *Journal of Vacuum Science & Technology* B 35: 040801. https://doi.org/10.1116/1.4983210.

Sila. 8 battery metrics that really matter to performance (30 March 2022). https://www.silanano.com/news/8-battery-metrics-that-really-matter-to-performance.

VanZwol, J. (2021). How does temperature affect your choice of Lithium UPS Battery? Schneider Electric website. https://www.datacenterdynamics.com/en/opinions/how-does-temperature-affect-your-choice-of-lithium-ups-battery/.

Notes

1 The Union of Concerned Scientist, REPORTS & MULTIMEDIA/EXPLAINER "Energy Storage How It Works and Its Role in an Equitable Clean Energy Future" Published 19 February 2015 Updated 4 October 2021.

2 Boeing, the FAA and the 787's Japanese battery maker shared the blame for the fire, which the NTSB determined was caused by a short circuit that led to thermal runaway in one of the battery's cells. https://www.flyingmag.com/news/cause-boeing-787-battery-fires-pinpointed-ntsb/.

3 Lithium-ion batteries, the primary choice for electric vehicles, are known for their potential to catch fire, although the incidences are rare. Last year, a Chevy Volt caught fire a few days after being crash tested. The problems are not limited to cars. Boeing's 787 Dreamliner was grounded after fires in the plane's lithium-ion battery. https://spectrum.ieee.org/energywise/transportation/advanced-cars/teslas-lithiumion-battery-catches-fire-.

4 "PREVAILING WAGE SCHEDULE FOR ARTICLE 8 PUBLIC WORK PROJECT," Project ID# 1000509999, 2020.

10

Data Collection

Key Chapter Points

- How to gather sufficient data to functionally improve PV plant and component durability, reliability, productivity, and production.
- Types of necessary and usable data (as addressed in Chapters 5–7).
- How to move from corporate sensitive data (due to IP/NDAs) to sharing industry data that benefits all stakeholders.
- Providing examples of a process to acquire, analyze, and share data across the industry.

Key Impacts

- Deliver a better understanding of the plant equipment's reliability and maintainability attributes.
- Stakeholders reap the benefits of better specifications driving the design of systems and lower lifecycle costs.
- The types and data essential for continual improvement are curated to improve RAMS attributes over time.
- Owners gain more profitable and productive extended life plants.

10.1 Introduction

> "If you don't look for PV plant problems and failures, you're probably not going to find them."[1]

The philosophy behind this chapter is to address the technical, financial, and even some political realities that tend to create adversarial camps and obfuscate and confuse the value of publicly available acquisition and distribution of data. The intent of this chapter is to address the importance and need for data collection and analysis to improve the reliability and delivery of solar-based power and its economics. Under the proposed best practices, the organizational operations and maintenance (O&M) team identifies PV power system (PVPS) operational problems and failures, acts on them as required to return the plant

Photovoltaic (PV) System Delivery as Reliable Energy Infrastructure, First Edition.
John R. Balfour and Russell W. Morris.

to full operational status and documents all actions needed to accomplish this. Industry experience in PV indicates that many PV professionals feel monitoring is a burden, just a contractual requirement with little or no benefit. This book presents several different FRACAS forms. This is meant to only provide the reader some suggestions.

The O&M team is tasked with monitoring energy collection, output, and performance, as well as providing scheduled and unscheduled maintenance per contract. They tend to be the principal gatherers of plant failure and performance data, and thus it is required that a sufficient budget be allocated for that task alone. That budget item must also be protected, i.e. not allowed to be shunted aside to make up for other short falls. Failure and performance data collection, analysis and curation are needed in order to correct RAMS-related issues that are identified early and resolved in a cost-effective manner.

It should be noted that almost every organization, owner, O&M, stakeholder, government (from local to national and international), etc., has certain data needed to satisfy legal and statutory requirements, but in different detail, type and to whom it is distributed. This is where the value of a database such as Microsoft™ SQL, MySQL™, Oracle™ Database, ACCESS™, SAP HANA™, and many more comes into play. As with any commercially available products, it is necessary to assess the ease of use, the type of data that the tool is optimized for, and the ability to generate the needed reports for all of the affected stakeholders. The tool(s) should also provide for automatic data RAMS analysis based on standard data collection elements and analysis processes.

Since many of the data elements for a Failure Reporting and Corrective Action System (FRACAS) tool can be held in lookup tables, the number of true variable fields (such as type, nomenclature, etc.) can use default selection drop-down menu items. Examples include; "Combiner; 8-Input, 5 kA, 1000 VDC, or Inverter; 100 kWAC, Skid Mounted." There is no requirement for specific vendor information. There is, however, always unique equipment for which anonymized data still points to a specific vendor. In these cases, it is recommended that dummy variables be used. The intent is to create a series of categories that allows for gathering critical information while legally protecting corporate names and intellectual property (IP).

The focus of this chapter is to address a major weakness in the present PV project model that fails to address the need for consistent and detailed data acquisition and analysis. The results of the analysis are the foundation for improvements to existing plants as well as improving new plant designs that assure future PVPS as reliable infrastructure.

To address the major questions, it is essential to be able to have usable information to act upon. In Chapter 6, we covered the FRACAS system of data acquisition for information to support reliability and maintainability attributes of the system and equipment. This relates to single or multiple failures that occur at a single point in time and are documented through the FRACAS process. Data from a single plant constitutes a single subset of data set that can be aggregated with other plant information to provide a data base of failure rates by component category, type (as appropriate), operating conditions, and site environments.

The questions data acquisition needed to be asked and answered are:

- "What are the operating and environmental conditions of failures at the time of occurrence and what does this tell us about the state of the plant?"
- "What is the plan for and investment needed in acquiring, analyzing, and curating a modest learning curve that will pay for itself many times over?"
- "The equipment/component how, what, when, where, and how bad of the failure?"

The fact is that developing a "data plan" during the initial project systems development is as important as the initial systems engineering (SE) for new and repowered plants. Its value, cost-wise, is that once established for one project, it is for all intents **SE/R**epowered™ and purposes identical for each follow-on plant for that developer, Engineering, Procurement, and Construction (EPC), and O&M. It is a clearly structured set of detailed requirements, processes, and procedures that owners and O&M personnel can be trained on. However, it is important to understand that this data acquisition does not just support SE or the design team and the other technical stakeholders! A successful data plan addresses the short, medium, and long-term needs of all stakeholders by identifying what metrics they need and issuing reports as requested or contracted. In other words, what the stakeholders do not know and understand can injure them and the project.

As noted in Chapters 1–5, the first step is to understand the system and its requirements by resolving many stakeholder requirements conflicts at specification prior to EPC bidding. Doing so results in a tenfold benefit in cost reduction and later cost avoidance. Practicing good PV system data collection and analysis pays benefits that exceed the costs of performing the tasks. The initial cost is the upgrading of the SCADA software (SW) to allow for improved fault detection and isolation and handling of the additional sensors needed to monitor the plant. Combined with connectivity with inverter status reporting, data analysis provides the understanding of what is the health and status of the system.

A small team of a skilled database designer, SE/RAMS engineer, O&M representative and the owner can take all of the project stakeholder data requirements and convert those to a workable system in less than three months.

One important aspect of this effort is that it can also feed data into the plant model, add data to the Typical Meteorological Year (TMY) database, provide feedback to the EPC and their OEMs, and provide the current status and trends of plant operation. Data acquisition and tracking are seldom looked at or understood by stakeholders to be a key tool for not just identifying anomalies, failures, and outages. It can be like a scalpel for determining causes and impacts to the system and its outputs and result in implementing improvements in the current plant while applying the information to develop more robust systems. Our emphasis is on understanding the system's health and condition, which goes beyond identifying failures or outages, while tracking performance characteristics, which supports the move toward preempting issues, lowering actual costs, and revenue risk reduction.

Over a century ago, the electrical utility industry determined that it was critical to determine what the reliability of components was and the level of repair necessary to keep them within clearly defined operational parameters. The emphasis was on the predictability of operations and "all in" cost. The outcome of this demand was the development of International Electronic and Electrical Engineers (IEEE) 493 Gold Book, now discontinued, which provides reliability and maintainability data for major utility power components.

As with utility and other infrastructure technology, the potential useful life of a PV plant can be 50 to over 100 years or more, *if that decision is part of the long-term business and revenue plan*. That is the foundation for the development of solid SE and RAMS work that also supports the realities and value of **SE/R**epowering™ planning and implementation in PV as infrastructure.

All too often, a system is purportedly designed for a 20- to 25-year life with a statement (not documented or contracted) that it can be upgraded or repowered with new technology at low facility cost at some point in the future. As previously noted in Chapters 4 and 5, these systems often barely last 7–10 years before there is significant degradation and or failures that are a prelude to a loss of capability to produce sufficient energy to meet contract. To have reliable infrastructure requires that we answer one fundamental question: *What is the actual health and condition of the system?*

10.2 Reducing Risk Begins with Data

The foundation for PV, as reliable infrastructure, relies on establishing fundamentally sound system design, and performing validation and verification of specification during EPC design and at commissioning and benchmarking.

Data needed to provide the assurance of compliance can be provided by the implementation of remote sensing and reporting and repair technician data entry to an O&M database. Benefits for the stakeholders include:

1. Understanding plant health, condition, and status,
2. Documenting equipment condition and trends,
3. Tracking maintenance times,
4. Supporting improved root cause analysis (RCA),
5. Documenting maintenance actions required on warranties,
6. Providing visibility into the cost (labor and equipment) needed for nonroutine maintenance, validates plant's equipment capability to generate power when demanded (Availability), and
7. Identifying O&M management costs.

This information effectively supports operating reductions in CAPEX, OPEX, and LCOE. For example, doing a simple financial analysis of squeezing 2%, 5%, or 10% more operational time out of a component over a 10- or 20-year period, dramatically changes the complexion of an LCOE. An added benefit is that the plant will be capable of producing more energy at a lower cost, and for the owner, it results in retaining the plant's commercial asset value. Given this, there is a lack of agreement on what is needed and what is possible.

So, what is driving this disparate industry view of what is and what is not possible? The simplest understanding occurs when one considers the need for but lack of acceptable industry-wide data metrics about the component and equipment RAMS attributes. Data based on agreed common metrics for RAMS, anonymized R&M data acquisition, and analysis. It must be handled through an *industry*-supported third-party storage and curation function. This benefits the whole industry, creating substantially better systems with superior performance and output.

10.2.1 Component and Equipment RAMS Data Sharing

The original need for a common RAMS database for utility electrical power equipment was recognized as far back as the 1960s. The IEEE started studying and gathering data from various industry sources in the 1970s with the result of the publication of the

"IEEE 493-1980 – IEEE Recommended Practice for the Design of Reliable Industrial and Commercial Power Systems." The data provided industry with a common nomenclature and standardized metrics. This has been expanded and improved a number of times, the latest of which is the IEEE Std 493-2007. As of November 2023, an IEEE standards committee is updating and revising the current 493 with expected publication in April 2027.

The IEEE 493 Gold Book provides a multi-decade summary of the average (or mean) failure rate for a large number of Power Industry components and equipment.

The providers of anatomized data would include stakeholders such as:

- Manufactures/suppliers (OEM),
- Third-party suppliers,
- EPCs,
- Supply houses,
- Project owners and/or asset managers,
- O&M providers.

The big missing ingredient after providing the information is anonymizing the participants. This would be based on setting up a series of anonymized accounts where only the participant and the industry-supported data management team have access to that data. This does require a certain level of security for the database as well as nondisclosure agreements from all personnel that have access to that database. All part-sharing participants/owners/buyers/providers are given highly secured random account numbers that provide no indication of a specific owner or project location.

10.2.2 Current Mandatory Reporting

As of 2023, the industry, through several organizations (IEC, North American Electric Reliability Corporation [NERC], or others) has standardized some elements of the information needed to understand the health, condition, and status of the equipment.

NERC's GADS maintains operating histories on more than 5000 generating units in North America. GADS is a mandatory power generation industry program for conventional (nonnuclear) generating units greater than or equal to 20 MW. Table 10.1 provides a description of the event identification and the event cause. What this does not provide is what about the event cause actually drove the downing event. Flooding, for example, may not actually cause the event, but a hole in the equipment caused by corrosion allowed the water to get into the equipment. The corrosion may have started when blowing sand eroded the paint away, exposing bare metal and allowing corrosion to occur. Cold weather conditions may be just temperature or an ice storm depositing 3 in. (7.12 cm) of ice on the module's surface. This results in approximately 14.3 lb/ft^2 (6.5 kg/m^2) additional load on the module and the structure. So, for a module measuring 7.2 ft by 3.43 ft or 2.22 m by 1.05 m, there is approximately 24.89 ft^2 or 2.31 $m^{2;}$ the module weight increases to 355 lb. or 161 kg. This may be sufficient to cause damage to the mounting hardware, cracks in the glass, or to break the module itself. Addressing weather as being "cold" does not provide the information needed to properly identify the stresses and what the design must accommodate.

At the current time, the amount of work time is voluntary. This is, however, a major component of understanding the true cost elements of any LCOE.

Table 10.1 GADS Table III-4: record layout of section C – primary cause of event (records 02 and 03).

Column ID	Number of columns	Starting position
Record 02		
A – Event Identification		
Record Code (required)	2	1
Utility (company) Code (required)	3	3
Unit Code (required)	3	6
Year (required)	4	9
Event Number (required)	4	13
Report Revision Code (voluntary)	1	17
Event Type (required)	2	18
C – Primary Cause of Event		
System/Component Cause Code (required)	4	20
Cause Code Amplification Code – (required for U1 events coming from in-service only; strongly recommended for all other events.)	2	24
Time Work Started (voluntary)	8	26

10.2.3 Proposed Data Format and Elements

The following is an example minimum set of data to be submitted:

- Date, Operating time in hours,
- Failure Mode Code,
- Failure Mechanism Code,
- Equipment Type (Inverter, Combiner, Module, Motor Controller, etc.),
- Class (1000 A/1000 VDC, 100 kWAC, 300 WDC, 400 WDC, etc.),
- Root Cause Analysis (RCA) result,
- Repair Code (Remove and Replace (R&R), Repair in Place (RiP), Clean),
- Repair time (Reporting to Work Order Closeout),
- Number of personnel required (Total needed to closeout Work Order),
- Environmental Zone (A – Coastal, B – Inland Desert, C – High Desert, D – Mountain >5000 ft (1.6 km)),
- Site Environmental data at the time of failure
  - o Altitude, km, ft,
  - o Temperature, °C/°F,
  - o Humidity, percent, and
  - o Weather Conditions (overcast, cloudy, raining, snowing, etc.),
- Repair Action Code (Return to Vendor for repair (RTVr) or replace (RTVp), Return to Vendor – Warranty (RTVw), Repair on Site (ROS), Repair at Shop (RAS), etc.), and
- Other data as determined by stakeholders.

The incorporation of repository data provides an early look at the RAM attributes of the proposed or existing plant being SE/**R**epowered™ during the concept, systems engineering, and design phases. The environmental data by area or region forms the basis of statistical analysis of the mean, high and low extremes, and the 95% confidence level calculations for specifications. This last data should be compared to data from the planned site to verify the appropriateness and applicability of the proposed design.

By type, for example, inverter, model number, operating time, and the total number of failures or performance challenges provide a baseline estimate. The limitation is that the data is comprised of an aggregate of manufacturers and thus only shows the mean or average of

Date	End Item	Vendor/Mfg	Resp Eng	Reliability Eng	Failure Rpt No.
/ /					

Failure Date	Plant	Op Time (hours)	Equipment Loc	Personnel
/ /		________.__		

Item Nomen	Serial No.	Part No.	Rev
Failure Description			

Failure Report/Date	Fault Detection Method	BIT		BITE	
Failure Report					
Corrective Action					

Corrective Action Approved	Y/N	Date / /	Management Approval

Group	Date	Signature	Group	Date	Signature
Quality			Finance		
Design			Software		
Systems			Material		
Reliability			Subcontracts		
Supplier			Customer		
Notes					

Figure 10.1 Example FRACAS Form for Data Collection.

a type. Reviewing the Failure Modes and Repair codes allows for looking for weaknesses in past systems and using them to work with the OEMs in particular to ensure that these areas have been addressed.

The form shown in Figure 10.1 is one implementation for data acquisition. The form can be built for a standard tablet, notebook, or laptop. The various cells would use drop-down menus with fixed responses but need to have an "other" categories for the occasional odd problems. The data covers the fault/failure, its source, and the actions taken to perform the RCA, responsible persons, and final closeout.

One other example is shown in Chapter 11, which used Postscript SQL™ with data stored on a cloud server – time scale™ and based on Python language. Since this a database-intensive application, there are possible other languages such as C++, C#, VB, NET, C, and Java™ script.

10.3 Shared RAMS Data

There are four standards that help identify strategies to move from a *high-risk cost and performance PV industry* to energy production with photovoltaic (PV) systems valued and delivering as *Reliable Energy Infrastructure.*

- The IEC Reliability Standard (63265),
- The IEC Availability Standard (63019),
- The IECRE PV system rating and certification standard (Began development 2020),
- The emerging reliability data from component and system failure reporting, and
- IEEE 493-2007 Gold Book (Inactive to be updated through IEEE PAR 493).

These standards and methodologies provide system owners and operators with increased revenues from more consistent energy production (increased equipment availability). It also reduces spare parts quantities, reduces overhead, and reduces maintenance unbudgeted time, energy, and materials. Further, there is reduced OPEX (fewer mobilizations for forced outages and fewer lifecycle replacements), thereby making projects more profitable. Increased profitability leads to sustainable industry growth.

An important element in establishing the level at which maintenance (scheduled and unscheduled) documentation is required is the Level of Repair Analysis (Chapters 5–7). This list identifies the equipment or component for which spares and repairs will take place in the event of a fault. Examples include but are not limited to:

- Modules,
- Combiners,
- Cabling,
- SCADA hardware,
- UPS,
- Inverter PC cards/assemblies/air conditioning,
- Inverter Container,
- Whole inverter (where excessive damage has occurred to major parts of the item), and
- Energy Storage Systems (ESS).

10.4 Stakeholders

The challenge with many current PV projects is that some stakeholders may not have been clearly identified, or their participation at critical points in the early project phases has not been solicited or adequately included. The challenge is identifying and bringing the stakeholders together to understand and communicate at project levels and stages not previously addressed. These discussions are supported by the aforementioned standards. However, success requires stakeholder participation in a voluntary opt-in anonymized program where the stakeholders share specific kinds of nonproprietary data for components, system health, condition, and performance. In addition, there is a need to provide some proprietary data that is anonymized to aid in equipment/component highest level of usable data. There is, however, no requirement to participate. It would be a voluntary program but nonparticipation may result in restricted access to any information that is supplied by actual participants. It is important to note that the NERC GADS program of data is required to be provided for all plants of 20 MW and greater. Participants reap the benefits of this previously unavailable information data set for end items and components.

Figure 10.2 is a summary of the major stakeholders currently in most of the industry. There are potentially a number of additional stakeholders that can be identified for a specific project. Identifying and engaging the stakeholders is the first step to creating a cohesive data-sharing process environment with procedures, requirements, recommendations, and the key ingredient – buy-in and participation. Although the standards will define the elements and processes, stakeholders have the choice of opting in or not.

As the market matures, the principal source of funding, insurers, financiers, and owners, have become focused on unreliable operations data and their sources to accurately assess the risk and value of plants at any time during a plant lifecycle. As they become more aware of the problem areas, their business responsibilities will require them to seek standards that more clearly define their participation in projects and potentially force the projects to follow their guidelines/requirements. Doing so will result in limiting finance and insurance based on existing widespread inconsistent bankability assumptions held by many past and current projects.

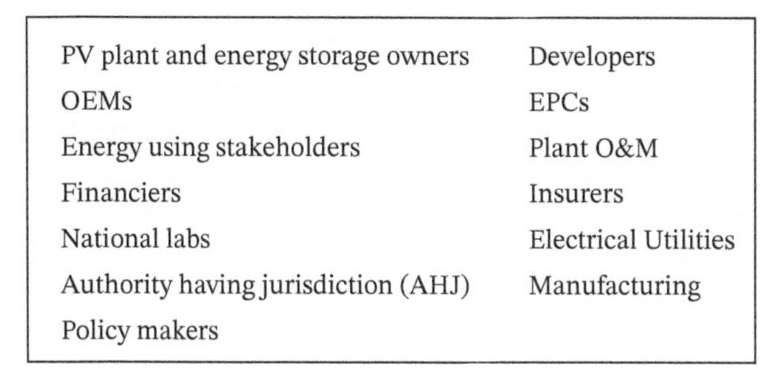

Figure 10.2 Major Stakeholders.

10.5 Anonymized Plant Data

To facilitate an industry data base and avoid legal entanglements, the data would be anonymized. This would be done, for example, by substituting specific traceable site information for climate zones, altitude, and weather information. While there is higher level (e.g. GADS) data sharing schemas in place, at this time they are seldom or not effectively complete or organized. A predefined set of climate zone information is shown in Appendix 10.A. Doing so provides usable and comparable data to make consistent industry-wide reliability determinations at specification prior to the EPC bidding and follows all the way through site restoration. This allows the EPCs to bid on the same plant design and provides greater clarity when making the selection of potential EPC candidates. Doing so provides a level playing field of comparative data contained in proposals. It will also allow owners and operators a clearer understanding of component lifecycles and the number of spares they will actually need.

Data that would specifically be excluded is:

- Organizational name,
- Project name,
- Project location,
- Model number,
- Vendor/supplier name,
- Any other information that would identify the specific supplier, site, or owner.

10.6 Stakeholder Business Case for Sharing Reliability Data

Component-based system health and condition drive Inherent and Operational Availability as a fraction of specified capability, which defines the energy production impact resulting in system revenue. Focusing strictly on performance metrics provides a false sense of understanding of risks and the cost of operation. This PV industry focus has led to misunderstanding and a false sense of security on the plant's actual health and condition and value. By not paying attention to the details, the commercial value of the plant plummets early in the plant life and, generally, just after the warranties have closed out.

Some of the issues of not having clear and concise data were addressed in the 2018 PAM paper, where we looked at performance impact and the need for standardized processes and data acquisition, public data storage, and analysis. This relationship was identified and clarified early on in a NASA paper published[2] in the 1970s. The paper focused on the importance of system availability as it relates to reliability and clearly illustrated the connectivity to performance. This relationship was cast aside during the development of later PV system financing requirements that shifted the focus primarily to system cost as the key metric ignoring what was taking place in the system itself in real-time.

For NASA, the bottom line came down to O&M as one of the most important budget items for life cycle cost control. Their focus was on building more robust (reliable) systems because of the cost of an extraterrestrial servicing mission. For example, the NASA mission to the Hubble Telescope to repair a manufacturing flaw (astigmatism), had a total price tag that was at least US$1.7 billion. NASA also at one point in time had a philosophy of "faster, better, cheaper." This approach, more often than not, led to faster and cheaper, but not better, as evidenced by some Mars mission failures[3] with the associated loss of science and more than 5 years to replace the mission elements with new spacecraft.

> NASA estimates the cost at between US$1.7 billion to US$2.4 billion. However, documentary support for portions of the estimate is insufficient. For example, NASA officials told us that the Hubble project's sustaining engineering costs run US$9 to 10 million per month, but they were unable to produce a calculation or documents to support the estimate because they do not track these costs by servicing mission.[4]

Although Hubble is an extreme example as it relates to data collection, reliability, availability, and data collection, the fascinating part was the gaps in the actual costs. These were not only higher, but after a 15-year period, NASA still had not completely determined all the costs.

Addressing the chain of issues associated with the true cost of low reliability or, more appropriately, the significant impact on the life cycle costs requires a structured approach beginning with the acquisition of data. For the IEC 63019 committee, it was resolved using the steps illustrated in Figure 10.3. The figure was further modified for the IEC 63019 Availability Standard but indicated that both performance and production cost are impacted by component health and condition.

It is important to note that the difference between the two graphics shift from talking about a performance impact to a production impact. The difference will only become clearly recognizable after the system has degraded below the system DC overbuild or more rapidly if energy storage is involved.

It should be noted that overbuild is needed to compensate for plant module and inverter degradation of 0.5–1.5% per year and to handle the failure of large sections of the plant. However, overbuild without monitoring masks plant health, condition, and status. The result will be the inability of the system and its components to meet their service life energy production demands.

It is noted that in many cases there is a rush toward quick plant system design selection, with EPC design, installation, and commissioning. All too often, this is completed by the

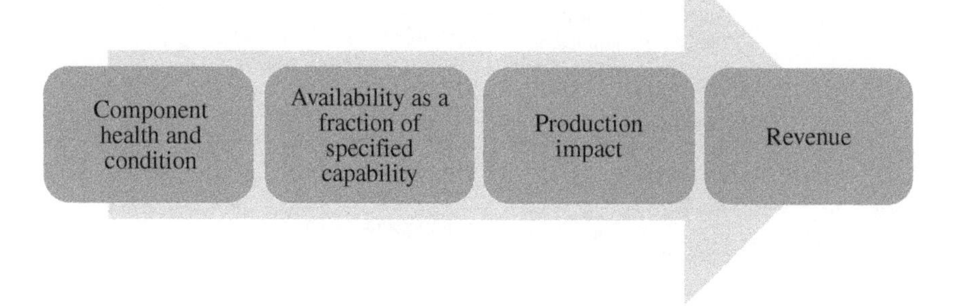

Figure 10.3 IEC 63019, Figure 2 – PVPS Component-to-Revenue Path.

EPC without adequate benchmarking or an appropriate design for monitoring needed by the owner, O&M, and other stakeholders.

The SE team should ensure that there is an effective monitoring system that allows for the highest level of maintenance of the system. This should be geared toward the long-term goals of a data and monitoring plan.

The goal is to build PV systems that are, over the service life, reliable, available, maintainable (testable), and safe (RAMS) energy-producing infrastructure that is manageable, measurable, and maintainable?

What seldom happens is that each of these segments identified in Figure 10.4 are monitored and compared to the initial benchmark. The more detailed and accurate the questions asked and answered prior to bidding will result in fewer cost and performance problems.

Has the modeling addressed the **SE/R**epowering™, RAMS, lifetime LCOE, and operations while being reflected in the questions below?

- Does the modeling reflect all of the system and site conditions?
- Does the benchmarking and acceptance testing indicate consistency, and if not, why?
- Does the monitoring confirm the other two, and if not, why? In other words, does system operational data and analysis reflect the initial model and results?
- Does the monitoring "zero in" on irregularities that indicate real or potential problems?
- Does the analysis address what is nominal versus what is potentially problematic and where?
- Does the O&M strategy and tactics planning leverage this information to control repair timing?
- Do the stakeholders understand what the monitoring and analysis are telling the project team?
- Are the monthly reports distributed and discussed on a regular basis?
- Or is critical information just buried in the reporting (archived)?
- Is there critical data that is being discarded or ignored due to lack of interest?
- Does all this monitoring and analysis get fed back into future designs to reduce problems?

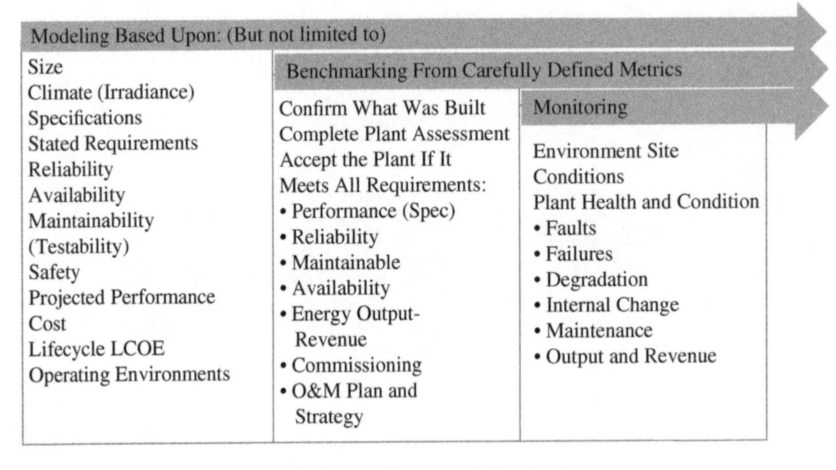

Figure 10.4 Process Flow From Concept to Field Operations.

If they do not understand the critical value of being able to measure, document, curate, and analyze the system's health, condition, and status, they will not have the information needed to achieve greater long-term success.

10.7 The Level Necessary to Control Costs and Improve PV Systems

Considering the data collection and analysis with the rest of PV system delivery elements requires a series of project plans that detail the picture of system development and project life. If the monitoring system or plan primarily reflects the needs of external interests (financiers and insurers), as often viewed today, stakeholders tend to get consistently incomplete results.

In order to determine what level of detail is necessary for data acquisition, project stakeholders need to quantify their goals and objectives (defining success and failure). To build PV systems as energy infrastructure, the preparation for bidding must be far more detailed and complete, requiring detailed data collection and analysis on all system facets.

However, from the buyer/owner's or the O&O's perspective, the goal is assumed and should be to focus on long-term operations and performance, with reduced O&M costs, to achieve improved profit. The side bonus comes in a reduction of wasted unbudgeted time and money and a reduction of "organizational angst."

10.8 Monitoring for Better Data, Security, and Plant Cost Control

To gain the full value of monitoring, data collection, and analysis, there must be a clear monitoring plan with long-term goals and a clear path toward implementation over time.

A well-planned and delivered monitoring system provides stakeholders the ability to see, measure and respond to, then clearly identify, and define the health and condition of the plant over time. Project stakeholders are already beginning to understand and apply this to existing and future systems. This will dramatically improve the value of that plant throughout its life and its ability to maintain output when it is bought, sold, refinanced, or reinsured (Figure 10.5).

Identification is the fundamental issue in controlling long-term PV plant costs and risks. Control begins with a clear understanding of the health and condition of the PVPS at all times during its lifecycle. This is fundamentally based on effective/accurate data collection and analysis, while making better decisions at specification prior to design and following a PV engineering systems process.

Figure 10.6 addresses the question of: What Are the "Data Plan" Goals and Objectives?

The goals and objectives of a data plan are to determine, define, collect, secure, curate, and analyze PV plant data. It also provides for "Anonymizing" data to an industry-shared data repository. This allows asset managers (and other stakeholders) to accurately observe, control and make the most cost-effective decisions to maintain plant/fleet health and condition and improve **SE/R**epowering™ design and operations decisions.

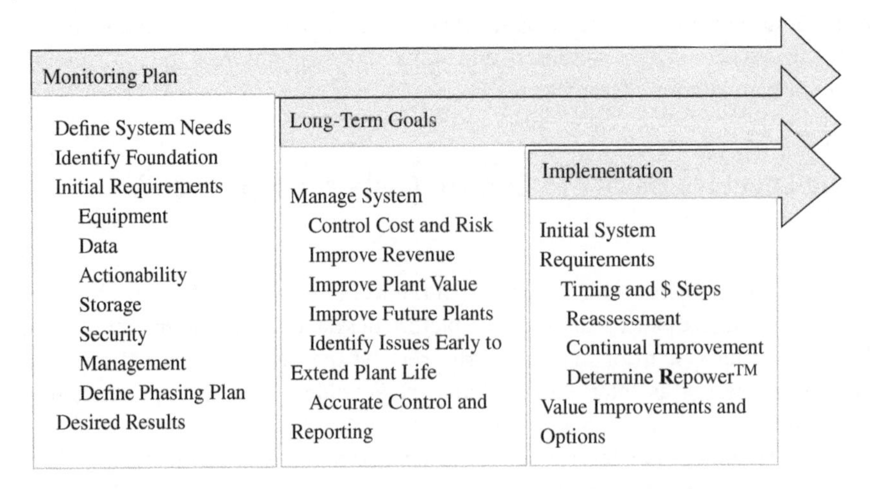

Figure 10.5 Lifecycle of Data Collection.

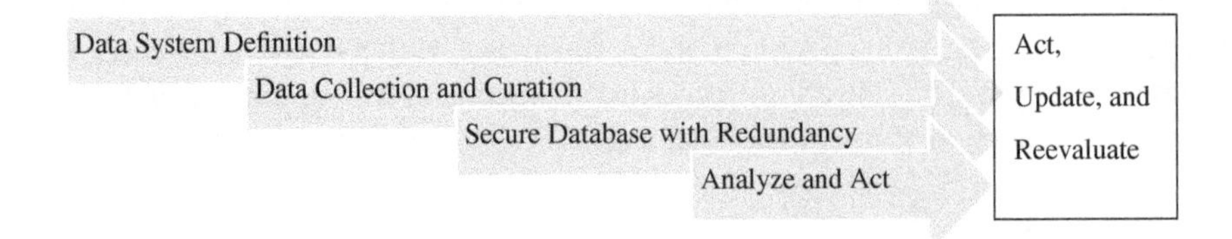

Figure 10.6 Data Plan Goals and Objectives.

Plans also need to be reviewed and updated periodically to stay in step with changes in technology and analytic techniques. This is especially as remote sensing improvements reach the market and costs are reduced. This approach allows the owners to define clearly their initial priorities while continuing to gather the information they will need to reevaluate the initial metrics and determine if the original data was sufficient to make their decisions.

The data analysis should include tracking the incremental changes that indicate that those changes had a beneficial or minimum impact on the plant's performance. Those changes indicate everything from faults, failures, anomalies, anticipated and unanticipated degradation, aging and maintenance issues, site and surrounding issues, among others. Most change is gradual or slow until it is triggered to be of consequence. Tracking those trends can result in carefully timed activity that reduces or eliminates system or subsystem downtime.

Reporting must focus on healthy affect system availability, reliability, performance, and output. In some instances, a failure is not the result of a single cause. The multiple causes leading to failure events can usually be identified if looked for through an RCA process.

Isolate issues, analyze, and take appropriate, timely, cost-effective action. Even if the team visually inspects the whole system, they will only see a very small portion of meaningful changes and challenges that impact the system over time. It is critical to clearly understand

and respond to issues that are problematic while understanding changes that are normal and within acceptable parameters.

Focusing on preemption with advanced sensing technology and analysis can often reduce not only the labor needed to address the situations but can eliminate or reduce downtime. This becomes incredibly important on a seasonal basis.

The simplest example of cost-effectiveness is to track module issues and have them replaced under warranty instead of finding them following failures afterward the fact when the warranty or company is gone. Untimely action results in all of those costs being shifted to a maintenance budget.

However, eliminating any eminent failure when it is most cost-effective to do so has a major impact over months and years. This is where experience becomes critical in the equation.

Compare data to modeling and benchmarking, document systemic changes, and review analysis on a timely and consistent basis to supply information for improving future system designs.

10.9 Data Analysis

Part of the value of data acquired from the lowest level of maintenance is the ability to identify patterns, link them, and allow users to understand what the information is telling the user. For the most part, this is not only identifying failure trends (i.e. increasing failures with time) but to associate descriptions, equipment, and consequences to improve understanding of plant status.

In PV, we look for a series of events or effects that indicate how a system, subsystem, or component is operating. If it is not operating to specification, why not? Nevertheless, when analyses are properly applied, there is a high probability of success in sorting out faults and failures that impact systems and raise costs.

By gathering, curating, and performing analyses of the data, the plant's performance, reliability, maintainability, and availability can be established. Data gathered on-site provides for an improved understanding of the site's natural, operating, and induced environments.

This provides the stakeholders with variations that they apply their own models that they must consider, including cost-effectiveness. For the plant systems engineers, EPCs, and designers, it flags those components or subsystems not designed for that environment or use.

The notion that PV failures are binary is inconsistent with the actual operations. By that, we mean that historically an inverter was available if it worked and unavailable if it did not. If the green light came on, it was available even if it was only operating at 10%. We have moved beyond that to address the capability of that inverter to produce 0% to 100% of its expected output. This becomes a far greater and more powerful metric.

Many, if not most, failures reduce the capability of the component to do its job before the actual total failure takes place. Making decisions that squeeze the most out of a component or set of components while still providing sufficient energy production is a simple financial consideration once you have sufficient information. By carefully picking and choosing the appropriate actions, losses can be reduced, budgets maintained, and revenues enhanced. More importantly, all of this addresses the issue of corporate angst. Project management and employees can spend a lot less time pointing fingers, thrashing about wasting time and energy, and trying to solve problems that should have been resolved before the EPC contract was signed.

In Chapters 3–5, we discussed and outlined the dataset that best gathers all of the information needed to address the performance, reliability, and maintainability data needed to understand the plant cost risks. In addition, it also allows for an assessment of the plant energy/power capability leading to the energy availability calculations in Chapter 6.

As a point of interest, this includes Met station and local weather data from original sources near the plant. To support plant capability assessment, it includes the ability to take optical (IR, visual, and UV), digital, and electrical data to determine if there are any irregularities that potentially lead to, or do indicate premature or unexpected failures or energy loss. It requires curation[5] of that data so it can be used later even if the analytics are not available at the beginning of plant life.

This is also important as one phase in monitoring system components and analysis. If your team is building a fleet, they may begin primarily with collecting and reviewing data acquisition system (DAS) data. However, as the fleet grows, all the information will be valuable across the fleet and for future plant(s), providing evidence for component improvements. With the line being blurred between DAS and SCADA systems analysis, this becomes more important. If data is properly collected and curated, a shift to a different monitoring platform allows for long-term use of all the plant performance and RAMS data.

When someone on the team says, "We already collect or have too much data!" the important questions are; "What data is being collected?" "What are we going to do with it?" "What is it telling us?" and "What are we missing?"

In all probability, the concern about too much data is an indicator that in spite of the value of that data, there is a lack of awareness of the benefits of understanding and using it more effectively. The goal is not to collect data for data's sake! The goal is to collect data that can be used over the plant's life cycle while including data that will become more valuable in time, especially as analytics processes and tools improve.

- One of the first issues with any data taken from United States infrastructure data, it does not do much good to collect and analyze data in itself because the value of that data will be long gone before it is needed, *unless you have defined a set of needs*. What is the long-term data plan?
- Is it being properly collected?
- Has it been properly screened and curated?
- How is it being stored? Is it on-site or in the cloud?
- Is it being adequately backed up, and is somebody checking to ensure it is?
- Is it being stored in such a manner that it will not be corrupted, stolen, damaged, or deleted?

These are questions that generally have not in the past been addressed by stakeholders. The results are that team members who understand the value of good data are sometimes ignored when talking about preserving that data. In some instances, even though they were not provided with the appropriate tools, they are blamed for the results.

10.9.1 Where Will the Responsibility for O&M and Data Decisions Reside?

We recommended owners begin initial plant operations with an in-house asset manager where possible. Even if part-time, their tasks will include reporting to and clearly informing stakeholders consistently. This requires the reporting process to be understandable to nontechnical stakeholders so there is an ability to explain what they are seeing, preferably graphically. Providing a trendline that links plant health and condition to anticipated and unanticipated costs allows a clearer understanding of cause-and-effect. A major goal is to eliminate unanticipated costs, which is the result of better planning in the feedback process. It does absolutely no good to collect this data and not use it.

The interests of the owner are different than the interests of a third-party operations or maintenance vendor. The third-party vendor has a limited contract with a limited budget and no responsibility outside of that contract. At the same time, the owner or asset manager can be responsible for ensuring sufficient accurate information to make substantively better decisions. If there is no one to administer those responsibilities and deliver on them, the quality of decision-making will deteriorate rapidly and will not generally be noticed until the cost of that deterioration becomes abhorrent.

10.10 Data Presentation

- How will different forms of the data be presented, and to whom?

As an EPC who provided a standard extended 5-year warranty to all our customers including (O&M), it was important to me as the company owner and EPC (J.R. Balfour) that we reviewed the data to determine if there were issues long before the owner would notice. It was also our documented practice to do a pre-end-of-five-year-warranty review.

There were substantive benefits to doing that end-of-warranty service. The most important was making sure that the systems were doing what they were supposed to do. However, there were two other advantages. Doing so allowed us to see if there were any issues that were caused by the equipment or by our installation and or maintenance of the equipment. Overtime, the lessons learned were incredibly valuable. Those lessons were important to the long-term needs of my company. The other issue was when we found faults, failures, and anomalies, we could replace them under warranty, and in some instances, when we were concerned a company was going out of business, we could address the problem while there was recourse.

We consider that excellent service. The fact that we included the O&M under warranty was a sign of both good faith, our strength, and quality of our work. It also contributed to not only enhancing our image, reputation, and marketing. It also saved us a fortune in unnecessary service calls.

…And yet, I found that the extra steps we took during PV system delivery that dramatically reduced O&M challenges and costs were continually being questioned internally by my staff, who would ask, "Why do we have to do all of this "extra stuff" that our competitors don't?" There was an ongoing desire to do what everyone else was doing even when we had the physical proof that what we were doing was more cost-effective. This is tough to deal with because many if not most people in the industry have no sense of history of what works and what doesn't and if unaddressed, raises costs. … even when there is proof many are loath to even consider it!

From an EPC and initial maintainer of these systems, many of the policies and procedures were very simple, straightforward, and basic. The minimal amount of time that we spent doing things right, resulting in better systems, resulted in a fraction of the cost of having to do a truck roll in-service. That was even before we fully understood the total cost of that truck roll, which included:

- The initial call to the office.
- The staff had time to determine what the problem was.
- If there was a problem that was not initially discovered (No Fault Found, NFF), which required us to take a closer look at the monitoring data.
- Scheduling that truck roll.
- Hoping the truck was sufficiently supplied to eliminate a second or third trip.
- Documenting the issues, their causes RCA, and solutions. (This was always a learning process that slowly improved.)
- Discussing it in staff meetings and then training.

Still, there was always resistance to this. We did not know of any competitors with the same practices, although they claimed to do so.

It was not until later that I discovered how ingrained this attitude was across the industry. This process of not dealing with real issues and falling back on the "other guy did not do it" excuse, along with the "it's too expensive" excuse, have horribly retarded our industries' progress and maturity. This is so that even when we had the proof in dollars and cents, it was still resisted!

The reason that I'm providing this detail is that there has been a tendency in the industry to do the absolute minimum and then express some form of pride in doing so. As we look at bigger and bigger systems, this tendency multiplies, costing the industry billions of dollars a year and putting a number of companies out of business. This is true for small companies and even companies valued at US$18 billion, yet it is all too often so avoidable.

The analysis available today in systems is far superior to what we dealt with in the past. Yet, leveraging those new capabilities requires determining what the analysis consists of, how and to whom it will be presented. Doing so must be part of key goals and objective discussions among the stakeholders very early on. If that discussion does not take places, the odds are that it will not take place until the plant has been sold, derated, possibly with a renegotiated PPA price, or may not take place at all because the plant has been closed and removed.

It is critical to address the whole series of technologies that are coming online in the industry that previously have been effective in other industries. Yet, without an effective platform to integrate them, they become an expense that does not generate a positive cash flow.

The proper plan and management of that process to drive results, is what makes it a success, not just buying equipment and collecting data. Starting with the right data plan and acquiring the appropriate technology is the very first critical step. This means that upper management must understand what is happening at the plant and in the fleet, then respond when unusual issues have been identified and raised. As we learned from a broad range of industry experience, the issues can clearly be identified long before they positively or negatively impact the bottom line. This requires planning and executing a series of steps that will allow that information to leverage performance and output. The existing industry model of hoping problems will go away, results in additional organizational angst. That angst tends to result in a form of organizational paralysis where effective decisions become difficult to make.

This results in substantive waste of unbudgeted time and effort while destroying organizational morale. We consider this a very poor management practice that is often blamed on others instead of being faced and resolved.

Each facet of stakeholder information access and delivery must address the most critical issues that are important to that stakeholder while outlining the impact on the whole project. That requires collecting the right information and then being able to properly present it in an understandable manner to the broadest group of technically skilled and unskilled people. Doing so properly dramatically reduces costs while improving revenues. Maximizing revenue and profit must become the critical goal, not just cutting costs.

10.11 Process

There are a series of well-known and successful steps that can be applied. If a recognized internal champion has the integrity, respect, and admiration of a broad enough group of organization members, this can usually be managed internally. That in itself requires a substantial level of managerial and inter personal relationship skills.

In many instances, an outside professional is needed that can navigate the muddy waters that have been stirred up over time. That third-party professional needs to apply a different approach. If they approach the process by working with the team to have the team identify the critical issues themselves and have the team provide them with solutions, the results are more likely to have a substantive buy-in. This results in an organizational win because the members of the team have solved the problem and, therefore, can take credit as a team.

However, as a warning, it is not unusual for management to hire somebody to solve problems without actually discussing, clearly defining the goals and objectives, including processes, procedures, and details. Further, as a consultant, all too often management does not wish or have the time, ability, capability, or interest to actually address what needs to take place. They become frustrated, often push the consultant into doing things that are not within the scope, and then blame the consultant if they don't get the results they anticipated or expected but never discussed.

Many of these problems can be avoided by addressing the following items:

1. Enlist management buy-in to a comprehensive data program prior to initial specification.
2. With active stakeholder involvement, define success and failure and the associated metrics when starting the project (Chapter 5).
3. O&O/Asset manager must collect and store site or portfolio data focusing on as much data acquisition as the project funding will initially allow (1–5 minute).

4. Institutionalize and initiate a series of planning processes that address the past, and present problems and the near and long-term goals and objectives. This is with the understanding that the specific requirements must be addressed primarily at specification prior to EPC design. Future decisions, such as **SE/R**epowering™, can be impacted by this planning.
5. Determine the steps that go into providing all of the essential equipment from the first project to the operations, maintenance, management, and sale of the whole fleet. This allows for a plan to phase in equipment as the organization and systems grow.

   Note: For the EPC and other stakeholders, the value may be in following this process with customers and stakeholders to include having access to the project data results over time. This gives them actionable information on component and subsystem reliability and the ammunition to present it to a potential client to help differentiate themselves. As part of their education, they begin to understand the details and challenges of specific equipment including issues related to product quality in their RAMS values.
6. Within the scope of the proposed work, clearly establish traceable cost and schedule for initiating milestones for each step. Monitor and track progress using risk assessment (Chapter 3) and initiate corrective action as soon as the risk is identified. That is: Don't wait for disaster to strike to fix the problem.
7. Establish operation and asset management at the beginning of the project to guide the development of SMART specifications, maintenance, cost controls and implement mid-course corrections in a timely manner. There must be someone within the organization with appropriate training and experience that will oversee the program/project processes and its details and insure they have been addressed. This responsible engineer/manager must be knowledgeable in all of the technology in the plant. They also must understand all project details to provide the organization with information needed to control costs thus improving profitability. This requires that they coordinate with the appropriate stakeholders their issues and concerns are adequately addressed. As a result, error reduction will dramatically improve value by offsetting common and unnecessary problems let tend to be overlooked in a rush to completion.
8. Save plant and field/supplier data in a *secured* cloud with *limited access to the project organizational personnel*. Having a firewalled and isolated non-Internet connection for security should be a serious consideration. For a number of reasons they include:
   - Direct hard storage provides the ability to continue to collect and store information even if the Internet or cloud is down.
   - Use data encryption to allow for rapid ability to update the database if one or the other is hacked or corrupted.
   - The ability to address communications outages from natural, or other reasons.
   - Reduces capability of cyberattack and or disruption.
   - It may be the cheapest insurance you will ever buy.
9. Secure data (controlled access and dissemination) to assure there is no corruption, loss, theft, or vandalism. Multiple or duplicate cloud storage of data can be compromised during a number of different phases of the project acquisition. Data curation must address both synchronization of cloud and hard storage for consistency and accuracy.

10. Clearly define all reporting including what the analysis indicates regarding the present/initial health and condition (benchmarking) of the plant changes and projections taking place over time. Include an updating process to always compare the present health and condition of the plant or system with the initial or last benchmarking and service life expectations.
11. Collect and be ready to provide all data necessary for analysis to support future budgeting and other decision-making. This requires not only having a data plan for the life of the system, and it includes following the plan and doing all of the necessary updates as new monitoring technology is implemented.
12. Determine what the actual subsystem degradation and failures are, carefully documenting actual failure modes and mechanisms (Chapters 5 and 6) while including an accurate and detailed incident report with documented RCA.
13. Provide clear documentation and training for the whole process that includes graphical explanations of how the system actually works, what it does, and how it impacts each phase of the system lifecycle. It is critical that the technical processes and systems show how costs can be controlled and risks can be reduced while improving revenue and asset value. The steps that are taken in the process will determine the real performance numbers while highlighting where risks arise and can be addressed.
14. Use the data acquired to recalculate and project future LCOE values on an annual basis. This delta (δ) LCOE is an early warning system for fiscal issues that can easily go undetected. This allows the O&O and other stakeholders to understand whether or not the plant is meeting original assumptions and expectations as modeled. This is part of an ongoing financial risk analysis that addresses the fiscal health, condition, and status of the plant by providing real numbers to the investors. This δ LCOE is a tool to understand the risk, while providing time to plan for corrective action when it is most effective. It delivers a real-time trend analysis of continually evolving cost model elements that identify the rate of change of fiscal health, condition, and status. Doing so gives the stakeholders an early warning system to provide far better control of the fiscal and physical conditions instead of just leaving everything to contractual default, liquidated damages, and substantive loss of asset value.

10.12 Implementation

Within the next few years data curation, analysis, and reporting will incorporate deep learning tools and algorithms. Indeed, many of these tools exist today, but not as effective as integrated packages. The current marketing hype, for example, calls this AI (artificial intelligence). A note on AI: There are and will be many claims as to the magic that can be done through AI long before it is actually a fact. The term AI implies that the item equipped with AI is self-learning and decision capable of changing it programming to handle problems outside the scope of its initial programming. In reality, the current "AI" is only as capable as its programmers. There is a learning curve to determine the value of the information coming out of "artificial intelligence" processes and systems that require humans to interpret and reprogram the system.

As indicated above, for a system, any system, to have a profitable and long service life requires understanding the operation of the system over time. The perspective here is that the plant has been designed with the capability to fault isolate to the lowest level of repair, at which point most failures will be observable as they occur. The PAM SE/RAMS approach and processes examine and identify defects, faults, failures, anomalies, rates of degradation, and other changes in the system performance grossly impacting LCOE, LCC, and O&M. This process generates data that, when properly analyzed, documents the performance and RAMS trends supporting making critical decisions about maintaining a reliable energy infrastructure element. Poor data collection or failure to analyze the available data results in some maintenance efforts being superfluous especially when they did not actually address "a" or "the" real reported performance issue or failure. This is where establishing an accurate RCA program becomes critical to solving problems. In doing so, it is important to always consider the fact that most issues are not the result of a single failure mechanism.

For example: Understanding the changes taking place between inverters, modules, or other equipment, will help address these issues:

- Determine if and when modules need cleaning.
- Define and determine the probable and evolving impacts and metrics for PID or LID.
- Detect degradation of the module caused by browning or discoloration.
- Assess changes regarding terminations that have been over or under torqued.
- Assess a broad range of induced failures as the result of damage in shipping or installation.
- Identify rising temperatures that indicate potential problems with IGBTs, capacitors, or other components or subsystems that can result in overheating, failure, or fire.
- Annual comparison of site conditions that could indicate erosion or slope slippage, or excessive under growth that can facilitate a fire hazard.
- Update changing climate, and local site conditions to support understanding potential impact to power generation.

Data gathered from current and previous projects can support owners needs for understanding the potential weaknesses, problems, forms of deterioration/degradation, and causes of failures in the system state prior to building their own. This information then can guide the specification design, build, and operation of a plant and control, mitigate, or eliminate performance problems earlier at a much lower cost.

Additional data collection from tools such as LiDAR, IR, UV, electroluminescence, and other emerging technologies currently used in other industries can provide further support for improved systems. These are incredibly valuable benchmarking and maintenance tools. As these technologies become more commonly used in the industry, their cost of acquisition and use will decrease, and the value of their analysis will rise. These new techniques will dramatically reduce maintenance labor while improving safety and accuracy and providing greater consistent clarity in metrics. Note: Many performance problems and concerns are not identified today due to masking[6] and lack of fault detection (FD) capability. Adding FD with automated reporting will result in rapid substantive change and system health and condition. This will include being able to produce the equivalent information remotely that normally requires on-site testing, for example, by opening a J box.

Do not trust your data until you have confirmed that it is accurate. Making decisions on faulty data significantly increase maintenance and logistics costs. As indicated in Chapter 6 there are FD/FI effects of false positive (failure reported but not real) and false negative (failure not detected) failure reporting. Just assuming that data is accurate will come back to haunt your organization for decades.

Implementation for data collection consists of seven critical steps:

1. Develop the data plan.
2. Set system and subsystem requirements for fault detection and isolation.
3. Monitor design, development, and installation of equipment.
4. Provide training on data acquisition of requirements by the O&M organization.
5. Perform data analysis and establish action plans based on failure trends or other anomalies.
6. Ensure transmission of anonymized data to a central data base repository.
7. Develop lessons learned data base and incorporate results of analysis for the plant for improvements or expansion to future sites.

10.13 The Monitoring Plan

A good monitoring plan effectively defines the system and stakeholders long-term needs. It thoroughly addresses numerous issues over the full life of the plant including addressing the critical issues of **SE/R**epowering™ plan. It addresses the vast majority, if not all, of the potential issues, defines them, clarifies their impact, and is measurable, explainable, traceable, and duplicatable. It provides historical documentation to compare issues and change over time while measuring both negative and positive change. Most importantly, it gives documented information to make hard-core business decisions.

In a phased approach, the focus is initially on identifying the foundational needs in the first phase of system operation following benchmarking and commissioning (Figure 10.7). When effectively administered, these identify the plant health and condition of the plant at start up. Without this essential step, hundreds, thousands and in some instances, tens of thousands of problems that should have never occurred, actively exist in a system signed off at commissioning and COD. Those issues are critical to resolve prior to taking full ownership of a plant, meaning that the punch list must be completed, and the retesting must be confirmed.

Defining the initial monitoring system requirements considers, evaluates, and determines the actual requirements for day 1 operations while carefully considering and understanding that the system will evolve and change. Determining how that evolution begins, continues and is fulfilled over time as a balance between cost and the ability to control future costs. Because this part of the process is linked with the **SE/R**epowering™ plan, many potential systemic issues are preempted, which more than offset the time and energy necessary to effectively develop these plans and their implementation. Yet, a further consideration is that these different plans, overlaid, provide their own sets of checks and balances toward ensuring greater planning and implementation consistency.

Figure 10.7 Monitoring Plan Steps.

During this phase, the initial equipment data requirements including the actionable selections regarding vendors, storage, security, analytics and monitoring management are identified. This includes defining the skills, tasks, requirements and duties of the asset manager/s who will oversee a single system on up to a fleet of systems. Even when a large organization has provided most of the O&M Service under contract, management-level operations must include their own asset "watchdog." Doing so ensures the monitoring and analysis is effectively and accurately reflecting the actual health, condition, and status of the plant while triggering the activities that maintain the plant or fleet and do so cost-effectively.

In most industry cases, this step is ignored initially, which defeats the whole reason for collecting data to improve the existing assets and future plants. It is critical to always keep the end in mind with a clear understanding that the initial system will evolve in parallel to a good data plan.

Desired results must be clearly identified and documented prior to taking the steps to ensure that these things will take place. As with all other elements of the PV system delivery process, "*If it is not clearly identified in the contract specifications and plans prior to EPC bidding, it will not be designed/built into the system and should not be expected!*"

Personally, I think some of my colleagues get tired of me saying this after over a decade, but it happens to be something that we should all keep in mind no matter how big our organization is or how smart we are.

10.13.1 Management Goals

These items must include, at the very beginning of the process, a clear understanding, determination, and acceptance of what the project team's long-term goals and objectives are. For example, under system management, a set of goals would include:

- A clear and concise set of requirements to assure that all data collected under the existing system tools will be available, transferable and usable in later steps of the system data plan,
- Using data collected to control cost and risk throughout the entire service life of the system,
- Assessing revenue improvement and plant value through financial analysis of data,
- Provide the automated tools and actions to identify, document, and deliver future requirements to ensure substantively better future plants,
- Ensuring that masked issues are identified, confirmed, and brought to light by effective data analysis, planning and actions,
- Using acquired data to confirm that the plant O&M is maintaining costs consistent with the originally projected LCOE.

This approach for systems management can lead to an extended service life at a lower cost primarily because implementing an effective data plan allows for better comprehensive decisions.

The initial system requirements, at a minimum, include:

- Clearly define monitoring equipment and analysis timing.
- Determine early on how data will be selected and stored to work across a variety of different monitoring and analytical systems.
- The basic budget for additional steps and equipment.
- Details for continual reassessment of the process.
- A focus on continual improvement.
- A functional set of processes to evaluate change and address it systemically.
- The additional goal is determining what and when to trigger and implementing the **SE/R**epowering™ and value improvements.

Preparatory steps include:

- Define system size and type (PV, PV with Storage, grid, other contributing energy technologies, microgrid, or other), level of data acquisition and access ports.
- Establish and review the **SE/R**epowering™ Plan, supporting data needs.
- Establish and review Operations and Maintenance plan, data acquisition method(s).
- Establish and review stakeholder information management processes, systems and reporting data requirements.

10.13.2 Establish Initial Data Requirements

Initial data requirements begin with defining the type of data to be acquired, how it will be handled, stored, and secured, and with great caution, how it will be distributed to the industry. Acquiring data is often gathered by a data acquisition system (DAS) and Met station and other reporting. This will be common in smaller systems and some commercial systems. If there is a Met station, the data must be compared to the original source of climate data for system modeling to determine the accuracy of the modeling and to adjust if there are variations.

Larger commercial industrial and utility systems may range from somewhat more complex DAS to a full-sized and utility-grade service SCADA system. The actual size of the SCADA/DAS is determined by the application, size, number of components or equipment monitored and the ability of some equipment to automatically report its own status such as inverters. The challenge comes when taking a look at the whole lifecycle of the system. Today, all of this data collection and analysis may seem excessive, however, as has been amply demonstrated over the last 40 years, necessary.

Considerations include but are not limited to:

1. What is the value of this data for initial and future use while considering the varied costs of not having it available later?
2. What are the environmental and climatic conditions that should be recorded?
3. What are the requirements for uninterrupted energy to the Collection, Analysis, Storage and Transmission elements of the system?
4. What are the access points for data transmission (hardline and/or uplink)?
5. For combined data and security, what does site security include?
6. When does the initial monitoring function begin?

A year or so before construction, ensure that MET data is being collected for comparison to modeling data. At this time the initial security may also be instituted. At construction, details and information to observe and protect the site, and document changes to the site, and any modifications to construction details must be included. During commissioning and at any annual benchmarking, the results of the testing and concurrent MET conditions must be documented.

(Note: The project team must clearly understand the differences between commissioning and benchmarking and their processes. This provides the foundation for calculating the present LCOE and the trend it is following.)

10.13.3 Define Assessment Strategy

Once data are collected, filtered, curated, and analyzed, it is necessary for the project or system management to determine what path the results will take. Completed analyses can have many output forms and formats. Stakeholders with a vested interest in this level of detail need to define the data set form and format they need in their reports and follow-up actions. This allows them to perform their own assessments and to coordinate any actions with the owner, O&M, finance, insurers, or other entities.

10.13.4 Identify Additional Data and Sensor Requirements

At the beginning of a new **SE/R**epowering™ event, there is a need to ensure that the most cost-effective technology has been employed to detect and report faults along with the concomitant data. This can be considered remote sensing, external sensing, or just adding additional sensors to known trouble spots in a system.

For projects with a planned service life of less than 20 years, most steps will be completed at commissioning and benchmarking while considering those incremental lifecycle changes that may occur with changes in customer needs and planning. The monitoring system itself will be verified and validated, and in some cases certified. This establishes the basis for data acquisition and analysis so that the system can be observed, tested, measured, and compared to expectations. This should occur regardless of whether it is a simple DAS or SCADA with expanded capability.

Data analysis and assessment of the health, condition, and status of the system, is used to determine if triggers have been reached and the need to begin planning and execution of **SE/R**epowering™ or upgrades to equipment. Data will also be assessed to identify any additional performance masking due to the granularity of the sensors and analysis. This will be some of the best money the stakeholders invest towards the future capability to generate electricity and its commercial asset value.

Commissioning must include a complete system inspection, benchmarking, completion of the punch list, and all warranty repairs indicated on that list. These are critical actions needed to establish the sign-off on the commercial operations date (COD) of record.

This process carries an additional cost, which is minor compared to the outsized costs being transferred to the operations and maintenance budget for failing to do so. This is one of the most overlooked sources of project risk and loss that is completely controllable and almost universally underestimated.

The benchmarking should be that demarcation point where the initial health, condition and status of the plant are carefully/accurately measured and certified. At this point, the system "should" be operating as specified, designed, and installed.

Future benchmarking is data that is compared against the original to assess reliability (MTBF, MTBCF), maintainability (MTTR, Mmax) and availability of the plant to provide power to the utility and eventually the customer. The plant COD should not be permitted until all of the items designated above have been cleared.

Although performance metrics based on IEC 61724-3 will be used in the process, that standard does not currently address system health, condition, and status. The application of IEC 63019 should be incorporated for availability metrics. IEC 63265 includes information to address the specifics of the plant reliability, and Orange Button, addresses others.

10.14 Warranty Issues

A major concern for the O&O, O&M as well as asset management is warranties. This makes detailed data acquisition during the first years of operation, while under warranty, critical to control costs. The collection of data beyond just the inverter, i.e. to the lowest level of maintenance, addresses the issues related to PC modules and other components. This supports claims for warranty replacement and repairs. This requires benchmarking and failure report data documenting equipment performance, anomalies, and failures.

Manufacturers establish a wide variety of starting points for their warranties. These include; manufacture date, date sold, and date installed. Warranties are biased toward the manufacturer. They have established their producer and consumer risk based on their reliability and life testing or field data. The risk for the user, is what does the contract say!

Aside from setting the specifics of what can or will be done by the manufacturers or OEM when a claim is filed, there are other elements that require the user (EPC, Owner, and O&M) to include in the purchase contract. These include:

1. Need for notification of product ceasing production at least 6 months before the planned termination date.
2. Need for legal notification of pending company bankruptcy, sale, or other disruption of company operation.
3. Fixed warranty duration to include clearly defined prorate tables and requirements.
4. List of required supporting evidence (failure and operating data, failure mode and mechanism codes (descriptions), photos (visual and IR, etc.), prior corrective actions or other repairs) to validate a claim.
5. Condition of the product on return.
6. Responsibility for costs of transportation and testing if the product has a no-fault found code.

As a side note, many companies use warranties as profit centers. The cost of testing can be very high and the responsibility of the user. It is possible, however, to renegotiate warranty provisions, with an expectation of a cost delta of the product.

Note: Cost of transportation and transportation loss insurance, potential loss of power production capability, cost of labor for handling and reinstallation, etc.

Warranties are often written for the OEM to mitigate quality risk. One example is that the equipment falls well short of the predicted MTBF of the equipment. This is a system risk commonly used in other industries because of the impact on O&M and logistics costs, but not fully implemented in PV as it is often undervalued.

The long-term impact of manufacturers receiving better data helps them improve their products. By providing data from multiple sources across the industry, products should improve at a moderately reasonable rate, which in turn benefits the whole industry.

Comparing inverters, combiners, transformers, and other equipment performance to established expectations based on detailed modeling has added benefits. It can impact the system's performance while helping to reduce maintenance costs.

Collecting and analyzing sensor data available today makes it possible to identify many potential problems early at low or even negligible cost. By including additional sensors and sensor types, which have been effectively used in other industries successfully, we can add to that capability to identify and document issues early, which give us cost control that we don't usually have or take a vantage of today.

10.15 Synthetic Data

Synthetic data is data that has been fabricated, developed, and inserted into the process because the initial data was damaged, incomplete, compromised, or missing. The problem with synthetic data is that it is all too often factored in as if it were real data. Resolving the issue of why it was damaged, incomplete, or missing is the first step: which needs to be properly addressed, not rushing into making up new faux data. Performing a Weibull analysis, for example, would require censoring the data (removing it from the sample population) as it does not reflect actual failures or deficiencies or real operating time information.

This data compromises advanced data collection systems with a focus on possible short-term anomalies that may or may not indicate developing problems. The mission of data in the near future will be expanded to address a number of systemic issues that, right now, are not measured or measurable.

If synthetic (made up data) data is going to be used, it should be clearly identified and considered suspect in providing accurate/real information. The point to consider is "What was missed during that time period and how will the team address that focusing on other irregularities in the data?"

In many instances, data loss, or damage is a serious situation that impedes the operators and maintainers of the system(s) from doing credible work that supports plant operations and performance. This also gives some impetus toward performing a high degree of maintenance on the monitoring system itself. The details must be carefully defined and included in O&M and data planning which, again, begins at specification and design.

It has not been unusual to find data that has been compromised because a lack of maintenance, rodents, or other pesky wildlife, fire, vandalism, or damage or results in data provided as intermittent data. The results are often that data is not taken seriously because it has been so compromised.

10.16 Conclusion

Effective data collection and analysis informs asset management, O&M, and all other stakeholders to improve specifications for cost control and maintain revenues. This is a requirement for effective documentation availability and an organizational process to control costs.

- This is pure and simple cost and risk control management that requires sufficient data.
- The planning process must focus on relevant and cost-driven aspects of data, asset management, operations, and effective maintenance control over the life of the system.
- Applying these processes will reduce the need for under-skilled technical professionals and highlights the need for higher level skill and training to better assess site realities.
- There is a cost–benefit to data collection and analysis that reduces operational cost, LCC, LCOE, energy production loss, and corporate angst.

Finally, organizations looking for the greatest value should become part of a network or group that shares reliable data supporting improvement across the industry. This creates a more vital and dynamic industry delivering PV infrastructure.

10.A Appendix

NOAA defines Climate as "the average weather conditions in a place over a long period of time—30 years or more. And as you probably already know, there are lots of different types of climates on Earth." https://scijinks.gov/climate-zones/

There are generally five (5) climate zones for the Earth. These are:

10.A.1 Tropical

In this hot and humid zone, the average temperatures are greater than 64 °F (18 °C) year-round, and there is more than 59 in. of precipitation each year.

10.A.2 Dry

These climate zones are so dry because moisture is rapidly evaporated from the air, and there is very little precipitation.

10.A.3 Temperate

In this zone, there are typically warm and humid summers with thunderstorms and mild winters.

10.A.4 Continental

These regions have warm to cool summers and very cold winters. In the winter, this zone can experience snowstorms, strong winds, and very cold temperatures – sometimes falling below −22 °F (−30 °C)!

10.A.5 Polar

In the polar climate zones, it's extremely cold. Even in summer, the temperatures here never go higher than 50 °F (10 °C)!

NOAA has the above definitions for North America and applies it globally.

Bibliography

IEC 62446-1 ED 2 (2020). *Photovoltaic (PV) systems – Requirements for testing, documentation and maintenance – Part 1: Grid connected systems – Documentation, commissioning tests and inspection.*

IEEE Std 493™ (2007). IEEE Recommended Practice for the Design of Reliable Industrial and Commercial Power Systems.

Type of Project: Revision to IEEE Standard 493-2007, Project Request Type: Initiation / Revision, PAR Request Date: 18 Sep 2023, PAR Approval Date: 08 Nov 2023, PAR Expiration Date: 31 Dec (2027).

Geoffrey T. Klise, Olga Lavrova, Renee Gooding (2018), SANDIA REPORT, SAND2018-1743. *Unlimited Release, PV System Component Fault and Failure, Compilation and Analysis*, Printed February 2018.

Balfour, J.R., Morris, R. (2018). Preemptive Analytical Maintenance (PAM), Introducing a More Functional and Reliable PV System Delivery Model – A Reliability Precursor Report, Academia.edu.

National Solar Radiation Database. (1992). National Solar Radiation Data Base User's Manual (1961–1990). [TMY2 format description.]

Patton, A.D. (1968). Determination and analysis of data for reliability studies. *IEEE Transactions on Power Apparatus and Systems* PAS-87: 84–100.

Sengupta, M.; Habte, A.; Kurtz, S.; et al. (2015). *Best Practices Handbook for the Collection and Use of Solar Resource Data for Solar Energy Applications*. NREL Report No. TP-5D00-63112. (PDF 8.9 MB).

TamizhMani, G. (2014). Reliability evaluation of pv power plants: input data for warranty, bankability and energy estimation model. *PV Module Reliability Workshop 2014*, Golden, CO.

Notes

1 Robert (Bob) Holsinger, formerly PV Solar/Fuel Cell Generation Supervisor at PG&E.

2 "Photovoltaic Power System Reliability Considerations," DOE/NASA/20370-79/19 NASA TM-79291.

3 Dillon R.L. and Madsen, P.M. (2015). Faster-better-cheaper projects: too much risk or overreaction to perceived failure?. *IEEE Transactions on Engineering Management* 62(2).

4 GAO, "Report to the Subcommittee on VA/HUD-Independent Agencies, Committee on Appropriations, U.S. Senate," November 2004.

5 Data curation is the organization and integration of data collected from various sources. It involves annotation, publication and presentation of the data such that the value of the data is maintained over time, and the data remains available for reuse and preservation. https://en.wikipedia.org/wiki/Data_curation

6 Balfour, J.R., Hill, R.R., Walker, A. et al. (2021). Masking of photovoltaic system performance problems by inverter clipping and other design and operational practices. United States: N. p. https://doi.org/10.1016/j.rser.2021.111067.

11

Operations and Maintenance (O&M)

11.1 Introduction

The concept of operations and maintenance (O&M) for Photovoltaic (PV) plants is a continual point of discussion and argument in the industry. Many PV professionals do not fully understand what O&M is, why it is important, what is required, how it must be practiced for effectiveness, and what the benefits are. From the perspective of this book, consider that: "Operations and Maintenance is about maintaining the health, condition, status and the asset value of the plant to meet its defined mission, and contracts over the planned life of the plant." Therefore, O&M covers a broad range of planning, specifications, time-based cause and effect considerations, requirements, education, training, and a full awareness of all of the stakeholders of what the consequences are of not addressing O&M seriously at system concept/specification.

Yet, before we can consider what should be included in the scope of a maintenance plan for photovoltaic systems, we must first consider why maintenance is important. For many, reducing the risk of litigation arising from safety and environmental issues and maximizing revenues from the project are at the top of their priority list.

The bankruptcy filing by PG&E (2020) underscores the importance of maintaining electrical equipment. A fire caused by neglecting proper maintenance is one of the largest threats to a company's profitability and reputation. Unfortunately, the number of fires caused by PV systems in wildfire-prone regions is increasing (Figure 11.1). If that were not enough, solar has caused its first documented wildfire in Northern California. While the number of PV fires in the US is still relatively low, this number is growing from year-to-year as installed equipment ages. However, unlike aged medium voltage power lines that were responsible for the Camp Fire (and others) that can have voltage removed at central distribution points like substations, there is no "stop-gap" allowing for the removal of voltage from hundreds of thousands of residential PV systems during times when wildfire risk is high. Thermal expansion and contraction of conductors occur in any electrical system – let alone one placed directly in the sun and subjected to extremely high temperatures in the afternoon and cold temperatures in the morning. It is only a matter of time until a significant amount of PV capacity is old, and the number of rooftop ignitions represents a larger portion of firefighting efforts.

As most residential and many commercial and utility PV systems are not maintained adequately, available data tells us to expect numerous fires in the future. This is especially true

Photovoltaic (PV) System Delivery as Reliable Energy Infrastructure, First Edition.
John R. Balfour and Russell W. Morris.

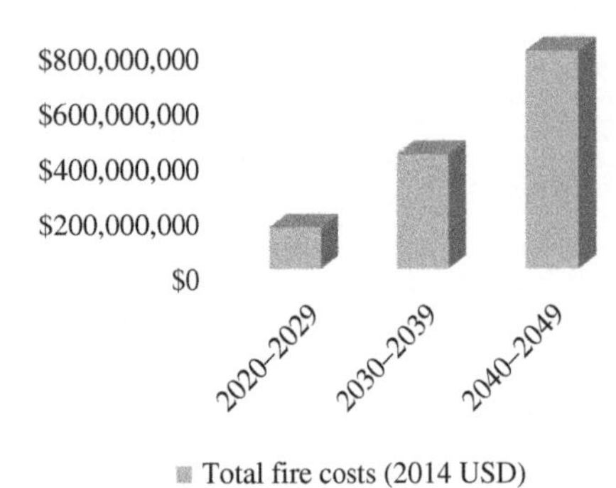

Figure 11.1 Forecast California Fires. (Source: Courtesy of Higher Powered LLC)

in western US states and probably other locations with similar conditions internationally, prone to wildfires. Forecasting historical data forward, the story is not pretty. In California alone, in addition to costing hundreds of millions of dollars, data indicates our industry will start maiming civilians annually in 2023, and in 2036, will start killing people regularly. Methodology to forecast PV sourced fires, deaths, and costs included in Appendix 11.A.

To keep ahead of this, it would be required to have quantitative field data that identifies common failure points in photovoltaic systems across a spectrum of operational and environmental profiles. Unfortunately, at this time, most data collection efforts have been geared toward understanding issues that may impact system production (revenue), and not failures while not or under addressing safety hazards, and failure to meet contractual requirements, excessive and or unexpected O&M costs, and not a planned **SE/R**epowering™. Data acquisition systems (DASs) are hooked up to inverters and simply cannot tell an operator if there is a hotspot in a PV module that has the potential to develop into a thermal event. To fully understand the risk of fire and the reliability of equipment, routine inspections have to occur, and standardized data must be collected by the boots on the ground maintainers in the field, and the operations team and included as part of the O&M plan.

From the perspective of this book, O&M is a key variable in reducing the potential for safety hazards while simultaneously maximizing revenues from system production. To elaborate a bit further on why this is important, this chapter will begin by discussing various reliability, availability, maintainability including testability, and safety (RAMS) topics from the perspective of O&M. Following this discussion, we will outline what should be contemplated during the specification phase of project development and what could be included in an O&M plan to successfully reduce unplanned risk of litigation and operational expenditures (OPEX).

Key takeaways in this chapter include:

- Impact and need for RAMS and data collection from the perspective of O&M.
- An outline of items to contemplate when formulating an O&M plan and cost model.
- Tradeoffs to consider when weighing the strengths and weaknesses of various O&M strategies (e.g. self-perform, regional third party, local electricians).
- What is essential to place under an operator's purview.
- What the maintainer's role is and how their scope will impact a project's OPEX.
- OPEX spent on O&M – reality versus a project's cost model.

11.2 Safety

Straying from convention, this chapter will discuss safety prior to other RAMS topics since plant safe operation and personnel safety while performing plant O&M should be at the top of everyone's priority list.

The textbook goal of safety is to prevent injuries or death, protect capital investments and prevent the release and exposure of hazardous materials or substances to the public to the environment. While many stakeholders might agree with this definition, they may not necessarily agree to the extent by which each of their organizations must participate or how to accomplish this goal. In fact, similar to the concept of O&M, the concept of safety is a point of continual discussion and argument in the industry. What is up for debate usually is not the goal of working safely but rather how to accomplish it.

Although not prescriptive or comprehensive for all tasks, we suggest safer plant O&M might be accomplished by:

1. Identification and prioritization of safety issues prior to dispatch,
2. Communicating and marking arc flash potential, voltages, and other known hazards to the field service team, and
3. Requiring field service to complete a job hazard analysis (JHA) and training to contemplate potential unknown hazards and protective measures.

In other words, understanding when to put personnel in a potentially hazardous situation, knowing the hazards that can be identified, and taking a moment to contemplate previously unidentified hazards and proper work controls prior to commencing a task.

11.2.1 Identification and Prioritization of Safety Issues

Supervisory control and data acquisition (SCADA) systems typically are not hooked up to PV modules (solar panels), strings, combiners, and cannot automatically flag safety issues such as hotspots to the plant operator. However, every mobilization (truck roll) is an opportunity for service providers to identify safety and working environmental concerns. Often, maintainers will in fact raise the flag if they see something. Yet, if there is not an aggressive organizational safety imperative, with a clearly defined way to input this site data into the operator's work management tool,[1] then recognition of these concerns will often get missed by decision-makers. In essence, if it is a handwritten report, that does not necessarily mean pertinent information will go anywhere beyond that particular piece of handwritten material (Figure 11.2).

Often, third-party service maintenance personnel provide the essential information needed to have a feedback loop of information. However, the disconnect occurs when this information is not standardized and then entered into a database by the operator. A more effective process may involve standardized nomenclature and electronic tablet reporting used by the maintainer when they are in the field (Figure 11.3).

Without a means to include field measurements, decision-makers cannot prioritize safety issues. Ensuring these observations get recorded into a Relational Database Management System (RDMS) may be an effective means to ensure the operator can dispatch according to priority and optimize the planning of resources. There are Computer Maintenance

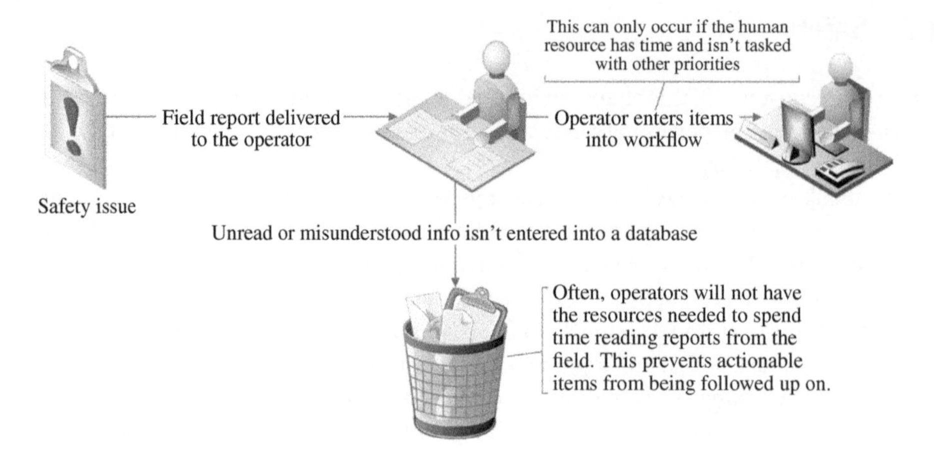

Figure 11.2 Source Higher Powered, LLC.

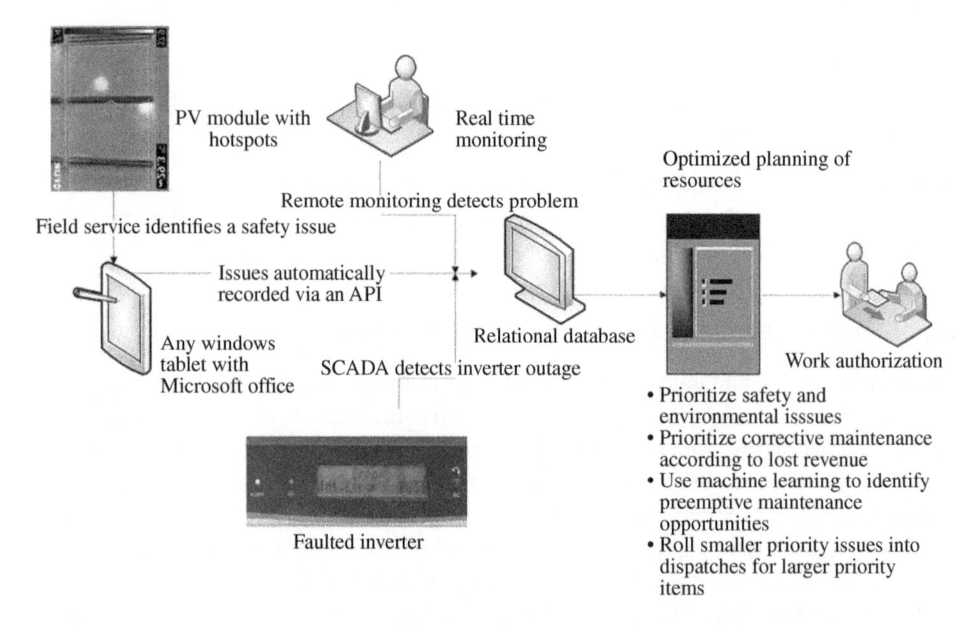

Figure 11.3 Illustration of PV Works CMMS workflow (Source: Higher Powered, LLC.).

Management Software (CMMS) platforms available that offer a tablet interface for field service, which may be an efficient means by which to capture field data.[2]

11.2.2 Arc Flash Potential

According to the Occupational Safety and Health Administration (OSHA), it is the employer's responsibility to provide a safe workplace.[3] Although this is true, it is easier said than done if employers do not have all the information needed in order to keep their employees safe. For instance, supposing the rated short circuit current belonging to a bank of energy storage devices was not shared with an O&M vendor? Their Engineering Safety Officer may not have all of the information needed to correctly:

- Calculate the potential for arc flash within a cabinet.
- Determine the appropriate amount of personnel protective equipment.
- Crew size needed for a particular task.

It is vital for success that enclosure arc flash potential and voltage, as well as other known hazards, is communicated to field service.

Although sharing of information is critical, as is recommended throughout this book (see Chapters 4–7), in today's world, where the potential for litigation is a real possibility, organizations tend to be cautious of sharing more information than necessary with external parties. This tendency to protect an organization from unwanted risk can hinder – not help – the apportion of crucial details needed for safe plant operations. For instance, a transformer manufacturer may be hesitant to label their product with arc flash stickers that define approach boundaries and Personnel Protective Equipment (PPE) for safe operation. Doing so may potentially expose them to the risk of a lawsuit if another company's employee is injured while following those procedures. Although there may be a modicum of logic supporting the manufacturer's hesitation, it is also likely that they, as well as the project developers, engineers, construction, and/or O&M company, would suffer damage to their reputations and/or loss of business if somebody was seriously injured or died resulting from an arc flash because they did not have enough information at their disposal to keep themselves safe.

Obtaining all pertinent information needed for arc flash calculations during project development could facilitate safer plant operations and potentially help a system owner have a more accurate O&M cost model. During the negotiation of equipment pricing is when the developer will have the maximum amount of leverage over a manufacturer. This is an optimum time to request all the information needed for arc flash calculations. An owner would then be able to share this information with their potential O&M vendors, who could in turn, provide more accurate price predictions and ensure their workers were safe once the plant was operational and an O&M contract had been awarded. It should therefore be common practice for the developer to share as much information as possible with the potential O&M vendors early in project development, i.e. at system concept through specification.

11.2.3 Job Hazard Analysis

Although a solar energy facility may have been built in accordance with the National Electric Code (NEC) NFPA 1, and certain hazards can be reasonably estimated, this does not mean an unforeseen hazard will not exist. For example, it is possible for a string (PV Source Circuit) of PV modules to have a continuous short circuit current of nine amps resulting from a fault and an over-current protection device (OCPD) (fuse) rated for 15 A that was sized correctly in accordance with NEC (NFPA 1, NFPA 70, NFPA 70B, NFPA 70E, NFPA 5000), that would never interrupt the circuit during the faulted condition. Given the amount of energy contained withing components of a PV system, to prevent themselves from being injured, field workers should take a moment to ponder what is not known about or what could go wrong with carrying out a task by completing a JHA.

The goal of assessing a particular task in the field for potential hazards is not to anticipate an infinite number of potential outcomes. That is, of course, impossible! Nor is it reasonable or effective to inundate the field service team with a mountain of paperwork. Instead, the objective should be to have the crew take a moment to assess the task for known hazards (e.g. dangerous voltages, work clearances, etc.), visualize the work ahead of themselves, anticipate potential problems or unknowns (e.g. pests, lack of ventilation, poor lighting, weather events, etc.), and ensure they have proper measures in place. Provided this step does not require filling out forms for hours, is easy to administrate, and is part of everyone's culture, the service team is far more likely to integrate it into their normal routine.

Since safety is the employer's responsibility according to OSHA and other requirements internationally, then at least partially, administration of the JHA should also fall under the purview of the employer. Technicians need to know what to expect before they head to the site. For instance, an employer should know if their crew is being dispatched to work on circuits with high or low voltages. They should also know what PPE should be available, worn, or used when working on those types of circuits and ensure that it is provided. If this part of the assessment was completed prior to dispatch, it would ensure only qualified personnel were sent and that they bring the correct tools and PPE needed for the job. It would also help limit paperwork for the field crew to an on-site assessment and review of the controls they need in place for the electrical work.

Completion of a JHA for every work activity does not necessarily need to be a time sink eating up a project's OPEX. Although a review of potential hazards and appropriate controls must be completed, time need not be spent on filling out all the fields on a form each time. Software automation can help reduce paperwork and increase the accuracy and reliability of information. For instance, if corrective maintenance (CM) was authorized on a central inverter, a maintenance management software program could be programmed to include the arc flash and PPE requirements onto the work ticket automatically. This type of functionality could save a project owner the expense of having a service technician (or their employer) fill out certain portions of a JHA. Doing so could save time and minimize repetition while showing respect for the workforce, their safety, and their professional needs.

11.3 Reliability

Per the Institute of Electrical and Electronics Engineers (IEEE), reliability is "the ability of a system or component to perform its required function under stated conditions for a specified period of time" (IEEE Standard Computer Dictionary 1991). This can have important implications for the O&M scope, plan, and strategy, since the period of time something is anticipated to be reliable (usually per information provided by the manufacturer) is typically longer than the period of time the manufacturer is obligated to replace or repair the equipment under warranty.

Manufacturers have a reputation for being overly optimistic when marketing and advertising the reliability of their products. There is a strong use of generalizations and allusions to "better than," or "best of class," or even "world's best!" Any customer interested in understanding reliability, with a high degree of accuracy, must opt for a more empirical, data-driven, approach, especially when involved with a portfolio of assets.

11.3.1 Reliability Data

There are a number of standards from the IEC on PV reliability, reliability methods, and data acquisition IEC 61078:2016, IEC 61703:2016, IEC TS 61586:2017, IEC TR 63292:2020, IEC 61163-2:2020, IEC 61123:2019, IEC TS 63265:2022. Reliability data is obtained by documenting field failures and operating conditions of like and similar equipment in the field and or from direct reliability testing (Chapter 6), which opens up multiple channels for risk identification, cost avoidance, and savings. For instance, an understanding of expected failures during a warranty period helps anticipate them following a warranty period. Using warranty data can anticipate early failures in the plant. Leaner maintenance strategies might also be achievable once historical data is available that quantifies which maintenance, completed at what frequency, will impact the ability of a component to function as required during its service life. However, an owner will not have this data unless measurements are included in the scope of O&M (Chapters 4–8).

Unfortunately, there are two major problems with how operators attempt to collect reliability data today:

(1) Many operators and or owners rely solely on their SCADA system to tell them what the problem is/was, based on event information. A fundamental assumption is that the data acquisition part of the SCADA is acquiring sensor data about the health and status of the system. Although potential causes may be listed for a type of failure by the manufacturer in a database, the actual root cause of the failure event cannot be determined until troubleshooting takes place in the field.[4] Unfortunately, most field service teams do not have a way to collect standardized data into their reporting tools that will feed back into the RDMS database that classified the outage in the first place. The resulting data may only indicate a component was replaced, which is insufficient. It may also be inaccurate since there is no validation for the cause of the problem determined by troubleshooting when cross-referenced with the initial problem identified by the SCADA system. From a data quality perspective, root cause analysis (RCA) can only be accomplished following troubleshooting.
(2) There is an issue with current industry practice that stems from where measurements are taken. As was just pointed out, it is often the SCADA system that identifies problems when they occur. This means project stakeholders that rely solely on the SCADA or other DAS to flag issues will only know about problems in a PV system where the SCADA or DAS is physically connected. For instance, failing or failed MC4 connectors used to interconnect modules would be missed entirely since this is not a point in a PV system typically monitored. Owners and operators will not have comprehensive reliability data unless a means to take additional (other than the SCADA or DAS) measurements is employed (Chapters 5–7 and 10).

Some owners recognize these challenges but are finding that solutions can be cost-intensive and if not properly managed, poorly executed. Certain CMMS solutions may have custom tablet software developed to collect the missing data, but these applications can be expensive (hundreds of thousands of dollars) and difficult, if not impossible, to deploy to third-party service personnel. Nevertheless, these obstacles are not insurmountable. At least one firm has developed a tablet-friendly data collection tool that is easily deployable to third-party field services.[5] Given the lucrative benefits achievable with reliability data, owners may prosper greatly by including the collection of, trending, curation, archival, and reporting of qualification tests, manufacturing, and field data in the O&M plan and scope. This also requires that the data form, format, and detail must be part of the initial SE and EPC (contractor) design.

11.4 Availability

How the word *availability* is defined and what equipment in a PV system it is applied to will always depend on the project's contract language. Chapter 8 provides the definitions for the various availability metrics. Although it is strongly recommended for system owners and developers to be early adopters of IEC TS 63019:2019 and to use the appropriate definitions, at the end of the day, what data operators collect to drive decision-making is always in the signed agreement. With this in mind, and as additional topics for availability, we will briefly discuss two topics: (i) Peer-based analysis and (ii) Inverter availability guarantee language. These are frequently included in an O&M agreement with little care given to the nuances in the commercial space or the impact that they can have on a project's risk and OPEX.

11.4.1 Peer-Based Analysis

Operators keen to understand the equipment availability of their commercial (or residential) assets must employ strategies different from what is used in the utility space. Whereas a peer-based, IEEE Std 762-2006 approach might yield some actionable information to utility operators, it is often not appropriate for a commercial solar energy facility.[6] Peer-based analysis only works if the DC build (number of, make, model, batch, tilt, azimuth, shading of, string sizes, etc.) of the PV modules and power conditioning units (inverters) (make, model, firmware, etc.) are always the same to provide points to measure for comparison. Sometimes a commercial site may only consist of a single central unit. In others, the makes and models of equipment installed may vary wildly. In fact, we have encountered a commercial facility with only four central inverters, each of which was a different make and model. Commercial operators (and contract language) would therefore do well to include an approach that does not solely rely on peer-based data. IEC 63019 is highly recommended.

11.4.2 Inverter Availability Guarantee

Borrego Solar 2019, a regional provider of O&M services for solar energy facilities, found that 25% of issues impacting system performance (excluding environmental) were related to inverter manufacturer warranty claims. Although the first major category of these instances were issues that were reportedly resolved in under five days, the next major subset of issues (The RMA process) took longer than 30 days to resolve. This type of response time puts system owners, O&M providers, and off-takers at the mercy of the inverter's original equipment manufacturer (OEM).

Historically, to reduce this type of risk, availability guarantees are clauses that have been included in O&M agreements. However, our industry is learning that certain things are beyond the O&M provider's control – proprietary part availability, OEM technician availability, and so forth. It may be more effective to place an administrative commercial backstop for production on the OEM's shoulders by having the OEM guarantee the availability of their equipment. This should not raise a project's development costs, as it is typical for this language to be negotiated and included in a contract (albeit with another party) anyway. Moreover, O&M providers are already providing services below cost (discussed later in this chapter) in an effort to stay competitive; they do not have pockets deep enough to keep their technicians servicing offline equipment if they are also dealing with liquidated damages associated with an availability guarantee caused by variables out of their control.

Notes:

1. This is not to suggest that an O&M provider should not have skin in the game. Required dispatch times and thresholds would be a great way to start.
2. Adoption of IEC 63019 (PHOTOVOLTAIC POWER SYSTEMS (PVPS) – INFORMATION MODEL FOR AVAILABILITY) will provide clarification of the necessary calculations and definitions for design and contractual purposes. The standard also clearly defines states of availability for comparison, which adds depth to communicating the system condition.
3. Availability language with an inverter OEM might include:
   (a) Equipment availability,
   (b) Supply chain for critical components (parts),
   (c) OEM technician dispatch criteria,
   (d) OEM technician dispatch response times, and so forth, and
   (e) A clear and concise glossary of definitions that are agreed to.

11.5 Maintainability

Although certain things, common sense in nature, should go without saying, real-life examples necessitate taking a moment to talk about the basic nature of keeping equipment serviceable and maintainable (Chapter 5). If a service technician cannot access an area, or bring needed materials, they cannot do their job. Equipment will remain offline if it cannot be accessed for servicing or if there's insufficient room for replacement (Chapter 7).

Although the NEC (NFPA 70) addresses sufficient working clearances in the vicinity of electrical enclosures, it does not necessarily provide for sufficient walking space between rows of PV modules installed on top of a residential, commercial, or industrial roof. If technicians do not have a path for walking, they may take it upon themselves to walk on top of PV modules. Silicon cells are very thin, brittle, and subject to micro cracking and crack propagation in this type of scenario. Moreover, PV modules are not designed to support the weight of an adult human. The force exerted by a pebble in a shoe or boot can cause the glass to shatter. Breaking through the glass can result in coming into contact with high voltage and current sufficient to result in injury or death.

> We have seen demonstrations at PV Industry conferences, where a demo array was built, and manufacturer representatives walked and rode bicycles on the array. The demo's intent was to send a message of how durable the product was. However, the message received was shockingly the wrong one.
>
> A group of experienced professional colleagues (with 10 or more years in the industry) gathered, stunned and dismayed at the real message that literally thousands of new industry members were receiving. We discussed how difficult it has been to get the industry to understand the fragility of modules in packaging handling, shipping, installing, and maintaining effective systems.
>
> This message of riding, walking, or carrying modules balanced on a hard hat undid a critical O&M set of messages. (John R. Balfour)

Another real-world example we have encountered was with string inverters that could not be replaced by field service personnel, as it would have required construction equipment (e.g. cranes, hoists, flat-bed trailers). Arrays of PV modules had been preassembled and mounted to their racking in a warehouse. These frames were then dropped in place, by a crane, on top of an in-situ support structure where string inverters had already been mounted. Unfortunately, sufficient clearance between the top of the string inverters and the bottom of the array was not available to allow for lifting a string inverter off the back mounting clip, should replacement ever become necessary. It was a simple, avoidable, and costly error that continues to hamper effective O&M.

Ignoring the maintainability and serviceability of equipment during system specification and development will result in higher risk and higher O&M costs. If a service tech with required support equipment cannot gain access to the vicinity of the task, it will not get done without potentially risking damage to equipment or injury. Should offline (normally revenue-generating) equipment not be serviceable with anything less than full-blown construction support equipment, it will either remain offline or a project's OPEX will skyrocket. Either way, the owner loses money. It would only be to a project's benefit to pay special care and consideration to the maintainability of the equipment beginning at specification during the project concept. Project stakeholders would do well to solicit feedback from potential maintenance team(s) early in, on, and continuously during project development as outlined in our **SE/R**epowering™ discussion and examples (see Chapter 4).

11.6 Testability

In PAM, a SE/RAMS approach for O&M must include agreed-upon testing, data collection, and curation procedures (Chapters 5–7 and 10) for the maintenance provider. Failure to clearly define and agree to which measurements are valid, what equipment to use, how to make the necessary measurements, and training needed for personnel may make performance evaluation or warranty claims difficult at best. It would clearly benefit all project stakeholders to formalize (contractually) testing during project development.

11.6.1 Module Warranty

Risk-adverse project stakeholders may want to ensure their warranty language includes a field test procedure for PV module performance.

It will not do any good to have a warranty in place by a Tier 1 manufacturer backed up by an insurance policy if a system owner cannot substantiate a large claim with evidence.

The expensive nature of laboratory analysis to collect evidence can be prohibitive and potentially could be avoided with an appropriate field test procedure.

Usually, the actual performance of a PV module must be compared to Standard Test Conditions (STC), which often necessitates sending a sufficiently large sample of modules to a third-party lab for testing. However, in reality, this is not necessary since an empirical approach taking appropriate measurements in the field may be used to determine module Fill Factor (FF) and thereby module efficiency[7, 8]:

$$FF = \frac{V_{oc} - \ln(V_{oc} + 0.72)}{V_{oc} + 1} \quad (11.1)$$

where V_{oc} = open circuit voltage normalized for temperature.

Notwithstanding science-based literature supporting field-based measurements, a module manufacturer may still require an owner to send modules to a laboratory for testing. High cost of removal (and reinstallation), potential damage during transit, and lost production revenues may all be undesirable to a project owner. It would be more to their fiscal benefit to negotiate a clearly defined field test procedure, confirmed to be achievable by the maintenance provider, that the manufacturer would agree to in writing at the time of the purchasing of the PV modules when they have the maximum amount of leverage.

11.6.2 Inverter Warranty

Unknown to many PV system stakeholders, an inverter's efficiency varies over time, from seconds to years (Chapter 2). From a DC, overbuild sometimes contributes to hiding this fact for at least a few years of operation. Once the inverter's output power drops low enough, it becomes apparent to owners they will need to replace the aging system components or accept lower revenues from production than was initially budgeted. This is easily preemptable with effective **SE/R**epowering™ planning and specification.

We have yet to encounter an inverter manufacturer that advertises the rate their equipment will decrease in power output from year-to-year. Equipment degradation is not defined for any specific usage, environment, or climate, resulting in high variability from site to site for the same equipment. We also have yet to encounter an inverter manufacturer that includes language for power degradation as part of their warranty obligations. Therefore, stakeholders need to negotiate the inclusion of the effects of aging and stress over the warranty period, again at the time of purchasing when there is the maximum amount of leverage. This language must include a field test procedure, confirmed to be easily performed by the maintenance provider, to verify an inverter's performance is as advertised from year-to-year.

11.7 Project Development

During the development of PV projects, there is often a push and pull between the business developers and the O&M team. This is a major reason we have defined the terminology "PV System Delivery versus PV Project Delivery." The discussion often goes, "If you could just give me that price, then you will be given the project!" Many times, the price is set first, and then the scope and logistics are figured out later. This can often be to the detriment of the project, its initial and future stakeholders, resulting in higher OPEX than was budgeted, lower power generation, less revenues from system production, and more risk from safety concerns. A more logical progression would be to first pick an O&M strategy, next define the scope of services, then request pricing from vendors to validate the number used in a project's cost model.

11.8 O&M Plan

Identification of what is needed to be successful for the operation and maintaining of a photovoltaic system is simply a process that needs to take place during the feasibility analysis of a project:

1. Select an O&M systems delivery philosophy and strategy.
2. Define the scope, roles, and responsibilities of the system operator.
3. Define the scope, roles, and responsibilities of the system maintainer.
4. Validate scope budget, pricing with providers and build a cost model for the O&M of a project.

In the following, we will discuss each component of this process.

11.8.1 O&M Philosophy and Strategy

Before the scope, roles, and responsibilities of services can be specified or defined, an O&M philosophy and strategy must first be selected. Should an owner opt to self-perform the O&M for the PV system(s), rely on regional third-party providers, or some other unbeknownst plan? We will discuss merits and drawbacks of each issue worth bearing in mind, as well as a potential third option that may have significant cost-saving potential. The philosophical approaches can be stated as selecting a basic operating approach, e.g. operate to failure, repair on major events, accumulate repairs to minimize truck rolls, etc. The strategy then further develops the philosophy into implementable form.

Further, a question to be answered: is the O&M a single company or organization, or is the operation of the plant being done by the owner or other party while maintenance is done by yet another? It is possible to have three or four alternatives, and the system engineering team, with the owner, must make the appropriate trade studies and selections based on local business conditions.

Many developers and system owners who have achieved some fleet scale experience believe a question worth pondering is whether they should continue to use regional third-party O&M providers or self-perform the O&M of their PV facilities. The assumption is that there will be greater control over OPEX in a self-perform scenario. With this in mind, system developers and owners must consider if the potential savings is worth the added risk of putting their own employees in situations and areas inherently more hazardous to life.

Each scenario is a gamble. Nobody has a crystal ball, and there is no way to know in advance if an employee will be injured while at work (presumably a very expensive outcome). On the other hand, a regional third-party service provider may have slow response times resulting in less revenue from system production. Either way, there are some expensive tradeoffs to contemplate in order to make an informed decision.

11.8.1.1 Self-Performing

Initially, one might suppose that self-performing the O&M would afford system owners tighter control of their OPEX. There would certainly be more insight into, and control of, planning and scheduling priorities. Technicians could be trained for and outfitted with the proper tools, test equipment, PPE, and everything else required for a particular service activity. Moreover, if the internal organization provided services at or close to the cost of those services, theoretically, they would be less expensive than an external organization. Anecdotally this seems like the best choice, but there are some weaknesses to consider as well.

Accounted correctly, the cost associated with self-performing the O&M of solar assets is an expensive proposition. Licensing, insurance (workers' compensation, general liability, etc.), real-time operations, technician training, employee benefits, vehicles, tools, equipment, etc., are all very real costs that are sometimes forgotten about at the project level. While third-party providers may be able to spread these expenses across all their clients, an owner choosing to self-perform would most likely need to justify these expenses based solely on the OPEX budget for their targeted portfolio of assets. For success in this process, the OPEX budget must reflect the realities of the system and all lifecycle considerations. This is seldom accomplished in the industry today.

Potential for loss of life or limb can result from hazards known to be in the field and must also be contemplated when weighing the pros against the cons of self-performing the O&M. Dangerous voltages, slip-trip-and-fall hazards, extreme temperatures, and pests (badgers, rattlesnakes, black-widow spiders, etc.) are all things a service technician might encounter in the field. The operation of safety devices such as interlocks should be verified prior to performing maintenance in that area. This adds a level of complexity and risk that should not be underestimated.

Notwithstanding the risk of opting to self-perform O&M, another factor must be considered. Business objectives can change. A company's stated goal of developing/owning utility solar projects one year may shift to only developing commercial projects the next. …Alternatively, an organization may choose to sell a portion of their solar assets to free up cash to invest elsewhere. Flexibility is sometimes a key component of success when market conditions can vary. Self-performing O&M may inhibit this flexibility (e.g. if a portion of the solar assets gets sold) since it takes a certain critical mass of projects to justify a service technician in a truck and all of the support equipment needed.

11.8.1.2 Third Party

Factoring the foregoing, it may seem third-party O&M would be the logical alternative. There may be less cost (vehicle, tools, test equipment, etc.) for some expenses that the individual projects would have to cover. Employees of the system owner would also be further removed from the dangers of field service activities. Additionally, managers would not have to worry about letting go of their employee technicians should their company decide to sell their solar assets. Nevertheless, there are some disadvantages to having third-party providers provide maintenance services for PV solar assets as well.

One potentially undesirable component of using third-party O&M is that project OPEX is spent on items not required for the O&M of a system owner's assets. Often, third-party providers will employ asset managers to liaise between the system owner's asset managers, their planner schedulers, and/or service technicians. Another example is the monitoring infrastructure (i.e. Real-time Operations Center [ROC]) frequently employed by third-party providers but also separately replicated, utilized, and maintained by system owners creating unnecessary and expensive redundancy. Finally, the maintenance service vendor is probably not interested in servicing one system owner's equipment exclusively. This means a decent amount of overhead is spent on marketing to other clients, which may sometimes be competitors. Although these types of expenses may be necessary for the O&M vendor to be in business (and grow their business), they are not necessarily needed for the O&M of a particular solar power plant.

Considering these tradeoffs, the selection of the appropriate methodology depends on the business strategy of the developer and/or system owner. For small utility sites and commercial projects, regional third party may be the way to go.

Conversely, a large utility might be another story. There is probably already a staff of full-time service technicians budgeted and paid for by the project. It does not matter if a company chooses to sell the project at a later date – the staff could be part of the transaction and follow the project (the project company is what will be bought or sold). However, the determination whether the staff is regional third party or employed by the system owner may depend on the owner's ability to manage field service personnel, or a regional third party's Key Performance Indicators (KPIs) (Mean Time to Repair, etc.). It is less about the overall business objectives (i.e., holding, or sale of, their solar assets) of the system owner for large utility and more about the ability to effectively manage the field service team.

Opting to self-perform the O&M, or to rely on regional third-party service providers requires the owner to understand their own business plans, strategies, and goals keeping in mind that sometimes the objectives can shift. Use of either of these choices when not aligned with corporate purposes will result in unnecessary project OPEX. Determination of what O&M strategy to employ early in the project development cycle will help mitigate unnecessary expenses and ensure a project is more profitable.

11.8.1.3 An Alternative Commercial PV O&M Strategy: Local Electricians

Contrary to a very prolific and popular assumption, the number of available strategies for O&M of PV facilities is not binary. At least one – and probably more – alternatives exist. Owners of commercial projects spread out across a large geographical footprint, interested in serious cost and risk reduction, might contemplate letting local properly trained, and experienced electricians provide field maintenance services.

Often when regional third-party O&M is employed, operations may become redundant and paid for from a project's OPEX twice. System owners learn, many times after losing hundreds of thousands of dollars, that third-party providers simply will not actively monitor their assets as aggressively as an owner might. Frequently, this results in the owner hiring their own employee to monitor the assets and ensure their service providers dispatch on a timely basis. In turn, the third party will sometimes employ an account manager to help set "reasonable" expectations from the owner, adding more unnecessary costs to a project's OPEX. Since owners are already adopting operation functions, it can be more cost-effective for them to simply hire maintenance providers (M versus O&M) and work with their dispatchers directly.

In many instances, the regional third party is also not licensed to provide electrical maintenance in all the jurisdictions for which they provide services. We have worked with, or are aware of, all the main regional third-party service providers and have yet to encounter one that is properly licensed in all but their home state and possibly a couple of others. A local electrician is more likely to carry licensing required by local law and would be a more suitable vendor should the owner wish to avoid unfavorable attention from the authority having jurisdiction (AHJ), which could result in steep fines.

More superfluous cost results from additional drive time to a commercial solar energy facility by a regional provider or an owner's employee than what would be necessary for an electrician that already lives or works nearby. Unnecessary wages and vehicle wear and tear could be reduced from a project's OPEX if the maintenance technician was already in the vicinity of the site. Spending money on excessive windshield time is needless.

Additionally, various specific service maintenance activities in the commercial space could be achieved without paying a premium for a technician certified by the inverter manufacturer. Forced outages in string inverters are usually resolved by cycling power to the unit or replacing the entire inverter at the direction or discretion of the equipment manufacturer. In either case, a licensed electrician is more than qualified to cycle power or connect and disconnect a wiring harness (and strings) along with the mounting hardware for a string inverter. It is usually not essential to pay for fully certified (by the inverter manufacturer) technicians to complete routine maintenance on string inverters.

Positive relationships with off-takers can be maintained by a good working relationship with facility personnel. The administration of the facility where the commercial PV plant is installed, may already have a "go-to" electrician they use and are happy with. If they know and trust that individual or company, they are more likely to accommodate site access requests in a timely and hassle-free manner. It is very frustrating (and costly) to service providers and owners when a technician arrives at the scheduled time only to find a locked perimeter fence or building with nobody on site to grant them access. Using the local electrician the facility folks are happy with, can reduce these headaches and keep the added cost of extra mobilizations down.

It is noteworthy to mention there is also an increased likelihood that generating units may have improved availability when using more local providers. It is analogous to using string inverters instead of one central inverter. Imagine two 250 kWac sites – one with a single central inverter, and another with 25–10 kWac inverters. Should a single unit trip off at each of these two sites, the impact on overall site availability (and production revenues) is very different (100% versus 4% of generating capacity). The same is true for vendor relationships. If a regional third-party provider does not do its job, the overall impact on a portfolio would potentially be much greater than if one local provider was not meeting its obligations. The cost and organizational angst to replace the regional service provider would be greater as well.

The tradeoff is that this type of process would require additional administrative measures and legwork up front. Using numerous electrical contractors based near the sites would necessitate the negotiation and execution of more contracts than what would be typically used between an owner and a single regional third-party provider. Administration, prioritization, planning, and scheduling, of work authorizations, via numerous contracts, would undoubtedly require a specialized tool set or system.[9] Nevertheless, the reduction in OPEX and the LCOE for the PV systems could certainly justify the added time and effort up front, required from system owners in order to use local electricians for the O&M of their PV projects.

After an O&M strategy has been identified, developing a detailed scope of services must be the next step in an O&M plan. However, with the ever-present need to achieve scale and generate income felt by businesses, system owners often forget that they – not just the lender – have an essential voice in what gets included in the O&M scope. It would well behoove system owners to take the time to identify the goals of O&M and use them to define and specify a scope of work for the O&M services of a project.

11.8.2 Operations Scope

Often the budgeted amount for the O&M of a project is based on multiple third party quotes provided in response to scope of works outlining preventive (preventative) maintenance (PM), scheduled maintenance (SM), condition-based maintenance (CBM), CM, and ancillary maintenance (AM) (Chapter 7). Unfortunately, these are all maintenance services and many times exclude the scope (and cost) of operations. Frequently, after a project gets built, the PV system owner realizes the need for active monitoring and dispatch of maintenance personnel. Foregoing proper consideration of this expense during project development jeopardizes a project's profitability.

Only after the distinction between operations (the O in O&M) and maintenance (the M in O&M) is recognized and a scope defined (specified) for operations can budgetary numbers be requested from vendors to assist with the development of a system cost model. Although the scope for the operator is ultimately specified by the PV system owner, the end goal should be to ensure the safe, efficient, and quality O&M of the solar energy facility – all the while maximizing revenues earned from system production. At a minimum, the PV system owner may want to incorporate the following in the scope outlined in an RFP issued to vendors:

1. Remote monitoring and technician dispatch to the maintenance provider.
2. Tracking and reporting of technician KPIs.
3. Tracking and reporting of maintenance provider KPIs.
4. Tracking and reporting of service maintenance planning and scheduling.
5. Tracking and reporting of equipment reliability.
6. Contract management (usually third-party maintenance providers).

7. Curation and analysis of service maintenance data and all documentation.
8. Oversee force majeure processes and procedures.
9. Coordination of outages and or maintenance activities with the utility or ISO (if required).
10. Curtailment of inverters via a SCADA (if required).
11. NERC CIP compliance (if required).
12. Other required data collection and reporting from a local through federal AHJ.

Note: An example Operations scope may be found in Appendix 11.B.

In addition to these items, there are other opportunities for the Operations team to add value to a project prior to commercial operation. The O&M team may be ideally qualified and suited for doublechecking the construction firm's work. It is also possible to refine dispatch priorities even further by correcting the performance model in accordance with measured weather data in real-time. As including these items in the Operations scope has the potential to increase the revenue from a project, we will elaborate a bit further about each of these topics before discussing maintenance.

11.8.2.1 Construction Oversight

Suppose during construction, corners are cut to save money. The EPC or construction company might profit at the expense and detriment of the system owners and other stakeholders. Installation errors may not show up till after the EPC warranty of, let us say, two years and then appear as a rash of problems demanding immediate repair. Without proper specification, clearly defined contractual obligations, and oversight by an owner's representative, a construction company's risk is relatively low, while the owner's risk for the life of the project can be unnecessarily high.

11.8.2.2 Substantial Completion Punch List

One opportunity for the operations team to provide construction oversight is when the construction of a project reaches substantial completion (SC). This milestone usually includes a punch list that documents all the construction-related issues that still need to be completed prior to final completion, which can greatly impact project delivery budgets.[10] An example previously cited by SunPower resulted in an estimated $60,000 in unplanned labor expenses.[11] Assuming the project's size was 100 MW AC, this cost would have already eaten over 10% of the budget for annual PM inspections.

Any cost issues not resolved prior to commercial operation and the completion of the EPC contract, move from the CAPEX to OPEX. Those issues should be resolved in the EPC CAPEX budget to control future expenditures.

This type of situation occurs all too often. *It is not to the EPC company's fiscal benefit to generate a comprehensive punch list.* For this reason, it would behoove a project owner to have another independent party such as a representative of the project's operations team (unless the EPC company is also the O&M provider) audit the EPC company's work. Independent plant review prior to declaring a site has reached SC can ensure everything has been added to the punch list as appropriate. The upfront cost of less than $10,000 would be money well spent compared to eating ten times this figure in OPEX down the road.

11.8.2.3 Capacity Testing

Another opportunity for the Operations team to provide construction oversight is during capacity testing. By addressing the importance of ASTM 2848-13 and IEC TS 61724-2:2016 – "Photovoltaic System Performance – Part 2: Capacity Evaluation Method Capacity Testing," key elements of commissioning and benchmarking provide additional data to support system startup and future O&M. Feeding the results into a real-time performance model can provide a baseline to support comparison of future annual evaluations of system health and condition.

11.8.2.4 System Performance

An understanding of performance is integral to understanding the evolving health, condition, and status of a PV system. Real-time weather will not be used during initial modeling and project conception. Sometimes there will be less irradiance than in a typical year – other times there will be more. A sampling of 39 sites by DNV GL determined that the average delta between predicted and actual energy is 3.1%.[12] As they point out, "It is imperative that modeled results are benchmarked (Chapter 2) against operational data to validate and refine approaches over time."[13] Without accounting for real-time conditions, it is impossible to understand the return on investment. To achieve this, the Operations scope should include a means to measure weather conditions and use this data to benchmark against the original performance model.

Benchmarking measured real-time data has some added fiscal benefits as well. Real-time irradiance, DC and AC power, temperature, and wind speed measurements can be plugged into a real-time performance model. These measurements need to be down to the string or lowest level of maintenance items to understand how much of the plant's capability has been affected. This would allow the operator to make more informed operational decisions instead of just simply choosing to mobilize when a central inverter trips offline. Of course, it may only make sense to mobilize for unplanned breakdowns if a certain amount of capacity is offline. However, if other issues such as offline strings or tracker issues can be identified and put in an administrative queue, this would allow for correction of these issues during the next mobilization. It would result in increased equipment availability and increased revenues from production.

Real-time performance monitoring and benchmarking are not necessarily a costly proposition. One bankable PV performance model, NREL's System Advisor Model (SAM), code is open source and available to software developers for free. Assuming a photovoltaic system has a DAS collecting and reporting hourly or more frequent weather, power, energy and failure event data can provide the inputs needed for real-time models. The cost to develop a real-time platform is not necessarily an added cost needed to be realized by PV system owners or developers as there are O&M providers, with rates competitive in the market, with this already integrated into their O&M standard offerings.

11.8.3 Maintenance Scope

The scope of work for the maintenance provider is ultimately specified by the PV system owner in accordance with PAM using RAMS methodologies. The end goal should be to ensure the safe, efficient, and defined quality of O&M for the PV facility, while maximizing revenues earned from system production. It is the same goal objective as plant operations. Although there is a distinction between O&M, they must work together to succeed.

In general terms, maintenance may be thought of as what occurs in the field. Sometimes this refers to activities that are "boots on the ground." More specifically, contractually, these actions often get segregated into three categories: PM including SM, CM, and AM. In the following, we will elaborate a bit more on what each of these categories should entail.

11.8.3.1 Preventive Maintenance (PM)

The objective of the PM scope should be an annual inspection and benchmarking intended to assess the overall system health, complete equipment manufacturer-prescribed maintenance, and identify issues requiring follow-up attention. At a minimum, recommended scope objectives include:

1. Prevention of safety issues.
2. Prevention of environmental issues.
3. Prevention of unplanned breakdowns.
4. An assessment of the overall system health, condition, status, and performance.
5. Identification of CM, and AM opportunities.
6. Replacement of disposable components such as air and oil filters, aged fuses and circuit breakers, and visual inspection for corrosion of metal structures.

Note: An example maintenance scope may be found in Appendix 11.C.

Yet, every site is different, and the uniqueness of the site must be considered and incorporated into the scope of an RFP issued to vendors. For example, we have encountered sites where waddles were required to be in good repair so they could ensure vernal pools essential to the continuation of endangered fairy shrimp did not dry out. We have also encountered sites where corrosion prevention painting on electrical enclosures was needed regularly since there were higher than normal concentrations of hydrogen sulfide in the atmosphere. As it is important to have an accurate project cost model, it is imperative site specific details are identified as early as possible during project development.

11.8.3.2 Scheduled Maintenance

SM are tasks that include the removal and replacement of parts on a regular basis to prevent those items from causing an unscheduled downing event. The majority of SM items are components that have a known limited life (time, cycles, overstress conditions, etc.). By careful tracking of the critical parameters, it is possible to perform this maintenance during a near coincident truck roll.

11.8.4 Corrective Maintenance (CM)

The CM scope should be strategically structured to address the eventuality of forced outages and emergency response. Directives "fix it – fix it now," and "just get it done" ringing in the ears of service technicians often underscore an owner's urgent need to get their revenue-generating unit back online. Unplanned breakdowns jeopardize a project's profitability if there is no mobilization plan, or budget in place. At a minimum, this plan should include:

1. An annual budget for unplanned breakdowns (e.g. $\frac{\$4}{kW_{DC}}$ for commercial plants[14]).
2. A budget for major inverter rebuilds or replacements (e.g. $\frac{\$0.10}{kW_{DC}}$ plus shipping, equipment, and labor).
3. Contractually defined dispatch times (e.g. >250 kW_{AC} be on-site within two business days).
4. An assessment of overall system health, status and performance: The maintenance service provider should know before leaving the site (preferably before arriving) how the site is performing (weather corrected) versus the modeled expectation. This will also enable them to ascertain any potential contributors to underperformance while on-site.

Never waste a mobilization!

5. The identification of additional CM and AM opportunities while on site (remember the mobilization is already paid for) that will not necessarily be flagged by a SCADA or DAS monitoring system:
   i. Safety issues (e.g. a hotspot on an infrared image),
   ii. Environmental (e.g. transformer oil leaking),
   iii. Major production issues (e.g. 500 kW_{AC} inverter with abnormal noise),
   iv. Minor production issues (e.g. soiled PV modules), and
   v. Potential warranty claims (e.g. delamination of PV module back sheet or cell cracking).

11.8.4.1 Opportunities for Process Improvement

Stakeholders are learning the administrative model where dispatch priorities determined solely by forced outages at the inverter level leave money on the table. Sophisticated organizations keen to get the most bang for their buck are refining their CM strategies to include preemptive and CBM.

11.8.4.2 Preemptive Maintenance

A subset of CM, preemptive maintenance corrects either a condition or small problem prior to it developing into a larger issue. For example, consider a single-axis tracker bearing in the Arizona desert. The manufacturer may only call for lubricating the bearing annually. However, data may indicate longer bearing life if it gets lubricated twice a year. Lubricating twice a year may therefore be more cost-effective than replacing the bearing at a higher frequency. However, data is often needed to justify the cost–benefit analysis of either scenario. That data is often not collected, inconsistent, or improperly curated. Another important question is when, according to fleet management priorities, to lubricate the bearing.

11.8.4.3 Condition-Based Maintenance

Supposing field service activities can be tracked in a SQL database, and assuming reliability data is properly acquired and processed, system owners would have the ability to really home in on when and where to complete CM items based on the condition of the component. Going back to the tracker bearing example in Section 11.8.4.2, imagine if statistical analysis determined it would be more cost-effective to replace the bearing but that it probably had another year of useful life. It would be beneficial for the bearing replacement to be in the service queue so it would not be forgotten – presumably as a lower-priority item. Now imagine seven months pass and a routine inspection is required at a nearby site. Would it not be more cost-effective if the work planning system raised the priority of the bearing replacement and rolled it into the same mobilization as the routine inspection?

Statistical analysis can help determine where and when the CM priorities are. This is the heart of CBM. Priorities could even be adjusted in real-time in response to real-time outages or the geographic deployment of field service personnel. Often the key limiting factor has been how many resources the operator could allocate to the analysis of data.

Very recent advances in machine learning may make it possible for organizations to analyze their field service and reliability data to identify priorities without over burdening internal labor resources. Driven to push machine learning to new heights, data scientists have come up with something called autogluon. It is powerful stuff. Imagine having a data set, and wanting to have it analyzed by not one, but all machine learning models. Then imagine selecting the models that turned out to be the most accurate in their predictions and creating an optimized predictive model that was an ensemble of the most accurate models that outperforms all the individual models. That is autogluon. That is not even the best part. The creators of autogluon wanted to keep it extremely simple for external users, so, they compiled all of these subprocesses into just three lines of python code[15,16,17]:

```
from autogluon import TabularPrediction as task
predictor = task.fit("train.csv", label = "class")
predictions = predictor.predict("test.csv")
```

Armed with statistical analysis, system owners can home in on which corrective and pre-emptive maintenance opportunities have priority. This would really help inform as to when it makes sense to spend money and when it does not.

11.8.5 Ancillary Maintenance (AM)

Often the AM scope is thought of as discretionary in nature, since the scope (module washing, vegetation removal, pest control, etc.) is usually provided by vendors other than the primary O&M provider. It is frequently an afterthought during project development, with little to no formal attention paid to the strategy or mobilization plan, which may be detrimental to a project. For example, the expedient removal of snow or ice following extreme winter conditions from an overburdened rooftop may be considered an emergency in terms of priority. Without a strategy in place and contract language with specific triggers defining priorities and dispatch times, a project owner may assume unnecessary risk in these types of scenarios. Furthermore, although this type of service vendor may not be savvy when it comes to electronics or solar-generating equipment, any mobilization authorized is already paid for, and the opportunity to take before-and-after photographs should never be wasted. They should be a standard operating procedure (SOP).

11.8.6 Pricing – Determining a Budget

Finally, once the scope of work has been fleshed out (and not before), one can begin to have an idea of what to budget for O&M. This is usually accomplished by providing the scope to O&M vendors and formal requesting pricing through RFPs and RFQs. However, current market conditions have added a layer of complexity, making it harder to understand how much money should actually be reserved for a project's OPEX. Should PV system owners wish to reduce the potential for high operational expenses, they should do diligence prior to building a project. This will allow for a better understanding of how O&M will impact the project's cash flows.

11.8.6.1 O&M Price Distortions

During initial project development, extra care is needed up front to ensure the O&M cost model is accurate and the forecasted yearly OPEX for a project is close to what actually gets or needs to be spent. In the "race to the bottom" numerous factors (e.g. missing portions of scope, invalid, absent licensing, etc.) have distorted the true price of O&M services. However, developers who invest a little time up front on diligence and scope specification and definition can overcome these challenges.

11.8.6.2 Scope

Solar project developers have seen the price of building solar drop significantly over the past decade.[18] As costs come down, the sentiment has been that the cost to operate and maintain these assets should also.[19] However, as labor and component prices actually continue to rise, the cost to provide services has also. To remain competitive, O&M providers have been simply eliminating portions of the scope from their contracts.[20]

By removing CM to address unplanned breakdowns from the scope of work, O&M providers can win contracts even in the current market conditions (5–6 dollars per kilowatt DC). The reality is that system owners will end up paying between 12 and 14 dollars per kilowatt DC. This is almost always the case and has a substantive negative impact on LCOE. Inverters will trip offline, and owners will invariably (as they should) opt to authorize the CM to work on a time and material basis in order to get the revenue meter spinning in their favor again. O&M vendors are familiar with how often forced outages occur for inverters and do not mind pretending the work will not be needed – it is not their project cost model, after all.

Does this mean that system owners should budget 14 dollars per kilowatt DC for the cost of O&M even though the market would say otherwise?

Maybe at first. However, in time, there are O&M providers that are utilizing smarter O&M strategies (such as tracking DC health and equipment reliability issues) that can drive cost reduction. Strategies such as careful vendor selection, clearly defined scopes, contracts at the portfolio (instead of project) level, and a distinction between O&M can help drive cost reduction toward 9 dollars per kilowatt DC. Although not as attractive as 5 dollars, a project's proforma can be profitable using this number. The key is that it is closer to what will actually be spent vs. a misdirection or misrepresentation that was told up front in order to get the project built. Also, consider that making better infrastructure grade decisions at specification will reduce faults, failures, and labor, thereby reducing real O&M costs.

As you have read through Chapters 4–8, we have raised the basic issue of building better, more reliable plants to deal with many of these long long-term yet more controllable costs, such as spares, warehousing, etc. Doing so allows stakeholders to determine the actual requirements more accurately for a plant as they are applied. This results in lower total lifecycle costs for O&M.

11.8.6.3 Licensing

Proper licensing is not solely about compliance with rules set forth by an AHJ. It is also about reducing risk to the system owner. It is easy to surmise that unlicensed personnel may also be untrained and or undertrained personnel. Certainly, it goes without saying that untrained people should not work in the vicinity of hazardous voltages. Workers' compensation claims resulting from death or injury can be very expensive. Moreover, any general liability insurance policy held by the maintenance service provider might be void if they were engaged in work requiring a license they did not possess. It would behoove risk-adverse system owners to ensure they are adhering to the regulations provided by the AHJ.

Notwithstanding the prudence inherent in proper licensing, many system owners might be surprised to learn their service provider is not licensed for electrical work in all the territories they provide services. This may be attributed to the time-intensive and costly nature of obtaining licenses in all of these places. New York for example has different licensing required for each city.[21] Nevertheless, many of these same places require a license for electrical work. In New York City, for instance, unlicensed electrical work is against the law.[22]

Lack of licensing may be causing O&M pricing to be artificially low. Examination (and generally test preparation) associated application/filing fees, and the wages of the qualifying party for time spent during the licensing process are all very real expenses that service providers should be paying for. Usually, licensing also requires bonding, which is yet another expense. Should a majority of vendors continue to evade these obligations, in addition to potentially breaking laws set forth by the AHJ, they further distort what the true cost is to deliver O&M of PV plants, which is unfair to their competitors and puts the system owner(s) at risk.

11.9 Conclusion

O&M of PV systems is not a trivial matter. Failure to adequately contemplate and address the challenges and nuances of O&M during the initial phases of the system development, jeopardize a project's profitability and expose system owners and downstream stakeholders to unnecessary risk. However, an appropriate strategy defined by a company's business objectives combined with a well thought out scope of services can help owners reign in their OPEX and reduce the number of unforeseen expenditures. When coupled with a safety-oriented culture, where all stakeholders participate, system owners have the ability to reduce the number of hazards they expose their employees or the public to and enjoy more revenue from system uptime.

11.A Appendix A: Photovoltaic Fires Calculation Methodology

Step 1: Determine the Number of PV Systems in California

First, we determined the population size. The State of California publicly shares data regarding the number of systems installed each year, their type (e.g. utility, commercial, and residential), and the overall total number of systems installed.[23] From this information, we can see the number of solar systems reported for all independently operated utility (IOUs) companies. For instance, in 2015, we know there were 144,478 newly installed systems and a total of 425,362 total installed systems. Once codified, we used these numbers to represent the overall population of photovoltaic systems installed in the State of California. Using the number of systems installed in California from 1996 through 2019, we forecasted the number of systems we anticipate to be installed in California in years to come.

Note: it is possible this process does not account for all off-grid photovoltaic systems installed in this region.

Step 2: Tally Number of Fires for California

We requested historical data provided by the U.S. Fire Administration and queried the data to determine the number and locations of photovoltaic-sourced fires every year. Data request is formatted as below:

```
SELECT Fireincident.INC_NO, Fireincident.INC_DATE, Fireincident.FDID,
Incidentaddress.LOC_TYPE, Incidentaddress.NUM_MILE,
Incidentaddress.STREET_PRE, Incidentaddress.STREETNAME,
Incidentaddress.STREETTYPE, Incidentaddress.STREETSUF, Incidentaddress.APT_NO,
Incidentaddress.CITY, Incidentaddress.STATE_ID, Incidentaddress.ZIP5,
Incidentaddress.ZIP4
FROM Fireincident INNER JOIN Incidentaddress ON (Fireincident.FDID =
Incidentaddress.FDID) AND (Fireincident.INC_DATE = Incidentaddress.INC_DATE)
AND (Fireincident.INC_NO = Incidentaddress.INC_NO)
WHERE (((Fireincident.EQ_POWER)="55"));
```

Notes:

1. We requested the data from the US Fire Administration and received it by mail.
2. The query results need to be checked for duplicate records.

We filtered the queried results for photovoltaic-sourced fires in California for 2015–2019.[24] This provided us with the number of fires per the number of installed systems in California.

Step 3: Forecast the Number of Photovoltaic System Sourced Fires in California

We assume that a majority of photovoltaic solar system fires will be caused by aging system components. Indeed, a case study in Japan of a hundred solar system fires found this to be the case and indicated a significant number of failures were attributed to components greater than or equal to seven years of age.[25] For our analysis, we assumed most fires would occur in systems at or older than seven years old, and adjusted the overall population size accordingly (e.g. number of fires in California in 2015 per number of installed systems in California in 2008) to determine an estimated number of fires per number of systems seven years old or older. This process was repeated for each year from 2015 to 2019 to establish

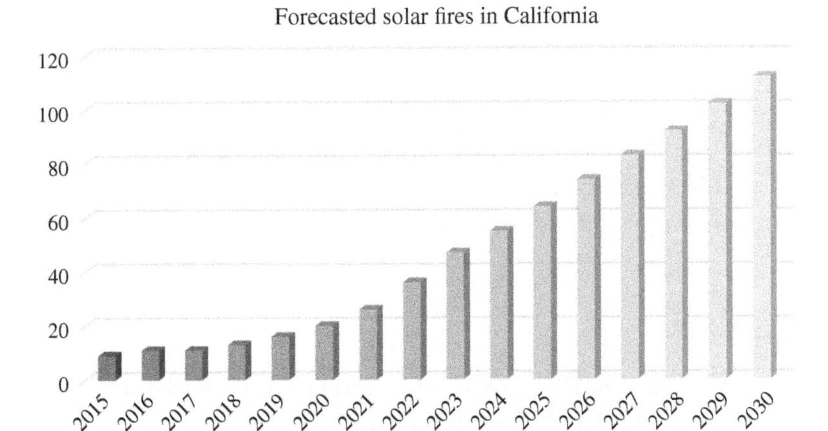

Figure 11.A.1 Forecast California Fires. (Source: Higher Powered LLC).

known values for the independent and dependent variables to forecast the number of California-based photovoltaic sourced fires anticipated in the future (Figure 11.A.1).
Notes:

1. These statistics only contemplate fires where a fire department was dispatched.
2. We understand PV-sourced fires are not always be counted as such[26] and therefore anticipate the actual value to be higher.

Step 4: Forecasting Civilian Casualties and Cost from Number of Fires
To better understand the overall impact, we combined the forecasted values with additional fire statistics.

The authors are aware of the fact that many companies distain the topic of system fires on their projects that, as a result, in many instances, field staff will not document or report them to their supervisors. Some of this is from our own personal experience.

The source of information regarding fires came either from others at the system site or by finding damaged parts in a truck or trash. When the entire staff was asked the question, "What happened and which project did this come from?", the staff would not answer the question. Part of the reason for this is a code of silence for errors made and the knowledge that other companies fire, embarrass, or harass employees for errors they made as a component of a culture of blame and punishment.

While some organizations are very proactive about all of these issues, it is not unusual for management to clearly indicate that, "We don't want to hear about problems!!!" This philosophy supports a lack of improvement that have an impact legally, financially, and on injury/deaths.

These should be used as learning and training tools and another industry reason to share data.

11.A.1 Rate of Deaths Per Fire

Referencing a 2014 report by the National Fire Protection Agency (NFPA), we learn there were 3275 US civilian deaths out of 1,298,000 reported fires.[27] As solar fires do not only happen at residences, and could happen elsewhere (e.g. outside), we used the number of deaths per all reported fire types to determine the average number of civilian deaths per fire as follows:

$$\frac{\text{3275 Civilian Deaths}}{\text{1,298,000 Fires}} = \frac{\text{0.002523112 Civilian Deaths}}{\text{Fire}}$$

When combined with the forecasted number of fires caused by solar systems in California, we estimate existing industry practices will kill civilians in California every year starting by 2036.

11.A.2 Total Cost Per Fire

From another report provided by the NFPA, we see there are more costs associated with fire than just property losses (e.g. disaster planning, preparing standards, etc.).[28] As this report provided an aggregate sum total (and breakdown) of costs of fires in the US in 2014, we used this value of $328.5 billion[29] with the total number of fires in the US in 2014[30] to determine the average total cost per fire

$$\frac{\text{\$328.5 Billion (2014 USD)}}{\text{1,298,000 Fires}} = \frac{\text{\$253,082 (2014 USD)}}{\text{Fire}}$$

Note: this value needs to be adjusted for inflation to better understand the current fiscal impact.

When combined with the forecasted number of fires caused by solar systems in California, this provided a forecasted total cost per fire.

11.B Appendix B: Operations Scope Example (Source: Courtesy of Higher Powered LLC)

11.B.1 The Operator's Role[31]

In general, the operator shall be responsible for the following activities connected with the operation of the commercial solar energy facilities:

1. Remote monitoring and technician dispatch to the maintenance provider.
2. Tracking and reporting of technician KPIs.
3. Tracking and reporting of maintenance provider KPIs.
4. Tracking and reporting of service maintenance planning and scheduling.
5. Tracking and reporting of equipment and plant reliability.
6. Contract management (usually third-party maintenance providers).
7. Reporting confirmation and acceptance.
8. Curation and availability of service maintenance documentation.
9. Oversee force majeure processes and procedures.
10. Coordination of service activities with the utility or ISO (if required).
11. Curtailment of inverters via a SCADA (if required).
12. NERC CIP compliance (if required).
13. Training and Education.
14. Remote Monitoring and Technician Dispatch.

11.B.1.1 Corrective Maintenance

The operator shall continuously remotely monitor the system via the DAS, and issue a CM work order to the maintenance planner in accordance with the following criteria (Table 11.B.1).

11.B.2 Preventive Maintenance

The operator will keep track of annual PM requirements and will assign a PM WO to the maintenance provider at least 50 days in advance of the maintenance completion "due" date. The operator will take reasonable steps to ensure the PM is scheduled at a time that will work for the offtaker as well as the third-party maintenance provider and confirm scheduling with each party in writing (typically email).

11.B.3 Ancillary Maintenance

The operator shall, from time-to-time, if the annual O&M budget permits, authorize AM by issuing an AM WO to the maintenance provider. Ancillary services (module washing, vegetation removal, snow removal, pest control, road, access repair, etc.) are discretionary in nature and will be evaluated on a case-by-case basis. The operator may work with, at the operator's sole discretion, other parties that may specialize in a specific trade appropriately suited for the AM scope, and need not necessarily work through the maintenance provider to facilitate the completion thereof.

11.B.4 Tracking and Reporting of Technician Key Performance Indicators (KPIs)

The operator shall track and report the mean time to repair (MTTR) for each service technician working to resolve CM issues further categorized according to type of problem and

Table 11.B.1 Priority number with time to issue work order.

Priority number (Included on work order)	Criteria	Time to issue a work order (WO)
1	Emergency response	Immediate (same day)
2	Urgent priority	Immediate (same day)
3	>= 250 kWac outage	Immediate (same day)
4	> 1 < 250 kWac inverter outage	Current business day (if on the weekend, next business day), or when another WO is issued for the same site in order to combine multiple service activities into the same mobilization
5	<=1 kWac forced outage and other issues	Current business day (if on the weekend, next business day), or when another WO is issued for the same site in order to combine multiple service activities into the same mobilization

location in the PV system in order to identify technician strengths and weaknesses and allow for the optimization of allocation of resources by assigning the correct technician(s) to the correct task(s).

11.B.4.1 Tracking and Reporting of Maintenance Provider KPIs

The operator shall track and report the MTTR for each service maintenance provider working to resolve CM issues further categorized according to the type of problem and location in the PV system in order to identify technician strengths and weaknesses and allow for the optimization of allocation of resources by assigning the correct maintenance provider(s) to the correct task(s).

Tracking and reporting of service maintenance planning and scheduling.

The operator shall track and report (weekly) all service maintenance activities in the cue and work with the service maintenance provider to ensure activities are scheduled according to the priority number assigned.

11.B.4.2 Tracking and Reporting of Equipment Reliability

The operator shall track, confirm and report (monthly) equipment reliability resulting from RCA for CM activities.

11.B.4.3 Contract Management of the Third-Party Maintenance Providers

The operator shall negotiate, and maintain in place, contracts with third-party maintenance providers to allow for the timely dispatch of service personnel. Should there ever arise a situation where the maintenance provider has failed to meet their dispatch criteria more than once in an annual period, the operator will take necessary steps, such as the identification and selection of a new maintenance provider, to ensure maintenance provider dispatch criteria are met.

11.B.5 Curation of Service Maintenance Documentation

11.B.5.1 Record Keeping

The operator shall track and maintain a record of all documentation of service activities connected with the O&M of the commercial solar energy facility. The operator shall provide copies of all service documents to the system owner following the completion of service activities.

11.B.5.2 Data Aggregation

URTD – Data Collection, Control, Curation and Exchange, Stakeholder Data IO.

11.B.5.3 Corrective Maintenance

Specific data (1. PV System Location 2. Problem Type 3. Contributing Factor 4. Actions Taken) shall be recorded for every CM activity for trending and analysis in a SQL database.
Ancillary maintenance

Specific data (1. PV System Location 2. Problem Type 3. Contributing Factor 4. Actions Taken) shall be recorded for every AM activity for trending and analysis in a SQL database.

11.B.5.4 Follow-up Issues

Every mobilization (CM, PM, and AM) affords the maintenance provider the opportunity to identify issues that require follow-up attention. Any issues requiring follow-up shall be recorded in the SQL database and included in the "cue" to be addressed during the next mobilization where appropriate, as defined by the "dispatch" criteria above.

11.B.5.5 Data Sharing

Developer The operator will provide, at the developer's request, equipment reliability data as reported from the SQL database to assist the developer with equipment selection for future projects in order to continue to install high-quality, reliable, solar energy facilities for their customers. The data provided to the developer is limited exclusively to output from the SQL database and will not include specific service activity documentation. The system owner/s shall have access to all system documentation from concept through site restoration.

Owner In addition to being provided with documentation for every service activity connected with the maintenance of the solar energy facility, the operator will provide, at the system owner's request, equipment reliability data as reported from the SQL database.

Industry From time-to-time, there may be opportunities that arise to share equipment reliability data:

(1. PV System Location,
2. Problem Type,
3. Contributing Factor,
4. Actions Taken) with industry that could help raise general awareness to the common types of photovoltaic system failures. Should such an opportunity arise, and we believe that it will add value to industry, we at our sole discretion, may opt to share aforementioned reliability but will take every reasonable effort to keep our partners' (clients, project developers, equipment manufacturers) anonymous.

11.C Appendix C: Maintenance Scope Example[32]

11.C.1 Introduction

The end goal is to ensure the safe, efficient, and quality O&M of the commercial or residential solar energy facility. Keeping with this goal, the maintenance provider shall deliver the following scope.

11.C.2 Maintenance

11.C.2.1 Safety

The number one priority is to ensure nobody gets hurt connected with the O&M of the solar energy facility. To that end, the service maintenance provider shall integrate a "safety first" paradigm into their culture and daily practices.

11.C.2.2 Site Access

Text notifications shall be provided to the operator to communicate the arrival to and safe departure from site for every service call.

11.C.2.3 Job Hazard Analysis

Field service personnel shall complete a JHA, or other such standard document used by the maintenance provider, and use appropriate Personal Protective Equipment (PPE) for every task. A copy (a photo is fine) of the filled-out document shall be included in every report detailing the service activities.

11.C.2.4 Training

Maintenance provider shall assign only qualified personnel who are trained for the assigned task.

[1]Example maintenance scope provided courtesy of Higher Powered, LLC.

11.C.2.5 Records

Record of safety training and completed JHA documentation shall be kept on file for at least four years and is subject to audit by the operator on request.

11.C.3 Preventive Maintenance (PM)

The preventive maintenance scope is an annual inspection intended to assess the overall system health, complete equipment manufacturer-prescribed maintenance, frequency, activity, and identify issues requiring follow-up attention.

Note: section headers with "if applicable" included in the text are applicable, and included in this scope of work, if that type of equipment is installed and included in the PV system. However, if that type of equipment is not installed or included in the PV system, that section is not included in this scope of work.

11.C.3.1 Administration

The annual PM will be completed by the maintenance provider following receipt of a PM WO from the operator authorizing the service of the system. Scheduling of service activities shall be within two weeks of the due date (included on the WO). The maintenance provider shall notify the operator of the scheduled date within two weeks of receiving a PM WO. WOs authorizing PM shall be issued to the maintenance provider at least 50 days in advance of the due date by the operator. Every reasonable effort will be made to coordinate the scheduling of the PV system maintenance with the site (host) facilities management for a day(s) that work for their team as well as the maintenance provider's team.

11.C.3.2 Measurement and Control Systems

Safety note: this procedure, order of operation, lock out tag out, arc potential, correct PPE, and appropriate crew size should be verified, documented and/or provided by the maintainer's electrical safety engineer based on the data published by the equipment manufacturer and installed equipment configuration.

Wearing appropriate PPE:

- □ Revenue meter(s),
  - □ OEM manual specified maintenance,
  - □ Meter calibration (at frequency [required by any applicable consideration of environmental conditions and agreements]),
- □ DAS or SCADA components,
  - □ Visual inspection,
  - □ Clean and inspect,
  - □ Any other OEM requirements,
- □ Meteorological station(s) (if applicable),
  - □ Clean the reference cell(s) (or irradiance sensor(s)) with a lint-free, nonabrasive cloth each time the site is visited for any type of service call,
  - □ Calibrate telemetry at the frequency specified by OEM,
  - □ Any other OEM requirements.

11.C.3.3 Gen-Tie and Switchgear Maintenance (If Applicable)

Safety note: this procedure, order of operation, lock out tag out, arc potential, correct PPE, and appropriate crew size should be verified and/or provided by the maintainer's electrical safety engineer based on the data published by the equipment manufacturer and installed equipment configuration.

Wearing appropriate PPE:

- □ Circuit breakers,
  - □ Check for the presence and condition of labeling,
  - □ Clean and inspect,
  - □ Thermal imaging and documentation of electrical terminations where accessible,
  - □ Manual operation (Note: the method of procedure for operation of breakers, and the order of operation, should be determined by the maintainer's electrical safety engineer),
  - □ Any other OEM requirements,
- □ Any other component of the PV system,
  - □ OEM requirements.

11.C.3.4 MV Transformer (If Applicable)

Safety note: this procedure, order of operation, lock out tag out, arc potential, correct PPE, and appropriate crew size should be verified and/or provided by the maintainer's electrical safety engineer based on the data published by the equipment manufacturer and installed equipment configuration.

Wearing appropriate PPE:

- □ External visual inspection,
- □ Nitrogen addition and bleed (as required),
- □ Check for the presence and condition of labeling,
- □ Temperature and pressure verification,
- □ Internal visual inspection,
- □ Thermal imaging (if the unit has an IR window),

- □ Oil sample and analysis (first year of operation, then every other year thereafter unless otherwise specified by the manufacturer),
- □ Electrical TTR testing (as required by the OEM manual or every three years),
- □ Any other OEM requirements.

11.C.3.5 AC Panel or Combiner Box (If Applicable)

Safety note: this procedure, order of operation, lock out tag out, arc potential, correct PPE, and appropriate crew size should be verified and/or provided by the maintainer's electrical safety engineer based on the data published by the equipment manufacturer and installed equipment configuration.

Wearing appropriate PPE:

- □ Visually inspect the exterior for enclosure integrity and any signs of corrosion.
- □ Check for the presence and condition of labeling.
- □ Visually inspect the door gasket material and seal condition.
- □ Visually inspect the interior of the enclosure for damage to, discoloration of, or corrosion on any internal component such as wire termination points, disconnect switches, breakers, wire insulation, fuse holders, etc. Also, visually inspect conduit entry points for adequate sealing material and/or mating.
- □ Look for any signs of water intrusion.
- □ Note the presence of insects and/or pests.
- □ Conduct IR scan of the enclosure and include any image(s) with hotspot(s) in the PM report if re-torquing (allowing for time to cool before operation) the termination point to OEM specifications does not clear the hotspot.
  NOTE: IR scans should be conducted with an imaging device that includes the measured temperature(s) in the image taken. Any images included will be annotated in such a way that records the equipment's location.
- □ Any other OEM requirements.

11.C.3.6 Shelter/Skid (If Applicable)

Safety note: this procedure, order of operation, lock out tag out, arc potential, correct PPE, and appropriate crew size should be verified and/or provided by the maintainer's electrical safety engineer based on the data published by the equipment manufacturer and installed equipment configuration.

Wearing appropriate PPE:

- □ Visual inspection,
- □ Clean and inspect air filters (replace if necessary or as per OEM requirements and site environment),
- □ Contribute to the general cleanliness of the shelter/skid, and
- □ Any other OEM requirements.

11.C.3.7 Energy Storage Device(s) (If Applicable)

Safety note: this procedure, order of operation, lock out tag out, arc potential, correct PPE, and appropriate crew size should be verified and/or provided by the maintainer's electrical safety engineer based on the data published by the equipment manufacturer and installed equipment configuration.

Wearing appropriate PPE:

- □ Visual inspection
  - □ Proper clearance between each battery cell,
  - □ Any bulging or leaking battery cells,
  - □ Presence of burn marks,
  - □ Any unaccounted for chemical residue,
- □ Measure and record in the PM report individual battery voltages and string voltages,
- □ Measure and record DC bus voltages,
- □ Measure and record ambient temperature,
- □ Measure and record individual battery ohmic readings,
- □ Measure and record all inter-unit and terminal connection resistances,
- □ Using an infrared camera, check for individual batteries or battery terminals 15 °C hotter than their counterparts.

11.C.3.8 Inverter(s)/Converters

Note: Describe power in (DC) and power out (AC) and how it is to be measured with varying climate/environmental conditions. Does it need to be taken at the inverter and the DAS or SCADA?

Safety note: this procedure, order of operation, lock out tag out, arc potential, correct PPE, and appropriate crew size should be verified and/or provided by the maintainer's electrical safety engineer based on the data published by the equipment manufacturer and installed equipment configuration.

Wearing appropriate PPE:

- □ Record inverter efficiency in the PM report

$$\text{efficiency}\,(\eta) = \frac{\text{Power Out}\,(P_{out})}{\text{Power In}\,(P_{in})} \tag{C.1}$$

- □ Listen for any unusual sounds from each inverter or any other component while operating and document.
- □ Check for the presence and condition of labeling.
- □ Visually inspect the inverter exterior and its anchoring,
- □ Visually inspect the interior of the inverter for damage to, discoloration of, or corrosion on any internal component such as bus bars, points of termination, contactors, insulation, door gasket material, din rail, etc.
- □ Look for any signs of water intrusion,
- □ Check for leaking or bulging capacitors (check temperatures).
- □ Clean inverter cabinet air vents in accordance with (IAW) the OEM manual.
- □ Clean and/or replace air filters IAW the OEM manual.
- □ Remove dirt and dust from interior of cabinet(s) using an anti-static vacuum, and IAW the OEM manual.
- □ Conduct an infrared scan (IR) of interior and termination points, and include any images of hot spots in the PM report.

NOTE: IR scans should be conducted with a calibrated imaging device that includes the measured temperature(s) (°F or °C) in the image taken. Any images included in the inverter will be annotated in such a way that records the equipment's specific location and ambient environment temperature (°F or °C).

- ☐ Check torque marks and re-tighten (ONLY if IR scan indicates there may be a loose connection or there is a visual indication the connection may be loose) to design specification torque IAW the OEM manual using a calibrated torque wrench or a calibrated torque screwdriver.

NOTE: if a termination is tightened, provide time for the termination to cool, then re-take the IR scan. If the hotspot is cleared, take the original IR image with the hotspot OUT of the "follow-up" section since follow-up is no longer required.

- ☐ Cycle DC, AC, and auxiliary power disconnect.
- ☐ If the inverter technology is "string" inverter, measure IV curves with an IV curve tracer of each string (PV source circuit) and include measurements in the PM report.
  NOTE: Should site global irradiance and ambient temperature measurements be available, it may be acceptable to measure the operating current and voltage for each string, provided this will yield all the measurements needed to submit a PV module warranty claim with the manufacturer.
- ☐ Any other OEM requirements

11.C.3.9 DC Combiner Box (If Applicable)

Safety note: this procedure, order of operation, lock out tag out, arc potential, correct PPE, and appropriate crew size should be verified and/or provided by the maintainer's electrical safety engineer based on the data published by the equipment manufacturer and installed equipment configuration.

Wearing appropriate PPE:

- ☐ Visually inspect the exterior and exterior for enclosure integrity and any signs of corrosion.
- ☐ Check for the presence and condition of labeling.
- ☐ Visually inspect the door gasket material and seal condition.
- ☐ Visually inspect the interior of the combiner box for damage to, discoloration of, or corrosion on any internal component such as wire termination points, disconnect switches, wire insulation, fuse holders, etc. Also, visually inspect conduit entry points for adequate sealing material and/or mating.
- ☐ Look for any signs of water intrusion.
- ☐ Note the presence of insects and/or pests.
- ☐ Conduct IR scan on a combiner and include any image(s) with hotspot(s) in the PM report if re-torquing (allowing for time to cool before operation) the termination point to OEM specifications did not clear the hotspot.
  NOTE: IR scans, whether handheld, aerial, or mobile, should be conducted with an imaging device that includes the measured temperature(s) in the image taken. Any images included in the report will be annotated in such a way that records the equipment's location.

- ☐ If operating current and voltage measurements were not already taken at the inverter(s), measure IV curves with an IV curve tracer of each string (PV source circuit) and include measurements in the PM report.
  NOTE: should site global irradiance and ambient temperature measurements be available, it may be acceptable to measure the operating current and voltage for each string, provided this will yield all the measurements needed to submit a PV module warranty claim with the manufacturer.
- ☐ Any other OEM requirements.

11.C.3.10 Trackers (If Applicable)

Safety note: this procedure, order of operation, lock out tag out, arc potential, correct PPE, and appropriate crew size should be verified and/or provided by the maintainer's electrical safety engineer based on the data published by the equipment manufacturer and installed equipment configuration.

Wearing appropriate PPE:

- ☐ Verify proper operation for the tracking unit.
- ☐ Check for the presence and condition of labeling.
- ☐ Inspect the condition of tracking actuators.
- ☐ Visually inspect mounting hardware.
- ☐ Any other OEM requirements.
- ☐ IR scan motor control boxes for hot or cold.

11.C.3.11 PV Modules and Other DC components

Safety note: this procedure, order of operation, lock out tag out, arc potential, correct PPE, and appropriate crew size should be verified and/or provided by the maintainer's electrical safety engineer based on the data published by the equipment manufacturer and installed equipment configuration.

Wearing appropriate PPE:

- ☐ Visually inspect the array to identify any broken PV modules (solar panels), loose components, etc.
- ☐ Check for the presence of module shading.
- ☐ Take IR images of any modules, and their *j*-boxes, belonging to strings with abnormal IV curves (measured at the combiner box or string inverter earlier in the procedure).
- ☐ For 100% of the system do the following: (see Chapter 3 on performance and safety defects)
  - ☐ Take IR images of PV modules;
  - ☐ Inspect the PV modules for signs of damage, degradation, delamination, soiling, corrosion, etc.;
  - ☐ Inspect mounting structure(s) condition;
  - ☐ Visually inspect mounting hardware and torque to manufacturer's specified values if needed;
  - ☐ Check condition of DC wires and connectors – look for signs of loose, pinched, worn, or cracked insulation (Table 11.C.1);
  - ☐ Mechanically spot check integrity of module mounting hardware.

□ Remove any PV module with any of the following safety issues from service (disconnect and bypass) and record for "follow-up" in the PM report.[33]

11.C.3.12 Grounds/Rooftop

□ Visually inspect
  □ Site drainage – ensure there is no signs of standing or pooling of water,
  □ Site security – verify that any fences, or other perimeter security measures are intact,
  □ Erosion – verify there is no erosion near or around any PV system equipment or perimeter fence,
  □ Signage is properly affixed, visible, and legible,
  □ Corrosion,
  □ Ground Slippage,
  □ Excessive undergrowth of vegetation, and
  □ Anything that does not appear to be normal or as expected
□ Mechanical maintenance (if applicable)
  □ Gates operate and lock correctly,
  □ Tracker operations per specification.

11.C.3.13 PM Report

A blank PM report will be attached by the operator to the emailed WO authorization to be completed during the PM. Every single item need not be completed to take credit for completion of the PM provided that the applicable scope is at least 90% completed. For example, a specific string inverter may be locked out and tagged out because it is

Table 11.C.1 Example PV module defects.

PV module location	Problem	PV module location	Problem
Backsheet	Peeling	Glass	Front glass crack
Backsheet	Delamination	Glass	Front glass shattered
Backsheet	Burn mark	Glass	Rear glass crack
Backsheet	Crack/cut under cell	Glass	Rear glass shattered
Backsheet	Crack/cut between cells	Junction box	Crack
Bypass Diode	Open circuit	Junction box	Burn
Cell	Interconnect arc tracks	Junction box	Loose
Cell	Hotspot over 20 °C (relative to avg. cell temp)	Junction box	Lid fell off
Frame	Grounding severe corrosion	Junction box	Lid crack
Frame	Grounding minor corrosion	Wires	Insulation cracked/ disintegrated
Frame	Major corrosion	Wires	Burnt
Frame	Joint separation	Wires	Animal bites/marks
Frame	Cracking		

Table 11.C.2 Job priority number with criteria.

Priority number (Included on work order)	Dispatch criteria	Time (from receipt of WO) to notify operator of schedule	Time (from receipt of WO) to be on site	Is overtime authorized?
1	Emergency response	ASAP (<24 hrs)	ASAP (<24 hrs)	Yes
2	Urgent priority	1 day	1 day	Yes
3	> =250 kWac outage	1 business day	2 business days	No
4	> 1 < 250 kWac inverter outage	2 business days	3 business days	No
5	<=1 kWac forced outage and other issues	6 business days	The next mobilization for other work	No

awaiting parts for repair. In this case, it would be noted as an exception in the "follow-up" portion of the report. Any other issues requiring attention should also be included in the "follow-up" portion of the report. These items will go into the queue to be addressed during the next mobilization. Only photographs, or IR images, of problems requiring attention, should be included in the "follow-up" portion. All other photographs taken during the PM should be saved to a shared folder location (URL provided in the work order).

11.C.4 Corrective Maintenance (CM) Scope

11.C.4.1 Corrective Maintenance (CM)

The corrective maintenance (CM) scope will be in the event of equipment outages and emergency response.

11.C.4.2 Administration

CM, sometimes coined "reactive repairs," will be completed by the maintenance provider following receipt of a CM WO from the operator authorizing the service of the system. Scheduling of service activities shall be in accordance with Table 11.C.2.

11.C.4.3 CM Work Order

The operator will include a general scope of work, any applicable screenshots from the DAS or photographs of the system, as well as a blank service report attached to the WO, which will be emailed to the maintenance provider to authorize the field service activity.

11.C.4.4 CM Field Service Scope

In addition to the scope of work included in the WO, the general purpose of a CM WO has four components:

1. Return the equipment to service (working with the OEM when necessary) as quickly as it is reasonable and safe to do so.
2. Call (or text from areas with reduced cell coverage) the operator to ensure the DAS is reporting that the previously offline equipment has been returned to service.

3. Identify additional issues with the system that can be followed up on later, and record these issues in the "follow-up" portion of the service report.
4. Act as a representative of the system owner to maintain a professional and positive relationship with the site (host) customer.

Note: For ease of procedural administration, these general instructions will be included in each WO, which should be shared with the service technician who will be doing the work.

11.C.4.5 Key Deliverables

- ◻ Provide notification of the SM data to the operator.
- ◻ Text notifications are provided to the operator to communicate the arrival to and safe departure from site.
- ◻ A service report documenting the field service activities.
- ◻ Before-and-after photographs of the primary service activity.
- ◻ The JHA, or other such document used by your organization, included in the service report.
- ◻ A visual and infrared (IR) camera inspection of all strings associated with any offline inverter (even if the inverter fault was inside the inverter). If the dispatch was not for an offline inverter, the visual, and IR inspection would be completed for 20% of the array. If any PV module safety defect (Table 11.C.1 in PM Section) is found, the module(s) will be immediately removed from service. Any issues identified will be included in the "follow-up" portion of the service report.
- ◻ If a return trip is required, the operator will need the updated schedule within two business days.
- ◻ If replacement parts are needed, the operator will need the order tracking number within two business days.
- ◻ If the CM service request must be placed on hold, for any reason, the operator will require an update email/text or a phone call, of the status at the end of day.

11.C.4.6 CM Service Report

A blank CM service report will be attached to the CM WO and will need to be filled out to document what field service activities were provided. The report will be touch screen and tablet friendly (works on desktops and laptops also) for any Windows device.

There are scenarios where one WO will require multiple trips to resolve the issue. For instance, following the initial investigation, the service technician may determine that they need to order parts and will need to return to install the parts when they arrive. Administratively, no new WO is needed when these situations arise – the WO has already authorized actions needed to resolve the issue. However, a service report and all the aforementioned key deliverables for each trip are required. The original blank report that was first issued with the WO may be used each time as a starting point (take care the correct trip number gets entered).

11.C.5 Ancillary Maintenance (AM)

The AM scope is for discretionary items and emergency response.

Often the operational mission may be regarded as the safe, efficient, maximization of revenues from system production. Within this framework, there are sometimes opportunities for maintenance providers to impact higher system production. For instance, if a service technician who just finished a six-hour job (leaving two remaining) noticed that the PV modules (solar panels) were dirty and had the time and material needed for washing, authorization to wash could be as simple as texting a photo of the dirty panels followed by a quick call to the operator. The SOP will be for the maintenance provider to look for additional opportunities to positively impact system production while already mobilized and on-site.

11.C.5.1 Administration

AM will be completed by the maintenance provider following receipt of an AM WO from the operator authorizing the service of the system. Scheduling of service activities shall be in accordance with Table 11.C.3.

11.C.5.2 Types of Ancillary Services

1. PV module (solar panel) washing – should be done at a time of day when the temperature of the water will not thermally stress the modules (e.g. early morning), the modules should be squeegeed dry, and the use of hard water or a pressure washer is not allowed.
2. Rooftop (if applicable) snow removal.
3. Solar Canopy (if applicable) snow removal.
4. Minor fence (if applicable) repairs.
5. Solar ground-mount (if applicable) weed abatement including mowing.
6. Landscape (if applicable) management.
7. Erosion (if applicable) management.
8. Solar Energy Facility (SEF) road (if applicable) maintenance.
9. Pest control.
10. Site tours – escort nonqualified personnel in the vicinity of the arrays for a visual tour serving as the safety coordinator for the visit and ensure that everyone in attendance participates in the JHA.

11.C.5.3 AM Service Report

A blank AM service report will be attached to the AM WO and will need to be filled out to document what field service activities were provided. The report will be touch screen and tablet friendly (works on desktops and laptops also) for any Windows device.

Table 11.C.3 Ancillary maintenance job priority.

Job priority number	Time (from receipt of WO) to notify operator of schedule	Time (from receipt of WO) to be on site	Is overtime authorized?
1	ASAP (<24 hrs)	ASAP (<24 hrs)	Yes
2	1 day	1 day	Yes
3	1 business day	2 business days	No
4	2 business days	4 business days	No

The requisite before-and-after photographs for AM report are a straightforward deliverable. For example, if the AM was grass mowing, the expectation would be for a photo of the tall grass (before) and a photo of the short, freshly mowed, grass (after).

Bibliography

IEEE Standard Computer Dictionary (1991). *A Compilation of IEEE Standard Computer Glossaries*. Institute of Electrical & Electronic.

Borrego Solar (2019). Thoughtful Inverter Procurement Can Prevent 25% of Lost Revenue: Inverter Warranty Management. Solar Risk Assessment, 2019.

IEC TS 63019 (2019). Photovoltaic power systems (PVPS) – information model for availability.

IEEE Std 762 (2006). Definitions for use in reporting electric generating unit reliability, availability, and productivity, pp. 1–75 (15 March 2007). https://doi.org/10.1109/IEEESTD.2007.335902.

IEC TS 63265 (2022). Photovoltaic power systems – reliability practices for operation.

NFPA 1. Fire Code. https://www.nfpa.org/News-and-Research/Resources/Emergency-Responders/High-risk-hazards/Photovoltaics-Systems.

NFPA 70. National Electrical Code®. https://www.nfpa.org/News-and-Research/Resources/Emergency-Responders/High-risk-hazards/Photovoltaics-Systems.

NFPA 70B. Recommended practice for electrical equipment maintenance (Chapter 33). https://www.nfpa.org/News-and-Research/Resources/Emergency-Responders/High-risk-hazards/Photovoltaics-Systems.

NFPA 70E. Standard for electrical safety in the workplace®. https://www.nfpa.org/News-and-Research/Resources/Emergency-Responders/High-risk-hazards/Photovoltaics-Systems.

NFPA 5000. Building construction and safety code®. https://www.nfpa.org/News-and-Research/Resources/Emergency-Responders/High-risk-hazards/Photovoltaics-Systems.

IEC 61078 (2016). Reliability block diagrams.

IEC 61703 (2016). Mathematical expressions for reliability, availability, maintainability and maintenance support terms.

IEC TS 61586 (2017). Estimation of the reliability of electrical connectors.

IEC TS 63265 (2022). Photovoltaic power systems – Reliability practices for operation.

IEC TR 63292 (2020). Photovoltaic power systems (PVPSs) – roadmap for robust reliability.

IEC 61163-2 (2020) provides guidance on RSS.

IEC 61123 (2019). Reliability testing – compliance test plans for success ratio.

XXX Condition Based Maintenance Plus, DoD Guidebook, May 2008.

Balfour, J., Hill, R., Walker, A. et al. (2021). *Masking of Photovoltaic System Performance Problems By Inverter Clipping and Other Design and Operational Practices*. Elsevier.

ASTM 2848-13 (2019). Standard test method for reporting photovoltaic non-concentrator system performance. PV Pro (18 February 2019). https://pvpros.com/astm-e2848-capacity-test-for-beginners/#:~:text=Capacity%20and%20performance%20ratio%20tests,Non%2DConcentrator%20System%20Performance%E2%80%9D.

Notes

1 Computer Maintenance Management Software (CMMS) will frequently include work tickets that are created based on forced equipment outages sensed by the SCADA or DAS but often not include an efficient means to initiate work flow from observations made by the boots on the ground maintainers who service the equipment in the field.

2 IT security protocols can make it challenging to deploy a CMMS tablet application when using multiple third party service providers. PV Works is a CMMS platform we're aware of that addresses this challenge and has a tablet application that is easily deployed when using multiple third party service providers.

3 UNITED STATES DEPARTMENT OF LABOR: Occupational Safety and Health Administration, "OSHA Worker Rights and Protections/Employer Responsibilities," [Online]. Available: https://www.osha.gov/as/opa/worker/employer-responsibility.html (accessed 15-March-2020).

4 A root cause of the failure event of the plant is not the same as the root cause of determining why and how a component failed.

5 PV Works, a Computer Maintenance Management Software (CMMS) developed by Higher Powered, LLC. collects field-based reliability data, is tablet friendly, and is easily deployed to third-party service personnel.

6 The use of string inverters or module level electronics such as micro-inverters may provide some opportunity for peer-based analysis in the commercial space.

7 C. Honsberg and S. Bowden, "PV EDUCATION.ORG," [Online]. Available: https://www.pveducation.org/pvcdrom/solar-cell-operation/solar-cell-efficiency (accessed 08-09-2018).

8 C. Honsberg and S. Bowden, "PV EDUCATION.ORG," [Online]. Available: https://www.pveducation.org/pvcdrom/solar-cell-operation/solar-cell-efficiency (accessed 08-09-2018).

9 PV Works, a Computer Maintenance Management Software (CMMS) developed by Higher Powered, LLC. has the capability of prioritizing, and the planning and scheduling of, numerous work authorizations sent to several third-party organizations.

10 SunPower, "Incomplete EPC Punch-listing Results in 1.2% Performance Loss in Year 1 Operations," Solar Risk Assessment: 2019, 2019.

11 SunPower, "Incomplete EPC Punch-listing Results in 1.2% Performance Loss in Year 1 Operations," Solar Risk Assessment: 2019, 2019.

12 DNV GL, "Narrowing the Performance Gap: Reconciling Predicted and Actual Energy Production," Solar Risk Assessment: 2019, 2019.

13 DNV GL, "Narrowing the Performance Gap: Reconciling Predicted and Actual Energy Production," Solar Risk Assessment: 2019, 2019.

14 This price doesn't account for sites that are very remote or have negligible generating capacity.

15 Requires the 64-bit version of python 3 and corresponding pip.

16 In command prompt to pip install the module, you'll run pip install mxnet autogluon.

17 https://auto.gluon.ai/stable/index.html.

18 Wood Mackenzie Power & Renewables, "Solar O&M Pricing has Dropped ~60% with More to Come," Solar Risk Assessment: 2019, 2019.

19 Wood Mackenzie Power & Renewables, "Solar O&M Pricing has Dropped ~60% with More to Come," Solar Risk Assessment: 2019, 2019.

20 Wood Mackenzie Power & Renewables, "Solar O&M Pricing has Dropped ~60% with More to Come," Solar Risk Assessment: 2019, 2019.

21 biz fluent, "Electrical License Requirements for New York," 26-09-2017. [Online]. Available: https://bizfluent.com/info-7969723-electrical-license-requirements-new-york.html (accessed 10-02-2020).

22 "NY Law Increases Penalties for Unlicensed Electricians," 04 2017. [Online]. Available: https://www.ecmag.com/section/your-business/ny-law-increases-penalties-unlicensed-electricians (accessed 10-02-2020).

23 Go Solar California, "California Distributed Generation Statistics," [Online]. Available: https://www.californiadgstats.ca.gov/charts/.

24 It was our experience that the most recent year of data wasn't published by the U.S. Fire Administration till a couple years had elapsed.

25 The Japan Times, "thejapantimes," 29-01-2019. [Online]. Available: https://www.japantimes.co.jp/news/2019/01/29/national/aging-bad-connections-behind-japans-home-use-solar-system-accidents/#.XNdKg3dFzcs.

26 PV Magazine, "There are – data missing – solar power fires per year," 22-08-2019. [Online]. Available: https://pv-magazine-usa.com/2019/08/22/there-are-solar-power-fires-per-year/.

27 National Fire Protection Association, "Fire Loss in the United States During 2014," NFPA, 2015.

28 Department of Industrial and Systems Engineering, University at Buffalo, "Total Cost of Fire in the United States," Fire Protection Research Foundation, 2017.

29 Department of Industrial and Systems Engineering, University at Buffalo, "Total Cost of Fire in the United States," Fire Protection Research Foundation, 2017.

30 National Fire Protection Association, "Fire Loss in the United States During 2014," NFPA, 2015.

31 Example Operation's Scope provided courtesy by Higher Powered, LLC.

32 Example maintenance scope provided courtesy of Higher Powered, LLC.

33 G. TamizhMani, *Risk Priority Number (RPN) for Warranty Claims,* Photovoltaic Reliability Laboratory, 2016.

Glossary

ACCESSIBILITY A measure of the relative ease of admission to the various areas of an item for the purpose of operation or maintenance.

ACTIVE REDUNDANCY Redundancy in which all redundant items operate simultaneously.

ADMINISTRATIVE TIME That element of delay time, not included in the supply delay time.

ALBEDO The measure of the diffuse reflection of solar radiation out of the total solar radiation and measured on a scale from 0, corresponding to a black body that absorbs all incident radiation, to 1, corresponding to a body that reflects all incident radiation.

AMBIGUITY The inability to distinguish which of two or more subunits of a product or item has failed.

AMBIGUITY GROUP The number of possible subunits of a product or item identified by built-in-test, or manual test procedures, which might contain the failed hardware or software component.

ANTHROPOMETRICS Quantitative descriptions and measurements of the physical body.

ARCHITECTURE OR SYSTEMS ARCHITECTURE of the field including all of the elements that make up the plant.

AVAILABILITY, INHERENT (Ai) A measure of availability that includes only the effects of an item design and its application and does not account for the effects of the operational and support environment. Sometimes referred to as "intrinsic" availability.

AVAILABILITY, OPERATIONAL The availability considering the MTBUM, spares availability, repair time, mean logistics delay times, etc., and factors outside of design control such as a utility demand to curtail or shut down due to a grid tie issue.

AVAILABILITY, RAW The current state of the plant equipment to produce power. The plant may be capable of producing more power but the equipment status indicates that not all of the equipment is available to do so.

AVAILABILITY A measure of the degree to which an item is in an operable and committable state at the start of a mission or operation when the item is called for at an unknown (random) time.

BUILT-IN-TEST (BIT) An integral capability of the mission equipment that provides an onboard, automated test capability, consisting of software or hardware (or both)

Photovoltaic (PV) System Delivery as Reliable Energy Infrastructure, First Edition.
John R. Balfour and Russell W. Morris.

components, to detect, diagnose, or isolate product (system) failures. The fault detection and, possibly, isolation capability is used for periodic or continuous monitoring of a system's operational health, observation, and diagnosis as a prelude to maintenance action.

CANNOT DUPLICATE (CND) The result related to receiving a call for maintenance (say by automated fault detection equipment), but when checked by maintenance staff and troubleshooting, the failure cannot be duplicated.

COMPONENT Within a product, system, subsystem, or equipment, a component is a constituent module, part, or item.

CORRECTIVE ACTION A documented design, process, procedure, or materials change implemented and validated to correct the cause of the failure or design deficiency.

CORRECTIVE MAINTENANCE (CM) All actions performed as a result of failure, to restore an item to a specified condition. Corrective maintenance can include any or all of the following steps: localization, Isolation, Disassembly, Interchange, Reassembly, Alignment and Checkout.

CRITICALITY A relative measure of the consequence and frequency of occurrence of a failure mode.

DEGRADATION A gradual decrease in an item's characteristic or ability to perform.

DELAY TIME That element of downtime during which no maintenance is being accomplished on the item because of either supply or administrative delay.

DERATING (a) Using an item in such a way that applied stresses are below rated values. (b) The lowering of the rating of an item in one stress field allows an increase in another stress field.

DETECTABLE FAILURE Failures at the component, equipment, subsystem, or system (product) level can be identified through periodic testing or revealed by an alarm or an indication of an anomaly.

DOWNING EVENT An event that causes an item to become unavailable to begin operation. (i.e., the transition from uptime to downtime).

DOWNTIME That element of time during which an item is in an operational inventory but is not in a condition to perform its required function.

ENVIRONMENT The aggregate of all external and internal conditions (such as temperature, humidity, radiation, magnetic and electrical fields, shock, vibration, etc.), whether natural, manmade, or self-induced, influences the form, fit, or function of an item.

FAILURE The event, or inoperable state, in which any item or part of an item does not, or would not, perform as previously specified.

FAILURE EFFECT The consequence(s) a failure mode has on the operation, function, or status of an item. Failure effects are typically classified as local, next higher level, and end.

FAILURE MECHANISM The physical, chemical, electrical, thermal or other process that results in failure.

FAILURE MODE AND EFFECTS ANALYSIS (FMEA) A procedure for analyzing each potential failure mode in a product to determine the results or effects thereof on the product. When the analysis is extended to classify each potential failure mode according to its severity and probability of occurrence, it is called a Failure Mode, Effects, and Criticality Analysis (FMECA).

FAILURE MODE The consequence of the mechanism through which the failure occurs, i.e., short, open, fracture, excessive wear.

FAILURE RATE The total number of failures within an item population, divided by the total number of life units expended by that population, during a particular measurement period under stated conditions.

FAILURE REPORTING AND CORRECTIVE ACTION SYSTEM (FRACAS) is the process by which failures are reported in a timely manner and analyzed with corrective actions devised and implemented to eliminate or mitigate the recurrence of failures.

FAILURE, CATASTROPHIC A failure that causes loss of the item, human life, or serious collateral damage to property.

FAILURE, CRITICAL A failure or combination of failures that prevents an item from performing a specified mission.

FALSE ALARM A fault indicated by BIT or other monitoring circuitry where no fault can be found or confirmed.

FAULT DETECTION (FD) A process that discovers the existence of faults.

FAULT ISOLATION (FI) The process of determining the location of a fault to the extent necessary to effect repair.

FAULT ISOLATION TIME The time spent arriving at a decision as to which items caused the system to malfunction. This includes time spent working on (replacing, attempting to repair, and adjusting) portions of the system shown by subsequent interim tests not to have been the cause of the malfunction.

FD/FI is Fault Detection/Fault Isolation capability of the system. Generally expressed as the probability of detection of a failure and the ability to isolate the failure down to the failed item 90% to 1 item, 95% to two items, and 97% to three or more items.

FAILURE MODES AND EFFECTS ANALYSIS (FMEA) is a process for assessing the effects of a single failure propagated from the point of failure to the next and system levels.

HAZARD FUNCTION The hazard function focuses on components, subsystems or systems failing, that is, on the failure event occurring.

HUMAN ENGINEERING (HE) The application of scientific knowledge to the design of items to achieve effective user-system integration (man-machine interface).

HUMAN FACTORS A body of scientific facts about human characteristics. The term covers all biomedical and psychosocial considerations; it includes but is not limited to, principles and applications in the areas of human engineering, personnel selection, training, life support, job performance aids, workloads, and human performance evaluation.

ISOLATION TO AN AMBIGUITY GROUP The percent of time that detected failures can be fault isolated to a specified ambiguity group of size n or less, where n is the number of replaceable items.

ITEM A general term used to denote any product, system, material, part, subassembly, set, accessory, etc.

LOWEST LEVEL OF REPAIR Lowest level of repair is lowest aggregate of components or items for which a repair action would be performed. For example, a combiner box may have a circuit breaker as the lowest level of repair while for the photovoltaics it would be the string.

LIFE CYCLE COST (LCC) LCC is defined as the total cost to the government of acquisition and ownership of a system over its full life. It includes the cost of development, acquisition, operation, support, and eventual disposal.

LIFE CYCLE COST (LCC) The sum of acquisition, logistics support, operating, and retirement and restoration expenses.

LIFE CYCLE PHASES Identifiable stages in the life of a product, from the development of the first concept to removing the product from service and disposing of it.

LIFE PROFILE A time-phased description of the events and environments experienced by an item throughout its life. Life begins with manufacture, continues during operational use (during which the item has one or more mission profiles), and ends with final expenditure or removal from the operational inventory.

LOGISTICS DELAY TIME The time between the demand on the supply system for a part or item to repair a product, or for a new product to replace a failed product, and the time when it is available.

MAINTAINABILITY The relative ease and economy of time and resources with which an item can be retained in, or restored to, a specified condition when maintenance is performed by personnel having specified skill levels, using prescribed procedures and resources, at each prescribed level of maintenance and repair. Also, the probability that an item can be retained in, or restored to, a specified condition when maintenance is performed by personnel having specified skill levels, using prescribed procedures and resources, at each prescribed level of maintenance and repair.

MAINTENANCE ACTION An element of a maintenance event. One or more tasks (i.e., fault localization, fault isolation, servicing and inspection) necessary to retain an item in or restore it to a specified condition.

MAINTENANCE EVENT One or more maintenance actions are required to effect corrective and preventive maintenance due to any type of failure or malfunction, false alarm, or scheduled maintenance plan.

MAINTENANCE TASK The maintenance effort necessary for retaining an item in, or changing/restoring it to a specified condition.

MAINTENANCE TIME An element of downtime that excludes modification and delay time.

MAINTENANCE All actions necessary for retaining an item in or restoring it to a specified condition.

MEAN CORRECTIVE MAINTENANCE TIME (Mct) Mct is the measured field maintainability attributes; see mean corrective maintenance time Mct above that includes delays.

MEAN DOWNTIME (MDT) The average time a system is unavailable for use due to a failure. Time includes the actual repair time plus all delay time associated with a repair person arriving with the appropriate replacement parts.

MEAN MAINTENANCE TIME A basic measure of maintainability taking into account maintenance policy. The sum of preventive and corrective maintenance times, divided by the sum of scheduled and unscheduled maintenance events, during a stated period of time.

MEAN TIME BETWEEN CRITICAL FAILURE (MTBCF) A measure of mission or functional reliability. The mean number of life units during which the item performs its mission or function within specified limits during a particular measurement interval under stated conditions.

MEAN TIME BETWEEN FAILURE (MTBF) A basic measure of reliability for repairable items. The mean number of life units during which all parts of the item perform within their specified limits during a particular measurement interval under stated conditions.

MEAN TIME BETWEEN MAINTENANCE (MTBM) A measure of reliability taking into account maintenance policy. The total number of life units expended by a given time is divided by the total number of maintenance events (scheduled and unscheduled) due to that item.

MEAN TIME BETWEEN MAINTENANCE ACTIONS (MTBMA) A measure of the product reliability parameter related to the demand for maintenance labor. The total number of product life units is divided by the total number of maintenance actions (preventive and corrective) during a stated period of time.

MEAN TIME BETWEEN REMOVALS (MTBR) A measure of the product reliability parameter related to the demand for logistic support: The total number of system life units divided by the total number of items removed from that product during a stated period of time. This term is defined to exclude removals performed to facilitate other maintenance and removals for product improvement.

MEAN TIME TO FAILURE (MTTF) A basic measure of reliability for nonrepairable items. The total number of life units of an item population divided by the number of failures within that population, during a particular measurement interval under stated conditions.

MEAN TIME TO REPAIR (MTTR) A basic measure of maintainability. The sum of corrective maintenance times at any specific level of repair, divided by the total number of failures within an item repaired at that level during a particular interval under stated conditions.

MISSION DURATION The time interval over which an analysis or analyses are being performed for most solar power fields. It is one year and contract life (20 years, for example).

MISSION PROFILE A time-phased description of the events and environments experienced by an item during a given mission. The description includes the criteria for mission success and critical failures.

MISSION RELIABILITY Mission reliability is the probability that the equipment will work as specified over a defined period of time and under stated operating/usage conditions. For general utilities, this is an annual metric.

MISSION TIME That element of uptime required to perform a stated mission profile.

MEAN LOGISTIC DELAY TIME (MLDT) is the mean logistic delay time that includes all transportation, ordering, handling, build, etc., for an item to be obtained for a repair

MAXIMUM MAINTENANCE TIME (Mmax) The maintenance maximum time to repair an item based on a confidence level assuming a log-normal distribution of the maintenance times to repair an item, then the Mmax at 90% is 1.65 times the MTTR.

MAINTENANCE MAN-HOUR/OPERATING HOUR (MMH/OPH) is the average maintenance burden to the plant per operating hour of the plant.

MEAN TIME BETWEEN CRITICAL FIALURE (MTBCF) is the mean time between failures that can take down a large portion or all of a field or can lead to significant damage to the equipment or can result in the plant being one fault away form a plant shutdown.

MEAN TIME BETWEEN FAILURE (MTBF) The time between failure for repairable items.

MEAN TIME BETWEEN FAILURE DORMANT (MTBF-D) MTBF dormant is the reliability of the component or subsystem as the result of being stored generally a degradation or chemical mechanism ex. corrosion.

MEAN TIME BETWEEN FAILURE INDUCED (MTBF-I) MTBF Induced is the mean time between failure caused or precipitated by damage induced by humans.

MEAN TIME BETWEEN FALSE ALARMS (MTBFA) MTBFA is the Mean Time Between False Alarms, as noted by the faults, reported where no fault actually exists.

MEAN TIME BETWEEN PLANT FAILURE (MTBPF) The interval of time between the plant's failure to meet the demand for a contracted energy

MEAN TIME BETWEEN REMOVAL (MTBR) The queuing number for the spares system I.e., a demand is made on the system to provide a spare

MEAN TIME BETWEEN SCHEDULED MAINTENANCE/PREVENTIVE MAINTENANCE (MTBSM/PM) MTBSM/PM The mean time between preventive or scheduled maintenance is part of the fundamental cost of maintenance of the field.

MEAN TIME BETWEEN UNSCHEDULED MAINTENANCE (MTBUM) The attribute used to estimate the cost of all maintenance, as it is the time between maintenance trips (personnel and materials) that are required to perform maintenance and return the plant to full operation.

MEAN TIME TO FAILURE (MTTF) The time between failure for non-repairable items.

MEAN TIME TO REPAIR (MTTR) is the time to repair an item as designed if the maintainer is at the site standing in front of the failed equipment and has all the tools, spares and test/support equipment needed to perform maintenance.

NO FAULT FOUND (NFF) The case where a part is removed for cause (failure) and sent back to the manufacturer for repair, except the manufacturer cannot find a failure or defect to repair. This has a direct impact on the cost of spares, maintenance, shipping and handling as well as testing and return charges from the manufacturer.

NON-DETECTABLE FAILURE Failures at the component, equipment, subsystem, or system (product) level that are identifiable by analysis but cannot be identified through periodic testing or revealed by an alarm or an indication of an anomaly.

NON-OPERATING TIME That time during which the product is operable according to all indications or the last functional test but is not being operated.

OFF-THE-SHELF FILL RATE Generally the probability (typically 80–90%) that the plant will have a spare available immediately when called for to perform a repair within the plant.

OPERABLE The state in which an item is able to perform its intended function(s).

OPERATIONAL ENVIRONMENT The aggregate of all external and internal conditions (such as temperature, humidity, radiation, magnetic and electric fields, shock vibration, etc.), either natural or manmade, or self-induced, that influences the form, operational performance, reliability, or survival of an item.

OPERATIONAL PROFILE How the equipment is used a flat array profile is that it is not capable of producing some minimum amount of power until 0937 on a specific day while it may be 0902 on another. A two-axis tracker has a profile that includes moving the panels to face the rising sun in the morning (parasitic or batter power) and producing some minimum power just after sunrise 0741; for example Đ, it then tracks (+/− accuracy) across the sky until sunset. In case of high winds, it goes into stow mode.

PERFORMANCE SPECIFICATION (PS) A design document stating the functional requirements for an item.

PROBABILITY OF LOSS OF PLANT (PLOP) Probability of loss of plant due to failure of the equipment from internal or external sources.

PREDICTED That which is expected at some future time, postulated on analysis of past experience and tests.

PREDICTIONS The results of reliability and maintainability analyses to provide an estimate of what the parts are expected to do. These are often assumed or estimated for modeling purposes for lifetime plant operation.

PREVENTIVE MAINTENANCE (PM) All actions performed to retain an item in a specified condition by providing systematic inspection, detection, and prevention of incipient failures.

QUALIFICATION TEST A test conducted under specified conditions, by or on behalf of the customer, using items representative of the production configuration to determine if item design requirements have been satisfied. Serves as a basis for production approval.

QUALITY FEEDBACK Information gathered during the manufacturing process that adjusts the builds as defects are found.

RELIABILITY BLOCK DIAGRAM (RBD) Graphically shows how the plant is connected including capability to detect and mitigate faults as well the effects of fault propagation if no protective equipment is designed in/installed.

REDUNDANCY The existence of more than one means for accomplishing a given function. Each means of accomplishing the function need not necessarily be identical. The two basic types of redundancy are active and standby.

RELIABILITY FEEDBACK Field information that is found as the result of a Root Cause Corrective Action (RCCA) process that adjusts design, supplier or even specification to correct the field defects in the future.

RELIABILITY (1) The duration or probability of failure-free performance under stated conditions. (2) The probability that an item can perform its intended function for a specified interval under stated conditions. (For nonredundant items, this is equivalent to definition (1). For redundant items, this is equivalent to definition of mission reliability.)

REPAIR TIME The time spent replacing, repairing, or adjusting all items suspected to have been the cause of the malfunction, except those subsequently shown by interim test of the system not to have been the cause.

REPAIRABLE ITEM An item which, when failed, can be restored by corrective maintenance to

REPLACEABLE ITEM An item, unit, subassembly, or part which is normally intended to be replaced during corrective maintenance after its failure.

REQUEST FOR PROPOSAL (RFP) A letter or document sent to suppliers asking to show how a problem or situation can be addressed. Normally the supplier's response proposes a solution and quotes a price. Similar to a Request for Quote (RFQ), the RFQ is usually used for products already developed.

REQUEST FOR QUOTE (RFQ) A solicitation for goods or services in which a company invites vendors to submit price quotes and bid on the job.

SCHEDULED MAINTENANCE Periodic prescribed inspection and servicing of products or items accomplished on the basis of calendar, mileage or hours of operation. Often included in preventive maintenance.

SINGLE-POINT FAILURE A failure of an item that causes the system to fail and for which no redundancy or alternative operational procedure exists.

STANDBY REDUNDANCY When some or all of the redundant items are not operating continuously but are activated only upon failure of the primary item performing the function(s).

STORAGE LIFE The length of time an item can be stored under specified conditions and still meet specified operating requirements. Also called shelf life.

SUBSYSTEM A combination of sets, groups, etc., which performs an operational function within a product (system) and is a major subdivision of the product. (Example: Data processing subsystem, guidance subsystem).

SUCCESS A favorable or desired outcome that the system provides power to the grid per specification and contractual requirements based on stakeholder wants, needs, and expectations.

SYSTEM SPECIFICATION Defines the system requirements and the overall hardware/software system design in top-level detail.

SYSTEM A composite of equipment and skills, and techniques capable of performing or supporting an operational role, or both. A complete system includes all equipment, related facilities, material, software, services, and personnel required for its operation and support to the degree that it can be considered self-sufficient in its intended operational environment.

TESTABILITY A design characteristic that allows an item's status (operable, inoperable, or degraded) to be determined and faults within the item to be isolated in a timely manner.

TIME Time is a fundamental element used in developing the concept of reliability and is used in many of the measures of reliability. Determining the applicable interval of time for a specific measurement is a prerequisite to accurate measurement. In general, the interval of interest is calendar time, but this can be broken down into other intervals daily, weekly, monthly, quarterly, and semi-annual.

- Inactive Time Includes time in storage or logistic pipeline
- Calendar Time
- Active Time
- Not-Operating Time
- Maintenance Time Delay Time
- Downtime
- PM Time Supply
- Delay Time

- Admin Time
- Uptime
- CM Time

TOTAL SYSTEM DOWNTIME The time interval between the reporting of a system (product) malfunction and the time when the system has been repaired and/or checked by the maintenance person and no further maintenance activity is executed.

UNSCHEDULED MAINTENANCE Corrective maintenance is performed in response to a suspected failure.

UPTIME That element of ACTIVE TIME during which an item is in condition to perform its required functions. (Increases availability and dependability). Service LIFE: The number of life units from manufacture to when the item has an unrepairable failure or unacceptable failure rate. Also, the period of time before the failure rate increases due to wear out.

VALIDATION Confirmation through the provision of objective evidence that the requirements for a specific intended use or application have been fulfilled. ISO/IEEE/IEC 15288.

VERIFICATION Confirmation through the provision of objective evidence that specified requirements have been fulfilled. ISO/IEEE/IEC 15288.

VERIFICATION TIME That element of maintenance time during which the performance of an item is verified to be a specified condition.

VERIFICATION The contractor effort to: (1) determine the accuracy of and update the analytical (predicted) data; (2) identify design deficiencies; and (3) gain progressive assurance that the required performance of the item can be achieved and demonstrated in subsequent phases. This effort is monitored by the procuring activity from date of award of the contract, through hardware development from components to the configuration item (CI).

WEAR OUT The process that results in an increase of the failure rate or probability of failure as the number of life units increases.

Index

Note: Page numbers followed by "n" denote endnotes.

a

Aberrations 180
Acceleration factor (AF) 340
Accessibility 367–371
Access™ 319
AC combiner box 508–509
Achieved availability 401–402
Acquisition costs 241, 243–244
Activation energy 339
Actual power capability (APC) 80, 82–84
Affordability 209
Ambiguity group 359, 371, 387n2
American Hydrogen Association (AHA) 139
American Society for Testing and Materials (ASTM) 346n13–14
Ancillary maintenance (AM) 351, 497, 503, 512–516
Arc flash potential 480–481
Arrhenius equation 339
Artificial intelligence (AI) 465
Asset valuation 185
Astronomical twilight 365
Authorities having jurisdiction (AHJ) 54, 141, 175, 190–191, 499
Automotive industry 170
Availability 192, 209, 389–391, 484
  achieved 401–402
  allocation 405
  annual solar fluence/irradiance 394–395
  calculation of 391
  capability and capacity 393–394
  component 391–395
  contractual 393
  customer demand 395
  energy 393, 396–397
  equipment-focused 393
  force majeure 394
  grid 404
  index 195
  information categories 395, 396
  inherent 399–400, 405
  inverter availability guarantees 484–485
  maintenance-driven 365–371
  measurement 391–395
  metrics 403–404
  O&M 405
  operational 400–401
  owner's responsibility 405–406
  peer-based analysis 484
  power 397–399
  raw 395, 402–403
  requirements 405–406
  spares 379
  specifications 392, 404–406
  standards 406
  third-tier vendors 405
  types of 395–403
Available maintenance time 364–365

Photovoltaic (PV) System Delivery as Reliable Energy Infrastructure, First Edition.
John R. Balfour and Russell W. Morris.

b

Bankability 172
Bankruptcy 107, 109, 159, 173
Bathtub curve 296–298
Battery 411–413
  augmentation costs 438
  basic types of 416
  calendar life 418
  cells 421–422
  charge rate 416
  comparison of 419, 420
  configuration 422
  cycle life 417, 419
  definition of 414
  degradation 432–433
  energy capacity 416–417
  energy density 417
  EV 440, 441
  failure 428
  fire suppression 428–429
  flow 421
  housing 427–428
  interface 423
  inverter 422
  lithium-ion 418
  long-duration 434
  maintenance of 374
  metrics 415–417
  misconceptions 434
  operating ambient 429
  powerpack 423
  primary 414
  racks 427
  reliability 432–433
  run time 417
  safety considerations 428
  secondary 414
  selection criteria 414–415
  shelf life 17, 419
  storage unit 423, 424
  swell rate 418
  technology costs 438
  thermal management 427–429
  toxicity 416
  warranties 439
Battery management system (BMS) 422, 426
Battery storage system (BSS) 137
Bayesian prediction techniques 283
Bayes Theorem 282–283
Bayes, Thomas 346n4
Benchmarking 134, 256–257, 269n34, 471, 494
  commissioning and 65–66
  external 68
  internal 68
  performance 68
  practice 68
  process 69
  report 69
Bidding process 27–28, 43
Bifacial photovoltaics 131
"Bigger Is Not Necessarily Better" (Bellini) 197
Bill of materials (BOM) 109, 186–188
Bird excrement 323
Bloomberg tier process 199–200
Boeing 787 battery 428, 442n2
Boeing 737 Max 25, 159
Brutal/brutalizing cost model 108
Budget, O&M 498–499
Bulk energy shifting 413
Bypass diodes 246

c

California fires 477, 478, 500–502
Cannibalization 162–163
Cannot duplicate (CND) 382
Capability 2, 28, 47, 78, 393–394
  actual power capability 80
  design power capability 79–80
  performance gap 82–84
  raw power capability 80–81
Capacitor 119, 120
Capacity 2, 28, 78, 393–394
  cumulative 143
  design 78–79
  testing 494
Capital expenditure (CAPEX) 25, 149, 151
Catch-22 13–15, 30
Cell efficiencies 121–123

Cell temperature 321
Central limit theorem (CLT) 298
Certification 30–31, 171, 172, 218, 286
Change 43
Charge rate (C-rate) 416
Circular economy 146
Civil twilight 365
Climate change 71–74, 177–184
Climate zones 474
Clipping 111, 135, 137, 139, 247, 248
Clouds 231
Codes 30, 51–54, 190–196
Commercial & industrial (C&I) 162–165
Commercial operation date (COD) 174, 256–257
Commissioning 96, 134, 216, 286, 471
  and benchmarking 65–66
  fractional 66
  installation and 219, 256
  monolithic 66
  primary function of 70
  process 70
  sectional 66
Communication 50–51, 63
Complementary metal oxide semiconductor (CMOS) 283
Compliance 196, 264
Component business model 116
Component selection 93–97
Component-to-revenue path 455
Component unavailability 397
Computerized maintenance management information system (CMMIS) 75
Computerized maintenance management system (CMMS) 75–76
Computer maintenance management software (CMMS) 480, 483, 517n1
Concept of Operation (ConOps) 230
Concept phase 228–229, 235
Concept to field operations, process flow 456
Condition-based maintenance (CBM) 496–497
Confidence interval (CI) 305–307
Confidence level (CL) 305–307, 376, 384
Conformance 196
Consensus 191
Constant failure rate 349
Construction oversight 493
Contracts 9–11
Contractual availability 393
Corporate-oriented business practices 51
Corrections reporting 71
Corrective actions 317, 318
Corrective maintenance (CM) 350, 482, 495–496, 498, 503, 504, 511–512
Corrosion 315–316, 449
Cost–benefit analysis 164
Cost control 18–21
Cost of ownership (COO) 188
Costs
  acquisition 241, 243–244
  ESS 429–431
  fire 478
  fixed 260
  hidden 258
  installation 430
  maintenance 355–357
  metrics 355
  O&M 430–431
  overrun 206, 207
  recycling 146
  spares 349
  technology 99–100
Crew size (CS) 366–367
Critical failure 135
Critical path method (CPM) 227
Culture 50
Cumulative DC energy 139
Cumulative PV capacity 143
Customer demand 395
Customer-generated maintenance 371, 373
Customers 287

d

Data 113, 216, 319
  accuracy 467
  anonymized 449, 454, 457
  curation 202n2, 464, 475n5
  decisions 461

Data (*contd.*)
encryption 464
environmental 451
format and elements 450–452
high-quality 130
loss/damage 473
presentation 461–463
reliability 483–484
security 464
synthetic 472–473
Data acquisition 446–447
Data acquisition system (DAS) 174, 177, 197, 478
Data analysis 459–461
Access 326
Excel™ 326
Praeto analysis 333–334
top-level 326, 331
Weibull++™ 326–333
Database 446
Data collection 197–202
FRACAS form 451, 452
IEEE 448–449
implementation 467
lifecycle 458
processes 463–465
tools 466
warranty issues 471–472
Data plan 447, 457–458
Data sharing 71–76, 505–506
DC combiner box 508–509
Decision approach 103–105
Degradation 87–88, 244, 245, 338, 366, 432–433
light, and elevated temperature induced 130
light-induced 130
module 128–130
potential-induced 131
sources of 128
Degradation rate (DR) 301–302, 390
Department of Energy (DOE) 105–106
Derating 337
Deregulation 31–32, 106, 107
Design capacity 78–79
Design considerations 96–97
Designer 141
Design FMEA 313
Design for maintainability (DFM) 356
Design margin 148n25
Design power capability 79–80, 82–84
Detailed design 218, 255
Discrimination 113
Disposal of equipment 142–144
Distressed plant **SE/R**epowering™ 172–174, 185
Documentation 114, 504–506
Dopants 143, 145
Downtime 401
Duck curve 241, 242, 412, 413
Dusk 365

e

Education 189–190
Efficiency
cell 121–123
module 122, 124–126
Electrical testing 70
Electric vehicles (EVs) 439–441
Electromagnetic interference/ electromagnetic compatibility (EMI/EMC) 209
Emissivity 321
Energy availability 393, 396–397
Energy demand 395
Energy loss 135, 137, 399
Energy loss budget 244–251
Energy planning and accounting 87
Energy storage devices 509
Energy storage system (ESS) 81, 86, 137, 373–374, 410–412. *See also* Battery
AC-coupled system 425–426
applications 412–414
assumptions 433
battery interface 423
components of 423–426
considerations 437–439
costs 430
installation cost 430
O&M 430–432

DC-coupled system 425
definition of 411
EVs 439–441
forms of 411–412
investment tax credit 426, 435
misconceptions 433, 435
PV + ESS 434, 435
reliability 432–433
storage architecture 423–424
in United States 412
value stacking 436–437
Engineering estimate 283
Engineering, procurement and construction (EPC) 18, 27, 29, 34, 43, 111, 112, 117, 217, 218, 225, 226, 285, 454
Entitlements 91n19
Entropy 92n39
Environmental conditions, site research 71–74, 177–184
Environmental engineering 209
Environmental factor 304
Environmental issues 183
Environmental profile 214, 248–253, 280, 281
Equipment delay time (EDT) 358
Equipment disposal 142–144
Equipment-focused availability 393
Equipment removal 142, 262
European Standards 269n38
E-waste disposal 146
Excel™ 326, 333, 348
Existing plant **SE/R**epowering™ 169–172, 184–185
Expected life of plant 234
Exponential probability density function 282
External benchmarking 68
External stakeholders 57–63

f

Failure 7, 96, 113, 115, 134, 135, 212–213, 223, 224, 229–230, 245, 246, 277, 290, 428, 459
Failure cause (mechanism) 316, 319–324
Failure effect 316, 317
Failure-free operation 279
Failure modes 315–316, 319–324
Failure modes and effects analysis (FMEA) 177, 223, 311, 376
application of 314–315
categories and subcategories 313
data field list for 314
design 313
development 313
failure category definitions 319
lifecycle application of 311, 312
procedure 315–317
process 313
RBD 315
RPN 317–318
service 313
system 313
types 313
updation 319
Failure rate (FR) 276, 277, 279, 280, 291, 303–307, 335
Failure reporting, analysis, and corrective action system (FRACAS) 74, 217, 257, 324–325, 446, 451
False negative 382
False positive 382
Fault detection (FD) 352, 357–358, 466
Fault detection/fault isolation 353, 381
Fault isolation (FI) 356, 359
Fault tree analysis (FTA) 177
Fault trees 311, 312
Fault verification (FV) 359, 377, 379
Federal Aviation Administration (FAA) 17
Fill factor (FF) 487
Financial institution risk 224
Financial risk 42
Financial viability 165
Fires, PV system 477–478, 500–502
Fire suppression 428–429
First-year failures 328, 330
Fiscal performance 87–89
Fixed costs 260
Flipping, system/project 33–34
Flow batteries 421
Fluence 394–395

Fly wheels 412
Forced outages 246–247
Force majeure 394
Fractional commissioning 66
Frame damage 115
F(*t*) 281
Funding 105–106, 108
Funding Opportunity Announcement (FOA, DOE) 106

g
GADS 449, 450, 453
GANTT chart 227
Garbage in–garbage out (GI–GO) 44
Generating unit 392
Gernet, Bryan 54, 55
Grading 109
Green energy 143, 271
Grid 106
Grid availability 272, 404
Ground coverage ratio (GCR) 369, 370, 387n7
Growth of the industry 105–111, 151

h
Hail 126–128, 180–182
Hardware design 285
Hazardous materials 263
Hazards 481–482
Heating, ventilation, and air conditioning (HVAC) 429, 430
High-latitude row separation 369
High-performance PV (HPPV) 58–60
High-quality data 130
Holistic issues 98
Hours of sunlight 231–233, 240–241
Hours of theoretical total production time (H_{ttp}) 396
Housing batteries 427–428
Hubble 455
Human engineering 209
Human factors 284
Hurricane Harvey 180
Hype (marketing) 100

i
IEEE 493 Gold Book 447, 449
Improper cleaning 367, 368
Inclination 387n6
Indentured parts list 380
Industry language 51–52
Information flow 194
Infrared emittance 322
Inherent availability 399–400, 405
Initial concept 214–215
Initial data requirements 469–470
Inspections 109
Inspectors 141
Installation 256
Institute of Electrical and Electronics Engineers (IEEE) 482
Insulated-gate bipolar transistor (IGBT) 99, 147n5
Insurability 165–166, 172
Insurance 212, 214–215, 226, 286
Insurance companies 192
Integrated circuits (ICs) 335, 336
Integrated system 150
Integrators 111–118
Intellectual property (IP) 55, 94
Internal benchmarking 68
Internal stakeholders 57, 58
International Electrotechnical Commission (IEC) 65, 91n21, 452
  IEC 61703 406
  IEC 61724-3 471
  IEC 61853-3 126
  IEC 61853-4 126
  IEC 62548 264
  IEC 63019 455, 471
  IEC 63265 471
  IEC TS 62915:2018(E) 203–204n20, 204n24
  IEC TS 63019 81, 389, 406
  IEC TS 63019:2019 192, 195, 395, 402, 484
  IEC TS 63265 81
  IEC TS 63397 126, 127
  IEC TS 92915: 2018 195
  modified revenue steam model 194

reliability 483
reliability testing 338
Interpolation 283
Inverters 110, 119, 134–142, 177, 179, 187, 243, 261, 291, 334, 484–485, 487, 509–510
Investment tax credit (ITC) 426, 435
IR scanning 383

j

Job hazard analysis (JHA) 481–482, 506

k

Key performance indicators (KPIs) 503–504
Knowledge 95, 111, 113

l

Landfilling 145–147, 262
Least-cost model 82, 116, 151, 159
Levelized cost of energy (LCOE) 19–21, 32, 90n1, 134, 150, 151, 234, 258–259, 347, 465
Lifecycle 76–77, 89, 96
Lifecycle cost (LCC) 25, 151, 235, 259, 273
Lifecycle cost optimization 25, 235, 237–238
Lifecycle plan 151
Life profile 288, 290
Light, and elevated temperature induced degradation (LeTID) 130
Light-induced degradation (LID) 130
Likelihood of occurrence 225
Limiting factors 201
Liquidated damages (LD) 86–87, 389–390
Lithium-ion battery 443n3
anode and cathode 420
power-focused 434
swell rate 417
types of 418
working of 419–420
Lithium iron phosphate (LFP) batteries 434
Local electricians 490–492
Logistics 348
Long-term energy production 185
Losses 247, 258
Lowest level of repair 278–279
Low-latitude row separation 368

m

Maintenance 203n3, 209, 257–258, 485–486, 506–516. *See also* Operations and maintenance (O&M)
ancillary 351, 497, 503, 512–514
attributes 348, 352, 353
available maintenance time 364–365
batteries 374
condition-based 496–497
corrective 482, 495–496, 498, 503, 504, 510–511
costs 355–357
customer-generated 371, 373
design for maintainability 356
documentation 504–506
emergency 365
energy storage 373–374
flow 357
equipment delay time 358
fault detection 357–358
fault isolation 359
fault verification 359
mean corrective maintenance time 361–362
mean logistic delay time 359
mean time till onsite 358
minimum maintenance time 360
overhead time 360
repair time 359
repair verification time 360
time to repair 360–361
work authorization delay 358
follow-up issues 506
hours 239–240
importance of 347–348
KPIs 503–504
metrics 364
O&M 492, 494–495
overtime and 357, 387n1
preemptive 496
preventive 350–351, 371–373, 385, 430, 495, 503, 507–513
priority classifications 352, 354–355
responsibility for 350
scheduled 350–351, 385, 430, 495

Maintenance (*contd.*)
software updates 351
specifications 384–386
team 389
times, calculation of 362–363
types of 350–355
unscheduled 350
Maintenance-driven availability 365–371
Manufacturers
Bloomberg tier process 199–200
Tier 1 102, 199–200, 487
Tier 2 102, 200
Tier 3 200
variability of quality 101–103
Manufacturing defect 328, 329
Manufacturing risk 100–101
Masked power 85–86
Masking 81, 85–87, 247
Maximum maintenance time (M_{max}) 384
Maximum power point tracking (MPPT) 110–111, 135–141, 148n24, 413
McMahill, Lanny 54, 55
Mean corrective maintenance time (M_{ct}) 348, 350, 361–362, 384, 401
Mean logistic delay time (MLDT) 348, 359, 385, 386
Mean time between critical failure (MTBCF) 276, 278, 281
Mean time between failure (MTBF) 97, 188, 235, 275, 277–280, 291, 292, 298, 301–303, 306, 348, 376, 377, 399–400, 405
Mean time between maintenance (MTBM) 277, 347, 402
Mean time between scheduled maintenance (MTBSM) 276, 278, 351
Mean time till onsite (MTTO) 358
Mean time to failure (MTTF) 275–278, 280, 285, 340–341
Mean time to issue work order (MTTIWO) 358
Mean time to repair (MTTR) 286, 350, 360–362, 384, 399–401, 405
Measure of effectiveness (MOE) 267n1
Mechanical testing 71
Median (50th percentile) time to repair (MRT(50%)) 364
Metrics 199–202, 274, 280, 287, 288, 290, 352, 364, 403–404
Microinverters 180
Minimum maintenance time (MMT) 360
Missing energy 40
Modified Vee diagram 216–217
Module(s) 98, 99, 117, 179, 512
capacity 2
defects 130–131
detection of 132–133
related to technology changes 126
sources of 115
degradation 128–130
efficiency 122, 124–126
recycling 185
cost of 146
in Europe 145
issues 146–147
need for 145
regulation of 145–146
in USA 145
reliability 272
warranty 487
Monitoring plan 457, 458, 467–468
additional data and sensor requirements 470–471
assessment strategy 470
initial data requirements 469–470
management goals 468–469
steps 469
Monolithic commissioning 66
Monte Carlo analysis 336
MV transformers 508

n

National Aeronautics and Space Administration (NASA) 170, 192–193, 206, 207, 454–455
National Centers for Environmental Information (NCEI) 183
National Electric Code (NEC) 53, 114, 140–141, 481

National Institute for Occupational Safety and Health (NIOSH) 368
National Oceanic and Atmospheric Administration (NOAA) 183, 473–474
National renewable energy laboratory (NREL) 311
Nautical twilight 365
Need(s) 215–216
New products 100, 114
Nickel-manganese-carbon (LFP) batteries 434
No fault found (NFF) 333, 382
Non-operating failures 277
Normal distribution 298–303, 307
North American electric reliability corporation (NERC) 223, 227, 272, 288, 449
NREL cell efficiencies 122, 123

o

Occupational Safety and Health Administration (OSHA) 480, 482
Ontology 36n9
Operating profile 215
Operational availability 400–401
Operational expenditure (OPEX) 25, 149–151
Operations and maintenance (O&M) 9–10, 26, 226, 257, 287, 389, 405, 446, 461. *See also* Maintenance
  budget 498–499
  concept 230–231
  fire risk 477–478
  licensing 499
  local electricians 490–492
  maintainability 485–486
  maintenance scope 494–497, 506–513
    ancillary 497
    condition-based 496–497
    corrective 495–496
    preemptive 496
    preventive 495
    scheduled 495
  operations scope 492–493, 502–506
    capacity testing 494
    construction oversight 493
    substantial completion punch list 493
    system performance 494
  philosophy and strategy 488–492
  price distortions 498
  project development 488
  safety 479–482
  scope 498–499
  self-performing 489
  team 493
  testability 486–487
  third-party 490, 491
  unsupported 30–31
Orange Button project 52
Organizational fat 22, 62
Original equipment manufacturer (OEM) 77, 119, 131, 187, 195, 485
Overbuild(ing) 33–34, 85, 86
Over-current protection device (OCPD) 481
Overhead time (T_{OH}) 360
Overtime 357, 387n1
Owners 287, 405–406, 505
Owners and operators (O&O)
  ancillary maintenance 351
  failure 335
Ownership
  issues 174
  new 172
  operations analysis and decisions 186

p

PerfectPower Inc. 61
Performance 494
  benchmarking 68
  capacity 2
  defects 131–134
  evaluation 71
  gaps 82–84
Personnel protective equipment (PPE) 365, 481, 482, 507–510
Phoenix, Arizona 178–180, 248–250
Phoenix Sky Harbor Airport 177, 251
Photoelectric cells 121
Photovoltaic (PV) arrays 264, 292

Photovoltaic Degradation Rates–An Analytical Review (Jordan & Kurtz) 128
Photovoltaic fires 500–502
Photovoltaic power system (PVPS) 40, 205, 210–212, 229, 289, 380, 455
Pi factors 335
Plant availability 395, 396
Plant capability 2, 390
Plant capacity 2
Plant lifecycle 76–77
Plant maintenance time 367
Plant overbuild 85, 86
Plant requirement 234–237
Plant retirement 261–262
Plant viability 163, 186
POI meter 80, 81, 96, 389, 391
Point of interconnection (POI) 47
Poisson distribution 387n12
Poisson process 275
Poisson, Siméon Denis 387n12
Potential-induced degradation (PID) 131
Power availability 397–399
Power generation model 238–239
Power grid 272
Power loss 258
Power purchase agreement (PPA) 31, 110, 174, 175, 407n4
Practice benchmarking 68
Praeto analysis 333–334
Preemptive analytical maintenance (PAM) 3, 152
  application of 17–18
  concept 24
  model 4
  SE/Repowering™ 175–176
Preemptive maintenance 496
Preventive maintenance (PM) 350–351, 371–373, 385, 430, 495, 503, 507–510
Price per kilowatt hour 32
Primary battery 414
Primary components 118–121
Problem solving 254
Process FMEAs 313
Production impact 195
Professional arrogance 113
Program evaluation and review technique (PERT) 227
Project delivery
  model 28–35
  process 21
Project development 488
Project management (PM) 287
Project management tools 227
Project phases
  detailed design 218
  installation and commissioning 219
  manufacturing, building and testing 218–219
  operation, upgrading, and **SE/R**epowering™ 219
  site restoration 219–220
  system concept 218
Project schedule constraints 45
Project value determination 31–33
Proprietary information (PI) 46, 94
P values 307
PV Cycle 145
PV module recycling
  cost of 146
  in Europe 145
  issues 146–147
  need for 145
  regulation of 145–146
  in USA 145
PV professionals 10, 20, 28, 35, 71, 76, 81, 110, 114, 115, 127, 136, 189, 383, 446
PVSyst 222
PVWorks 517n2
Pyranometer 40, 394
Pyrheliometer 40, 394

q

Q-factors 335
Quality 98

r

Rain 180, 252–253, 322
Raw availability 395, 402–403
Raw power capability 80–84
Reactive repowering threshold 81–82

Rechargeable batteries 414
Recycling 144–147, 262–263
Redundancy 277
Reflectance of material *(R)* 321
Regulatory requirements 214
Relational database management system (RDMS) 479
Reliability 18–19, 32, 106, 108, 114, 192, 209, 271
  analytical/statistical 280
  data 483–484
  definition 276, 482
  descriptive 295
  design 282
  drivers 292–293
  ESSs 432–433
  failure 290
  IEC standards 483
  importance of 272–274
  issues 273
  mathematics 295
    bathtub curve 296–298
    failure rate 303–307
    normal distribution 298–303
    Weibull 295–296
  metrics 274, 280
  module 272
  predictions 334–337
  predictive 280, 295
  program plan 293–295
  semiconductor 339–340
  specifications 292
  synthetic data 282–283
  tasks 293–294
  testing 338–341
  test time 340–341
  tracking and reporting of 504
Reliability availability maintainability safety (RAMS) 10, 24, 46, 99, 105–108, 116, 193, 448–449
Reliability block diagrams (RBDs) 307
  FMEA 315
  functional diagram 307–309
  system-level 309, 310
  unit generator level diagram 309
Relocation plant **SE/R**epowering™ 174–175, 185
Removal of equipment 142
Repair, lowest level of 278–279
Repairable item level 37n15
Repair time (T_R) 359
Repair verification time (T_{RV}) 360
Report(ing)
  benchmark 69
  corrections 71
Repowered™ 156, 260
Repowering™. *See also* **SE/R**epowering™
  plan 159
  vs. repowering 67
Request for proposal (RFP) 34, 273
Request for quote (RFQ) 34, 255, 273
Resource Conservation and Recovery Act (RCRA) 263
Restricted substances 263
Restriction on Hazardous Substances (RoHS) 263–264
Retirement, plant 261–262
Revenue 164, 186
Risk 222
  management 217, 222–227
  manufacturing 100–101
  matrix 223
  perception 29–30
  reduction 160–161
  shedding 46
  technology 97–99
Risk priority number (RPN) 130, 317–318
Root cause analysis (RCA) 49, 52, 120, 286, 325, 351
R*(t)* 281
Run to failure 356–357

S

Safety 37n28, 52–57, 209, 253–254, 286, 479, 506
  arc flash potential 480–481
  employer's responsibility 480, 482
  issues, identification and prioritization 479–480
  job hazard analysis 481–482

Sampling 66
Schedule 215
Scheduled maintenance (SM) 350–351, 385, 430, 495
Scheduling 207
Scope creep 254
Seattle, Washington 240
Secondary battery 414
Sectional commissioning 66
Security engineering 209
Semiconductor materials 262
Semiconductor reliability 339–340
Semiconductor technologies 99
SE/Repowering™ 151–152, 219, 261
  cannibalization 162
  C&I 162–165
  concept 167–169, 184
  distressed 172–174, 185
  effective 151–154, 160
  existing 169–172, 184–185
  financial viability 165
  impacts of 163–166
  improved and higher asset valuation 163
  insurability 165–166
  long-term energy production 163
  need for 150
  operations analysis and decisions 164
  overview of 153
  PAM 175–176
  plan 213–214
  planning elements 175–176
  plant viability 163
  process 152–156
  RAMS 176–184
  relocation 174–175, 185
  revenue 164
  risk reduction 160–161
  system long-term considerations 154–155
  triggers and hooks 154
  types 161, 166–175
Service FMEAs 313
Service life 214, 234
Shading 369, 370
Shelter/skid 509
Shoulder 135
Silicon solar cells, degradation sources 128
Siloing 56, 63, 108, 113
SIMILAR (state, investigate, model, integrate, launch, assess, and reevaluate) 220–221
Single point failures (SPFs) 277
Site access 506
Site capability survey (SCS) 63–64
Site restoration 219–220, 262
Site-specific environmental data 71–74, 96, 177–184
SMART (specific, measurable, assignable/achievable, realistic and/or relevant, time-related, and traceable) 221–222, 268n6
Soft costs 186
Software (SW)
  defects 286
  design 285
  updates maintenance 351
Soil conditions 253
Soiling 322
Sol-air temperature 177
Solar cell technology 95
Solar Energy Industries Association (SEIA) 263
Solar fluence 394–395
Solar irradiance 240–241, 394–395
Solar noon 387n8
Solar power generation system 228
Solar waste 145
Solders 264
Spares 188, 374, 387n3
  calculation 374–379
  storage and availability 379
Sparing 36n6
Special support equipment (SSE) 383
Specifications 239
Speculative PV 150
SQL database 505
Stakeholders 4, 6, 15–17, 36n10, 156
  agreement 230
  benefits 448

engagement 453
external 57–63
functions of 288
internal 57, 58
long-term 202n1
major 453
needs 185–196, 284–287
RAMS-centric view of 284
reliability 284
requirements 213–214
success and failure criteria 229–230
unidentified 254
Standards 30–31, 51, 52, 81, 99–100, 126, 146, 171, 172, 190–196, 264, 269n38, 406, 452
Standard test condition (STC) 7, 77–78, 121, 487
Substantial completion (SC) punch list 493
Success 8–9, 212, 229–230
Sun Edison bankruptcy 159
SunPower 493
Sun up/down 407n1
Supervisory control and data acquisition (SCADA) 177, 197, 210, 285, 309, 324, 346n5, 479, 483
Suppliers 286–287
failure 335
models 335
Support costs 243–244
Surface mount components 99
Sustainable development 212
Synthetic data 49, 472–473
Synthetic reliability data 282–283
System concept 218
System delivery 21–24, 216–217
System design 87–89, 114, 255
System flipping 33–34
System FMEA 313
System-level RBD 309, 310
System lifecycle optimization 235, 237–238
System Operational Concept (OpsCon) 230
System owner 141–142
System performance 195
System resilience 209
Systems engineering (SE) 107, 112, 284–285, 447. *See also* **SE/Repowering**™
cost overruns *vs.* percent of budget spent 206, 207
definition of 205
function 210, 213
NASA 207
need for 206
problems 254
process 208, 210–218
references 209
risk management 217, 222–227
specialty engineering skill sets 209
team 206–207, 209–210, 217, 219, 456
tools 208, 220–227
Systems engineers 219

t

TamizhMani (Mani), Govindasamy 130, 131
Taxonomy 36n8, 51–52
Technology
cost of 99–100
fatigue 18, 45–47, 97, 116, 196–197
present state of 97
risk 97–99
Testability 379
design 381
function of 380–381
percent coverage 381–382
requirements 382–383
special test equipment 383
specifications 384–386
Testing 69, 218–219, 255–256, 486–487
capacity 494
electrical 70
hail 126–127
mechanical 71
reliability 338–341
Thermal conductivity and resistance 321, 322
Third-party O&M 490, 491
Third-party vendor 461
Tier 1 manufacturers 102, 199–200, 487
Time 230

Time to failure (TTF) 279, 327
Time to issue work order 503
Time to repair (TTR) 360–361. *See also* Mean time to repair (MTTR)
Trackers 509
Training 190, 506
Transistor-transistor-logic (TTL) 283

u

Uncertainty 245
Universal real-time data (URTD) 71, 505, 506
 CMMS 75–76
 data sources and users 75
 implementation of 74–76
 site environment 71–74
Unprofessional conduct 113
Unscheduled maintenance 350
Upgrading 219, 259–261
Uptime 401
Usage profile 215, 231
User profile 215, 247
Utility economics 164–165
Utility industry 107, 109

v

Value engineering 25
Value stacking 435–437
Variability in manufacturing 101–103
Vee diagram, modified 216–217
Verification 196
Verification matrix 385

w

Warranties 28–29, 117, 170, 439, 471–472, 487
Waste 264
Waste from Electrical and Electronic Equipment (WEEE) 145
Weak parts 328, 329
Wearout failures 329–332
Weibull distribution 295–296
Weibull++™ tool 326–333, 473
Wildfires 477–478
Wind 252
Wind damage 322, 323
Work authorization delay 358
WUNDERGROUND® 251

z

Zero likelihood of occurrence 225–227
z-score 306, 307